ORGANIC CHEMICALS in the AQUATIC ENVIRONMENT

Distribution, Persistence, and Toxicity

Alasdair H. Neilson

LEWIS PUBLISHERS

Boca Raton   Ann Arbor   London   Tokyo

Library of Congress Cataloging-in-Publication Data

Neilson, Alasdair H.
Organic chemicals in the aquatic environment: distribution, persistence, and toxicity/Alasdair H.
Neilson.
    p.   cm.
  Includes bibliographical references and index.
  ISBN 0-87371-597-7
    1. Aquatic organisms–Effect of water pollution on. 2. Organic water pollutants–
Environmental aspects. I. Title.
  QH545.W3N45 1994
  574.5'263—dc20   94-22460
                                                      CIP

© 1994 by CRC Press, Inc.
Lewis Publishers is an imprint of CRC Press

No claim to original U.S. Government works
International Standard Book Number 0-87371-597-7
Library of Congress Card Number 94-22460
Printed in the United States of America   2 3 4 5 6 7 8 9 0
Printed on acid-free paper

PREFACE

This book stems from the author's experience with a variety of problems of the fate, distribution, and toxicity of organic compounds in the aquatic environment. It became increasingly clear that the procedures for investigating these problems crossed the traditional boundaries of organic and analytical chemistry, microbiology, and biology and after many years this resulted in the idea of selecting the relevant aspects of these and writing the present book. Environmental problems have become increasingly complex and environmental impact studies should be based as far as possible on incontrovertible scientific facts. In this book the basic issues of chemical analysis, distribution, persistence, and ecotoxicology have therefore been discussed although emphasis has been placed on microbial reactions with which the author is most familiar. The final chapter attempts to address the issue of environmental hazard assessment and to construct a strategy for carrying it out. The necessary input and the difficulties of achieving a rigorous synthesis of the various elements are illustrated with specific examples. Throughout the book an attempt has been made to include a wide range of structurally diverse compounds as illustration, and a mechanistic approach to biodegradation and biotransformation has been adopted. At the same time, the limitations in this book should be clearly appreciated: it is not designed for the specialist in any of the traditional disciplines, although it is hoped that the level of detail is acceptable to those who seek discussions of a range of environmental issues. The book is by no means comprehensive, but a list of references for those seeking further detail is provided at the end of each chapter.

THE AUTHOR

Alasdair H. Neilson, Ph.D. is Principal Scientist at the Swedish Environmental Research Institute in Stockholm. Dr. Neilson studied chemistry at the University of Glasgow and obtained his Ph.D. in organic chemistry in Cambridge where he was elected into a Research Fellowship at Christ's College. Dr. Neilson has carried out research in organic and theoretical chemistry, and has held academic and industrial positions in chemistry before turning to microbiology. His research interests have varied widely, and have encompassed studies of nitrogen fixation, carbon and nitrogen metabolism in algae, and various aspects of environmental science including biodegradation, environmental organic chemistry, and ecotoxicology. These studies have resulted in over forty publications and review articles. Dr. Neilson is a member of the American Chemical Society, the American Society for Microbiology, the American Crystallographic Association, the Society for General Microbiology, the Society for Applied Bacteriology, the Society for Ecotoxicology and Chemistry and the Arctic Institute of North America.

TABLE OF CONTENTS

5. Persistence: Experimental Aspects

6. Pathways of Biodegradation and Biotransformation

ACKNOWLEDGMENTS

When all is said and done, it remains to thank all those who have contributed in many different ways to the making of this book. It is a pleasure to express my deep gratitude to the following.

My teachers in the Universities of Glasgow, Cambridge, and Oxford and the University of California, Berkeley — including many who are no longer with us — for illustrating by example the rigors of scientific enquiry.

The late Percy W. Brian, F.R.S., for patiently directing my faltering footsteps into microbiology many years ago.

Lars Landner, my former chief at IVL, for his interest and support during the critical years when our studies on environmental hazard assessment procedures were initiated.

My collaborators, Ann-Sofie Allard, Per-Åke Hynning, Marianne Malmberg and Mikael Remberger, not only for their scientific contributions, but also for their friendship and tolerance over many years.

Ann-Sofie Allard for valuable comments on Chapters 4 and 5 that resulted in greater clarity, as well as for her patience and expertise in producing the numerous figures from sometimes erratic drafts; Per-Åke Hynning for bringing to my attention numerous relevant references on analytical procedures that have been incorporated into Chapter 2; Marianne Malmberg for perceptive comments on Chapter 7 and for removing some of the glaring taxonomic errors. The Knut and Alice Wallenberg Foundation and its executive director Professor Gunnar Hoppe for providing instrumentation which opened new horizons to our research.

Reviewers of our publications in a variety of journals who have helped to maintain our scientific integrity and diminish our sense of isolation from the greater world of science.

My good and long-standing friends, the Søndmør families in Øversjødalen, Norway for providing the tranquility during which important chapters and the final revision of this book were written.

Ewa Berg, Margaretha Eriksson, and Ole Jensen without whose support and encouragement this book would never have materialized.

CHAPTER 1

Introduction

It is convenient historically to date the beginning of popular concern about the environment with the publication of Rachael Carson's *Silent Spring* in 1962. Although this book made no pretense at a scientific exposure of the potential dangers of pesticides, the date happily coincided with the coming of age of the analytical instrumentation needed to provide a firm scientific foundation for the unease which was expressed.

The succeeding years have seen an increasing degree of sophistication in the analysis of environmental issues and a clear appreciation of the central role of chemical analysis. Indeed, since the birth of quantitative chemistry, there has been an almost continuous search for more sensitive and selective methods of analysis. For example, it was the invention of the humble Bunsen burner almost 150 years ago which provided a non-luminous flame that made possible atomic emission spectroscopy, and before the turn of the century this had contributed to the discovery of no less than 12 new elements. Within the last 30 years or so, the revolution in organic analysis has been no less spectacular: application of gas chromatographic and high-resolution liquid chromatography have become routine, and structural identification using mass spectrometry and nuclear magnetic resonance are commonplace.

For organic compounds, evaluation of their environmental impact, based on data for acute toxicity to fish and estimation of biochemical oxygen demand, has been steadily replaced by the results of experiments on sub-acute and chronic toxicity, sophisticated investigations of microbial metabolism, and an appreciation of the true complexity of partitioning processes. It may conservatively be stated that during the 1960s there slowly emerged a new interdisciplinary activity — environmental science.

1.1 ORIENTATION

This book attempts to provide discussions on the partitioning, the persistence, and the toxicity of organic compounds within the wider context of environmental hazard assessment. The presentations are restricted to organic compounds and virtually exclusively to the aquatic environment. Although the most detailed analysis has been given to aspects of microbiology which have been the main interest of the author during many years, these discussions that extend over three chapters have been buttressed by a brief section dealing with chemical analysis which is a cornerstone of all aspects of environmental activity and short sections devoted to problems of partitioning and ecotoxicology. Yet none of the chapters stands in isolation: problems in partitioning are intimately bound to aspects of metabolism, assessment of toxicity requires understanding of bioavailability and partition, and mechanisms of association are highly relevant to microbial metabolism.

There are certain threads which run throughout this series of essays: the importance of the transformation of xenobiotics, the association between xenobiotics and components of environmental matrices, and the dynamics of ecosystems. By implication, therefore, environmental hazard assessments should take into account not only the original xenobiotic but also its possible transformation products together with the actual state — free or associated — in which the compound exists in the environment. It should be clearly appreciated that both the persistence and the toxicity of xenobiotics are critically determined by the extent and reversibility of the interactions between low-molecular weight xenobiotics and polymeric material in the environment. Indeed, the importance of dynamic interactions at both the physico-chemical and the molecular levels cannot be too strongly emphasized.

Although a comprehensive study on the hazard of chemicals in the aquatic environment has been presented and illustrated by application to three pure compounds (Landner 1989), it is important to appreciate the difference between the emphasis in that study and that developed in the present one. Landner's book deals in depth with a wide range of biological factors at different levels — though microbiological and chemical aspects are also included. The present study is, by contrast, however, essentially chemical and mechanistic in approach. An attempt has been made in Chapters 7 and 8 to indicate some of the concepts that may profitably be translated — albeit with appropriate modification — from human toxicology and epidemiology to ecotoxicology. A number of non-specialist books provide valuable and interesting ideas and a good background to the general area of hazard evaluation, even though their emphasis is on radiological hazard and carcinogenesis in man.

1.2 LITERATURE CITED

One aim of this book was to provide a forum for presenting an overview of the issues that are basic to producing an environmental hazard assessment of organic compounds discharged into the aquatic environment. This evaluation requires the application of expertise in the analysis of the compounds, knowledge of their distribution among the various environmental compartments and their dissemination, and appreciation of the factors that determine their persistence and toxicology. An attempt has been made throughout to illustrate these principles with concrete examples, even though no effort has been made to provide a comprehensive account of any single group of compounds. It was hoped at the same time that the reader would obtain some appreciation of the complexities in the design and interpretation of the relevant experiments which are being carried out in the laboratory. This has, therefore, necessitated some degree of compromise both in the depth of the presentation and in the number of references to the literature.

It was decided at the outset that it was clearly impossible to provide citations for every statement. On the other hand, controversial points or possibly less well-known facts which have not yet reached the textbooks deserve citation of their sources.

References are almost invariably given to the primary literature which has been subjected to the scrutiny of peer review. It is therefore assured that even when the present author's interpretations should prove faulty — and this is inevitable — a solid and reproducible basis of fact is available to the critical reader. As a result, although the number of references is considerably greater than had been visualized at the outset, they represent merely an eclectic selection from a vast literature. The choice of references may, however, seem quixotic. Unduly historical, and recent though incomplete relevant references have, no doubt, been omitted, but the writer can assure the authors of these that there is no malice in the selection. Some older work has been cited when this has led to lasting concepts, though other early work may be difficult to evaluate by the standards of today; no doubt work at the cutting edge of current research will rightly require modification and extension in the future. Whereas no consistent attempt has been made to trace the earliest relevant references, it can no longer necessarily be assumed that the most recent publications cite all the appropriate earlier studies. The author has, therefore, exercised the privilege of selection. It is too much to hope that an adequate compromise has been reached, and without doubt, important studies have been omitted, and errors of interpretation have been made. In the last analysis, all that can be done is to offer humble apologies and hope that the damage done is not too serious. The author hopes he may take refuge in the reply given to a lady enquiring why, in his great dictionary, Dr. Johnson had defined pastern as the knee of a horse: "Ignorance, madam, pure ignorance". There are a large number of books and review articles dealing with the microbiological aspects of the material presented here, and a list of some of these is given at the end of Chapter 4 which attempts to provide an overview of the problem of persistence predominantly — though not exclusively — from a microbiological point of view.

1.3 PERSONAL BIAS

It is inevitable that the coverage of the areas included for discussion is uneven. The selection of material for inclusion is, therefore, substantially determined by personal bias, and some of these prejudices should be taken into account.To quote Dr. Johnson, discussing the state of learning in the present author's native country: "Their learning is like bread in a besieged town: every man gets a little, but no man gets a full meal." This is certainly true of the discussions presented in this book.

1. There is a huge literature on organochlorine compounds since these have awakened serious and well-merited concern. Although, in many cases, investigations concerned with these compounds have yielded principles of general application, an attempt has been made to redress the balance by including illustrative examples using non-chlorinated compounds. The coverage of, for example, DDT, PCBs, and chlorinated dibenzo-1,4-dioxins does not, therefore, reflect the substantial research effort that has been directed to these compounds. It is hoped, however, that — apart altogether from the specific interest in individual compounds — the general principles that have been developed in the course of such investigations have clearly emerged. The other large class of compounds that has been extensively studied are the PAHs, but it is pointed out in several places that

conclusions drawn from studies with neutral compounds such as PAHs and many of the organochlorines may not be directly applicable to compounds with polar substituents such as hydroxyl, amino, or ketonic groups. It is particularly important to underscore the great structural diversity of agrochemicals and pharmaceutical products that have been developed against specific biological targets using the increasing sophistication of methods available to the synthetic organic chemist. Whereas agrochemicals have attracted attention on account of their potential adverse effects on non-target organisms and concern with their persistence, the same intensity of effort has not so far been directed to pharmaceutical compounds. Although this presumably reflects the much smaller quantities that are involved, the widespread distribution of some of them — including, for example, antibiotics and steroids — merits attention.

2. Some of the material may possibly seem exotic. Apart from the predilection of the author for the curious, there are serious scientific reasons for introducing these issues.

   • Valuable concepts may emerge from the results of isolated experiments whose relevance in a wider context was not appreciated at the time they were published and which, therefore, have not attained sufficient prominence.

   • Areas that have attracted intense research effort may not have done so exclusively on account of their scientific interest, but as a result of economic or even social pressure. The remaining uncertainty surrounding the hazard associated with exposure to 2,3,7,8-tetrachlorodibenzo-1,4-dioxin (U.S. EPA 1993) in spite of massive international effort illustrates both the influence of public opinion and the intrinsic difficulty of the hazard assessment process.

1.4 SCOPE OF THE WORK: TANGENTIAL AREAS NOT INCLUDED

A number of important areas that are relevant have inevitably been omitted, and in order to provide perspective and avoid misunderstanding, it seems worth drawing attention to some of them. In some cases, the principles enunciated in this book are directly relevant, and an attempt has been made to indicate this.

Modeling — There is at least one major area of activity pertaining directly to the aquatic environment for which the reader will seek in vain. The complexity of environmental problems and the availability of personal computers has led to extensive studies on models of varying sophistication. A discussion and evaluation of these lies well beyond the competence of an old-fashioned experimentalist. This gap is left for others to fill, but attention is drawn to a review that covers recent developments in the application of models to the risk assessment of xenobiotics (Barnthouse 1992) and a book (Mackay 1991) that is devoted to the problem of partition in terms of fugacity — a useful term taken over from classical thermodynamics. Some superficial comments are, however, presented in Chapter 3 (Section 3.5.5) in an attempt to provide an overview of the dissemination of xenobiotics in natural ecosystems. It should also be noted that pharmacokinetic models have a valuable place in assessing the dynamics of uptake and elimination of xenobiotics in biota, and a single example (Clark et al. 1987) is noted parenthetically in another context in Chapter 3 (Section 3.1.1). In similar vein, statistical procedures for assessing community effects are only superficially noted in Chapter 7 (Section 7.4).

Eukaryotic organisms — Although the metabolic potential of fungi and algae is referred to in Chapter 4 (Section 4.3.4 and Section 4.3.5) as well as in various parts of Chapter 6, the main thrust of this account is directed to reactions mediated by bacteria. This is due only partly to personal prejudice but mainly to scientific conviction: on the one hand, the author has limited experience with fungi; on the other, in most natural aquatic systems, bacteria and algae dominate the microbial flora. It is hoped, however, that the salient differences between the metabolism of xenobiotics by prokaryotic and eukaryotic organisms have clearly emerged.

Biotechnology — Biotechnology with the aim of producing organic chemicals by biological reactions is not systematically discussed, although a few comments are made in Chapter 6 in the context of biotransformation reactions, and occasional examples are also cited throughout the same chapter. Biotechnology is often interpreted to include biological waste treatment, so that one important application of studies in biodegradation is in the design and optimization of biological systems for the treatment of industrial waste. Attention is, therefore, directed to a review that draws attention to the possible application of organisms not traditionally considered in this context (Kobayashi and Rittman 1982). Biological waste treatment has not been discussed at all here since most of the fundamental physico-chemical principles are not included in this book. For similar reasons, and in view of many unresolved issues, the important issue of bioremediation or biorestoration of contaminated areas is not specifically discussed, although the contamination of groundwater by leachate from landfills clearly interfaces with that of the aquatic environment. Instead, attention is directed to some reviews of the problems in this difficult area of application (Madsen 1991; Edgehill 1992; Thomas and Lester 1993; Wilson and Jones 1993). The application of this technology has also achieved prominence in the remediation of oil spills (Atlas 1991). On the other hand, many critical — and, in some cases, unresolved — issues that are directly relevant have been taken up in the appropriate chapters of this book: partition into the solid phase, into biota, and into the atmosphere, together with the reversibility of these processes (Chapter 3); persistence to biotic and abiotic attack and the formation of metabolites (Chapters 4 and 6); and toxic effects (Chapter 7). One important issue which may be of dominant significance in some of these artificial environments — and which is clearly not so in the aquatic environment — is the water concentration which may be a factor that limits the growth and metabolism of the relevant organisms.

Some other aspects of biological treatment systems are clearly germane to the principles discussed in this book, and attention is briefly drawn to the following examples:

1. Binding of xenobiotics and their metabolites to the sludge phase may occur so that, although acceptably clean effluents may be attained before discharge into receiving waters, the problem of finding environmentally acceptable solutions to disposal of the sludge may still remain.
2. Biotransformation or only partial degradation may occur so that an evaluation of the functioning of the system cannot be based solely on an analysis of the

components of the effluents before treatment. During treatment, compounds with totally different physico-chemical properties and toxicity may be produced.

3. Most industrial effluents consist of a complex mixture of individual components, and the degradation of these may present formidable problems in metabolic regulation of which some simple examples are given in Chapters 4 and 6. This is addressed in the context of toxicity in Chapter 8 (Section 8.3.1).

4. Care must be exercised not to confuse the aims of the investigation. For example, anaerobic reactors that are developed to treat wastewater with the object of producing methane may not degrade recalcitrant xenobiotics for which inoculation with specific microorganisms may be necessary (Ahring et al. 1992).

5. There exist unresolved issues in the application of genetically engineered organisms for biological treatment (a) from the potential risk of dissemination of the organisms and (b) from the problem of maintaining these organisms in a mixed bacterial population in competition with other organisms.

6. Addition of nutrients, such as nitrogen and phosphorus, must be carefully adjusted so that excess is not discharged into receiving waters, while for effluents with high concentrations of readily degraded substrates, advantage might be taken of nitrogen-fixing bacteria to diminish or even eliminate additions of combined nitrogen (Neilson and Allard 1986).

Terrestrial systems — Two areas which interface with aquatic systems are not discussed since they lie beyond the expertise of the author and, indeed, merit separate and detailed treatment. Nevertheless, many of the principles are common to those discussed in this book so that some brief comments on these areas seems justified.

1. There is no discussion of terrestrial systems in spite of the fact that there is considerable interest and concern over the fate of agrochemicals in the terrestrial environment and over their possible effect on non-target organisms. These problems are considered only parenthetically here, though the transport of agrochemicals is governed by the principles of distribution discussed in Chapter 3, their fate by those given in Chapters 4 and 6, and their toxicology by those in Chapter 7. Extensive results on the persistence of agrochemicals has also provided gratuitous support for the principles of elective enrichment that are discussed in Chapter 5.

2. Considerable interest has arisen in the environmental problems associated with the disposal of solid waste; again, this interfaces closely with the aquatic environment through leaching of organic compounds from landfills — both as solutes and as particulate material — into water courses, rivers, and lakes. As noted above, many of the principles that are discussed in this book in other contexts are directly relevant, and specific attention is drawn to the following:

   • The association between monomeric compounds and naturally occurring polymeric material in environmental matrices is a recurring theme in this book, and its significance for the analysis of xenobiotics, its importance in transport processes, and its critical role in determining the accessibility of xenobiotics to biota including microorganisms are discussed in Chapter 2 (Sections 2.2.4 and 2.2.5), Chapter 3 (Section 3.2.3), and Chapter 4 (Section 4.5.3).

- The wider environmental significance of metabolites produced by microbial action from the original compounds is discussed in Chapter 6 (Section 6.11.5).
- In the context of bioremediation, the stability of introduced microorganisms and their ability to withstand competition from endemic ones are cardinal issues that have not received final resolution.
- The differences in the design of experiments to determine the toxicity of xenobiotics to terrestrial organisms is a result both of the organisms involved and of the possibility that the mobility of some organisms, such as earthworms, may simply enable them to avoid exposure to toxicants.

The atmosphere — It will soon become apparent that separate consideration of the atmospheric and aquatic phases is hardly justified, but only two examples which support this view will be used as illustration.

1. It has been repeatedly demonstrated that organic compounds — even those of apparently low volatility — may be transported via the atmosphere over great distances and may, therefore, be encountered in samples of precipitation and particulate deposition collected from areas remote from the point of initial discharge. This is discussed further in Chapter 3 (Section 3.5.3).
2. The atmosphere is not an inert medium, and compounds discharged into it may be transformed by photochemical reactions or by interaction with atmospheric constituents, such as oxygen, or the oxides of sulfur and nitrogen. This is discussed briefly in Chapter 4 (Section 4.1). These transformation products may then subsequently enter the aquatic and terrestrial environments in the form of both particulate matter and precipitation.

For these reasons, a thorough discussion of the role of atmospheric process ought to be presented if the aquatic environment is to be adequately discussed. This is, however, a monumental task far beyond the competence of the author; all that can be offered instead are brief comments at the appropriate places in the various chapters.

Natural microbial metabolites—Although this account is directed almost exclusively to xenobiotics, and a discussion of biosynthetic reactions mediated by microorganisms therefore lies beyond its assigned limits, a few brief comments may be justified in light of an upsurge of interest in naturally occurring halogenated compounds. Attention is, therefore, drawn to the following specific issues:

1. Biosynthetic reactions carried out by microorganisms produce a structurally diverse range of chlorinated and brominated compounds (Strunz 1984; Gribble 1992), while some higher plants produce fluorinated compounds. Brominated compounds are widely distributed in marine environments as the metabolites of higher organisms — in particular, sponges — and their possible therapeutic application has resulted in intense activity (Munro et al. 1987). Attention has also been drawn to the possibility of exploiting the synthetic capability of marine bacteria for producing compounds of pharmaceutical interest (Austin 1989) and to the use of chlorinated phenylpyrroles produced by *Pseudomonas cepacia* for controlling fruit spoilage fungi (Roitman et al. 1990).

2. Halogenated compounds are produced both by *de novo* synthesis involving direct incorporation of halide ion and by transmethylation reactions (Chapter 6, Section 6.11.4). The former include reactions mediated by haloperoxidases in the presence of hydrogen peroxide and halide ion, and these enzyme systems have wide biosynthetic capability (Neidleman and Geigert 1986), including oxidative dimerization resulting, for example, in the formation of 2,3,7,8-tetrachlorodibenzo-1,4-dioxin from 2,4,5-trichlorophenol (Svensson et al. 1989).

3. A few naturally occurring compounds, such as methyl chloride, and halogenated phenols are both industrial products and anthropogenic, although the relative quantitative contribution of these natural products on a global basis cannot be estimated reliably. In addition, it seems unlikely that the more highly chlorinated phenols — and in particular, for example, 2,4,5-trichlorophenol — are produced in significant concentrations by such reactions (Wannstedt et al. 1990).

4. Although the distribution, persistence, and toxicity of these halogenated compounds — some of which such as the chlorinated grisans, the chlorotetracyclines, and chloramphenicol are antibiotics — are subject to the same principles as those outlined in this book, this aspect has seldom been examined. One exception that may serve as an illustration is the debromination of naturally occurring bromophenols by bacteria under anaerobic conditions (King 1988).

Since the quantitative significance of almost none of these microbial metabolites has been evaluated and the toxicology of only very few has been investigated — invariably, in other contexts — it seems justified to accord a low priority to such problems in light of more urgent issues. On the other hand, there is no doubt that the production of polycyclic aromatic compounds during the diagenesis of di- and triterpenes and which may be microbially mediated is quantitatively significant and could seriously compromise non-specific analysis for aromatic compounds. This is noted again in Chapter 2 (Section 2.4.2).

The effect of xenobiotics on microbially mediated processes — The discussion throughout this book is directed to the degradation and transformation of xenobiotics by microorganisms. The converse issue — the inhibitory effect of xenobiotics towards microbial processes that are not directly involved in degradation, but may play an important role in natural geochemical cycles — is not explored. This is of enormous importance, particularly in the terrestrial environment where non-target organisms may be subjected to exposure to agrochemicals. Some of these organisms have an autotrophic way of life and utilize CO_2 as principal or exclusive source of carbon — rather than a heterotrophic metabolism that is characteristic of those organisms that are primarily responsible for degadation of xenobiotics. The autotrophic organisms include algae, ammonia-oxidizing bacteria, and many thiobacilli. In the context of degradation, however, the relevant microorganisms are generally sufficiently resistant to the xenobiotics to preclude serious inhibition at the concentrations involved. Some useful examples of the reactions that may be involved are found in a review (Smit et al. 1992) that discusses the hazard of genetically manipulated organisms released into the terrestrial environment.

REFERENCES

Ahring, B.K., N. Christiansen, I. Mathrani, H.V. Henriksen, A.J.L. Macario, and E.C. de Macario. 1992. Introduction of a de novo bioremediation ability, aryl reductive dechlorination, into anaerobic granular sludge by inoculation of sludge with *Desulfomonile tiedjei. Appl. Environ. Microbiol.* 58: 3677-3682.

Atlas, R.M. 1991. Microbial hydrocarbon degradation — bioremediation of oil spills. *J. Chem. Tech. Biotechnol.* 52: 149-156.

Austin, B. 1989. Novel pharmaceutical compounds from marine bacteria. *J. Appl. Bacteriol.* 67: 461-470.

Barnthouse, L.W. 1992. The role of models in ecological risk assessment: a 1990's perspective. *Environ. Toxicol. Chem.* 11: 1751-1760.

Clark, T.P., R.J. Norstrom, G.A. Fox, and H.T. Won. 1987. Dynamics of organochlorine compounds in herring gulls (*Larus argentatus*): II. a two-compartment model and data for ten compounds. *Environ. Toxicol. Chem.* 6: 547-559.

Edgehill, R. 1992. Factors influencing the success of bioremediation. *Aust. Biotechnol.* 2: 297-301.

Gribble, G.W. 1992. Naturally occurring organohalogen compounds — a survey. *J. Nat. Prod.* 55: 1353-1395.

King, G.M. 1988. Dehalogenation in marine sediments containing natural sources of halophenols. *Appl. Environ. Microbiol.* 54: 3079-3085.

Kobayashi, H., and B.E. Rittman. 1982. Microbial removal of hazardous organic compounds. *Environ. Sci. Technol.* 16: 170A-183A.

Landner, L. (Ed.) 1989. *Chemicals in the aquatic environment. Advanced hazard assessment.* Springer-Verlag, Berlin.

Mackay, D. 1991. *Multimedia environmental models. The fugacity approach.* Lewis Publishers, Chelsea, Michigan.

Madsen, E.L. 1991. Determining in situ biodegradation. Facts and challenges. *Environ. Sci. Technol.* 25: 1663-1673.

Munro, M.H.G., R.T. Luibrand, and J.W. Blunt. 1987. *In Bioorganic marine chemistry,* Vol. 1, (Ed. P.J. Scheuer), pp. 93-176. Springer-Verlag, Berlin.

Neidleman, S.L., and J. Geigert. 1986. *Biohalogenation: principles, basic roles and applications.* Ellis Horwood, Chichester, England.

Neilson, A.H., and A.-S. Allard. 1986. Acetylene reduction (N_2-fixation) by Enterobacteriaceae isolated from industrial wastewaters and biological treatment systems. *Appl. Microbiol. Biotechnol.* 23: 67-74.

Roitman, J.N., N.E. Mahoney, W.J. Janisiewicz, and M. Benson. 1990. A new chlorinated phenylpyrrole antibiotic produced by the antifungal bacterium *Pseudomonas cepacia. J. Agric. Food Chem.* 38: 538-541.

Smit, E., J.D. van Elsas, and J.A. van Veen. 1992. Risks associated with the application of genetically modified microorganisms in terrestrial ecosystems. *FEMS Microbiol.* Revs. 88: 263-278.

Strunz, G.M. 1984. Microbial chlorine-containing metabolites pp. 674-773. In *CRC Handbook of microbiology,* Vol. V, 2nd. ed. (Eds. A.I. Laskin and H.A. Lechavalier). CRC Press, Boca Raton, Florida.

Svensson, A., L.-O. Kjeller, and C. Rappe.1989. Enzyme-mediated formation of 2,3,7,8-tetrasubstituted chlorinated dibenzodioxins and dibenzofurans. *Environ. Sci. Technol.* 23: 900-902.

Thomas, A.O., and J.N. Lester. 1993. The microbial remediation of former gasworks sites: a review. *Environ. Technol.* 14: 1-24.

U.S. EPA 1993. Interim report on data and methods for assessment of 2,3,7,8-tetrachlorodibenzo-*p*-dioxin risks to aquatic life and associated wild life. EPA / 600 / R-93 / 055. Office of Research and Development, Washington DC 20460.

Wannstedt, C., D. Rotella, and J.F. Siuda. 1990. Chloroperoxidase mediated halogenation of phenols. *Bull. Environ. Contam. Toxicol.* 44: 282-287.

Wilson, S.S., and K.C. Jones. 1993. Bioremediation of soil contaminated with polynuclear aromatic hydrocarbons (PAHs): a review. *Environ. Pollut.* 81: 229-249.

CHAPTER 2

Analysis

SYNOPSIS

Chemical analysis is an integral part of all environmental investigations. Brief attention is directed to aspects of sampling in order to avoid interference from possible artifacts, to various procedures for extraction and concentration of analytes, and to the importance of cleanup procedures before identification and quantification. Procedures for the analysis of water and sediment samples and for biota are outlined, and some of the reactions used for derivatizing functional groups are summarized. Methods for the identification of components of environmental samples are briefly summarized, and attention is directed to the application of non-destructive procedures such as infrared and nuclear magnetic resonance spectroscopy. The importance of access to authentic reference compounds is emphasized. Gas chromatographic procedures for quantification are briefly discussed, and attention is directed to recent developments, including the use of chiral support phases and various detector systems. The use of supercritical fluids for the extraction of samples and as mobile phases for chromatography is briefly noted, together with the potential application of immunologically based assays. Attention is drawn to the range of analytes — including transformation products — that may be encountered in environmental samples and to some of the problems in the analysis of commercial mixtures and complex effluents. There is an inherent indeterminacy in assessing the recoverability of analytes from naturally aged sediment samples as a result of dynamic associations between analytes and macromolecular components in environmenal matrices.

INTRODUCTION

It is appropriate to begin this book with a discussion of chemical analysis which lies at the heart of almost all environmental investigations whether they are devoted to monitoring the distribution of a xenobiotic, evaluating its persistence and toxicity, or determining its partition among environmental matrices. Analytical support should be incorporated into all of these programs, and its central role should be clearly appreciated at the planning stage; otherwise, the conclusions that are drawn from the results of the investigation may be equivocal. These comments should not, however, be interpreted as negating the fundamental contributions that analytical chemistry itself has made, not least in illuminating the role of metabolites and transformation products and in revealing the global distribution of hitherto unsuspected compounds.

There have been revolutionary instrumental developments during the last 35 years or so, and these have completely altered the scope and possibilities of environmental research. The following examples may be given as illustration:

1. After early concern over the accumulation of DDT in biota, analysis of DDT and its metabolites was carried out by nitration and colorometric measurement after treatment with alkali. This was superseded by gas chromatography some fifteen years later, and this development facilitated more rapid analysis, greater precision, and simultaneous unambiguous analysis of DDT metabolites such as DDE and DDD.
2. Although PCBs were first synthesized in 1881 and introduced into industrial use as electrical insulators in the 1920s, they were detected in environmental samples only some thirty-five years later in 1966 by Jensen (Jensen 1972; Jensen et al. 1972). This discovery was facilitated by the development of the electron-capture detector in gas chromatography, and it also illustrates an early example of the application of mass spectrometry to the tentative identification of environmental contaminants. Since that date, PCBs have become recognized as virtually universal contaminants of environmental samples from all parts of the world.
3. The interest of the synthetic organic chemist in stereospecific synthesis has resulted in the need for methods for the analysis of chirons. This has led to the development of both chiral reagents and chiral supports for GC and HPLC analysis. The application of these methods to environmental samples is also beginning to draw attention to the environmental input of enantiomers without established biological activity as well as to their possible persistence.
4. The identification of xenobiotics at the concentrations existing in environmental contaminants and of metabolites formed during laboratory experiments in biodegradation and transformation has been completely revolutionized by the availability of modern instrumentation for structure determination. For example, infrared and nuclear magnetic resonance spectroscopy and mass spectrometry have been widely used, and their sensitivity has been facilitated by Fourier transform techniques for signal processing.
5. The detection and quantification of extremely low concentrations of chlorinated dibenzo-1,4-dioxins in environmental samples would not have been posssible without the development of high resolution mass spectrometers. As a dividend, the instrumentation thereby developed has been adapted to other analytes and has significantly lowered both their limits of detection and the level of quantification.
6. The increased sensitivity of NMR instrumentation has made possible the application of ^{13}C NMR to studying the pathways and kinetics of microbial reactions in cell suspensions, and these procedures have provided direct evidence for the structure of intermediate metabolites without the need for their isolation. Techniques have also been developed to overcome the negative magnetogyric ratio of nuclei, such as ^{15}N and ^{29}Si (Morris and Freeman 1979), and these procedures have been applied to a number of environmental problems that were previously inaccessible.

Today, few modern laboratories engaged in environmental studies lack facilities for carrying out analysis using gas chromatography, high performance liquid chromatography, or mass spectrometry, and many have access at least to nuclear magnetic resonance instrumentation. Most of these instruments are coupled to systems for automatic injection of samples and to data systems for processing the output signals and reducing background interference. This makes possible the analysis of extremely small amounts of material, and this single advantage can hardly be overestimated since it is seldom possible to isolate environmental contaminants in sufficient quantity for identification by conventional chemical procedures.

At the outset, brief mention should possibly be made of sum parameters since these have traditionally dominated environmental analysis. Some of these, such as total organic carbon (TOC), dissolved organic carbon (DOC), particulate organic carbon (POC), and their nitrogen and phosphorus analogues, have been included in conventional water quality crtiteria and are still useful in wide-ranging monitoring, for example, in oceanography. With the upsurge of interest in halogenated xenobiotics, extensive effort has, therefore, been directed to devising comparable methods for organohalogen compounds. Analysis for total PCBs, which depends on estimation of decachlorobiphenyl after chlorination (Lin and Hee 1985) or by dechlorination with LiAlH$_4$ (de Kok et al. 1981) or of halogenated compounds in general by reduction with dispersed sodium followed by potentiometric analysis (Ware et al. 1988), have been used but are now primarily of historical interest although they might find application to other chemically stable analytes. Methods have also been developed for analyzing total organic chlorine, bromine, or iodine by neutron activation analysis (Gether et al. 1979) or for non-specific halogen by coulometry or potentiometric titration after combustion (Sjöström et al. 1985). None of these procedures merit inclusion in contemporary environmental hazard assessment programs since it has been consistently demonstrated that differences in the persistence, the toxicity, and the phase partitioning among, for example, PCB congeners make the use of sum parameters quite inappropriate. Such parameters should be replaced by analysis for the concentrations of specific compounds. Although sum parameters have been used quite extensively in some monitoring programs (Martinsen et al.1988), they are really justifiable only when it can be established that a restricted range of structurally similar compounds is present.

The discussion presented here makes no attempt at being comprehensive, nor does it provide experimental details of standardized methodologies. This account is not, therefore, directed to the professional analyst, nor is it intended to serve as a handbook, although some general comments on laboratory practice are made in Section 2.5. The needs of the professional are fulfilled by several complementary books (Keith 1988, 1992; Heftmann 1992). A few examples may be given to illustrate the substantial limitations in the present account:

1. There is no systematic presentation of analytical procedures developed by the U.S. EPA, nor of those suggested by the OECD, although attention is directed to a valuable critique of EPA procedures (Hites and Budde 1991).
2. No attempt is made to evaluate, for example, the range of columns which may advantageously be used in gas-liquid or high-resolution liquid chromatography, those used for solid-phase concentration of analytes from water samples, or commercially available columns for cleaning up samples.

3. There is no discussion of technical aspects of mass spectrometry nor of the various procedures that are finding increasing application. Attention is directed to a review (Barceló 1992) that contains references to a number of books dealing with these specialized topics.

Experimental details may be found in reviews dealing with specific groups of compounds including polycyclic aromatic compounds (Bartle et al. 1981), chlorinated hydrocarbons (Hale and Greaves 1992), PCBs (Duinker et al. 1991; Lang 1992; Schmidt and Hesselberg 1992; Creaser et al. 1992), and pesticides in general (Barceló 1991). Increased interest in aromatic nitro compounds has led to development of procedures for their analysis (Lopez-Avila et al. 1991), and some of the material presented here is covered in a review dealing specifically with marine organic pollutants (Hühnerfuss and Kallenborn 1992). Attention is also directed to special issues of the *Journal of Chromatography* (1993: 642 and 643) that are devoted to a wide spectrum of procedures that have been applied to the analysis of a wide range of analytes in environmental samples.

In broad outline, the following steps will be incorporated into virtually all environmental investigations: (1) sampling from a predetermined site, (2) extraction and cleanup of the sample, (3) identification of the components, and (4) their quantification. The objective of the investigation will naturally determine the degree of sophistication of the procedures that will be applied. It should also be appreciated that unforseeable difficulties may arise during the investigation, and it is important to resolve these at as early a stage as possible. Attempts will be made in this chapter to discuss all of the steps outlined above, although no effort has been made to provide a comprehensive account of any single aspect. Instead, this chapter will attempt to erect signposts along the wayside and to provide an overview of analytical problems and a perspective on their application to specific problems in partitioning, estimating persistence, and evaluating toxicity. It is hoped, however, that some of the pitfalls awaiting the unwary have been revealed and that at least some of the major unresolved issues have emerged. The available literature is both specialized and extensive, and no attempt has been made to provide either a complete or a systematic coverage.

2.1 SAMPLING

Sampling from laboratory experiments — for example, on partitioning, toxicity, or on degradation and transformation — generally presents few problems: the concentrations of analytes are relatively high, only small volumes of samples are required, and these systems are relatively homogeneous. Samples may be frozen after collection if they are not analyzed immediately. On the other hand, the magnitude of the problem with field material is substantially greater since the systems are less homogeneous and the concentrations of the analytes may be low. Although no attempt at a comprehensive treatment will be attempted here, it cannot be too strongly emphasized that the quality of an investigation is critically determined by the care devoted to the selection of samples, their collection, and their preservation before analysis. The relative costs of the sampling and the subsequent analyses should be kept in proportion. The following issues merit brief notice:

1. Attention should be paid not only to the cleanliness of the sample containers but also to their composition; use of plastic containers is attractive on account of their physical inertness and robustness, but their use will inevitably contaminate the sample with plasticizers such as phthalate esters. Glass containers are, therefore, generally to be preferred for samples of water and sediment. Problems arising from sorption of analytes to glass surfaces may exist and may be significant for particular groups of compounds; in the case of sterol analysis, this has been circumvented by silylation of glassware to which the analyte is exposed (Fenimore et al. 1976). Samples of biota should be wrapped in aluminium foil rather than any kind of plastic material.

2. Water samples present fewer problems than other matrices, though if low concentrations of xenobiotics are to be analyzed, it may be necessary to use high-volume samplers and process the samples on board the ship. One issue which should be recognized — even if it cannot be resolved — is that presented by samples having particulate material. This is discussed in greater detail in Sections 2.2.4 and 2.2.5. Freezing of such samples may bring about alterations in their chemical composition and, hence, in their toxicity (Schuytema et al. 1989).

3. Sediment sampling may present a number of problems including the patchiness of the area being investigated (Downing and Rath 1988; Brandl et al. 1993) and the difficulty of obtaining a truly representative sub-sample from a possibly heterogeneous bulk sample. These problems may be particularly severe where monitoring is directed to providing a historical record of deposition. Ambiguities may arise (Sanders et al. 1992) where the sediment surface is non-uniform or where diffusion and bioturbation are significant. There is probably no single optimal procedure for preserving sediment samples after collection; for chemical analysis, although freezing may eliminate or at least minimize the possibility of chemical reactions occurring after sampling, analytes may be released from sediment-xenobiotic associations during subsequent thawing. Preservation at 4°C will generally minimize microbial alterations of the analyte, but the addition of inhibitors of microbial activity, such as mercuric chloride or sodium azide, may introduce problems during subsequent analysis (Section 2.2.4), and this should be avoided.

4. Particular care should be exercised in the collection of samples of fish and attempts made to minimize the inevitable stress of capture; this is particularly critical for analysis of compounds such as steroid hormones or for assays of enzymatic activity (Munkittrick et al. 1991; Hontela et al. 1992).

5. There has been increased interest in samples from remote areas such as the Arctic and Antarctic. These areas are, however, increasingly visited both by tourists and by scientific expeditions so that particular care should be exercised in the conclusions drawn from the results of such sampling. Atmospheric input may not be the only source of xenobiotics, and truly undisturbed localities are becoming increasingly rare throughout the world.

6. For the sake of completeness, attention is drawn briefly to procedures for passive sampling. For example, xenobiotics may be concentrated from soil by passive diffusion into solid adsorbents or into organic solvents in dialysis membranes (Zabik et al. 1992). Desorption of the solid supports, such as C-18 bonded silica or XAD resins, is then carried out for analysis of the xenobiotics. Xenobiotics have been concentrated from aquatic systems into dialysis membranes containing

solvents such as hexane, and this procedure has been suggested as simulating uptake by biota (Södergren 1987; Huckins et al. 1990a). In view of the discussion in Chapter 3 (Section 3.1), this should, however, be regarded as merely a qualitative evaluation. Although these procedures are valuable for demonstrating the *presence* of compounds, they are of only limited applicability to hazard assessment for which the *concentrations* of the analyte are clearly needed.

2.2 EXTRACTION AND CLEANUP

After collection and transfer to the laboratory, the samples have to be analyzed. Two factors combine to magnify the problem: (1) the low concentration of the analyte which generally necessitates concentration and (2) interference from other compounds that may occur in much greater concentration and which necessitate purification or cleanup procedures to reduce their concentration. These factors have resulted in increasing demands upon the skill of the analyst and in the need for the application of sophisticated procedures for the pretreatment of samples before identification and quantification. The analysis will, therefore, generally encompass at least the following steps: extraction, concentration, and cleanup with or without derivatization, identification, and quantification. A range of matrices — samples of water, sediment, and biota, ranging from microbial cultures to higher organisms — may be submitted for analysis, each presenting its own special problem, so that procedures for extraction and cleanup will vary considerably. An attempt will be made to present only an outline of the methods that have been widely used and to summarize some of the problems that may be encountered.

2.2.1 Solvents

Organic solvents are used at all stages of the analytical procedures — for extraction, during cleanup, and in identification and quantification. It is, therefore, appropriate to preface this section with general comments on some of the important issues that should be addressed. The purity of the solvents is of cardinal importance: the initial extracts from the samples which may have volumes of tens of milliliters are generally concentrated to volumes of the order of hundreds of microliters, and volumes in the microliter range are then used for the final analysis. Concentration procedures should be designed to minimize the loss of compounds with low boiling points and appreciable volatility. For this reason, solvents such as dichloromethane, pentane, or diethyl ether or *t*-butyl methyl ether are frequently used, while a less volatile solvent such as *iso*octane may be added as a "keeper" to retain a liquid phase. A gas atmosphere of N_2 is generally maintained during concentration to minimize oxidation of sensitive components in the sample. There is, unfortunately, no single extraction solvent which is optimal for all analytes, and the choice depends both upon the nature of the matrix and upon the structure of the analyte. Solvents with limited capacity for dissolving water and ranging in polarity from hexane and dichloromethane through toluene or benzene to diethyl and *t*-butyl methyl ethers and ethyl acetate may be usefully exploited, though no single one of these is invariably optimal. For some purposes such as extraction of wet sediment samples, water-miscible solvents such as methanol, propan-2-ol, tetrahydrofuran, acetonitrile, or dimethylformamide may be applicable; the analyte is then subsequently partitioned into a water-immiscible phase. The major drawback to the use of solvents such as ethyl acetate — and to a lesser extent, the dialkyl ethers — for extracting aqueous solutions is the high solubility of water in these solvents. Such extracts should, therefore, be dried with anhydrous sodium sulfate or preferably by azeotropic distillation before attempting derivatization with water-sensitive reagents such as acid anhydrides, acid chlorides, or *iso*cyanates.

Most common solvents are now available commercially in a high state of purity; these have extremely low residue levels, and they can be purchased as glass-distilled quality to obviate this step in the laboratory. Care should be exercised in their storage after purchase to prevent contamination with volatile laboratory chemicals and especially with standards used for analysis. Particular caution should be exercised in the use of solvents which contain stabilizers: for example, 4-*t*-butyl-2-methylphenol in diethyl ether or cyclohexene in dichloromethane. Dichloromethane is recommended for many extraction procedures on account of its volatility and the simplification in extraction resulting from its greater density than water. Reactions of the analyte with the cyclohexene inhibitor may, however, result in the production of artifacts such as halogenated cyclohexanes and cyclohexanols (Campbell et al. 1987; Fayad 1988), and the presence of iodocyclohexanol has been associated with low recoveries of phenolic compounds using EPA standard procedures (Chen et al. 1991). Attention is also drawn to the reaction of methanol with carbonyl compounds that may result in the formation of acetals or ketals (Hatano et al. 1988) or of esters from chlorinated acetic acids (Xie et al. 1993). Care should, therefore, be exercised in the analysis of such compounds particularly by HPLC for which methanol is a widely used mobile phase. For the extraction and analysis of compounds containing reactive carbonyl groups — especially aldehydes — their chemical reactivity clearly precludes the use of solvents with active methylene groups such as acetone, acetonitrile, dimethylsulfoxide, or nitromethane.

2.2.2 Cleanup Procedures

Extracts from environmental samples are seldom sufficiently free of contaminants that they can be analyzed directly, since the high-resolution columns used for GC or HPLC analysis must not be overloaded with compounds other than the analyte. Pretreatment of the sample before GC, GC-MS, or HPLC analysis is, therefore, often a critical step in the analysis of extracts from complex matrices, and this is particularly important when the concentrations of the interfering compounds greatly exceed that of the analyte. These compounds may then seriously compromise the analytical procedure. The cleanup procedures that will be used subsequent to the initial extraction depend both on the nature of the matrix and on that of the analyte — in particular, the sensitivity of the analyte to chemical reagents such as strong acids (Bernal et al. 1992) or bases (Vassilaros et al. 1982) or oxidizing agents including molecular oxygen (Wasilchuk et al. 1992). All of these may bring about significant chemical changes in the analytes and should, therefore, be avoided unless the stability of the analyte towards them has been unequivocally established. For example, although the quantification of chlorinated dibenzo-1,4-dioxins and some of the PCB congeners is facilitated by taking advantage of the great stability of these compounds to chemical reagents such as sulfuric acid for removing interfering compounds, comparable cleanup procedures are clearly not applicable to more sensitive compounds. Less drastic cleanup procedures should, therefore, be developed, and these should be adapted to the chemical structure and to the reactivity of the analyte.

A number of different cleanup procedures for environmental samples have been effectively employed, and the following may be used as illustration:

1. Advantage may profitably be taken of classical chromatographic separation methods using open columns of silica gel, alumina, florisil, or activated charcoal, although it should be realized that these do not generally provide complete separation of structurally similar compounds. Gel-permeation chromatography has been effectively used for separating classes of compounds such as toxaphene (polychlorinated bornanes and bornenes) and PCBs (Brumley et al. 1993) and has increasingly been used for the removal of interfering lipid material; this is further discussed below.

2. Counter-current chromatography (CCC) has been extensively used for the isolation of natural products (Foucault 1991) and has many features applicable to the fractionation of environmental samples before analysis. A good example is the resolution of intractable mixtures of polyphenolic compounds in plant extracts (Okuda et al. 1990), and this procedure might be effective in the analysis of organic constituents of soils and organic-rich sediments. One additional advantage is the elimination of ambiguities resulting from possible alteration in the structure of the analyte during gas chromatographic separation (Nitz et al. 1992). Many features of CCC are identical to those involved in HPLC, and the main limitation probably lies in the availability of suitable immiscible solvent systems, although ternary systems, such as hexane/acetonitrile/t-butyl methyl ether, or even quaternary systems, such as hexane/ethyl acetate/methanol/water, have been used effectively.

3. Separation of the desired analyte may be achieved by taking advantage of the formation of specific complexes. Examples include those between Ag^+ and $C=C$ bonds for separating alkenes or alkenoic acids from their saturated analogues (Christie 1989) and of clathrate complexes between urea and normal saturated aliphatic hydrocarbons for separating these from branched-chain and other hydrocarbons (Richardson and Miller 1982).

4. Specific derivatives may be used for removing interfering compounds, and ideally the desired analyte should be recoverable, for example, by hydrolysis of semicarbazones formed between aldehydes or ketones and the Girard reagents (Wheeler 1968). This reagent has been employed for the effective removal of high concentrations of interfering chlorovanillins during the analysis of low concentrations of chlorocatechols (Neilson et al. 1988).

Two serious problems frequently arise: the presence of elementary sulfur in sediments samples and of lipids in samples of both sediments and biota. Fortunately, satisfactory solutions have been found for both of them.

1. Anaerobic sediments frequently contain large amounts of elementary sulfur (S_8) which may cause serious interference during gas chromatographic analysis. The traditional method that is applicable to chemically stable analytes is treatment with activated copper, but an elegant and milder procedure in which elementary

sulfur is converted into water-soluble thiosulfate by reaction with tetrabutyl ammonium sulfite has been developed (Jensen et al. 1977). This method cannot, however, be applied to samples containing sensitive compounds such as phenols; in such cases, the sulfur may be removed from the phenol acetates by silica gel chromatography and elution with cyclohexane (Allard et al. 1991).

2. Samples of biota — and to a lesser extent, sediment — will almost invariably contain substantial amounts of lipid material which is often dominated by relatively high-molecular weight triglycerides that may seriously interfere with analysis of the desired analyte. The choice of procedure for the removal of lipids depends on a number of factors including the nature of the analyte, the amounts of lipid that have to be removed, and the number of samples that have to be processed. For highly stable compounds such as PAHs and related compounds, samples have been submitted to alkaline hydrolysis before further cleanup (Vassilaros et al. 1982). This is not, of course, possible with more sensitive analytes, and a number of different procedures have been used for the removal of lipid material. These procedures generally take advantage of the differences either in the molecular weight and configuration of the analytes and the lipids or in the sorbent properties of the analytes and the lipids. Some of the most widely used procedures include the following:

- Gel-permeation chromatography (Stalling et al. 1972) which is essentially a fractionation dependent on differences in molecular weight: instrumental improvements to avoid the problem of tailing triglyceride peaks have been described and may be particularly valuable for application to on-line systems (Grob and Kälin 1991).
- Dialysis using membranes of polyethylene and methylcyclopentane as solvent (Huckins et al. 1990b): this general procedure has been examined for samples containing 50g of lipid or more and has proved an attractive simplification of conventional procedures (Meadows et al. 1993).
- On-line columns of silica gel provided care is taken not to overload the columns (Grob et al. 1991).
- Use of HPLC and columns of silica gel with gradient elution after saponification of the glycerides (Hennion et al.1983) or of HPLC using a Nucleosil 100-5 column (Petrick et al. 1988).
- Use of supercritical CO_2 or CO_2/methanol and silica gel or aluminium oxide (France et al. 1991): the use of supercritical fluids is discussed in a more general context in Section 2.2.5.

2.2.3 Toxicity-Directed Fractionation

The foregoing fractionation procedures are essentially empirical, and the individual fractions must be analyzed and their components identified and quantified. Clearly, this could be a substantial undertaking, and the results might not necessarily justify the expenditure of effort. Many investigations are directed to determining the toxic components in the samples. By analogy with the use of biological assays in isolation procedures for biologically active substances such as vitamins and steroid hormones, efforts have, therefore, been made to direct the fractionation of environmental extracts on the basis of the results of assays for toxicity. The Ames test using mutant strains of *Salmonella typhimurium* has been used most widely both on account of the ease and rapidity with which it may be carried out and also because concentrates may be assayed as solutions in, for example, dimethyl sulfoxide so that problems with solubility are readily overcome. This procedure may be illustrated by its use in an investigation of marine sediments (Fernández et al. 1992) or combined with assays for aryl hydrocarbon hydroxylase activity in a study of the organic components in automobile exhausts (Alsberg et al. 1985). A wider range of toxicity assays to marine organisms has been used to direct the fractionation of high molecular weight components of bleachery effluents, though a direct correlation of toxicity with molecular weight was apparently not possible (Higashi et al. 1992). Other types of parameters could also be used to direct the fractionation and separate the compounds of primary interest. For example, use could be made of reversed phase HPLC to separate and isolate compounds with P_{ow} values indicative of significant bioconcentration potential.

2.2.4 Specific Matrices: Water Samples

For homogenous water samples, there are generally few serious problems, and several procedures for concentration may be used:

1. Solvent extraction of the analytes may be carried out with suitable water-immiscible solvents using either batch or continuous procedures.
2. For compounds with appreciable vapor pressure, purging systems may be used followed by collection of the compound on Tenax columns from which the analyte may subsequently be thermally desorbed or — alternatively, for less volatile compounds — eluted with methanol.
3. For less volatile compounds, closed-loop stripping has been used in which the analytes are collected on a column of activated carbon from which they are then eluted with carbon disulfide. Attention should especially be drawn to an evaluation of this system for a range of analytes (Coleman et al. 1983) and in particular to the conclusion that for quantification, recovery efficiencies must be evaluated using matrices spiked with the analyte of specific interest.
4. The analyte may be sorbed directly from the aqueous phase, for example, on XAD columns (Dietrich et al. 1988), which are then eluted with a suitable solvent such as methanol, acetonitrile, dichloromethane, or diethyl ether.
5. Increasing use has been made of solid phase sorption systems incorporating, for example, C18 silica columns from which the analyte is eluted with ethyl acetate or benzene (Junk and Richard 1988). Use of discs incorporating C18 bonded silica facilitates the processing of large sample volumes (Barceló et al. 1993). Alternatively, advantage may be taken of ion-exchange resins followed by elution with salt solutions or phosphate buffer (Arjmand et al. 1988), and a combination of reversed phase C18 and ion-exchange columns has been incorporated into an automated system for the analysis of triazine metabolites in samples of soil and water (Mills and Thurman 1992). A wide range of ion-exchange columns is commercially available, and all of these systems have the advantage of using only small volumes of organic solvents: this aspect has achieved increasing prominence in attempts to diminish the use in analytical laboratories of solvents, particularly chlorinated ones. A microsystem using optical fibers, which are then used for direct injection into GC-MS systems, has

been suggested for concentrating and analyzing volatile aromatic hydrocarbons (Arthur et al. 1992).

Samples may be spiked with internal standards to simplify calculation and eliminate small errors in pipetting and injection, or surrogate standards may be employed where, for example, incomplete extraction of the analyte is unavoidable. Samples with particulate matter may present quite serious problems, and it may be desirable to remove particles by centrifugation and examine this fraction by procedures applicable to solid phases which are discussed in Section 2.2.5. Tangential-flow high-volume filtration systems have been used for analysis of particulate fractions (>0.45 µm) where the analytes occur in only low concentration (Broman et al. 1991). Attention has already been drawn to artifacts resulting from reactions with cyclohexene added as an inhibitor to dichloromethane. It has also been suggested that under basic conditions, Mn^{2+} in water samples may be oxidized to Mn(III or IV) which in turn oxidized phenolic constituents to quinones (Chen et al. 1991). Serious problems may arise if mercuric chloride is added as preservative after collection of the samples (Foreman et al. 1992) since this has appreciable solubility in many organic solvents; its use should, therefore, be avoided.

Water samples may contain appreciable amounts of particulate matter or dissolved organic carbon or colloidal material, and all of these may form associations with the analytes and affect their recoverability. For these reasons, discrepancies may arise between the concentrations of analytes determined by liquid-liquid extraction and those obtained by sorption on polyurethane or XAD resins (Gómez-Belinchón et al. 1988). Empirical procedures have been developed (Landrum et al. 1984) for fractionating samples to assess the relative contribution of the associations of xenobiotics with the various organic components, while sediment traps for collection of particulate matter have been extensively used in investigations in the Baltic Sea where appreciably turbid water may be present (Näf et al. 1992).

2.2.5 Specific Matrices: Sediments and Soils

The analysis of solid matrices such as sediments and soils almost invariably presents more serious problems than those encountered with water samples, and incorporation of cleanup steps or fractionation by silica gel or alumina chromatography will generally be needed to remove high concentrations of interfering compounds that are almost always present in such samples.

Conventional procedures — Before extraction, soil and sediment samples may be dried by freeze-drying — provided that volatile compounds are not to be analyzed — or by mixing with anhydrous sodium sulfate and extraction in a Soxhlet apparatus. If wet samples are to be analyzed directly, acetonitrile, propan-2-ol, or ethanol may be employed first, and these may be valuable in promoting the chemical accessibility of substances sorbed onto components of the matrices. The analyte may then be extracted into water-immiscible solvents and the water phase discarded. Alternatively, if the analyte is sufficiently soluble in, for example, benzene, the water may be removed azeotropically in a Dean & Stark apparatus and the analyte then extracted with the dry solvent. Analytes may, however, be entrapped in micropores in the soil matrix so that recovery of even the volatile 1,2-dibromoethane required extraction with methanol at 75°C for 24 h (Sawhney et al. 1988).

If a large number of samples are to be processed, Soxhlet-based procedures may be unduly cumbersome. As an alternative, attention has, therefore, been given to the use of sonication in a variety of solvents although this procedure should clearly be avoided where the analytes are sensitive to oxidation which may be exacerbated by cavitation. Sonication has been effectively used for recovering neutral priority organic pollutants from a number of matrices (Ozretich and Schroeder 1986) and for four nitrated explosives from soil (Jenkins and Grant 1987). It should be recognized, however, that the recovery efficiencies for some compounds may be quite low compared with alternative procedures.

Supercritical fluids — Increased effort has been directed to the application of supercritical fluids for extraction. Both static and dynamic extraction procedures have been developed, and supercritical fluids such as methanol for the extraction of bound pesticide residues (Capriel et al.1986) or of carbon dioxide/methanol for the cleanup of organochlorine pesticides in fatty tissue (France et al. 1991) have been used. In the latter study, attention was drawn to the importance of using carbon dioxide of suitable purity. It has emerged from several investigations that only poor recovery of some analytes can be achieved using carbon dioxide without the addition of polar additives — including such apparently drastic reagents as formic acid for the extraction of resin and fatty acids from sediment samples (Lee and Peart 1992). Somewhat conflicting results on the recoverability of a number of analytes have been reported, so that caution should be exercised against uncritical application of this attractive procedure. For example, whereas one extensive study in which a total of 88 pesticides were analyzed in spiked samples of sand showed considerable variation in recoverability and the recovery of polycyclic hydrocarbons in certified samples was generally poor (Lopez-Avila et al. 1990), another in which CO_2 modified with 3% methanol was used showed good recoveries that were comparable with those from conventional procedures using sonication treatment or Soxhlet extraction (Snyder et al. 1992). Another study that included PCBs showed good agreement between Soxhlet extraction and dynamic supercritical fluid extraction of soil samples with the added advantage that no cleanup step was needed with the latter (van der Velde et al. 1992). It, therefore, appears that some procedural details remain incompletely resolved and that, as with all extraction procedures, due attention should be directed both to the nature of the analyte and to that of the matrix being analyzed. A combined extraction using CO_2 and acetylation with acetic anhydride and triethylamine has been used for the analysis of chlorophenolic compounds in soil (Lee et al. 1992) and air-dried sediment samples (Lee et al. 1993). This is both rapid and gives results comparable with those using conventional methods of steam distillation.

Use of acidic CO_2 is clearly undesirable for the extraction of basic compounds, and N_2O has been used effectively for a number of amines (Mathiasson et al. 1989; Ashraf-Khorassani and Taylor 1990). Similarly, CO_2 was ineffective for extraction of 1-nitropyrene from diesel exhaust particulate matter, although this could be accomplished effectively using the freon $CHClF_2$ (Paschke et al. 1992). In conclusion, it seems safe to state the obvious: no single supercritical fluid is likely to be optimal for the extraction of structurally diverse analytes.

The combination of supercritical fluid chromatography and supercritical fluid extraction with standard GC and LC techniques offers an attractive extension to conventional procedures (Greibrokk 1992), and this will probably receive increased attention in the analysis of samples containing a large number of components.

Association: the recoverability of "bound" analytes — In the light of increasing evidence for the existence of "bound" residues that is discussed more fully in Chapter 3 (Section 3.2.3), relatively drastic procedures have been used for the analysis of both soil and sediment samples. Difficulties experienced in the recovery of even such apparently unreactive compounds as lindane have led to the suggestion of a procedure using BF_3 in methanol (Westcott and Worobey 1985), and extraction of glyphosate (N-phosphonomethylglycine) from contaminated soils was facilitated by the use of alkaline extraction (Miles and Moyne 1988) which appears to be a viable procedure for soils up to 56 d after spiking (Aubin and Smith 1992). These observations are relevant to the issue of "aging" that is particularly important for agrochemicals applied to the terrestrial environment. The principles are, however, equally relevant to aquatic sediments containing xenobiotics that have been deposited over many years. For example, application of chemical treatment has been used to make an empirical division between "free" (extractable with simple solvents) and "bound" (accessible only after chemical treatment such as methanolic alkali) concentrations of chloroguaiacols in contaminated sediments (Remberger et al. 1988): this distinction is, of course, both relative and pragmatic. In addition, it has emerged that there were clear differences in the relative recoverability of chloroguaiacols and chlorocatechols, and this was plausibly attributed to different types of association between the analytes and the sediment phase.

It cannot, therefore, be emphasized sufficiently that there is no such thing as the absolute concentration of a given analyte in an environmental sample. Significant alterations in recoverability occur after deposition (aging); measured concentrations are empirical and subject to an inherent indeterminacy.

2.2.6 Specific Matrices: Biota

Procedures for the extraction of aquatic biota — particularly fish — have been extensively developed and generally involve grinding with anhydrous sodium sulfate followed by extraction with a suitable solvent. An attractive alternative (Birkholz et al. 1988) has used homogenization of fish tissue with solid carbon dioxide; after removal of CO_2 by sublimation at -20°C, a finely divided powder is obtained, and this may either be extracted directly or freeze-dried and subjected to Soxhlet extraction. As is the case for samples of soil and sediment, supercritical fluid extraction has also been used for samples of biota, and the results have been compared with those using standard procedures: supercritical fluid extraction has found substantial application in the analysis of food products. After extraction, advantage may be taken of any of the procedures noted in Section 2.2.2 for removing interfering contaminants such as lipids which will generally be present in high concentration. For highly stable compounds such as chlorinated dibenzo-1,4-dioxins, quite drastic cleanup stages including, for example, treatment with sulphuric acid may be incorporated. In contrast to this, and to the situation with "bound" residues in soil or sediment samples, analytes in samples of biota may be present as conjugates — generally as metabolites of the compounds to which biota were initially exposed. Many of these are sensitive to acidic or basic reagents so that mild procedures for their analysis are obligatory. Further details of some of the reactions involved in the metabolism of xenobiotics by higher organisms are given in Chapter 7 (Section 7.5). Illustrative examples are the glycoside and sulfate conjugates of phenols which may be cleaved with acid or enzymatically or the glutathione conjugates which may be reduced with Raney nickel. Even neutral compounds may be tightly associated with lipid components of fish, and the extractability of both PAHs and related compounds (Vassilaros et al. 1982) and PCBs (de Boer 1988) is significantly improved by preliminary alkali treatment. Problems of accessibility may also be encountered during the analysis of cultures containing microbial cells which may effectively bind the analyte; this may be especially severe with Gram-positive bacteria for which treatment with acetonitrile combined with sonication has been found to be highly effective (Allard et al. 1985).

2.2.7 Specific Matrices: Atmospheric Deposition

This is not the place to discuss air sampling which is a highly sophisticated discipline. Since, however, organic compounds in the atmosphere are scavenged by rain and by sorption onto particulate matter, some of these compounds will ultimately enter the aquatic and terrestrial environments. Brief comments on procedures for collecting wet and dry deposition are, therefore, relevant especially since considerable alteration in the distribution of congeners of, for example, chlorinated dibenzo-1,4-dioxins may occur after emission into the atmosphere and before reaching the ultimate sinks (Czuczwa and Hites 1986; Koester and Hites 1992). Samples of rain may be filtered to remove particulate matter and then analyzed by procedures used for water samples, athough a considerable degree of concentration may be necessary if the concentrations of the analytes are low. Alternative procedures for the collection of dry deposition have been described using either inverted frisbee samplers or flat plates, both of which are coated with a thin layer of mineral oil to improve collection efficiency. Samples are then removed from the collectors by wiping with glass wool, and these are extracted by procedures used for dried soil or sediment samples (Koester and Hites 1992).

2.3 PROCEDURES INVOLVING CHEMICAL REACTIONS: DERIVATIZATION

Introduction — For GC or HPLC quantification or GC-MS identification, neutral compounds are, of necessity, analyzed without further chemical treatment. For compounds with functional groups such as hydroxyl, amino, carboxylic acid, or reactive carbonyl groups, however, it may be convenient to prepare suitable derivatives. This has a long tradition in organic chemistry highlighted by the use of crystalline phenylosazones of carbohydrates to prepare and separate otherwise intractable mixtures of non-crystalline carbohydrates (Fischer 1909). In the present context, there are several advantages in using such procedures:

- A degree of selectivity — and thereby purification — is automatically introduced.
- It is possible to prepare volatile compounds with shorter GC retention times than their precursors.
- By preparation of halogenated derivatives, advantage may be taken of the enhanced sensitivity of the EC detection system in GC analysis.

- The lifetime of GC columns is prolonged by avoiding exposure to reactive compounds such as phenols, carboxylic acids, or reactive amines.
- Reactive compounds such as phenyl ureas are thermally unstable and are advantageously derivatized before GC analysis (Karg 1993).
- For HPLC, advantage may be taken of the high sensitivity of fluorescent detection systems.

For GC analysis, several of these advantages may be combined by preparing heptafluorobutyrate esters of phenols or trichloroethyl esters of carboxylic acids. Although trimethylsilylation is widely used and gives derivatives with structurally valuable mass spectra, the trimethylsilyl ethers and, particularly, esters are often relatively unstable. The choice of derivatives should, therefore, take into consideration their suitability for the cleanup procedures that may subsequently be employed.

Derivatization may be carried out at any stage of the analysis: on the initial sample, on sample extracts, or on the GC column itself. Generally, cleaner samples for analysis are obtained when extracts are used, since direct derivatization of a soil or sediment sample will generally result in large numbers of interfering peaks on the chromatograms due to reaction of the reagent with organic components of the solid matrix as well as with the desired analytes.

Derivatization of the untreated sample may not be universally applicable, but attention is drawn to examples where this has been examined and successfully applied.

1. Methylation of 2,4-dichlorophenoxyacetic acid and 3,6-dichloro-2-methoxy-benzoic acid in soil samples was examined using trimethylphenylammonium hydroxide prior to supercritical extraction with CO_2 (Hawthorne et al. 1992). Although recoveries of these compounds were apparently limited by competition for the reagent of components of the soil matrix, this procedure possesses obvious advantages for polar analytes such as carboxylic acids and is capable of extension to other analytes.
2. Analysis of chlorophenolic compounds in sediment samples has been carried out by *in situ* acetylation followed by solvent extraction (Xie et al. 1985) or by combined supercritical fluid extraction with carbon dioxide and derivatization (Lee et al. 1992).
3. Low concentrations of volatile aldehydes in air samples have been analyzed by collection on porous glass impregnated with 5-dimethylaminonaphthalene-1-sulfohydrazide with which they react to form hydrazones that were analyzed on-line using HPLC with fluorescence detection (Nondek et al. 1992).

Specific procedures for derivatization — There is voluminous literature on procedures for derivatization though, in practice, only relatively few have achieved extensive application. The most valuable derivatives combine a number of features including chemical stability, physical properties which enable their incorporation into cleanup steps such as open column chromatography before analysis, and a high response to the detector system. Extensive compilations are available (Knapp 1979; Blau and Halket 1993) to which reference may be made. Attention should also be drawn to an extensive range of reagents with varying selectivity for silylating functional groups (Pierce 1979). Only a few illustrative examples of commonly used reactions are given here:

- Acylation of phenols, amines, and alcohols
- Esterification of carboxylic acids and sulfonic acids
- Formation of silyl ethers of alcohols and phenols
- Formation of ureas from alcohols by reaction with *iso*cyanates
- Formation of oximes from aldehydes and ketones.

Attention should also be directed to some additional procedures:

1. Single-stage reduction and trimethylsilylation of a number of hydroxylated naphthoquinones and anthraquinones has been carried out using *N*-methyl-*N*-trimethylsilyltrifluoroacetamide in the presence of NH_4I (Bakola-Christianopoulou et al. 1993).
2. The use of bifunctional reagents to derivatize several functional groups simultaneously:
   - di-*t*-butyldichlorosilane derivatization of bifunctional analytes such as 1:2-diols, 2- hydroxybenzoates, and catechols (Brooks and Cole 1988).
   - phosgene for 2-hydroxyamines related to catechol amines (Gyllenhaal and Vessman 1988).
3. The use of chiral reagents. The use of these greatly facilitates the analysis of the enantiomers of compounds that contain functional groups and is essentially an application of the traditional procedure for resolving optically active compounds. Most of the reagents are chiral compounds with reactive groups such as acyl chlorides, chloroformate esters, *iso*cyanates, or amines that are suitable for the preparation of derivatives by conventional procedures. The advantage of such procedures is that standard chromatographic columns may be used. Their application to the analysis of carboxylic acids may be illustrated by two examples:
   - The analysis of chiral propionic acid anti-inflammatory drugs and herbicides was carried out by converting the acids into their acid chlorides with thionyl chloride and then reacting them with *R*-1-phenylethylamine (Blessington et al. 1993).
   - For the analysis of isomers of the synthetic pyrethroid insecticide permethrin, enantiomeric amides were prepared using carbonyldiiminazole and a range of chiral amines of which *R*(-)- or *S*(+)-amphetamine gave baseline separation of the four isomers (Taylor et al. 1993).

It should also be appreciated that problems may arise even when functional groups are to be derivatized using apparently straightforward reactions. For example,

1. The susceptibility to hydrolysis of derivatives of nitrophenolic compounds (Hynning et al. 1989).
2. The difficulty of acetylating phenolic groups which have carbonyl groups in the para position, for example, vanillin without the use of pyridine as a catalyst (Neilson et al. 1988).

3. The difficulty of esterifying sterically hindered carboxylic acids such as the C_{18} carboxylic acid group in dehydroabietic acid by conventional procedures using alcohols and an acid catalyst. Diazoalkanes may be used or esterification carried out with pentafluorobenzyl bromide (Lee et al. 1990).

4. Diazomethane reacts rapidly and quantitatively with acidic hydroxyl groups such as carboxylic acids, though less readily even with phenols containing electron-attracting groups such as chlorine, and hardly at all with unsubstituted phenols. 2,2,2- trifluorodiazomethane is even more selective and does not react even with carboxylic acids, though it forms esters with the much more acidic sulfonic acids (Meese 1984).

Procedures using ion-pair alkylation and esterification have been effectively used in a number of cases, and a few illustrative examples may be given.

1. Esterification of chlorophenoxy acids and of pentachlorophenol has been assessed using tetrabutylammonium hydroxide and methyl iodide in methanol and shown to be as effective as the standard procedure using diazomethane (Hopper 1987). A further development has examined the application of methanol/ benzyltrialkylammonium for simultaneous extraction and methylation of 2,4-dichlorophenoxyacetic acid in soil (Li et al. 1991); the quaternary ammonium hydroxide may, in this case, also play a significant role in releasing the "bound" analyte.

2. Tetrabutyl ammonium hydroxide and dimethyl sulfate in dichloromethane have been used to methylate a number of aromatic hydroxyacids and hydroxyaldehydes (Ramaswamy et al. 1985).

3. An on-column methylation of dimethylphosphorothioate in urine samples used trimethylanilinium hydroxide in acetone (Moody et al. 1985).

Finally, attention is directed to procedures for less commonly encountered compounds for which specific procedures may advantageously be applied: one example is the analysis of substituted benzoquinones. For halogenated benzo-1,2-quinones, an analytical procedure was used in which these compounds reacted specifically with diazomethane to give methylenedioxy compounds (Remberger et al. 1991b). Similarly, analysis of some benzo-1,4-quinones (Remberger et al. 1991a) required reduction with dithionite since conventional reducing agents, such as ascorbate, were insufficiently active. In neither case would conventional procedures have been satisfactory. These examples illustrate the value of using chemical methods directed to specific functional groups as an alternative to exclusive reliance on traditional procedures that have been used for alcohols, phenols, or carboxylic acids.

It should be recognized that some derivatives, such as the O-acetates of phenols, give mass spectra with extremely weak — or even lacking — parent ions, so that it may be useful to prepare a range of derivatives, some of which may provide more informative mass spectra (Wretensjö et al. 1990). In addition, the possible thermal instability of compounds during passage through GC columns may present serious limitations to application of this technique; at the expense of resolution, shorter columns may sometimes be used to overcome this problem.

The same principles have been applied — although to a considerably lesser extent — to HPLC analysis for which underivatized samples are most frequently used. For compounds without UV absorption, however, it may be convenient to prepare derivatives with suitable absorption or fluorescence for the detector system. A few illustrative examples may be given:

1. Alcohols may be converted into benzoate esters (Fitzpatrick and Siggia 1973) or into fluorescent derivatives by reaction with 9-fluorenylmethyl chloroformate (Miles and Moyne 1988).
2. Amines may be condensed with 1,2-phthaldialdehyde (Lindroth and Mopper 1979).
3. Thiols may be alkylated with 3,6,7-trimethyl-4-bromomethyl-1,5-diazabicyclo-[3.3.0]-octa-3,6-diene-2,8-dione (Velury and Howell 1988).

In summary, almost all reactions that have traditionally been used for preparing derivatives in organic chemistry have been effectively applied — modified as necessary to increase their sensitivity to the appropriate detector systems.

2.4 IDENTIFICATION AND QUANTIFICATION: BASIC DEFINITIONS

It is highly desirable to avoid any misunderstanding of these operational terms; they are, therefore, defined as they will be used in this account.

Identification — Use of the term clearly implies that it has been unambiguously established that the structure of an unknown compound is identical to that of an authentic standard. Comparisons are most frequently made using infrared, mass, or nuclear magnetic resonance spectroscopy. These procedures are discussed in more detail below.

Quantification — Quantification is the determination of the concentration of a compound in a given matrix. For water samples, concentration units in terms of volume are universally used, but for other matrices alternatives to volume have to be used for normalization. For soils and sediments, results are usually expressed relative to the total wet (or dry) weight — or preferably to organic carbon determined, for example, by loss after combustion of acidified samples. For biota, the total wet (or dry) weight may be used — or the fat or lipid content. Limits of detection are determined by the background levels, i.e., noise levels, in the instrumentation used. As a working rule, the detection limit may be taken as three times, and the quantification limit as ten times, the standard deviation of background levels (Keith et al. 1983).

2.4.1 Experimental Techniques: Identification

Several identification procedures are available, but two cardinal criteria should always be fulfilled:

1. Comparison should be made using several parameters — this practice is hallowed by historical application — combustion analysis of three independent

derivatives for the establishment of the empirical formula of unknown compounds or the establishment of identity by comparison of the R_f values in paper chromatographic systems using three different solvent systems, and
2. The obvious fact is that samples of authentic compounds must be available; in many cases — and particularly for metabolites — these may not be commercially available and must, therefore, be synthesized by the laboratory carrying out the investigation. This is discussed in more detail below.

Identification most often relies upon comparison of the gas chromatographic retention time, together with the mass spectrum, or of the nuclear magnetic resonance spectrum with those of authentic reference compounds. Use of GC or LC retention times alone is not adequate, although for GC this may be an acceptable compromise procedure if several derivatives are available. Use of only the mass spectrum is clearly unacceptable since isomers generally provide essentially identical spectra. Naturally, compounds whose structures are unknown may be encountered; this presents a much greater challenge since comparison is not then possible. Determination of the structure of such compounds must, therefore, rely upon interpretation of the mass spectrum, or the NMR spectrum, or ultimately the results of an X-ray diffraction study; the last has been effectively applied to the resolution of some important problems including, for example, the absolute stereochemistry of the *cis*-toluene dihydrodiol produced from toluene by *Pseudomonas putida* F1 (Kobal et al. 1973), the structure of 3,4,5-trichloroguaiacol (Lindström and Österberg 1980), and the structure of important congeners in technical chlordane (Dearth and Hites 1991). It should be noted that X-ray structure analysis has become significantly more accessible with the advent of modern high-speed computing facilities and that its application is possibly limited only by the need for single crystals of the pure compounds.

Mass spectrometry — Identification of components in environmental samples and in samples from laboratory studies of biodegradation and biotransformation is generally based on the application of mass spectrometry coupled to either GC or LC systems. For environmental samples which may contain only small amounts of the relevant compounds, mass spectrometry is particularly attractive in view of the extremely small amounts of samples — of the order of nanograms — which are required. There is an important additional advantage: since the mass spectrometer can be interfaced with GC or LC systems which incorporate separation procedures, pure samples are not required. Probably most investigations have used electron impact ionization, but the greater sensitivity of chemical ionization is, in some instances, attractive and may enable use of underivatized samples (de Witt et al. 1988). An application of negative ion chemical ionization for quantification is noted below. Ideally, comparison should be made between the complete mass spectrum of the analyte and that of the reference compound. If very low amounts of the analyte are available, it may not be feasible to obtain a complete spectrum, and it may be necessary to carry out the comparison by monitoring selected ions determined from the mass spectrum of the authentic compound. Many compilations of suitable ions for a range of widely encountered environmental contaminants have been provided.

It is appropriate to draw attention briefly to a few specific issues.

1. Isomers generally yield essentially identical electron-impact mass spectra although halogenated biphenyls carrying substituents at both the 2- and 2'-positions yield characteristic spectra in which loss of one halogen atom yields an intense ion fragment (Sovocool et al. 1987), and aromatic dinitro compounds substituted in the *ortho-* or *peri-*positions display unique fragmentation pathways (Ramdahl et al. 1988).

2. Interest in bromine-containing fire retardants and contaminants, such as brominated dibenzo-1,4-dioxins and dibenzofurans, has led to increased application of both negative chemical ionization (Buser 1986) and electron impact (Donnelly et al. 1987) procedures.

3. Current developments in mass spectrometry are directed to the application of alternative ionization methods for sensitive molecules, to the use of tandem systems (Durand et al. 1992), and to the development of interfaces with chromatographic systems (Arpino 1990; Niessen et al. 1991; Arpino et al. 1993). Particle beam LC/MS systems, which provide spectra compatible with conventional EI spectra and enable analysis of compounds unsuitable for GC analysis, seem promising (Kim et al. 1991; Behymer et al. 1990), while the inherent difficulties with LC interfaces may be overcome in some cases by use of supercritical fluid chromatographic systems (Arpino 1990). For some kinds of compounds, however, LC/MS systems using fast atom bombardment or electrospray may be more suitable.

4. A computer program has been developed to use the mass-selective detector as a chlorine-selective detector (LaBrosse and Anderegg 1984; Johnsen and Kolset 1988), but it does not appear to have been widely evaluated. Attention should be directed to three situations which may give rise to error in its application:

   - Compounds from which HCl is eliminated during electron impact so that no characteristic isotope pattern is observed
   - Compounds containing elements other than chlorine, such as bromine or sulfur, which compromise the isotope pattern
   - Compounds lacking chlorine atoms but which fortuitously give clusters of peaks with distributions resembling or identical to those expected for molecules with chlorine substituents.

Whereas none of these problems is unique to use of such detector systems, they clearly underscore the care that should be exercised by the conscientious mass spectroscopist.

Ideally, authentic samples of pure compounds should be examined under identical operating conditions, but frequent use is made of published spectra or of those contained in libraries of spectra that are directly coupled to the data system of the mass spectrometer. Although these compilations contain mass spectral data for a large number of compounds, care in their use should be exercised since operational conditions, such as the amount of the compound used for analysis and the accelerating voltage, may differ from those used in a given laboratory. Where library MS spectra are not available, the relevant compounds must be prepared by the laboratory itself, and this may be a demanding exercise. Examples of the effort required may be illustrated by the synthesis of the 209 PCB congeners (Mullin et al. 1984), the 75 congeners of chlorinated dibenzo-1,4-dioxins, or the complete set of chlorinated guaiacols and catechols. Without access to these reference compounds, none of the studies dependent on their availability could have been accomplished. On the other hand, the mass spectra of even fewer microbial or other metabolites or transformation products currently exist in library catalogues so that these compounds must almost always be synthesized. A few examples may be given as illustration:

1. Various hydroxylated tri- and tetrachlorodibenzofurans have been synthesized (Burka and Overstreet 1989) since, as conjugates, these could plausibly be expected to be mammalian metabolites of 2,3,7,8-tetrachlorodibenzofuran. A number of methoxy, dimethoxy, and dihydroxy derivatives of 2,3,7,8-tetrachlorodibenzo-1,4-dioxin have also been synthesized for similar reasons (Singh and Kumar 1993).
2. Possible metabolites of dichlorop-methyl (methyl 2-[4-(2,4- dichlorophenoxy)-phenoxy]propanoate) produced by plant-resistant species have been synthesized (Tanaka et al. 1990) in order to facilitate identification and quantification.
3. Decarboxylation products of dehydroabietic acid and 12,14-dichlorodehydro-abietic acid have been synthesized (Hynning et al. 1993) to enable the conclusive identification of the 18-*nor* and 19-*nor* compounds in environmental samples of fish and sediment.
4. A series of hydroxy-, nitrooxy-, and dinitrooxyalkanes that are produced by reactions between alkenes and hydroxy and nitrate radicals were synthesized to develop procedures both for their collection from atmospheric samples and to facilitate their identification and analysis (Muthuramu et al. 1993).
5. Three trichlorodiphenyl sulfides were synthesized and their structures determined by ^{1}H and ^{13}C NMR, and these were used for comparison with compounds in pulp mill effluents and stack gas samples (Sinkkonen et al. 1993).
6. Dimethyl sulfones are ultimate mammalian metabolites of PCBs produced by a series of reactions involving arene oxides and glutathione conjugates and have been synthesized to enable their conclusive identification and quantification (Bergman and Wachmeister 1978; Haraguchi et al. 1987).

Purely analytical work may, therefore, profitably be supplemented with competent — and sometimes extensive — organic synthetic activity. As noted in Section 2.5, however, great care must be taken to ensure that samples for analysis do not become contaminated in the laboratory with compounds synthesized in very much larger — in many cases, gram — quantities.

Non-destructive procedures — In contrast to GC-MS procedures, NMR methods are non-destructive though they require access to essentially pure samples. In addition, relatively large amounts of samples may be required especially if advantage is to be taken of natural levels of ^{13}C in the samples. NMR has, however, been effectively used for many years in identifying microbial metabolites where relatively large amounts of sample can readily be obtained. Its application to environmental samples is more restricted though a good illustration is the determination of the structures of the two principal components of toxaphene congeners isolated from biota (Stern et al. 1992). ^{1}H NMR showed these to be octachloro and nanochloro congeners, both of which display chirality although the enantiomers were not resolved in this study (Figure 2.1). Resolution of enantiomers by NMR may be achieved using a range of chiral shift reagents that are complexes of the rare earth

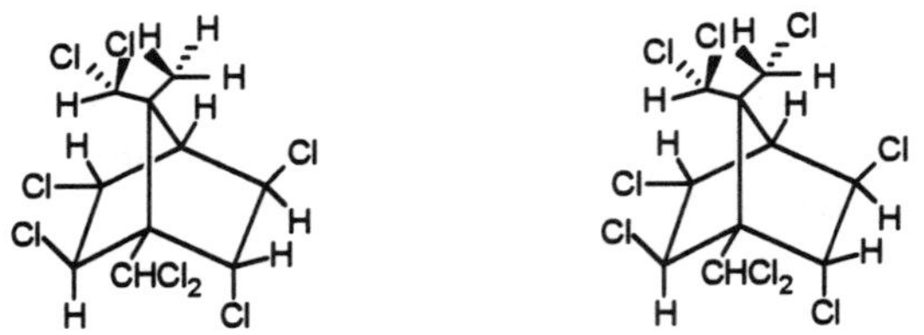

Figure 2.1 Structures of octachloro- and nonachlorocamphanes.

trivalent cations europium, praseodymium, and ytterbium with enantiomeric *R* and
S-camphor. When, however, only small quantities of sample are available and where
several compounds are present, the use of chromatographic methods that have
already been discussed is clearly advantageous. NMR has been increasingly used for
in vivo investigations of microbial degradation since samples containing cell suspensions and the ^{13}C-labelled substrates can be examined directly and the kinetics of the
appropriate resonance signals monitored. Further details of the application of NMR
to metabolic problems using both ^{13}C and ^{19}F are given in Chapter 5 (Sections 5.5.3
and 5.5.4). An example of the application of ^{13}C NMR to study the interaction of 2,4-
dichlorophenol with humic acid is given in Chapter 3 (Section 3.2.3).

    More restricted has been use of the ^{15}N, ^{17}O, and ^{31}P nuclei, but a few illustrative
examples may be cited:

1. The application of ^{15}N NMR to the structural determination of humic material
   is noted in Chapter 3 (Section 3.2.3).
2. Limited application of ^{17}O NMR has been made to a restricted range of chlorinated aromatic compounds (Kolehmainen et al. 1992).
3. ^{31}P NMR has been used to examine the effectiveness of pentachlorophenol on
   the energy metabolism of abalone (*Haliotis rufescens*) (Tjeerdema et al. 1991).

    Use of infrared spectra for identification seems to have declined, possibly because
of the need for pure and relatively large amounts of material. With the availability
of interface systems for GC and FTIR, and increasing interest in SCF FTIR interfaces
(Bartle et al. 1989), the situation may change. Use of FTIR methods is particularly
attractive for compounds containing carbonyl groups, such as aldehydes, ketones,
esters, lactones, and quinones, which have highly characteristic C=O stretching
frequencies well separated from other absorptions. A good example of the combined
use of MS and FTIR is provided by a study of the bacterial metabolites produced
from methylbenzothiophenes (Saflic et al. 1992). This is a particularly favorable case
since the relevant absorptions are strong compared to others in the spectra so that
effective use could be made of IR absorptions both for the S-O stretching frequencies
of the sulfoxides and for the C=O stretching frequencies of the benzothiophene-2,3-
diones. As with mass spectra, substantial libraries of IR spectra are available though,
once again, these should be used with due attention to the conditions under which the
spectra were obtained.

    Where the chirality of the product is of interest, optical rotatory dispersion and
circular dichroism may be effectively used, for example, in the determination of the
configuration of the dihydrodiol metabolites formed during fungal metabolism of
7,12-dimethylbenz[a]anthracene (McMillan et al. 1987) and in combination with

HPLC using a chiral stationary phase (Section 2.4.2) to resolve details of the metabolism of phenanthrene by several fungi (Sutherland et al. 1993). This is discussed further in Chapter 6 (Section 6.2.2). Reference has already been made to the use of X-ray diffraction.

2.4.2 Experimental Techniques: Quantification

In many cases, the identity of the analyte will be known; nonetheless, it is highly desirable that this be confirmed to avoid the possibility that an interfering compound has fortuitously the same GC retention time as that of the desired analyte. Indeed, many protocols that are now advocated use mass spectrometric systems so that this control is automatically incorporated.

Any measuring system may, in principle, be used for quantification provided that it produces a linear response between the output signal and the concentration of the analyte over a suitable range. Greatest use has probably been made of gas-liquid chromatography (GC) and high pressure liquid chromatography (HPLC) systems interfaced with appropriate detector systems; chromatographic systems may incorporate temperature gradients (GC) or gradients in the composition of mixed mobile phases (HPLC). Use of supercritical fluids (Section 2.2.5) incorporating, for example, carbon dioxide offers a promising alternative technique (Schoenmakers and Uunk 1987) to conventional GC or HPLC. For GC analysis, it should be appreciated, however, that the use of a single type of column packing or of a single detector system can never be optimal for all kinds of compounds. For quantification, translation of the detector response into amounts, and thereby concentrations, clearly requires access to authentic compounds. These will, however, also be needed for unambiguous identification in the first place so that they will generally be available. Attention should also be drawn to increasing interest in capillary zone electrophoresis that is applicable to ionizable compounds including carboxylic acids and quaternary nitrogen compounds. Its potential may be illustrated by application to the analysis of phenoxyacids and the effective separation of the enantiomers of phenoxypropionic acids (Nielen 1993) and of explosive residues containing 2,4,6-trinitrotoluene transformation products (Kleiböhmer at al. 1993).

For GC systems, it is convenient to note briefly developments in the two basic components — the column and the detector system.

GC columns — The introduction of capillary columns brought about a revolution in the level of resolution, and these columns are now almost universally used. Nonetheless, no single column is able, for example, to resolve all of the 209 PCB congeners so that the application of two columns in tandem with separate ovens and detector systems has been advocated. Fraction cuts from one column are quantitatively transferred to the other (Duinker et al. 1988). The value of the procedure has been graphically illustrated by the possibility of unambiguously quantifying toxic congeners such as 3,3',4,4'-tetrachlorobiphenyl (IUPAC No. 77) or 2',3,4,4',5-pentachlorobiphenyl (IUPAC No. 123) which, otherwise, co-elute with other congeners. Analytical protocols for the analysis of the toxic PCB congeners IUPAC 77, 126 (3,3',4,4',5-pentachlorobiphenyl) and 169 (3,3',4,4',5,5'-hexachlorobiphenyl) using conventional cleanup procedures which are compatible with those used for chlorinated dibenzo-1,4-dioxins and dibenzofurans have, however, been developed (Harrad et al. 1992). Use of parallel columns after splitting the sample prior to injection has been used for a range of aromatic nitro compounds (Lopez-Avila et al. 1991).

Another development is the use of chiral support phases. Determination of the chirality of analytes may be extremely important since the generally significant difference in the biological activity of enantiomers is well established. Striking examples are provided by (*R*)-asparagine that is sweet whereas the (*S*) enantiomer is bitter, or D-penicillamine that is antiarthritic whereas the L-enantiomer is toxic. In general, only one of the enantiomers of many compounds displays the desired biological activity, and this means that up to half of many compounds that have traditionally been used as rodenticides, insecticides, fungicides, and herbicides is biologically inactive towards the target organism and that this, therefore, contributes unnecessarily to the environmental burden of xenobiotics. Chiral support phases have, therefore, found important applications in a number of investigations. Detailed accounts have been provided of the methods for selected compounds, together with valuable chromatographic details (Armstrong et al. 1993). A few examples will be used to illustrate the potentially great scope for the application of these columns.

1. A stationary phase consisting of permethylated heptakis (2,3,6-tri-O-methyl)-β-cyclodextrin diluted with polysiloxane was used for the analysis of hexachlorocyclohexane isomers (Müller et al. 1992). Although the α isomer is optically active, the synthetic product is racemic; nonetheless, the ratio of the enantiomers in environmental samples may deviate considerably from unity (Mössner et al. 1992). Application of the same chiral support phase revealed similar deviations in the composition of epimeric octachloro- and nonachloro-chlordanes from samples of biota including an Adélie penguin (*Pygoscalis adeliae*) from Ross Island Antarctica (Buser et al. 1992a; Buser and Müller 1992). It seems almost certain that biological activity mediated either by bacteria or by higher organisms is the explanation for these deviations from a ratio of 1:1, although care should be exercised in the interpretation of enantiomeric ratios close to unity since there is evidence that some of the products from the chlorination of camphene are not racemic and contain a slight enantiomeric excess of some congeners (Buser and Müller 1994). A range of chiral stationary phases based on cyclodextrins has also been used for the resolution of enantiomeric hydrocarbon components which are of interest as biomarkers (Armstrong et al. 1991).

2. Stereospecific synthesis is of enormous importance in biotechnology in view of the generally divergent biological effects of enantiomers. A few examples of this are given in Chapter 6 (Sections 6.11.1 and 6.11.2), and only one illustration will be given here. The indanols formed by microbiological oxidation of indan with a strain of *Escherichia coli* containing the toluene dioxygenase genes from *Pseudomonas putida* F1 were separated after conversion into their *iso*propylurethanes on an XE-60-(*S*)-valine-(*S*)-phenylethylamide fused silica column (Brand et al. 1992).

GC detector systems — A variety of detection systems has been used employing flame ionization, electron capture — which is particularly sensitive to most halogenated compounds — N/P-sensitive, and mass selective detectors. It should be emphasized that the EC response of a compound is dependent not only upon the number of halogen atoms but also upon their position in the molecule; estimates of response based on different isomers may, therefore, be totally unreliable. Illustrative examples may be found among isomers of PCBs, trichlorophenols, and monochlorocatechols. Increasingly, Hall detector systems have been used because of their combined sensitivity to organochlorine compounds and their selectivity. In a convenient splitting system, this detector may be combined with a traditional FID system (Dahlgran 1981).

Although FID and EC detection systems have been by far the most widely used, the mass selective detector is highly attractive and is becoming incorporated into many standardized protocols; mass spectral comparison with authentic compounds — and thereby identification — is simultaneously made available. These quantification procedures may make use of labelled standards incorporating either ^{13}C or ^2H. Suitable compounds may not, however, be commercially available and may have to be synthesized; in such cases, synthesis of ^2H-labelled compounds may be more readily and economically accomplished. Selected ion monitoring has also been increasingly used — particularly for the quantification of chlorinated dibenzo-1,4-dioxins in environmental samples in which the concentrations may be extremely low — and its application with negative ion chemical ionization has been effectively applied to analysis of PCB congeners in fish (Schmidt and Hesselberg 1992) and toxaphene in arctic biota (Bidleman et al. 1993).

Procedures for the analysis of organic compounds of elements such as As and Sn lie beyond the scope of this book. For the sake of completeness, however, it may be noted that although analysis of organic As compounds has widely used atomic absorption detection systems, these have been complemented for organic As compounds by mass spectrometric procedures (Christakopolous et al. 1988) and for organic Sn compounds by procedures that have taken advantage of both flame photometric and mass spectrometric systems (Jackson et al. 1982).

HPLC systems — Quantification may also use HPLC systems, though the sensitivity is somewhat less that that attained by optimal GC procedures using EC detector systems. Probably GC and HPLC systems should be regarded as complementary, and both are invaluable in the appropriate applications. Ultraviolet or fluoresence detector systems are most generally used, although for specific groups of compounds such as catechols electrochemical detectors have proved valuable. Fluorescence has been extensively used in monitoring polycyclic aromatic hydrocarbons in environmental samples. Caution should be exercised since it has been shown that the major polycyclic aromatic components of polychaete samples were not anthropogenic. These fluorescent compounds have been identified as alkylated octahydrochrysenes putatively originating from triterpenes by microbial dehydrogenation of rings A and B (Farrington et al. 1986). Conclusive identification of the analytes should, therefore, be incorporated into a more specific protocol for HPLC analysis. This is only one example of a much wider spectrum of transformations of di- and triterpenes to aromatic compounds that take place in sediments after deposition; as a result, a structurally wide range of polycyclic compounds containing one or more aromatic rings has been isolated from sediments, oils, coal, and amber (Wakeham et al. 1980; Tan and Heit 1981; ten Haven et al. 1992). Conclusive identification of such compounds in samples of sediment and biota by MS or NMR is, therefore, necessary. HPLC has also been applied to the separation of the enantiomers of 2-(2,4-dichlorophenoxy)propionic acids after microbial degradation

(Ludwig et al. 1992) using α_1-AGP as the chiral stationary phase. Increasing interest in the separation of enantiomers that has been noted above will certainly encourage the further application of such methods.

2.4.3 Application of Immunological Assays

The preparation of environmental samples for analysis is often complex, time-consuming, and expensive. Attempts have, therefore, been made to circumvent this by using immunological assays. Enzyme-linked immunosorbent assays (ELISA) have been extensively applied to screening a number of environmental matrices and samples of various foods for the presence of agrochemical residues (Vanderlaan et al. 1988; Van Emon and Lopez-Avila 1992). Nonetheless, although these have been applied increasingly to a number of pesticides and to chlorinated dibenzo-1,4-dioxins, there are a number of technical problems including their relatively low sensitivity, their lack of specificity, and interference from other compounds. An ELISA system has been evaluated using a polyclonal antibody system for the analysis of triazine herbicides, and this was effective at the low concentrations (< ca. 2 µg/L) that are encountered in field situations. Confirmation of the results by GC-MS is still necessary, however (Thurman et al. 1990). This was clearly illustrated in this case since the structurally related triazines ametryn, prometon, and related compounds displayed significant cross-reactivity. On the other hand, atrazine metabolites give much lower response. The ELISA system has been effectively applied to monitoring a group of triazine herbicides in water courses over a wide area in the corn- and soybean-growing areas of the United States (Thurman et al. 1992). Further development could make this an even more attractive method for monitoring concentrations of a range of xenobiotics in large numbers of samples. Procedures for sample preparation, though considerably less complex than those used for conventional chemical analysis, appear to merit further development.

2.5 GENERAL COMMENTS: ORGANIZATION AND PRECAUTIONS

General issues — In spite of the fact that this is not intended to be a handbook, some very brief comments on what may be considered organizational aspects of laboratory practice may not be out of place. These may be considered within the wider context of precautions that should be exercised in all analytical laboratories. First of all, attention is drawn to the important principles of quality control, to procedures for numerical analysis of data, and to the important issue of documentation, all of which have been covered succinctly in a review (Keith et al. 1983) and more extensively in books (Keith 1988, 1991). It cannot be too strongly emphasized that the analyst is part of a team and that he or she should play an active part in both the planning and the execution of the proposed investigation; thereby, many pitfalls — and unnecessary irritation — may be avoided. For example, some conflict may arise over the number of samples required to answer the specific questions that are posed, and resolution of this issue should take priority in planning discussions. In addition, the level of accuracy should be decided at the outset and care taken that sufficient samples are available for duplicates to be preserved for re-analysis, if necessary — and that these are preserved in an acceptable manner.

It has already been noted in Section 2.4.1 that authentic compounds may need to be synthesized by the laboratory which also carries out analysis. Whereas this is a convenient — and indeed, desirable — activity, the possibility of cross-contamination should not be underestimated. It is clearly hazardous for an analyst working with possibly nanograms of a compound to be in the close vicinity of a synthetic chemist producing gram quantities of the same substance. These two operations should, therefore, be separated, and particular care should be given to the design of, and access to, the facility in which synthesis is carried out (Alexander et al. 1986). The operation of the ventilation system should be foolproof, and residues from synthetic activity must not enter communicating waste systems either for water or air. A good example of the effort which may be needed to trace the source of contaminants is provided by the occurrence of 1,3,6,8- and 1,3,7,9-tetrachlorodibenzo-1,4-dioxin in a laboratory for dioxin analysis; these compounds had not been synthesized in the laboratory but gained access through the use of a phenol-based cleaning solution in which these compounds were a minor impurity (Alexander et al. 1986).

Laboratory practice — A valuable review (Duinker et al. 1991) has summarized some problems encountered during analysis of PCBs, and these are of sufficient general importance to merit brief summary.

1. The purity of standards should be checked, and for GC this should include not only EC detection systems but FID systems to detect the presence of non-halogenated impurities.
2. The purity of solvents should not be taken for granted, and if redistillation is practiced, this should be carried out in a stream of N_2 to eliminate contamination by volatile laboratory contaminants. Cross-contamination is a potentially serious problem, and steps should be taken to ensure that samples, standards, and solvents are not exposed to this hazard.
3. Procedures for presenting numerical data should be critically reviewed and attempts made to determine the error in the analytical results.
4. Optimum conditions in operating GC equipment, including technical aspects of injection procedures, should be maintained and checked periodically. Although this may seem self-evident, clear evidence of its importance — and neglect — emerged from the results of an inter-laboratory study of the analysis of PCBs and agrochemicals in environmental samples (Alford-Stevens et al. 1988).
5. Attention should be directed to the important question of toxic hazards associated with some reagents that may be used. If the user is not familiar with the hazards, how to cope with accidental spillage, and emergency procedures, these compounds should not be used. With proper knowledge, good experimental practice, and adequate ventilation systems, the careful experimenter may, however, safely handle such compounds. Without attempting any numerical comparison of hazard, the most dangerous compounds commonly encountered are aliphatic diazo compounds, such as diazomethane and diazoethane, dimethyl and diethyl sulfate, dicyclohexylcarbodiimide, and phosgene. Exposure to all solvents — particularly aromatic solvents, including benzene — should be minimized, and solvents such as diethyl ether and dioxan, which may form explosive peroxides under some conditions, should be stored in darkness. In addition, these solvents generally contain peroxide inhibitors. The synthetic

chemist is exposed to a much wider range of hazardous compounds, but such discussion lies well beyond the scope of this brief summary.

Flexibility in operation: an open approach — Standardized procedures have been developed for a number of classes of analyte, but these should not be followed blindly. It has been clearly demonstrated that problems of recovery may be encountered due to the occurrence of unprecedented chemical reactions, for example, in the analysis of phenols in water samples (Chen et al. 1991). Disappointing agreement between laboratories in an intercalibration study in which PCBs in water samples were analyzed could be attributed both to numerical errors in calculation and to unacceptable operation of gas chromatographs (Alford-Stevens et al. 1988). Similar variability has been encountered during analysis of fish samples for a number of organochlorine compounds, and it has been suggested that for regulatory purposes a range of concentrations should be set rather than an absolute value (Miskiewicz and Gibbs 1992).

Problems encountered in the analysis of soils and sediments have already been noted, and it is clear that after deposition many xenobiotics are bound to organic constituents and are, therefore, not accessible by simple solvent extraction. The problem of alteration in the degree of accessibility after deposition (aging) is unresolvable. Plausible chemical reactions between the xenobiotic and organic components of the matrix are discussed in Chapter 3 (Section 3.2.3). The extent of their reversibility is generally unpredictable, so that increasing attention has been directed to purely empirical procedures for the analysis of these "bound" residues. It should be emphasized that the degree of recoverability of an analyte from a given matrix cannot validly be assessed from the results of short-term spiking experiments and that an inherent degree of indeterminacy in the concentrations of xenobiotics in such matrices must be accepted. For this reason, reference samples of various matrices and containing a variety of analytes for example, PAHs, neutral organochlorine compounds, agrochemicals, and including chlorophenols have been prepared to provide more realistic material.

Automation has truly revolutionized analysis, but the contribution of the skilled analyst has increased rather than diminished. The occurrence of extraneous peaks in chromatograms due to unexpected compounds should not be merely accepted blindly, but their structures should be verified. A good example is the occurrence in sediment samples of a compound which had the same GC retention time as PCB 77 (3,3′,4,4′-tetrachlorobiphenyl) but which on further examination was revealed as a C_{22} alkane (Mudroch et al. 1992). Pitfalls in interpreting mass spectra should be realized, and identification that relies solely on library searches should be accepted with caution (Swallow et al. 1988). The meaning of the term "identity" should be critically interpreted and the qualifier "tentative" added, if necessary, to avoid possible misunderstanding.

The spectrum of analytes — Attention should always be directed to the presence in environmental samples of compounds not previously encountered. The occurrence of metabolites and transformation products may require the development of specific procedures for their quantification, and these may be radically different from those used for their precursors. This may also necessitate substantial synthetic organic activity to provide authentic compounds (Section 2.4.1). In addition, the occurrence of such transformation products in environmental matrices complicates the evaluation of data from interlaboratory analysis of samples if appropriate attention has not been directed to this possibility. Among the more important sources of these novel analytes are:

- Chemical transformations during combustion, whereby the products are released into the atmosphere and then enter the aquatic and terrestrial environment in the form of precipitation or particulate matter, e.g., soot
- Reactions in the atmosphere including photochemical transformations
- Biotransformation reactions in aquatic and terrestrial systems
- Abiotic and biotic transformations in the sediment phase.

Further details of reactions mediated by microorganisms will be found in Chapter 4 (Sections 4.1.1 and 4.1.2), throughout Chapter 6, and by higher biota in Chapter 7 (Section 7.5). Only a brief summary will be given here to illustrate some of the widely occurring chemical and photochemical reactions together with a few examples of biotransformation. The transformation of naturally occurring terpenes has already been referred to in Section 2.4.2.

1. The formation during combustion of a range of chlorinated aromatic compounds including polychlorinated benzenes, styrenes, and naphthalenes (Yasahura and Morita 1988) and chlorobenzoic acids (Mowrer and Nordin 1987).
2. The production of nitroaromatic compounds by both combustion and by photochemical reactions of polycyclic hydrocarbons with oxides of nitrogen and oxygenating free radicals (Nielsen et al. 1983).
3. The formation of ketonic and quinonoid derivatives of aromatic hydrocarbons during combustion (Alsberg et al. 1985; Levsen 1988). These transformation products have been identified, for example, in marine sediment samples (Fernandéz et al. 1992) and in fish from a contaminated river (Vassilaros et al. 1982).
4. The identification of steranes and triterpene hydrocarbons in vehicle particulate emissions and that probably arise from lubricants (Rogge et al. 1993).
5. Unsaturated carboxylic acids are photooxidized to C_8 and C_9 dicarboxylic acids that have been detected in atmospheric particle samples and in recent sediments (Stephanou 1992).
6. Tris(4-chlorophenyl)methanol whose source has not been determined but has been shown to have a global distribution in birds and marine mammals (Jarman et al. 1992).
7. Biotransformation reactions of the original xenobiotic may result in compounds with fundamentally different physical properties and toxicities. The apparently ubiquitous distribution of halogenated anisoles (Wittlinger and Ballschmiter 1990) attests to the importance of O-methylation reactions that are discussed in Chapter 6 (Section 6.11.4).
8. The demonstration and partial identification of a wide range of methyl sulfone substituted PCBs as metabolites in samples from grey seals (*Halichoerus grypus*) serves as further confirmation (Buser et al. 1992b) of a previously established (Jensen and Jansson 1976) metabolic pathway. Both sulfone and sulfoxide metabolites of pentachlorobenzene have been identified in samples of parsnips and were presumed to have been translocated from the soil (Cairns et al. 1987).

Multi-component commercial products and effluents — Greatest attention has been directed in the preceding sections to single compounds since these represent an analytically accessible problem and serve to illustrate most of the principles to which this chapter is devoted. It is frequently the results of such analyses that are used in environmental hazard assessments which require data on distribution, toxicity, and persistence. In practice, however, many commercial products are complex mixtures of which PCBs are probably the most conspicuous. These are, however, by no means the only examples, and the following may be noted: polychlorinated naphthalenes and terphenyls, chlorinated paraffins, and the chlorinated monoterpenes, toxaphene and chlordane. All of these present serious problems for their quantification and raise the specter of sum versus congener-specific analysis which has aroused so much controversy in the case of PCBs. Attention may also be directed to creosote and to alkylphenol polyethoxylates and their halogenated analogues that have been the object of extensive studies (Ahel et al. 1987; Stephanou et al. 1988; Wahlberg et al. 1990). Many of these compounds have apparently global distributions (Chapter 3, Section 3.5) and, therefore, present potentially serious environmental problems. The cardinal issue in their identification and quantification in environmental samples is, of course, the availability of authentic standard compounds. This has been discussed in Section 2.4.1. Whereas authentic standards of PCBs, PCDDs, and PCDFs are commercially available, increased interest in the distribution of toxaphene has drawn attention to the need for authentic samples of both chlorinated bornanes and bornenes since — as with PCBs — substantial alteration in the distribution of the various congeners has generally occurred after discharge into the environment (Bidleman et al. 1993).

Industrial effluents almost always contain a wide range of compounds, often by-products of manufacture, and their analysis may present formidable problems both in identification and quantification. An illustrative example of site-specific contaminants is provided by the recovery from samples of sediment and fish in the Niagara River - Lake Ontario system of a benzophenone, a difluorodiphenylmethane and several biphenyls — by-products from the manufacture of 4-chloro(trifluoromethyl)benzene — all of which contained chloro and trifluoromethyl substituents (Jaff and Hites 1985). In addition, these effluents may contain polymeric material with which monomers are associated. A classic — and possibly unique — example is the plethora of compounds produced during the manufacture of bleached pulp by conventional technologies using molecular chlorine; these effluents contain both low molecular weight components and high molecular weight "chlorolignin" (Neilson et al. 1991). A recent investigation (Jokela and Salkinoja-Salonen 1992) has revealed a new facet of the composition of such effluents. Analysis of the molecular weight distribution of these compounds by size exclusion chromatography (SEC) and by ultrafiltration has revealed significant differences in the results obtained by the two procedures. Micelle formation during ultrafiltration, which is broken by dilution, and association phenomena in aqueous solutions that are analyzed by size exclusion chromatography led to significantly revised estimates of the molecular weight distribution of the components. In contrast to previous estimates, ca. 85 to 95% of the organochlorine constituents have molecular weights < 1000. The importance of these observations, and of associations in general, should, therefore, be critically evaluated in analyzing complex effluents.

2.6 CONCLUSIONS

Quantitative data are essential components of virtually all environmental investigations, but the range of chemical compounds involved and the different matrices in which xenobiotics may be encountered make generalizations on analytical procedures extremely hazardous. Whereas it is relatively straightforward to take advantage of modern instrumentation and methodology and to apply widely accepted procedures for detection, identification, and quantification, there are several fundamental difficulties to which specific attention should be directed and possible solutions sought:

1. Procedures for preparation of samples before analysis are of cardinal importance and particular emphasis should be placed on procedures for removal of interfering substances without destruction or alteration of the analyte.
2. Attention should be given to the possibility that the compounds are not present in the "free" state but are associated with macromolecules, both soluble and insoluble organic matter, and inorganic material. Analytical procedures should attempt to take this into account using the appropriate empirical fractionation protocols.
3. There are inherent dangers in assessing recoverability from the results of spiking experiments, and the significance of aging in natural sediments should be appreciated even if it cannot be quantitatively evaluated.
4. Attention should be directed to developments in structural determination including new techniques in mass spectrometry, the increasing accessibility of X-ray methods, and particularly the value of non-destructive methods such as NMR.
5. All natural systems are dynamic, and appropriate attention should be directed to kinetic processes, such as sorption and desorption, and to the significance and complexities arising from abiotic transformation and the metabolism of xenobiotics by biota.

REFERENCES

Ahel, M., T. Conrad, and W. Giger. 1987. Persistent organic chemicals in sewage effluents. 3. Determinations of nonylphenoxy carboxylic acids by high-resolution gas chromatography/mass spectrometry and high-performance liquid chromatography. *Environ. Sci. Technol.* 21: 697-703.

Alexander, L.R, D.G. Patterson, G.L. Myers, and J.S. Holler. 1986. Safe handling of chemical toxicants and control of interferences in human tissue analysis for dioxins and furans. *Environ. Sci. Technol.* 20: 725-730.

Alford-Stevens, A.L., J.W. Eichelberger, and W.L. Budde. 1988. Multilaboratory study of automated determinations of polychlorinated biphenyls and chlorinated pesticides in water, soil, and sediment by gas chromatography/mass spectrometry. *Environ. Sci. Technol.* 22: 304-312.

Allard, A.-S., M. Remberger, and A.H. Neilson. 1985. Bacterial O-methylation of chloroguaiacols: effect of substrate concentration, cell density, and growth conditions. *Appl. Environ. Microbiol.* 49: 279-288.

Allard, A.-S., P.-Å. Hynning, C. Lindgren, M. Remberger, and A.H. Neilson. 1991. Dechlorination of chlorocatechols by stable enrichment cultures of anaerobic bacteria. *Appl. Environ. Microbiol.* 57: 77-84.

Alsberg, T., U. Stenberg, R. Westerholm, M. Strandell, U. Rannug, A. Sundvall, L. Romert, V. Bernson, B. Petterson, R. Toftgård, B. Franzén, M. Jansson, J.Å. Gustafsson, K.E. Egebäck, and G. Tejle. 1985. Chemical and biological characterization of organic material from gasoline exhaust particles. *Environ. Sci. Technol.* 19: 43-50.

Arjmand, M., T.D. Spittler, and R.O. Mumma. 1988. Analysis of Dicamba from water using solid-phase extraction and ion-pair high-performance liquid chromatography. *J. Agric. Food Chem.* 36: 492-494.

Armstrong, D.W., G.L. Reid, M.L. Hilton, and C.-D. Chang. 1993. Relevance of enantiomer separations in environmental science. *Environ. Pollut.* 79: 51-58.

Armstrong, D.W., Y. Tang, and J. Zukowski. 1991. Resolution of enantiometic hydrocarbon biomarkers of geochemical importance. *Anal. Chem.* 63: 2858-2861.

Arpino, P. 1990. Coupling techniques in LC/MS and SFC/MS. *Fresenius J. Anal. Chem.* 337: 667-685.

Arpino, P.J., F. Sadoun, and H. Virelizier. 1993. Reviews on recent trends in chromatography/ mass spectrometry coupling. IV. Reasons why supercritical fluid chromatography is not so easily coupled with mass spectrometry as originally assessed. *Chromatographia* 36: 283-288.

Arthur, C.L., L.M. Killam, S. Motlagh, M. Lim, D.W. Potter, and J. Pawliszyn. 1992. Analysis of substituted benzene compounds in groundwater using solid-phase microextraction. *Environ. Sci. Technol.* 26: 979-983.

Ashraf-Khorassani, M., and L.T. Taylor. 1990. Nitrous oxide versus carbon dioxide for supercritical fluid extraction and chromatography of amines. *Anal. Chem.* 62: 1177-1180.

Aubin, A.J., and A.E. Smith. 1992. Extraction of [^{14}C]glyphosate from Saskatchewan soils *J. Agric. Food Chem.* 40: 1163-1165.

Bakola-Christianopoulou, M.N., V.P. Papageorgiou, and K.K. Apazidou. 1993. Gas chromatographic-mass spectrometric study of the reductive silylation of hydroxyquinones. *J. Chromatogr.* 645: 293-301.

Barceló, D. 1991. Occurrence, handling and chromatographic determination of pesticides in the aquatic environment. *Analyst* 116: 681-689.

Barceló, D. 1992. Mass spectrometry in environmental organic analysis. *Anal. Chim. Acta* 263: 1-19.

Barceló, D., G. Durand, V. Bouvot, and M. Nielen. 1993. Use of extraction disks for trace enrichment of various pesticides from river water and simulated seawater samples followed by liquid chromatography-rapid-scanning UV-visible and thermospray-mass spectrometry detection. *Environ. Sci. Technol.* 27: 271-277.

Bartle, K.D., M.L. Lee, and S.A. Wise. 1981. Modern analytical methods for environmental polycyclic aromatic compounds. *Chem. Soc. Rev.* 10 113-158.

Bartle, K.D., M.W. Raynor, A.A. Clifford, I.L. Davies, J.P. Kithinji, G.F. Shilstone, J.M. Chalmers, and B.W. Cook. 1989. Capillary supercritical fluid chromatography with Fourier transform infrared detection. *J. Chromatogr. Sci.* 27: 283-292.

Behymer, T.D., T.A. Bellar, and W.L. Budde. 1990. Liquid chromatography/particle beam/ mass spectrometry of polar compounds of environmental interest. *Anal. Chem.* 62: 1686-1690.

Bergman, Å., and C.A. Wachtmeister. 1978. Synthesis of methylthio- and methylsulphonyl-polychlorobiphenyls via nucleophilic aromatic substitution of certain types of polychlorobiphenyls. *Chemosphere* 7: 949-956.

Bernal, J.L., M.J. Del Nozal, and J.J. Jiménez. 1992. Some observations on clean-up procedures using sulphuric acid and florisil. *J. Chromatogr.* 607: 303-309.

Bidleman, T.F., M.D. Walla, D.C.G. Muir, and G.A. Stern. 1993. Selective accumulation of polychlorocamphenes in aquatic biota from the Canadian Arctic. *Environ. Toxicol. Chem.* 12: 701-709.

Birkholz, D.A., R.T. Coutts, and S.E. Hrudey. 1988. Determination of polycyclic aromatic compounds in fish tissue. *J. Chromatogr.* 449: 251-260.

Blau, K., and J.M. Halket (Eds.) 1993. *Handbook of derivatives for chromatography.* John Wiley & Sons, New York.

Blessington, B., N. Crabb, S. Karkee, and A. Northage. 1993. Chromatographic approaches to the quality control of chiral propionate anti-inflammatory drugs and herbicides. *J. Chromatogr.* 469: 183-190.

Brand, J.M., D.L. Cruden, G.J. Zylstra, and D.T. Gibson. 1992. Stereospecific hydroxylation of indan by *Escherichia coli* containing the cloned toluene dioxygenase genes from *Pseudomonas putida* F1. *Appl. Environ. Microbiol.* 58: 3407-3409.

Brandl, H., K.W. Hanselmann, R. Bachofen, and J. Piccard. 1993. Small-scale patchiness in the chemistry and microbiology of sediments in Lake Geneva, Switzerland. *J. Gen. Microbiol.* 139: 2271-2275.

Broman, D., C. Näf, C. Rolff, and Y. Zebühr. 1991.Occurrence and dynamics of polychlori-nated dibenzo-*p*-dioxins and dibenzofurans and polycyclic aromatic hydrocarbons in the mixed surface layer of remote coastal and offshore waters of the Baltic. *Environ. Sci. Technol.* 25: 1850-1863.

Brooks, C.J.W., and W.J. Cole. 1988. Cyclic di-*tert.*-butylsilylene derivatives of substituted salicylic acids and related compounds. A study by gas chromatography-mass spectrom-etry. *J. Chromatogr.* 441:13-29.

Brumley, W.C., C.M. Brownrigg, and A.H. Grange. 1993. Determination of toxaphene in soil by electron-capture negative ion mass spectrometry after fractionation by high-perfor-mance gel permeation chromatography. *J. Chromatogr.* 633: 177-183.

Burka, L.T., and D. Overstreet. 1989. Synthesis of possible metabolites of 2,3,7,8-tetrachlorodibenzofuran. *J. Agric. Food Chem.* 37: 1528-1532.

Buser, H.-R. 1986. Selective detection of brominated aromatic compounds using gas chroma-tography/negative chemical ionization mass spectrometry. *Anal Chem.* 58: 2913-2919.

Buser, H.-R., and M.D. Müller. 1992. Enantiomer separation of chlordane components and metabolites using chiral high-resolution gas chromatography and detection by mass spectrometric techniques. *Anal. Chem.* 64: 3168-3175.

Buser, H.-R., and M.D. Müller. 1994. Isomeric and enantiomeric composition of different commercial toxaphenes and of chlorination products of (+)- and (–)-camphenes. *J. Agric. Food Chem.* 42: 393-400.

Buser, H.-R., M.D. Müller, and C. Rappe. 1992a. Enantioselective determination of chlordane components using chiral high-resolution gas chromatography-mass spectrometry with application to environmental samples. *Environ. Sci. Technol.* 26: 1533-1540.

Buser, H.-R., D.R. Zook, and C. Rappe. 1992b. Determination of methyl sulfone-substituted polychlorobiphenyls by mass spectrometric techniques with application to environmental samples. *Anal. Chem.* 64: 1176-1183.

Cairns, T., E.G. Siegmund, and F. Krick. 1987. Identification of several new metabolites from pentachloronitrobenzene by gas chromatography/mass spectrometry. *J. Agric. Food Chem.* 35: 433-439.

Campbell, J.A., M.A. LaPack, T.L. Peters, and T.A. Smock. 1987. Gas chromatography/mass spectroscopy identification of cyclohexene artifacts formed during extraction of brine samples. *Environ. Sci. Technol.* 21: 110-112.

Capriel, P., A. Haisch, and S.U. Khan. 1986. Supercritical methanol: an efficaceous technique for the extraction of bound pesticides from soil and plant samples. *J. Agric. Food Chem.* 34: 70-73.

Chen, P.H., W.A. VanAusdale, and D.F. Roberts. 1991. Oxidation of phenolic acid surrogates and target analytes during acid extraction of neutral water samples for analysis by GC/MS using EPA method 625. *Environ. Sci. Technol.* 25: 540-546.

Christakopolous, A., B. Hamasur, H. Norin, and I. Nordgren. 1988. Quantitative determination of arsenocholine and acetylarsenocholine in aquatic organisms using pyrolysis and gas chromatography-mass spectrometry. *Biomed. Environ. Mass Spectrom.* 15: 67-74.

Christie, W.W. 1989. Silver ion chromatography using solid-phase extraction columns packed with a bonded-sulfonic acid phase. *J. Lipid Res.* 30: 1471-1473.

Coleman, W.E., J.W. Munch, R.W. Slater, R.G. Melon, and F.C. Kopfler. 1983. Optimization of purging efficiency and quantification of organic contaminants from water using a 1-L closed-loop stripping apparatus and computerized capillary column GC/MS. *Environ. Sci. Technol.* 17: 571-576.

Creaser, C.S., F. Krokos, and J.R. Startin. 1992. Analytical methods for the determination of non-*ortho* substituted chlorobiphenyls: a review. *Chemosphere* 25: 1981-2008.

Czuczwa, J.M., and R.A. Hites. 1986. Airborne dioxins and dibenzofurans: sources and fates. *Environ. Sci. Technol.* 20: 195-200.

Dahlgran, J.R. 1981.Simultaneous detection of total and halogenated hydrocarbons in complex environmental samples. *J. High Res. Chromatogr. Chromatogr. Commun.* 4: 393-397.

de Boer, J. 1988. Chlorobiphenyls in bound and non-bound lipids of fishes: comparison of different extraction methods. *Chemosphere* 17: 1803-1810.

de Kok, A., R.B. Geerdink, R.W. Frei, and U.A.Th. Brinkman. 1981. The use of dechlorination in the analysis of polychlorinated biphenyls and related classes of compounds. *Int. J. Environ. Anal. Chem.* 9: 301-318.

de Witt, J.S.M., C.E. Parker, K.B. Tomer, and J.W. Jorgenson. 1988. Separation and identification of trifluralin metabolites by open-tubular liquid chromatography/negative chemical ionization mass spectrometry. *Biomed. Environ. Mass Spectrom.* 17: 47-53.

Dearth, M.A., and R.A. Hites. 1991. Complete analysis of technical chlordane using negative ionization mass spectrometry. *Environ. Sci. Technol.* 25: 245-254.

Dietrich, A.M., D.S. Millington, and Y.-H. Seo. 1988. Specific identification of synthetic organic chemicals in river water using liquid-liquid extraction and resin adsorption coupled with electron impact, chemical ionization and accurate mass measurement gas chromatography-mass spectrometry analysis. *J. Chromatogr.* 436: 229-241.

Donnelly, J.R., W.D. Munslow, T.L. Vonnahme, N.J. Nunn, C.M. Hedin, G.W. Svocool, and R.K. Mitchum. 1987. The chemistry and mass spectrometry of brominated dibenzo-*p*-dioxins and dibenzofurans. *Biomed. Environ. Mass Spectrom.* 14: 465-472.

Downing, J.A., and L.C. Rath. 1988. Spatial patchiness in the lacustrine environment. *Limnol. Oceanogr.* 33: 447-458.

Duinker, J.C., D.E. Schulz, and G. Petrick. 1988. Multidimensional gas chromatography with electron capture detection for the determination of toxic congeners in polychlorinated biphenyl mixtures. *Anal. Chem.* 60: 478-482.

Duinker, J.C., D.E. Schulz, and G. Petrick. 1991. Analysis and interpretatiion of chlorobiphenyls: possibilities and problems. *Chemosphere* 23: 1009-1028.

Durand, G., P. Gille, D. Fraisse, and D. Barceló. 1992. Comparison of gas-chromatographic-mass spectrometrtic methods for screening of chlorotriazione pesticides in soil. *J. Chromatogr.* 603: 175-184.

Farrington, J.W., S.G. Wakeham, J.B Livramenta, B.W. Tripp, and J.M. Teal. 1986. Aromatic hydrocarbons in New York Bight polychaetes: ultraviolet fluorescence analysis and gas chromatography/gas chromatography-mass spectrometry analysis. *Environ. Sci. Technol.* 20: 69-72.

Fayad, N.M. 1988. Gas chromatography/mass spectroscopy identification of artifacts formed in methylene chloride extracts of saline water. *Environ. Sci. Technol.* 22: 1347-1348.

Fenimore, D.C., C.M. Davis, J.H Whitford, and C.A. Harrington. 1976. Vapor phase silylation of laboratory glassware. *Anal. Chem.* 48: 2289-2290.

Fernández, P., M. Grifoll, A.M. Solanas, J. M. Bayiona, and J. Albalgés. 1992. Bioassay-directed chemical analysis of genotoxic compounds in coastal sediments. *Environ. Sci. Technol.* 26: 817-829.

Fischer, E. 1909. *Untersuchungen über Kohlenhydrate und Fermente (1884-1908)* p.138-238. Verlag Von Julius Springer, Berlin.

Fitzpatrick, F.A., and S. Siggia. 1973. High-resolution liquid chromatography of derivatized non-ultraviole absorbing hydroxy steroids. *Anal. Chem.* 45: 2310.-2314.

Foreman, W.T., S.D. Zaugg, L.M. Faires, M.G. Werner, T.J. Leiker, and P.F. Rogerson. 1992. Analytical interferences of mercuric chloride preservative in environmental water samples: determination of organic compounds isolated by continuous liquid-liquid extraction or closed-loop stripping. *Environ. Sci. Technol.* 26: 1307-1312.

Foucault, A.P. 1991. Countercurrent chromatography. *Anal. Chem.* 63: 569A-579A.

France, J.E., J.W. King, and J.M. Snyder. 1991. Supercritical fluid-based cleanup technique for the separation of organochlorine pesticides from fats. *J. Agric. Food Chem.* 39: 1871-1874.

Gether, J., G. Lunde, and E. Steinnes. 1979. Determination of the total amount of organically bound chlorine, bromine and iodine in environmental samples by instrumental neutron activation analysis. *Anal. Chim. Acta* 108: 137-147.

Gómez-Belinchón, J.I., J.O. Grimalt, and J. Albalgés. 1988. Intercomparison study of liquid-liquid extraction and adsorption on polyurethane and amberlite XAD-2 for the analysis of hydrocarbons, polychlorobiphenyls, and fatty acids dissolved in seawater. *Environ. Sci. Technol.* 22: 677-685.

Greibrokk, T. 1992. Recent developments in the use of supercritical fluids in coupled systems. *J. Chromatogr.* 626: 33-40.

Grob, K., and I. Kälin. 1991. Towards on-line SEC-GC pesticide residues? The problem of tailing triglyceride peaks. *J. High Resol. Chromatogr.* 14: 451-454.

Grob, K., I. Kaelin, and A. Artho. 1991. Coupled LC-GC: the capacity of silica gel (HP)LC columns for retaining fat. *J. High Res. Chromatogr.* 14: 373-376.

Gyllenhaal, O., and J. Vessman. 1988. Phosgene as a derivatizing reagent prior to gas and liquid chromatography. *J. Chromatogr.* 435: 259-269.

Hale, R.C., and J. Greaves. 1992. Methods for the analysis of persistent chlorinated hydrocarbons in tissues. *J. Chromatogr.* 580: 257-278.

Haraguchi, K., H. Kuroki, and Y. Masuda. 1987. Synthesis and characterization of tissue-retainable methylsulfonyl polychlorinated biphenyl isomers. *J. Agric. Food Chem.* 35: 178-182.

Harrad, S.J., A.S. Sewart, R. Boumphrey, R. Duarte-Davidson, and K.C. Jones.1992. A method for the determination of PCB congeners 77, 126 and 169 in biotic and abiotic matrices. *Chemosphere* 24: 1147-1154.

Hatano, T., T. Yoshida, and T. Okuda. 1988. Chromatography of tannins III. Multiple peaks in high-performance liquid chromatography of some hydrolyzable tannins. *J. Chromatogr.* 435: 285-295.

Hawthorne, S.B., D.J. Miller, D.E. Nivens, and D.C. White. 1992. Supercritical fluid extraction of polar analytes using in situ chemical derivatization. *Anal. Chem.* 64: 405-412.

Heftmann, E. (Ed.). 1992 *Chromatography. Fundamentals and Applications of Chromatography and Related Differential Migration Methods. Part A: Fundamentals and Techniques.* 5th edition. Elsevier Science Publishers, Amsterdam.

Hennion, M.C., J.C. Thieblement, R. Rosset, P. Scribe, J.C. Marty, and A. Saliot. 1983. Rapid semi-preparative class separation of organic compounds from marine lipid extracts by high-performance liquid chromatography and subsequent quantitative analysis by gas chromatography. *J. Chromatogr.* 280: 351-362.

Higashi, R.M., G.N. Cherr, J.M. Shenker, J.M. Macdonald, and D.G. Crosby. 1992. A polar high molecular mass constituent of bleached kraft mill effluent is toxic to marine organisms. *Environ. Sci. Technol.* 26: 2413-2420.

Hites, R.A., and W.L. Budde. 1991. EPA's analytical methods for water: the next generation. *Environ. Sci. Technol.* 25: 998-1006.

Hontela, A., J.B. Rasmussen, C. Audet, and G. Chevalier. 1992. Impaired cortisol stress response in fish from environments polluted by PAHs, PCBs, and mercury. *Arch. Environ. Contam. Toxicol.* 22: 278-283.

Hopper, M.L. 1987. Methylation of chlorophenoxy acid herbicides and pentachlorophenol residues in foods using ion-pair alkylation. *J. Agric. Food Chem.* 35: 285-289.

Huckins, J.N., M.W. Tubergen, and G.K Manuweera. 1990a. Semipermeable membrane devices containing model lipid: a new approach to monitoring the biavailability of lipophilic contaminants and estimating their bioconcentration potential. *Chemosphere* 20: 533-552.

Huckins, J.N., M.W. Tubergen, J.A. Lebo, R.W. Gale, and T.R. Schwartz. 1990b. Polymeric film dialysis in organic solvent media for cleanup of organic contaminants. *J. Assoc. Off. Anal. Chem.* 73: 290-293.

Hühnerfuss, H., and R. Kallenborn. 1992. Chromatographic separation of marine organic pollutants. *J. Chromatogr.* 580: 191-214.

Hynning, P.-Å., M. Remberger, and A.H. Neilson. 1989. Synthesis, gas-liquid chromatographic analysis and gas chromatographic-mass spectrometric identification of nitrovanillins, chloronitrovanillins, nitroguaiacols, and chloronitroguaiacols. *J. Chromatogr.* 467: 99-110.

Hynning, P.-Å., M. Remberger, A.H. Neilson, and P. Stanley. 1993. Identification and quantification of 18-nor- and 19-norditerpenes and their chlorinated analogues in samples of sediments and fish. *J. Chromatogr.* 643: 439-452.

Jackson, J.-A. A., W.R. Blair, F.E. Brinckman, and W.P. Iverson. 1982. Gas-chromatographic speciation of methylstannanes in the Chesapeake bay using purge and trap sampling with a tin-selective detector. *Environ. Sci. Technol.* 16: 110-119.

Jaffe, R., and R.A. Hites. 1985. Identification of new, fluorinated biphenyls in the Niagara River-Lake Ontario area. *Environ. Sci. Technol.* 19: 736-740.

Jarman, W.M., M. Simon, R.J. Norstrom, S.A. Burns, C.A. Bacon, B.R.T. Simoneit, and R.W. Risebrough. 1992. Global distribution of tris(4-chlorophenyl)methanol in high trophic level birds and mammals. *Environ. Sci. Technol.* 26: 1770-1774.

Jenkins, T.F., and C.L. Grant. 1987. Comparison of extraction techniques for munitions residues in soil. *Anal. Chem.* 59: 1326-1331.

Jensen, S. 1972. The PCB story. *Ambio.* 1: 123-131.

Jensen, S., and B. Jansson. 1976. Methyl sulfone metabolites of PCB and DDE. *Ambio.* 5: 257-260.

Jensen, S., A.G. Johnels, M. Olsson, and G. Otterlind. 1972. DDT and PCB in herring and cod from the Baltic, the Kattegat and the Skagerak. *Ambio. Spec. Rep.* 1: 71-85.

Jensen, S., L. Renberg, and L. Reutergård. 1977. Residue analysis of sediment and sewage sludge for organochlorines in the presence of elemental sulfur. *Anal. Chem.* 49: 316-318.

Johnsen, S., and K. Kolset. 1988. The mass-selective detector as a chlorine-selective detector. *J. Chromatogr.* 438: 233-242.

Jokela, J,K., and M. Salkinoja-Salonen. 1992. Molecular weight distributions of organic halogens in bleached kraft pulp mill effluents. *Environ. Sci. Technol.* 26: 1190-1197.

Junk, G.A., and J.J. Richard. 1988. Organics in water: solid phase extraction on a small scale. *Anal. Chem.* 60: 451-454.

Karg, F.P.M. 1993. Determination of phenylurea pesticides in water by derivatization with heptafluorobutyric anhydride and gas chromatography-mass spectrometry. *J. Chromatogr.* 634: 87-100.

Keith, L.H. (Ed.). 1988. *Principles of environmental sampling*. American Chemical Society, Washington, D.C.

Keith, L.H. 1992. *Environmental sampling and analysis: a practical guide*. Lewis Publishers, Chelsea, Michigan.

Keith, L.H., W. Crimmett, J. Deegan, R.A. Libby, J.K. Taylor, and G. Wentler.1983. Principles of environmental analysis. *Anal. Chem.* 55: 2210-2218.

Kim, I.S., F.I. Sasins, R.D. Stephens, J. Wang, and M.A. Brown. 1991. Determination of chlorinated phenoxy acid and ester herbicides in soil and water by liquid chromatography particle beam mass spectrometry and ultraviolet spectrophotometry. *Anal. Chem.* 63: 819-823.

Kleibőhmer, W., K. Cammann, J. Robert, and E. Mussenbrock. 1993. Determination of explosives residues in soils by micellar electrokinetic capillary chromatography and high performance liquid chromatography. *J. Chromatogr.* 638: 349-356.

Knapp, D.R. 1979. *Handbook of analytical derivatization reactions*. John Wiley and Sons, New York.

Kobal, V.M., D.T. Gibson, R.E. Davis, and A. Garza. 1973. X-ray determination of the absolute stereochemistry of the initial oxidation product formed from toluene by *Pseudmonas putida* 39/D. *J. Amer. Chem. Soc.* 95: 4420-4421.

Koester, C.J., and R.A. Hites. 1992. Wet and dry deposition of chlorinated dioxins and furans. *Environ. Sci. Technol.* 26: 1375-1382.

Kolehmainen, E., K. Laihia, J. Knuutinen, and J. Hyőtyläinen. 1992. ^{1}H, ^{13}C and ^{17}O NMR study of chlorovanillins and some related compounds. *Magn. Reson. Chem.* 30: 253-258.

LaBrosse, J.L., and R.J. Anderegg. 1984. The mass spectrometer as a chlorine-selective detector I. Description and evaluation of the technique. *J. Chromatogr.* 314: 83-92.

Landrum, P., S.R. Nihart, B.J. Eadie, and W.S. Gardner. 1984. Reverse-phase separation method for determining pollutant binding to Aldrich humic acid and dissolved organic carbon of natural waters. *Environ. Sci. Technol.* 18: 187-192.

Lang, V. 1992. Polychlorinated biphenyls in the environment. *J. Chromatogr.* 595: 1-43.

Lee, H.-B., and T.E. Peart. 1992. Supercritical carbon dioxide extraction of resin and fatty acids from sediments at pulp mill sites. *J. Chromatogr.* 594: 309-315.

Lee, H.-B., T.E. Peart, and J.M. Carron. 1990. Gas chromatographic and mass spectrometric determination of some resin and fatty acids in pulpmill effluents as their pentafluorobenzyl ester derivatives. *J. Chromatogr.* 498: 367-379.

Lee, H.-B., T.E. Peart, and R.L. Hong-You. 1992. In situ extraction and derivatization of pentachlorophenol and related compounds from soils using a supercritical fluid extraction system. *J. Chromatogr.* 605: 109-113.

Lee, H.-B., T.E. Peart, and R.L. Hong-You. 1993. Determination of phenolics from sediments of pulp mill origin by *in situ* supercritical carbon dioxide extraction and derivatization. *J. Chromatogr.* 636: 263-270.

Levsen, K. 1988. The analysis of diesel particulate. *Fresenius Z. Anal. Chem.* 331: 467-478.

Li, C., L.M. Markovec, R.J. Magee, and B.D. James. 1991. A convenient method using methanol/benzyltrialkylammonium reagents for simultaneous extraction and methylation of 2,4-dichlorophenoxyacetic acid in soil, with subsequent analysis via gas chromatography. *J. Agric. Food Chem.* 39: 1110-1112.

Lin, J.M., and S.S. Que Hee. 1985. Optimization of perchlorination conditions for some representative polychlorinated biphenyls. *Anal. Chem.* 57: 2130-2134.

Lindroth, P., and K. Mopper. 1979. High performance liquid chromatographic determination of subpicomole amounts of amino acids by precolumn fluorescence derivatization with o-phthaldialdehyde. *Anal. Chem.* 51: 1667-1674.

Lindström, K., and F. Österberg. 1980. Synthesis, X.-ray structure determination, and formation of 3,4,5-trichloroguaiacol occurring in kraft mill spent bleach liquors. *Can. J. Chem.* 58: 815-822.

Lopez-Avila, V., J. Benedicto, E. Baldin, and W.F. Beckert. 1991. Analysis of classes of compounds of environmental concern: I. nitroaromatic compounds. *J. High Resol. Chromatogr.* 14: 601-607.

Lopez-Avila, V., N.S. Dodhiwala, and W.F. Beckert. 1990. Supercritical fluid extraction and its application to environmental analysis. *J. Chromatogr. Sci.* 28: 468-476.

Ludwig, P., W. Gunkel, and H. Hühnerfuss. 1992. Chromatographic separation of the enantiomers of marine pollutants. Part 5: Enantioselective degradation of phenoxycarboxylic acid herbicides by marine organisms. *Chemosphere* 24: 1423-1429.

Martinsen, K., A. Kringstad, and G.E. Carlberg. 1988. Methods for determination of sum parameters and characterization of organochlorine compounds in spent bleach liquors from pulp mills and water, sediment and biological samples from receiving waters. *Water Sci. Technol.* 20[2]: 13-24.

Mathiasson, L., J.Å. Jönsson, and L. Karlsson. 1989. Determination of nitrogen compounds by supercritical fluid chromatography using nitrous oxide as the mobile phase and nitrogen-sensitive detection. *J. Chromatogr.* 467: 61-74.

McMillan, D. C., P.P. Fu, and C.E. Cerniglia. 1987. Stereoselective fungal metabolism of 7,12- dimethylbenz[a]anthracene: identification and enantiomeric resolution of a K-region dihydrodiol. *Appl. Environ. Microbiol.* 53: 2560-2566.

Meadows, J., D. Tillitt, J. Huckins, and D. Schroeder. 1993. Large-scale dialysis of sample lipids using a semipermeable membrane device. *Chemosphere* 26: 1993-2006.

Meese, C.O. 1984. 2,2,2-trifluorodiazoethane: a highly selective reagent for the protection of sulfonic acids. *Synthesis* 1041-1042.

Miles, C.J., and H.A. Moyne. 1988. Extraction of glyphosate herbicide from soil and clay minerals and determination of residues in soils. *J. Agric. Food Chem.* 36: 486-491.

Mills, M.S., and E.M. Thurman. 1992. Mixed-mode islation of triazine metabolites from soil and aquifer sediments using automated solid-phase extraction. *Anal. Chem.* 64: 1985-1990.

Miskiewicz, A.G., and P.J. Gibbs. 1992. Variability in organochlorine analysis in fish: an interlaboratory study and its implications for environmental monitoring and regulatory standards. *Arch. Environ. Contam. Toxicol.* 23: 45-53.

Moody, R.P., C.A. Franklin, D. Riedel, N.I. Muir, R. Greenhalgh, and A. Hladka. 1985. A new GC on-column methylation procedure for analysis of DMTP (*O,O*-dimethyl phosphorothioate) in urine of workers exposed to fenitrothion. *J. Agric. Food Chem.* 33: 464-467.

Morris, G.A., and R. Freeman. 1979. Enhancement of nuclear magnetic resonance signals by polarization transfer. *J. Amer. Chem. Soc.* 101 760-762.

Mössner, S., T.R. Spraker, P.R. Becker, and K. Ballschmiter. 1992. Ratios of enantiomers of alpha-HCH and determination of alpha-, beta-, and gamma-HCH isomers in brain and other tissues of neonatal northern fur seals (*Callorhinus ursinus*). *Chemosphere* 24: 1171-1180.

Mowrer, J., and J. Nordin. 1987. Characterization of halogenated organic acids in flue gases from municipal waste incinerators. *Chemosphere* 16: 1181-1192.

Mudroch, A., R.J. Allen, and S.R. Joshi. 1992. Geochemistry and organic contaminants in the sediments of Great Slave Lake, Northwest Territories, Canada. *Arctic* 45: 10-19.

Mullin, M., C.M. Pochini, S. McCrindle, M. Romkes, S.H. Safe, and L.M. Safe. 1984. High-resolution PCB analysis: synthesis and chromatographic properties of all 209 PCB congeners. *Environ. Sci. Technol.* 18: 468-476.

Müller, M.D., M. Schlabach, and M. Oehme. 1992. Fast and precise determination of alpha-hexachlorocyclohexane enantiomers in environmental samples using chiral high-resolution gas chromatography. *Environ. Sci. Technol.* 26: 566-569.

Munkittrick, K.R., C.B. Portt, G.J. Van Der Kraak, I.R. Smith, and D. A. Rokosh. 1991. Impact of bleached kraft mill effluent on population characteristics, liver MFO activity, and serum steroid levels of a Lake Superior white sucker (*Catostomus commersoni*) population. *Can. J. Fish. Aquat. Sci.* 48: 1371-1380.

Muthuramu, K., P.B. Shepson, and J.M. O'Brien. 1993. Preparation, analysis, and atmospheric production of multifunctional organic nitrates. *Environ. Sci. Technol.* 27: 1117-1124.

Neilson, A.H., A.-S. Allard, P.-Å Hynning, and M. Remberger. 1991. Distribution, fate and persistence of organochlorine compounds formed during production of bleached pulp. *Toxicol. Environ. Chem.* 30: 3-41.

Neilson, A.H., A.-S. Allard, P.-Å. Hynning, and M. Remberger. 1988. Transformations of halogenated aromatic aldehydes by metabolically stable anaerobic enrichment cultures. *Appl. Environ. Microbiol.* 54: 2226-2236.

Nielen, M.W.F. 1993. (Enantio-)separation of phenoxy acid herbicides using capillary zone electrophoresis. *J. Chromatogr.* 637: 81-90.

Nielsen, T., T. Ramdahl, and A. Björseth. 1983. The fate of airborne polycyclic organic matter. *Environ. Health Perspect.* 47: 103-114.

Niessen, W.M.A, U.R. Tjaden, and J. van der Greef. 1991. Strategies in developing interfaces for coupling liquid chromatography and mass spectrometry. *J. Chromatogr.* 534: 3-26.

Nitz, S., M.H. Spraul, F. Drawert, and M. Spraul. 1992. 3-butyl-5,6-dihydro-4*H*-isobenzofuran-1-one, a sensorial active phthalide in parsley roots. *J. Agric. Food Chem.* 40: 1038-1040.

Nondek, L, D.R. Rodler, and J.W. Birks. 1992. Measurement of sub-ppb concentrations of aldehydes in a forest atmosphere using a new HPLC technique. *Environ. Sci. Technol.* 26: 1174-1178.

Näf, C., D. Broman, H. Pettersen, and Y. Zebühr. 1992. Flux estimates and pattern recognition of particulate polycyclic aromatic hydrocarbons, polychlorinated dibenzo-*p*-dioxins, and dibenzofurans in the waters outside various emission sources on the Swedish Baltic Coast. *Environ. Sci. Technol.* 26: 1444-1457.

Okuda, T., T. Yoshida, T. Hatano, K. More, and T. Fukuda. 1990. Fractionation of pharmacologically active plant polyphenols by centrifugal partition chromatography. *J. Liquid Chromatogr.* 13: 3637-3650.

Ozretich, R.J., and W.P. Schroeder. 1986. Determination of selected neutral priority organic pollutants in marine sediment, tissue, and reference materials utilizing bonded-phase sorbents. *Anal. Chem.* 58: 2041-2048.

Paschke, T., S.B. Hawthorne, and D.J. Miller. 1992. Supercritical fluid extraction of nitrated polycyclic aromatic hydrocarbons and polycyclic aromatic hydrocarbons from diesel exhaust particulate matter. *J. Chromatogr.* 609: 333.340.

Petrick, G., D.E. Schulz, and J.C. Duinker. 1988. Clean-up of environmental samples by high-performance liquid chromatography for analysis of organochlorine compounds by gas chromatography with electron-capture detection. *J. Chromatogr.* 435: 241-248.

Pierce, A.E. 1979. *Silylation of organic compounds.* Pierce Chemical Company, Rockford, Illinois.

Ramaswamy, S., M. Malalyandi, and G.W. Buchanan. 1985. Phase-transfer-catalyzed methylation of hydroxyaromatic acids, hydroxyaromatic aldehydes, and aromatic polycarboxylic acids. *Environ. Sci. Technol.* 19: 507-512.

Ramdahl, T., B. Zielinska, J. Arey, and R.W. Kondrat. 1988. The electron impact mass spectra of di- and trinitrofluoranthrenes. *Biomed. Environ. Mass Spectrom.* 17: 55-62.

Remberger, M., P.-Å. Hynning, and A.H. Neilson. 1988. Comparison of procedures for recovering chloroguaiacols and chlorocatechols from contaminated sediments. *Environ. Toxicol. Chem.* 7: 795-805.

Remberger, M., P.-Å. Hynning, and A.H. Neilson. 1991a. 2,5-dichloro-3,6-dihydroxybenzo-1,4-quinone: identification of a new organochlorine compound in kraft mill bleachery effluents. *Environ. Sci. Technol.* 25: 1903-1907.

Remberger, M., P.-Å. Hynning, and A.H. Neilson. 1991b. Chlorinated benzo-1,2-quinones: an example of chemical transformation of toxicants during tests with aquatic organisms. *Ecotoxicol. Environ. Saf.* 22: 320-336.

Richardson, J.S. and D.E. Miller. 1982. Identification of dicyclic and tricyclic hydrocarbons in the saturate fraction of a crude oil by gas chromatography/mass spectrometry. *Anal. Chem.* 54: 765-768

Rogge, W.F., L.M. Hildemann, M.A. Mazurek, G.R. Cass, and B.R.T. Simoneit. 1993. Sources of fine organic aerosol. 2. Noncatalyst and catalyst-equipped automobiles and heavy-duty diesel trucks. *Environ. Sci. Technol.* 27: 636-651.

Saflic, S., P.M. Fedorak, and J.T. Andersson. 1992. Diones, sulfoxides, and sulfones from the aerobic cometabolism of methylbenzothiophenes by *Pseudomonas* strain BT1. *Environ. Sci. Technol.* 26: 1759-1764.

Sanders, G., K.C. Jones, J. Hamilton-Taylor, and H. Dörr. 1992. Historical inputs of polychlorinated biphenyls and other organochlorines to a dated lacustrine sediment core in rural England. *Environ. Sci. Technol.* 26: 1815-1821.

Sawhney, B.L., J.J. Pignatello, and S.M. Steinberg. 1988. Determination of 1,2-dibromoethane (EDB) in field soils: implications for volatile organic compounds. *J. Environ. Qual.* 17: 149-152.

Schmidt, L.J., and R.J. Hesselberg. 1992. A mass spectroscopic method for analysis of AHH-inducing and other polychlorinated biphenyl congeners and selected pesticides in fish. *Arch. Environ. Contam. Toxicol.* 23: 37-44.

Schoenmakers, P.J., and L.G.M. Uunk. 1987. Supercritical fluid chromatography — recent and future developments. *Eur. Chromatogr. News* 1: 14-22.

Schuytema, G.S., A.V. Nebeker, W.L. Griffis, and C.E. Miller. 1989. Effects of freezing on toxicity of sediments contaminated with DDT and endrin. *Environ. Toxicol. Chem.* 8: 883-891.

Singh, S.K., and S. Kumar. 1993. Synthesis of oxygenated derivatives of 2,3,7,8-tetrachlorodibenzo-*p*-dioxin. *J. Agric. Food Chem.* 41: 1511-1516.

Sinkkonen, S., E. Kolehmainen, K. Lalhia, J. Kolstinen, and T. Rantio. 1993. Polychlorinated diphenyl sulfides: preparation of model compounds, chromatography, mass spectrometry, NMR, and environmental analysis. *Environ. Sci. Technol.* 27: 31319-1326.

Sjöström, L., R. Rådeström, G.E. Carlberg, and A. Kringstad. 1985. Comparison of two methods for the determination of total organic halogen (TOX) in receiving waters. *Chemosphere* 14: 1107-1113.

Snyder, J.L., R.L. Grob, M.E. McNally, and T.S. Oostdyk. 1992. Comparison of supercritical fluid extraction with classical sonication and Soxhlet extractions for selected pesticides. *Anal. Chem.* 64: 1940-1946.

Sovocool, G.W., R.K. Mitchum, and J.R. Donnelly. 1987. Use of the '*ortho* effect' for chlorinated biphenyl and brominated biphenyl isomer identification. *Biomed. Environ. Mass Spectrom.* 14: 579-582.

Stalling, D.L., R.C. Tindle, and J.L. Johnson. 1972. Cleanup of pesticide and polychlorinated biphenyl residues in fish extracts by gel permeation chromatography. *J. Assoc. Off. Anal. Chem.* 55: 32-38.

Stephanou, E.G. 1992. α,ω-dicarboxylic acid salts and α,ω-dicarboxylic acids. Photooxidation products of unsaturated fatty acids, present in marine aerosols and marine sediments. *Naturwiss.* 79: 128-131.

Stephanou, E., M. Reinhard, and H.A. Ball. 1988. Identification and quantification of halogenated and non-halogenated octylphenol polyethoxylate residues by gas chromatography/mass spectrometry using electron ionization and chemical ionization. *Biomed. Environ. Mass Spectrom.* 15: 275-282.

Stern, G.A., D.C.G. Muir, C.A. Ford, N.P. Grift, E. Dewailly, T.F. Bidleman, and M.D. Walla. 1992. Isolation and identification of two major recalcitrant toxaphene congeners in aquatic biota. *Environ. Sci. Technol.* 26: 1838-1840.

Sutherland, J.B., P.P. Fu, S.K. Yang, L. S. von Tungeln, R.P. Casillas, S.A. Crow, and C.E. Cerniglia. 1993. Enantiomeric composition of the *trans*-dihydrodiols produced from phenanthrene by fungi. *Appl. Environ. Microbiol.* 59: 2145-2149.

Swallow, K.C., N.S. Shifrin, and P.J. Doherty. 1988. Hazardous organic compound analysis. *Environ. Sci. Technol.* 22: 136-142.

Södergren, A. 1987. Solvent-filled dialysis membranes simulate uptake of pollutants by aquatic organisms. *Environ. Sci. Technol.* 21: 855-859.

Tan, Y.L., and M. Heit. 1981. Biogenic and abiogenic polynuclear aromatic hydrocarbons in sediments from two remote Adirondack lakes. *Geochim. Cosmochim. Acta* 45: 2267-2279.

Tanaka, F.S., R.G. Wien, R.G. Zaylskie, and B.L. Hoffer. 1990. Synthesis of possible ring-hydroxylated metabolites of Diclofop-methyl. *J. Agric. Food Chem.* 38: 553-559.

Taylor, W.G., D.D. Vedres, and J.L. Elder. 1993. Capillary gas chromatographic separation of some diastereomeric amides from carbonyldiimidazole-mediated microgram-scale derivatizations of the acid moiety of permethrin insecticide. *J. Chromatogr.* 645: 303-310.

ten Haven, H.L., T.M. Peakman, and J. Rullkötter. 1992. Early diagenetic transformation of higher-plant triterpenoids in deep-sea sediments from Baffin Bay. *Geochim. Cosmochim. Acta* 56: 2001-2024.

Thurman, E.M., M. Meyer, M. Pomes, C.A. Perry, and A.P. Schwab. 1990. Enzyme-linked immunosorbent assay compared with gas chromatography/mass spectrometry for the determination of triazine herbicides in water. *Anal. Chem.* 62: 2043-2048.

Thurman, E.M., D.A. Goolsby, M.T. Meyer, M.S. Mills, M.L. Pomes, and D.W. Kolpin. 1992. A reconnaisance study of herbicides and their metabolites in surface water of the midwestern United States using immunoassay and gas chromatography/mass spectrometry. *Environ. Sci. Technol.* 26: 2440-2447.

Tjeerdema, R.S., T. W.-M. Fan, R.M. Higashi, and D.G. Crosby. 1991. Effects of pentachlorophenol on energy metabolism in the abalone (*Haliotis rufescens*) as measured by *in vivo* ^{31}P NMR spectroscopy. *J. Biochem. Toxicol.* 6: 45-56.

Van der Velde, E.G., W. de Haan, and A.K.D. Liem. 1992. Supercritical fluid extraction of polychlorinated biphenyls and pesticides from soil. Comparison with other extraction methods. *J. Chromatogr.* 626: 135-143.

Vanderlaan, M., B.E. Watkins, and L. Stanker. 1988. Environmental monitoring by immunoassay. *Environ. Sci. Technol.* 22: 247-254.

Van Emon, K.M., and Lopez-Avila, V. 1992. Immunochemical methods for environmental analysis. *Anal. Chem.* 64: 79A-88A.

Vassilaros, D.L., P.W. Stoker, G.M. Booth, and M.L. Lee. 1982. Capillary gas chromatographic determination of polycyclic aromatic compounds in vertebrate fish tissue. *Anal. Chem.* 54: 106-112.

Velury, S., and S.B. Howell. 1988. Measurements of plasma thiols after derivatization with monobromobimane. *J. Chromatogr.* 424: 141-146.

Wahlberg, C., L. Renberg, and U. Wideqvist. 1990. Determination of nonylphenol and nonylphenol ethoxylates as their pentafluorobenzoates in water, sewage sludge and biota. *Chemosphere* 20: 179-195.

Wakeham, S.G., C. Schaffner, and W. Giger. 1980. Polycyclic aromatic hydrocarbons in recent lake sediments — II. Compounds derived from biogenic precursors during early diagenesis. *Geochim. Cosmochim. Acta* 44: 415-429.

Ware, M.L., M.D. Argentine, and G.W. Rice. 1988. Potentiometric determination of halogen content in organic compounds using dispersed sodium reduction. *Anal. Chem.* 60: 383-384.

Wasilchuk, B.A., P.W. Le Quesne, and P. Vouros.1992. Monitoring cholesterol autoxidation processes using multideuteriated cholesterol. *Anal. Chem.* 64: 1077-1087.

Westcott, N.D., and B.L. Worobey. 1985. Novel solvent extraction of lindane from soil. *J. Agric. Food Chem.* 33: 58-60.

Wheeler, O.H. 1968. The Girard reagents. *J. Chem. Ed.* 45: 435-437.

Wittlinger, R., and K. Ballschmiter. 1990. Studies of the global baseline pollution XIII. C_6-C_{14} organohalogens (α and g [sic]-HCH, HCB, PCB, 4,4'-DDT, 4,4-DDE, cis- and trans-chlordane, trans-nonachlor, anisols) in the lower troposphere of the southern Indian Ocean. *Fresenius J. Anal. Chem.* 336: 193-200.

Wretensjö, I., L. Svensson, and W.W. Christie. 1990. Gas chromatographic-mass spectrometric identification of the fatty acids in borage oil using the picolinyl ester derivatives. *J. Chromatogr.* 521: 89-97.

Xie, T.-M., E. Abrahamson, E. Fogelqvist, and B. Josefsson. 1986. Distribution of chlorophenols in a marine environment. *Environ. Sci. Technol.* 20: 457-463.

Xie, Y., D.A. Reckhow, and R.V. Rajan. 1993. Spontaneous methylation of haloacetic acids in methanolic stock solutions. *Environ. Sci. Technol.* 27: 1232-1234.

Yasahura, A., and M. Morita. 1988. Formation of chlorinated aromatic hydrocarbons by thermal decomposition of vinylidene chloride polymer. *Environ. Sci. Technol.* 22: 646-650.

Zabik, J.M., L.S. Aston, and J.N. Seiber. 1992. Rapid characterization of pesticide residues in contaminated soils by passive sampling devices. *Environ. Toxicol. Chem.* 11: 765-770

CHAPTER 3

Partition: Distribution, Transport, and Mobility

SYNOPSIS

The dissemination of a xenobiotic after discharge into the aquatic environment is determined by its partition between the water, sediment, and atmospheric phases and its potential for concentration in biota. These processes determine both the biological impact of the xenobiotic and the extent of its dissemination. Procedures for determining the partition of xenobiotics into biota are discussed, and attention is drawn to complicating factors including the association of xenobiotics with macromolecules in the aquatic phase and the important interdependence of metabolism and bioconcentration. Surrogate procedures for evaluating bioconcentration potential that use physico-chemical partition coefficients are outlined, and their intrinsic limitations are pointed out; the inability of such systems to take into account the important issues of metabolism in biota and the structure of biological lipid membranes is noted. Procedures for determining the distribution of xenobiotics between aqueous and sediment phases are presented. The desorption of xenobiotics from the sediment phase is discussed, and a brief account is given of interaction mechanisms between xenobiotics and components of solid matrices. Attention is drawn to the phase heterogeneity of the water mass in many natural systems and to the role of both particulate and dissolved matter in the distribution and dissemination of xenobiotics in lakes and rivers. Brief comments are devoted to the partitioning of xenobiotics between the aquatic phase and the atmosphere and to the significance of atmospheric transport on a global scale. A discussion of monitoring strategies is presented, together with brief comments on biomagnification and some of the complexities involved. It is emphasized throughout that partitioning involves a complex set of molecular interactions, that these are reversible to varying degrees, and that attention should be directed both to the structure of the xenobiotic and to the ecosystem to which the results are to be applied. Equations used for correlating partition coefficients with physico-chemical parameters have been presented, and some of their limitations have been noted.

INTRODUCTION

With the availability of suitable analytical procedures, the next question that should be addressed is the distribution of xenobiotics among the various phases after their discharge into the environment. This information provides a basis for deciding upon the ultimate fate of these compounds — particularly those that are not readily degradable — whose dissemination, persistence, and toxicity have aroused the greatest environmental concern. The distribution of xenobiotics is determined, on the one hand, by physico-chemical equilibria and on the other, by chemical or biologically mediated reactions, some of which may result in essentially irreversible associations between the xenobiotic and organic or inorganic components of the aquatic and sediment phases. The distribution of xenobiotics is, therefore, a function of many interacting factors, and it is to a discussion of these that this chapter is devoted. The most detailed discussions are devoted to the aquatic and sediment phases, although attention is also directed to the atmosphere on account of its established significance in the global dissemination of many xenobiotics. In this chapter the term "aquatic phase" will be taken to include the water phase together with biota (e.g., algae and fish) and particulate material (seston), while the term "aqueous phase" will be applied in a more restricted sense to the water phase alone.

The partitioning of organic compounds between the aqueous and the sediment phases, and between the aqueous phase and particulate matter including algae, is important for a number of rather different reasons:

1. It determines the *exposure of biota* to a potential toxicant initially discharged into the aqueous phase (Section 3.2.2) and the extent to which it is justifiable to correlate observed biological effects with measured concentrations of the toxicant.
2. Has a significant bearing on the *persistence* of a xenobiotic which is discussed in greater detail in Chapter 4 (Section 4.5.3).
3. An assessment of the *ultimate fate* of xenobiotics — and of putative metabolites — requires estimates of their concentration and distribution in all environmental compartments.
4. The *dissemination* of xenobiotics (Section 3.5) initially discharged, for example, into the aquatic phase may take place in several of the phases — within the water mass including suspended particulate matter, in the sediment phase to which the compound is sorbed, or via the atmosphere — and alterations in the structure of the xenobiotic may take place during transport within all of these phases.

Possibly greatest attention has, however, traditionally been directed to the concentration of organic compounds from the aqueous phase into biota. This effort has been motivated by the consistent recovery of many compounds of industrial interest such as PCBs and the more persistent agrochemicals such as DDT (and its metabolite DDE), mirex, and aldrin from samples of fish, birds, and marine mammals such as seals, whales, and polar bears. In a few cases, a plausible correlation has been established between injury to biota and exposure to a toxicant, and this is discussed in a wider perspective in Chapter 8 (Section 8.2). Only two examples of such correlations will, therefore, be given here as illustration:

1. Exposure of bottom-dwelling fish to concentrations of PAHs in contaminated sediments in Puget Sound, WA, and the incidence of disease including hepatic neoplasms (Malins et al. 1984).

2. Exposure of fish-eating herring gulls (*Larus argentatus*) in the Great Lakes and the incidence of porphyria in the gulls (Fox et al. 1988).

However, even though exposure of biota to xenobiotics does not necessarily result in toxification of these organisms, the possibility that such compounds could thereby enter the food chain and could, therefore, ultimately be consumed by the final predator — man — has awakened serious concern over the dissemination of such compounds. A good example is provided by the concern over possible adverse effects on human health including reproduction that could result from the consumption of fish from the Great Lakes that may be heavily contaminated with organochlorine compounds including PCBs (Swain 1991).

It should be appreciated that the concentration of a xenobiotic in biota is a dynamic process and represents a balance between uptake and elimination and that, as discussed in Sections 3.1.2 and 3.1.3, elimination may involve both the unchanged xenobiotic and its metabolites. Depuration, therefore, provides both a mechanism for the detoxification of the xenobiotic and its return — either unaltered or in the form of metabolites — to the aquatic phase and a means of its dissemination within the water mass. This aspect is discussed in Section 3.5.2.

Aquatic ecosystems are highly heterogeneous and comprise at least three apparently distinct phases: the aqueous phase, seston, the sediment phase, and the biota. None of these phases should, however, be considered as an independent entity. For example, probably most sediments have a rich biota consisting of microorganisms together with a spectrum of higher organisms, such as oligochaetes and amphipods, and the exposure of sediment-dwelling biota to toxicants is significantly determined by exposure to the interstitial water in the sediment phase. The distribution of an organic compound initially discharged into the aquatic environment is therefore exceedingly complex and is determined by the dynamics of a number of partition processes between (1) the aquatic phase and biota including microalgae, higher plants, invertebrates, and fish; (2) the aquatic phase and the sediment phase; and (3) the sediment and sediment-dwelling biota.

Almost all of these involve potentially reversible partitions, all of which should be taken into consideration; they may be mediated, for example, by chemical desorption processes from the sediment phase or by depuration and elimination from biota. It should also be appreciated that few — if any — of these distributions are in true equilibrium; this fact should be borne in mind especially in extrapolating the results of laboratory experiments to natural ecosystems. In addition, the situation is complicated by the fact that none of these phases is truly homogeneous. Even the aquatic phase is heterogeneous and often contains particulate matter, including inorganic material and both soluble and insoluble organic matter originating from aquatic biota and terrestrial plants. In addition, components of the sediment phase may have originated from atmospheric transport and deposition; the quantitative importance of all these distribution processes has, therefore, received increasing attention. As a result, intensive investigations have increasingly been directed to factors whose quantitiative significance had not been fully appreciated previously. A few examples may be given to illustrate some of the important issues:

1. The sorption to particulate matter in the water column and the dynamics and resuspension of surficial sediments.

2. The role of dissolved organic matter in the water column, accompanied by an increased appreciation of the important distinction between truly dissolved and finely divided particulate matter that may be colloidal.
3. The significance of interstitial water both in mediating exposure particularly to sediment-dwelling biota and in diffusion of xenobiotics into the water mass.
4. The importance of partitioning between the aquatic phase and the atmosphere even for compounds with relatively low volatility and the role of the atmosphere in mediating the long-distance transport of xenobiotics.

These factors have focused attention on important new aspects of the phase partitioning of organic substances and have, indeed, often revealed complexities that have merited intensive investigation and resulted in new perspectives. It is appropriate to note an increased awareness of possible limations in extrapolating data from laboratory studies to the natural environment. Two simple examples may be used as illustration — both involving PCB and laboratory-determined values of P_{ow}.

1. Partitioning between the aquatic and particulate phase in New Bedford Harbor was not strongly correlated with values of P_{ow} and revealed the importance of temperature (Bergen et al. 1993).
2. Studies of partitioning between the aquatic phase and algae have revealed that in natural ecosystems equilibrium is not reached in growing populations of algae so that use of P_{ow} values is not justified (Swackhamer and Skoglund 1993). This may, indeed, have wider implications and is discussed in greater detail subsequently.

It is important not to be left with the impression that biota and sediments function primarily as sinks for xenobiotics. A number of mechanisms exist for their elimination from these phases including metabolism and depuration in biota (Chapter 7, Section 7.5) and desorption from the sediment phase (Section 3.2.2). Elimination from biota may also depend on diffusion mechanisms when the biota are in intimate contact with another phase; for example, elimination of 2,3,3'-trichlorobiphenyl, DDE and γ-hexachlorocyclohexane from larvae of the midge *Chironomus riparius* was generally greater in sediments with higher organic content, and a significant correlation was found between the rate of elimination and the octanol/water partition coefficient of the compounds (Lydy et al. 1992).

3.1 PARTITIONING INTO BIOTA: UPTAKE OF XENOBIOTICS FROM THE AQUEOUS PHASE

3.1.1 Direct Measurements of Bioconcentration Potential

Outline of experimental procedures — For aquatic organisms, bioconcentration is the accumulation of a chemical from the aqueous phase; exposure takes place only via the water, although the compound may exist either in the dissolved form or associated with dissolved organic material. It is, therefore, distinguished from bioaccumulation that includes all modes of uptake including that of particulate matter. This is discussed more fully in a later section.

Bioconcentration factors (BCF) may be calculated by either of two procedures; the basic assumption in both is that the uptake and depuration are governed by first-order kinetics, although deviations may occur that may be accounted for by the induction of enzymes for metabolism of the xenobiotic. In practice, a number of additional factors may be involved including the toxicokinetics of different organs and possible interference from growth of the test organism if the compound is only slowly accumulated. In one method, concentrations in the biota and in the surrounding medium are measured after a steady state has been reached, and the ratio of the two concentrations is used to obtain the BCF value. In the other, rates of uptake and elimination of the xenobiotic are measured, and the ratio is used to calculate concentrations in the biota. The BCF is then calculated from these values. The possible complications resulting from metabolism of the test compound are discussed in Section 3.1.3.

In laboratory experiments using fish, exposure takes place primarily by uptake through the gills directly from the aquatic phase (Pärt 1990), and the bioconcentration factor may be estimated by either or both of the procedures outlined above. Both procedures have been evaluated in experiments in which guppy (*Poecilia reticulata*) were exposed to a series of organophosphorus pesticides that are metabolized only slowly. It was shown that there was a linear relation between the BCFs and the ratios of the uptake and elimination rates within the logarithmic range of 2.6 and 4.7 (de Bruijn and Hermens 1991). Although in this case the two procedures produced essentially identical results, some discrepancy would be expected if the compounds were metabolized to a significant extent and the metabolites were subsequently eliminated from the fish. This is discussed more fully in Section 3.1.3.

Exposure to the xenobiotic generally extends over a period of days or weeks but even for up to several months; either semi-static or flow-through systems may be used, and analytical control of the concentrations of the test substrate should be maintained. After exposure, fish are generally maintained in a xenobiotic-free environment to allow excretion of toxicants or their metabolites to take place. A variety of different fish including rainbow trout (*Oncorhynchus mykiss* syn. *Salmo gairdneri*), fathead minnows (*Pimephales promelas*), guppy (*Poecilia reticulata*), zebra fish (*Brachydanio rerio*), and medaka (*Oryzias latipes*) have been employed, even though it has been clearly established that fish have highly effective metabolic potential for a wide range of compounds (Section 3.1.3 and Chapter 7, Section 7.5.1) and that this metabolic potential varies with the species. Different BCF values may, therefore, be found in experiments using different fish; for example, for a restricted range of chlorobenzenes using fathead minnows, green sunfish (*Lepomis cyanellus*), and rainbow trout, experimental values for rainbow trout were the lowest (Veith et al. 1979a), and this might plausibly be correlated with their established metabolic capability. It should also be appreciated that the disposition of xenobiotics within the organisms may differ significantly. For example, a number of neutral organochlorine compounds are accumulated in the central nervous system (CNS) of cod (*Gadus morhua*) but not in that of rainbow trout, and for hexachlorobenzene it has been shown that, whereas it is the xenobiotic itself that is present in the CNS system, it is metabolites that are found in cerebrospinal fluid (Ingebrigtsen et al. 1992). Low concentrations of the test compound are generally employed, and particular care should be exercised with compounds displaying even subliminal toxic effects at the concentrations used during exposure. For example, the value of 39000 for the BCF of 2,3,7,8-tetrachlorodibenzo-1,4-dioxin at the concentration where rainbow trout were least affected may well be too low, since the corresponding value of the less toxic 2,3,7,8-tetrachlorodibenzofuran increased from 2455 at a concentration of 3.93 ng/L to 6049 at a concentration of 0.41 ng/L (Mehrle et al. 1988).

Virtually any aquatic organism may, of course, be used, and, for example, common mussels (*Mytilus edulis*) have been used for investigating the uptake of a restricted range of neutral organochlorine compounds (Ernst 1979), the crustacean *Daphnia pulex* for the uptake of azaarenes (Southworth et al. 1980), and freshwater mussels (*Anodonta anatina*) for the uptake of chlorophenolic compounds (Mäkelä et al. 1991).

The design of the uptake experiments themselves and the analytical determinations are straightforward: specific analysis may be carried out for the compounds being examined (together with their metabolites) or advantage may be taken of, for example, ^{14}C-labeled substrates. A number of important limitations in the numerical significance of the values obtained in laboratory studies have been pointed out (Oliver and Niimi 1985), and these are worth emphasizing:

1. Uptake of the xenobiotic may be so slow that the length of exposure is insufficient to attain a steady state.
2. The molecules may be too large for uptake, for example, via the gills of fish, so that BCF values are negligible, and uptake in the environment is dominated by uptake via the food. This is discussed in greater detail below.
3. The xenobiotic is metabolized by the biota, and this results in erroneously low concentrations in the biota and, hence, low BCF values. The interdependence of bioconcentration and metabolism in fish is considered in Section 3.1.3 and in a wider context in Section 3.5.2, while additional details of metabolism by fish are given in Chapter 7 (Section 7.5.1).

These limitations have been systematically explored for 34 halogenated compounds in rainbow trout, and they were shown to be particularly relevant to making realistic predictions of the concentrations in wild biota. Indeed, the striking incidence of DDE (the principal transformation product of DDT) in environmental samples is consistent with its bioconcentration in field samples to a degree greatly exceeding that predicted from laboratory measurements (Oliver and Niimi 1985).

Significant differences in measured BCF values may also result from the design of the experiments and from the inevitable biological variability in the test organism. For example, BCF values for 2,3,4,5-tetrachloroaniline in guppy (*Poecilia reticulata*) increased with increasing exposure time or increasing concentration of the test compound (de Wolf et al. 1992b), and log BCF values on a lipid basis for the same trichloroaniline isomer obtained in the same laboratory over a period of time using different strains of guppy ranged from 2.61 to 3.21 for 2,3,4-trichloroaniline and from 2.88 to 3.40 for 2,4,5-trichloroaniline (de Wolf et al. 1993). The significant role of the lipid content of the test organism is discussed in detail later.

It cannot, therefore, be too strongly emphasized that all of these considerations should be critically evaluated in discussions of bioconcentration potential.

The role of particulate matter and uptake via food — The inhomogeneity of many water masses is well established, so that attention has been directed to the role of particulate matter in binding xenobiotics and to its significance in determining their uptake and their bioavailability. The term bioaccumulation includes all transport routes including exposure to the xenobiotic in food and particulate matter, although the numerical difference between factors for bioconcentration and bioaccumulation will generally not be large except for highly hydrophobic compounds. The influence of organic matter in any form — dissolved or particulate — should, however, be kept in mind. For example, the bioconcentration factors of chlorobenzenes were reduced when guppy (*Poecilia reticulata*) were exposed to these compounds in sediment suspensions (Schrap and Opperhuizen 1990), and even dissolved organic carbon may also form associations with xenobiotics and, thereby, diminish their bioavailability (Landrum et al. 1985).

It has become increasingly realized that the exposure of biota to xenobiotics may occur not only from the dissolved state but also to a significant degree through consumption of particulate matter in sediments or in the water mass. Indeed, this exposure route may be dominant for demersal fish and for sediment-dwelling organisms. Two examples that support the role of particulate matter in determining exposure to xenobiotics may be given as illustration.

1. It has been estimated that in New Bedford Harbor, MA, sediment is responsible for 83% of the body burden of tetrachlorinated PCBs in winter flounder and for 42% in lobster (Connolly 1991).
2. In the clam *Macoma nasuta*, mass balance studies showed that the major route of uptake of hexachlorobenzene was via the gut from ingested solids (60-80%), with contributions of around 10% for other routes including interstitial and overlying water (Boese et al. 1990).

It is important to examine in a little more detail the significance of uptake via food, although biomagnification in a wider context in field situations is discussed in Section 3.5.4. As a general rule, it has been accepted that uptake via food rather than in the disssolved state directly from the water mass is the dominant exposure route for compounds with log $P_{ow} > 5$ (Thomann 1989; Connolly and Pedersen 1988). This alternative exposure route is not, however, generally taken into consideration in laboratory studies since it is difficult — and indeed, may be impossible — to distinguish between direct uptake via the food and simultaneous desorption from the food that results in direct uptake from the aquatic phase. Evidence that bioconcentration via contaminated food is the principal route of uptake for poorly water-soluble compounds has, however, been clearly demonstrated for PCBs. Two simple illustrations will be given here:

1. Lake trout (*Salvelinus namaycush*) in Lake Michigan have concentrations of PCBs which are three to four times higher than in alewife (*Alosa pseudoharengus*) which is preyed upon by older fish (Thomann and Connolly 1984).
2. PCB concentrations in the food chain of a small freshwater lake in Holland increase in the higher trophic levels (van der Oost et al. 1988). In the latter study, attention is also directed to the important issue of differences in the distribution of PCB congeners at the various trophic levels. An exhaustive study undertaken over a complete life cycle of guppy (*Poecilia reticulata*) has revealed important details of such processes (Sijm et al. 1992), while the wider issue of biomagnification including additional factors and the issue of cotransport with lipid material is discussed briefly in Section 3.5.4.

Whereas similar conclusions on the significance of uptake by food have been drawn from the results of a laboratory study with chlorinated dibenzo-1,4-dioxins using juvenile rainbow trout and fathead minnows (Muir and Yarechewski 1988), laboratory experiments with PAHs using rainbow trout did not reveal significant accumulation through dietary exposure, apparently as a result of poor absorption efficiency from the diet and rapid elimination of the xenobiotics (Niimi and Dookhran 1989). The results of these experiments with PAHs probably do not, however, exclude the significance of this exposure route for demersal fish that are exposed to high concentrations of these compounds in contaminated sediments.

Collectively, the results of these studies clearly illustrated that attention should be directed both to the feeding habits and to the physiology of specific organisms and that there may exist serious limitations in the application of models attempting universal application and which fail to take these into account.

The molecular size and shape of xenobiotics, and the role of lipid content of biota — Increasing evidence points to the specific role of lipids in determining bioconcentration potential, and two different kinds of situations may be clearly distinguished. It should be clearly appreciated, however, that the term lipid is used for a class of structurally diverse compounds united by a single physico-chemical property (solubility in organic solvents). They include, for example, neutral glyceryl triesters and glyceryl galactosides, zwitterionic phosphate diesters of glycerol and ethanolamine, and diesters of glycerol and inositol.

Some compounds, such as hexabromobenzene, octachlorodibenzo-1,4-dioxin, and tetradecachloroterphenyl, are not accumulated by fish, presumably due to the size and configuration of the molecules (Bruggeman et al. 1984). Such compounds have been termed superhydrophobic since they have values of log $P_{ow} > 6$, but it has been shown, on the other hand, that many of these compounds have — possibly unexpectedly — only low lipid solubility and that this *decreases* with increasing P_{ow} (Chessels et al. 1992). In the case of decachlorobiphenyl, it has been suggested that only ca. 3% of the substrate in the aqueous phase is available to guppy (*Poecilia reticulata*), and this results in a BCF value that is between ten- and one hundred-fold lower than would be predicted on the basis of the P_{ow} value of the compound (Gobas et al. 1989).

By definition, hydrophobic compounds would be expected to accumulate in lipid material so that variations in the lipid content of biota may be an important determinant both in laboratory experiments and in feral populations. For example, in a study on the uptake of hexachlorocyclohexanes and of pentachlorophenol by the mussel *Mytilus edulis* and by the polychaete *Lanice conchilega*, variations in the lipid content of the test organism could have introduced serious errors (Ernst 1979), and the BCF values for a number of organochlorine compounds increased linearly with lipid content to maxima at ca. 5% (Tadokoro and Tomita 1987). The lipid content of fish is also a function of their age, and in lake trout, for example, the lipid content increased from 7% in the age group 3 to 5 years to 16% in those in the age group 7 to 10 years (Thomann and Connolly 1984). The lipid content may also be subject to seasonal variation, and its significance in determining the half-lives of a number of organochlorine compounds in herring gulls (*Larus argentatus*) has been examined; analyses for lipid content and for plasma concentrations were then incorporated into a two-compartment model to describe the dynamics of clearance of the compounds (Clark et al. 1987). It may, therefore, be generally valuable to express BCF values based on lipid concentration (Mackay 1982) as well as on wet weight. These observations have directed attention to the whole question of the role of lipids, and this is discussed further in Section 3.1.2 in the context of surrogate procedures for estimating bioconcentration potential. In particular, the *structure* of biological lipid membranes should be taken into consideration.

In summary, all of these results illustrate the care that must be exercised in predicting the concentrations of xenobiotics in natural biota from values of bioconcentration factors assuming that uptake takes place exclusively from the water mass. The various factors which may seriously compromise the interpretation of measurements of bioconcentration potential are:

- Bioconcentration may occur by uptake of particulate material (bioaccumulation).
- The compound may be "bound" to dissolved components of the aquatic or sediment phases so that it is not freely accessible to biota.
- Transport into biota is not passive and must be evaluated in its relation to metabolism (Section 3.1.3).
- There may exist intrinsic limitations to transport into cells due to steric effects or the mere size and shape of the xenobiotic molecule.

Concentration of xenobiotics into algae — The preceding discussion has dealt almost exclusively with bioconcentration by fish or invertebrates, but transport into photosynthetic organisms also merits attention. Algae are the primary producers in aquatic systems and, therefore, play a key role in the food chain and in the transport of xenobiotics into higher trophic levels and — after their death — into the sediment phase. An interesting study (Swackhamer and Skoglund 1993) investigated the bioaccumulation of a range of PCB congeners into a strain of *Scenedesmus* sp. under different laboratory conditions and into field phytoplankton at different seasons of the year. Two important conclusions could be drawn:

1. In laboratory experiments at 11°C when growth was slow and when the length of exposure was 3d or more, the lipid-normalized bioaccumulation factor was a linear function of P_{ow} with a slope of unity for congeners with log $P_{ow} < 7$. When comparable experiments were carried out at 20°C in growing cells, the linearity was observed only for congeners with log $P_{ow} < 5.5$.
2. In the experiments using field material, for samples collected in both summer and winter, there was linearity for all congeners up to those with log P_{ow} of 8, but whereas the slope for winter samples was 0.93, that for summer samples was only 0.4.

These results show that equilibrium conditions do not prevail in growing populations of algae; the use of P_{ow} values to assess the bioaccumulation of hydrophobic xenobiotics into phytoplankton under growth conditions is not, therefore, justified.

Concentration of xenobiotics in higher plants — Uptake into aquatic plants may also be important and presents another redistribution pathway for xenobiotics. The uptake of a few pesticides has been investigated using the aquatic plant *Hydrilla verticillata* (Hinman and Klaine 1991), though only low levels of attrazine, chlordane, and lindane were accumulated. These plants have, moreover, only low levels of lipids, and this is consistent with the role of lipid material in determining uptake. Even these low levels of bioconcentration should, however, be taken into consideration in lakes with high densities of such plants. The situation for terrestrial plants is considerably more complex since uptake involves a number of partitioning processes including soil/water, water/plant roots, and atmosphere/leaves. For these reasons, a simple relation between bioconcentration factors and P_{ow} is not to be expected and this is confirmed by results using different plants and different xenobiotics (Scheunert et al. 1994).

3.1.2 Surrogate Procedures for Evaluating Bioconcentration Potential

Introduction — The concept of bioconcentration is derived from that of distribution coefficients in physical chemistry. In these, the equilibrium concentrations of a compound distributed between two phases are measured, for example, between water and a water-immiscible solvent such as hexane. If partitioning were a passive reaction, direct physico-chemical measurements of the partition between an aquatic phase and a suitable model for the biological membrane would be possible. It would, therefore, be attractive to measure distribution coefficients in a chemically defined system and to seek a correlation between the values found and those obtained by direct measurements in biota.

The octan-1-ol/water partition as a surrogate — A commonly used system measures — directly or otherwise — partition between water and octan-1-ol to derive the distribution coefficient (P_{ow}) and then applies an empirical formula to translate these values into bioconcentration factors (BCF) using a range of benchmark compounds. As would be expected, the numerical relationships depend on the organism used so that different equations result. Some equations that have been used for different organisms are the following (Mackay 1982; Hawker and Connell 1986):

- Fish $\log BCF = \log P_{ow} - 1.320$
- Molluscs $\log BCF = 0.844 \log P_{ow} - 1.235$
- Daphnids $\log BCF = 0.898 \log P_{ow} - 1.315$

As implied in Section 3.1.1, it should be noted that such equations cannot, however, be applied to uptake by aquatic plants — or ideed, other biota — with low lipid content (Hinman and Klaine 1992).

Accurate values of P_{ow} are clearly necessary for the ultimate calibration of all surrogate systems, but in practice, direct measurements of P_{ow} by the traditional shake-flask method are seldom used. Particularly for compounds with low water solubility, experimental difficulties may arise from problems in phase separation without carryover, sorption to glass surfaces, or from formation of emulsions. All of these introduce serious uncertainties into the concentrations in the appropriate phases and may, consequently, lead to substantial errors in the estimates of partition coefficients. The problem is particularly acute for compounds with extremely low solubility in water, such as the chlorinated dibenzo-1,4-dioxins for which widely varying values have been reported (Marple et al 1986; Shiu et al. 1988). For such compounds, use of a generator column has been advocated (De Voe et al. 1981; Woodburn et al. 1984). In essence, the following steps are carried out (1) a solution of the test substance in octanol is equilibrated with water, and the concentration in the octanol phase is determined, (2) the octanol phase is loaded onto a column packed with Chromosorb W, and (3) octanol-saturated water is pumped through the column and the solute collected in a Sep-Pak cartridge for analysis.

A possibly more expedient procedure is the slow-stirring method that has been applied to a structurally diverse range of hydrophobic xenobiotics (de Bruijn et al. 1989), and it was shown that there was generally good agreement with the values obtained from HPLC or the generator-column procedures.

The octanol water partition coefficient — and, hence, the bioconcentration potential — has also been correlated with the aqueous solubility, though the experimental determination of the latter for poorly water-soluble compounds also presents some problems. The following relations have been proposed (Mackay 1982):

Liquids $\log P_{ow} = 3.25 - \log X_l$ where X_l is the molar solubility (mol.m^{-3}) in water.

Solids $\log P_{ow} = 3.25 - \log X_s + 2.95 (1-T_m/T)$ where X_s is the molar solubility of the solid, T_m is the melting point, and T the ambient temperature.

A detailed thermodynamic discussion has been presented (Miller et al. 1985).

There are a number of basic questions which must also be addressed in the application of such surrogate procedures; among the most fundamental are the choice of the water-immiscible solvent and the neglect of metabolic transformation of the test compound.

Glycerol trioleate has been used in an attempt to simulate lipid membranes and to take into account some of the solvent associations plausibly occurring in biota; an impressive direct correlation was observed between $\log P_{tw}$ (P_{tw} is the partition coefficient between glycerol trioleate and water) and BCF values in rainbow trout expressed on a lipid basis, and these results were used to support the view that the bioconcentration of non-polar hydrophobic xenobiotics is significantly determined by their lipid solubility (Chiou 1985). This conclusion is further supported by the results of an extensive examination of a series of highly hydrophobic compounds which do not demonstrate a high potential for bioconcentration (Chessels et al. 1992). A few words of caution: it has been shown by careful thermodynamic analysis of rates of uptake that lipid membranes of biota differ significantly in structure from octanol/water interfaces (Opperhuizen et al. 1988), and attention has already been drawn to the low lipid solubility of some superhydrophobic compounds (Chessels et al. 1992).

Clearly, surrogate systems cannot take into account the metabolic activity of biological systems. Although the extent of biotransformation may be restricted for some classes of compounds, it is unlikely to be totally absent, and it has been suggested that for highly lipophilic compounds which have a low rate of physico-chemical elimination, the total rate of elimination may be significantly affected even by low rates of biological elimination (de Wolf et al. 1992a). The intrusion of metabolism results in lower concentrations in biota than would be predicted on the basis of the linear relationships between values of BCF and P_{ow} that have been noted above (de Wolf et al. 1992a). Such discrepancies have been observed for fish with compounds of diverse structure including trichloroanilines (de Wolf et al. 1993), chloronitrobenzenes (Niimi et al. 1989), and azaarenes (Southworth et al. 1980, 1981). For the azaarenes, the expected correlation was shown to hold for *Daphnia pulex* that apparently did not metabolize the compounds (Southworth et al. 1980).

Care should, therefore, be exercised in extrapolation of the results from all surrogate methods to assessing the uptake of xenobiotics in natural biota from the water phase.

Alternative surrogate procedures — A number of other surrogate procedures have been developed. These include the use of reverse-phase thin-layer chromatography (TLC) (Bruggeman et al. 1982; Renberg et al. 1985) to measure relative mobilities (R_m) or high-pressure liquid chromatography to measure capacity factors (Veith et al. 1979b); these values are then correlated with experimentally established P_{ow} values for standard compounds, and the correlation is then used to calculate values of P_{ow} for the unknown compounds. The TLC procedure is extremely easy to carry out but is essentially restricted to neutral compounds, and correlations for a range of structurally diverse compounds must be carried out with caution since appreciably different relations between R_m and P_{ow} exist for different classes of compounds such as PAHs and chlorinated compounds. Use of the reverse-phase HPLC system is highly flexible, since it can also be applied to ionizable compounds such as carboxylic acids, phenols, and amines. The partition coefficients relate to the unionized compounds that are the principal forms in which these compounds are transported into biota, even though their concentration may be low in comparison with the dissociated states at physiological pH values. Acidic compounds, such as highly chlorinated phenols or many carboxylic acids, have pK_a values < 7, and aqueous solutions of compounds with pK_a values < 6 contain only ca. < 10% of the free acid at neutral pH values. Although the influence of toxicity on pH has been examined (Neilson et al. 1990), its effect on bioconcentration has been less extensively explored. There is evidence, however, (Pärt 1990) that for some compounds both the unionized (free) and the dissociated forms may be accumulated. In the application of these methods, difficulties may emerge due to the absence of suitable detection methods, for example, in quantification of compounds lacking, for example, ultraviolet absorbance, fluorescence, or groups suitable for electrochemical detection.

3.1.3 Interdependence of Bioconcentration and Metabolism

It appears plausible to extrapolate to biological systems the concept of partitioning between two phases — representing the aquatic phase to which biota are exposed and a water-immiscible phase representing the lipid membranes. This simplification fails, however, to take into account a number of significant factors. The limitations concerning lipids have been briefly noted in Section 3.1.2, but there is an additional and frequently invalid assumption that is not always sufficiently appreciated. In fish, although the structure of the compound will generally remain unaltered during the partitioning between the aquatic phase and outer biological membanes (gills and skin), this will seldom be the case after transport into the organisms. Most of them have developed the capability of metabolizing many — if not most — xenobiotics, and some striking examples exist. For example, even compounds such as 1,3,6,8-tetrachlorodibenzo-1,4-dioxin can be metabolized by fish (Muir et al. 1986), although this is only one facet that may account for the apparently low bioconcentration potential of this compound. The same situation prevails in more complex natural systems. For example, whereas mirex may pass through the food chain *Daphnia magna—Lepomis macrochirus* without apparent metabolic change (Skaar et al. 1981), alterations in the relative distribution of the PCB congeners as they are transported through the food chain cod-seal-polar bear (Muir et al. 1988) clearly suggest metabolism by the terminal predator. Many xenobiotics are toxic to biota so that their metabolism in higher organisms generally serves as a mechanism for their detoxification and elimination. Some of these reactions are described in more detail in Chapter 7 (Section 7.5), and for fish at least, metabolism frequently results in the transformation of the xenobiotic to water-soluble compounds which are then excreted. The concentrations of the xenobiotic in the organism are, therefore, determined by the rates of metabolic processes as well as by the kinetics of bioconcentration. Elimination of the hydrophobic organophosphate insecticide chlorpyrifos (*O,O*-diethyl-*O*-[3,5,6-trichloropyridyl]phosphorothioate) from guppy (*Poecilia reticulata*) is accomplished almost exclusively by metabolism (Welling and de Vrise 1992). In channel catfish (*Ictalurus punctatus*), the major metabolite in urine and bile is the glucuronide conjugate of 3,5,6-trichloropyridinol, while the parent chloropyrifos is strongly bound to blood proteins (Barron et al. 1993). The significant role of metabolism has already been discussed in Section 3.1.2 in the context of surrogate procedures for assessing bioconcentration potential. There is, however, no sharp line dividing compounds that may be metabolized and those that are more persistent; these classes of compounds differ only in the *magnitude* of their rates of transformation. Whereas, for example, polychlorinated benzenes are only slowly metabolized by fish and, therefore, provide a suitable example of "inert compounds" (de Wolf et al. 1992b), it is probably safe to assume that, in the absence of evidence to the contrary, most xenobiotics can be metabolized by the fish species that are widely used for evaluating bioconcentration potential. Although details of the metabolism of xenobiotics by higher biota are given in Chapter 7 (Section 7.5), it may be useful to summarize here some of the groups of compounds for which metabolism should be taken into consideration in interpreting the results of experiments on bioconcentration: (a) PAHs, azaarenes, and thiaarenes; (b) phenols, anilines, and benzoates; and (c) chloronitrobenzenes.

Further consideration suggests that it may, indeed, be inappropriate to assess independently the apparently separate issues of bioconcentration and metabolism. Experiments on bioconcentration are generally designed so that exposure is continued until an apparent steady state is achieved; the test organisms are then generally maintained in the absence of the test compound to evaluate clearance by depuration. It would, therefore, be particularly attractive to combine these investigations with metabolic studies in which the nature of the metabolites is identified, though this seems only seldom to have been carried out.

Two general conclusions may be drawn from the foregoing discussion. First, the interpretation of data from experiments on bioconcentration of xenobiotics should recognize possible complications from the effects of metabolism and excretion, and second when aquatic organisms, such as leaches, mussels, or crustaceans, that are assumed to display limited metabolic potential for xenobiotics are used for monitoring purposes, interpretation of the data should consider the possibility of metabolism and excretion after initial exposure to the toxicant.

The role of metabolism in the wider context of the detoxification and elimination of xenobiotics from biota is discussed in Chapter 7 (Section 7.5), and its potential role in the dissemination of xenobiotic metabolites in Section 3.5.

3.1.4 Cautionary Comments

It may be questioned whether, in a dynamic system, the concept of bioconcentration as currently defined is experimentally accessible, although pragmatically the concept is certainly valuable provided that its limitations are sufficiently appreciated. Although it has been shown that there are some inherent ambiguities in the concept of bioconcentration potential, direct estimates can be made using fish, and surrogate procedures are well developed though care should be exercised in the interpretation of the results and especially in extrapolating these to calculate the concentrations of xenobiotics in natural biota where uncertainties about the exposure route may exist. Correlations between bioconcentration potential and various parameters, such as water solubility, octanol-water partition, and relative mobility on reversed-phase chromatographic systems, has been demonstrated and may be considered satisfactory if agreement within a power of ten is achieved. For more refined analysis of field material and for an assessment of potential public health hazard, however, a much greater degree of certainty may be required. It should also be pointed out that there appears to be a striking difference in the relative concentrations of different types of xenobiotics in sediments and fish; for example, in samples collected in the same area, the sediment/fish ratio was 200 for PAHs but only 0.05 for the organochlorine compounds (Malins et al. 1984). It would be plausible to attribute the difference to the fact that the PAHs are more readily metabolized by fish than the relatively recalcitrant organochlorine compounds. These issues are discussed in the context of biomagnification in Section 3.5.4.

An experimentally and interpretatively serious problem emerges with complex mixtures which are probably typical of many industrial effluents. The question immediately arises: if the nature of the compounds is unknown, how can measurements be made of partition? There are several strategies — each of them with inherent difficulties. Probably the least objectionable is the obvious one of first identifying the components of the mixture and then determining their partition coefficients. An equally acceptable — and possibly more realistic — procedure would be to fractionate the mixture into groups of compounds with putative bioconcentration potential having values of $\log P_{ow} > 3$. This could be carried out, for example, using HPLC, followed by identification of the relevant compounds. The least attractive method is to quantify unknown compounds using a surrogate with no established relation to the compounds in question. This procedure has been used on account of the ease with which it can be carried out, but its adoption seriously increases the number of links in the chain between measurements of partition coefficients and estimations of bioconcentration potential and, thereby, seriously jeopardizes estimates of bioconcentration potential.

3.2 PARTITIONING BETWEEN THE AQUATIC AND SEDIMENT PHASES

Introduction — The partitioning of compounds from the aqueous phase into biota is not the only significant process which occurs after the initial discharge of

Table 3.1 Examples of Classes of Xenobiotics Recovered from Sediment Samples

Hydrocarbons	
Polycyclic aromatic hydrocarbons	Prahl and Carpenter 1983
Alkylated aromatic hydrocarbons	Peterman and Delfino 1990
Chlorinated aromatic compounds	
Chlorobenzenes	Pereira et al. 1988
Polychlorinated biphenyls	Swackhamer and Armstrong 1986
Polychlorinated dibenzo-1,4-dioxins	Czuczwa and Hites 1986; Macdonald et al. 1992
Chlorinated guaiacols and catechols	Remberger et al. 1988
Nitrogen-containing aromatic compounds	
Azaarenes and aromatic nitriles	Krone et al. 1986
Oxygenated aromatic compounds	
2,4-dipentyl phenol	Carter and Hites 1992
Polycyclic quinones and ketones	Fernandez and Bayona 1992
Aliphatic carboxylic acids	
C_8 and C_9 dicarboxylic acids	Stephanou 1992

xenobiotics into aquatic systems; partitioning of xenobiotics from the aqueous phase into the sediment phase may be of equal significance, and its significance is attested by the structural range of organic compounds that have been recovered from contaminated sediments (Table 3.1). Attention is also drawn to the aromatic trifluoromethyl compounds noted in Chapter 2 (Section 2.5). Many of these compounds such as PAHs, PCBs, and PCDDs are widely distributed and only selected — and more or less random references — have been provided here. Some of these compounds certainly enter ecosystems as a result of long-distance transport but, irrespective of their origin, the sediment phase may clearly function as a highly effective sink for these compounds.

There are a number of important environmental consequences resulting from the partitioning of xenobiotics into the sediment phase (1) sediment-dwelling biota and demersal fish may be exposed to these compounds and the recovery of, for example, PAHs, polycyclic thiaarenes, and azaarenes (Vassilaros et al. 1982) from fish clearly demonstrates their progress through the food chain; and (2) since the sediment phase is not static but is subjected to the effect of currents and tides, the sediment phase may act as an effective transport system. These are discussed in greater detail in Section 3.5, and the important issue of the bioavailability of sediment-sorbed xenobiotics is discussed in Section 3.2.2.

3.2.1 Outline of Experimental Procedures

Natural sediments vary widely both in their physical structure and in their chemical composition. In addition to the inorganic matter that is universally present in the sediment phase, many shallow-water sediments contain appreciable amounts of humic material together with other organic matter originating from terrestrial plants. Furthermore, sediments in the neighborhood of industrial discharge often contain high concentrations of organic matter originating from manufacturing processes. This heterogeneity should be carefully evaluated in interpreting the results of experiments on partitioning and on the degree of recoverability of xenobiotics from natural sediments. Although it is customary to normalize partition data to the organic content of the sediment, using a relation

$$K_p = f_{oc} \cdot K_{oc}$$

where f_{oc} is the fractional organic carbon in the sediment and K_{oc} represents the partition to "generic" organic matter, it should be appreciated that the chemical structure of components of the sediment may play a critical role; the use of total organic carbon may, therefore, be misleading. For example, the sorption of toluene and trichloroethene to soil was dependent on the nature of specific organic components (Garbarini and Lion 1986), and the partition of pyrene to dissolved organic humic material was influenced by its structure and was dependent on factors other than the total content of organic carbon (Gauthier et al. 1987). The same structural dependence holds for association between xenobiotics and the organic constituents of interstitial water, and for marine samples, unexpectedly high sorption may be due to the high lipid content (Chin and Gschwend 1992). On the other hand, for aquifer samples with a low content of organic carbon, there was no correlation with the organic carbon content (Stauffer et al. 1989). Care should, therefore, be exercised in comparing the results of partitioning using sediments that have predominantly mineral components with those containing substantial amounts of structurally diverse organic components.

Direct measurements of sediment/water partition — The experimental determination of partition coefficients in laboratory experiments is, in principle, straightforward: it involves mixing samples of the sediment and of the aqueous phase until a steady-state is reached — generally within 24h — followed by analyses of the phases after separation. Azide may conveniently be added to inhibit bacterial transformation of the xenobiotic during the experiment. Attention should, however, be directed to an important factor that may seriously compromise the results: after equilibration, the aqueous phase may contain dissolved organic material from the sediment phase, and this may compromise estimates of the truly dissolved concentrations of the xenobiotic. This problem may be especially acute in determining the partition of compounds with extremely low water solubility such as 1,3,6,8-tetrachlorodibenzo-1,4-dioxin (Servos and Muir 1989), and it should also be clearly appreciated that the values of partition coefficients obtained in this way cannot take into account the significant alterations (aging) that occur after deposition and that may be of cardinal significance (Section 3.2.2).

Surrogate procedures — Extensive use of surrogate procedures has been used for estimating BCFs, and this is also the case for K_{oc}. Similarly, it should be appreciated that some implicit assumptions are made. First, that the compounds are neutral and hydrophobic, and do not react with the sediment phase, and second, the partitioning is determined by the organic carbon content of the sediment. Detailed descriptions of two surrogate procedures have been provided (Karickhoff 1984) so that only an outline of these is required here.

1. On the basis of water solubility and, for solids, thermodynamic parameters such as the entropy of fusion and the gas constant, two equations have been suggested that take into account the gross chemical structures of the compounds

$$\log K_{oc} = -0.9211 \log X_s - 1.405 - 0.00953 (T_m-25) \qquad \text{PAHs}$$
$$\log K_{oc} = -0.83 \log X_s - 0.93 - 0.01 (T_m-25) \qquad \text{Organochlorines}$$

where X_s is the molar solubility in water and T_m is the melting point in °C.

2. Cogent arguments have been presented to support the existence of a correlation between K_{oc} and P_{ow}, and again substantially different equations had to be applied to different structural classes of compound

$$\log K_{oc} = 0.72 \log P_{ow} + 0.49 \qquad \text{Simple benzenoid compounds}$$
$$\log K_{oc} = \log P_{ow} - 0.317 \qquad \text{PAHs}$$

Application of either procedure generally yields acceptable predictions for values for K_{oc} (Karickhoff 1981), though critical attention should be directed to the important limitations inherent in the method particularly when using empirically derived values of P_{ow}. As is generally the case, the correlations are least reliable for extreme values of P_{ow} (> ca. 10^6).

3.2.2 Reversibility: Sorption and Desorption

Introduction — It is well established that many compounds after introduction into the environment are not readily accessible to chemical recovery. This does not necessarily imply, however, that they are of no environmental significance; the degree to which they are desorbed and, therefore, become accessible to biota is a central issue that has implications both for the toxicity of xenobiotics and for their resistance to microbial attack. Several general considerations are worth noting.

1. There is substantial literature showing that a substantial fraction of agrochemicals introduced into the terrestrial environment is not recoverable by standard chemical procedures and is apparently bound irreversibly to either organic or inorganic components of the soil matrix (Bollag et al. 1983; Lee 1985; Ou et al. 1985; Smith 1985) or physically inaccessible through inclusion in micropores (Steinberg et al. 1987).
2. Laboratory experiments on sorption have shown that even over a short period of time sorption may be irreversible or exhibit hysteresis. Illustrative examples are provided by chlorophenols in sediment fractions (Isaacson and Frink 1984) and a number of chlorinated alkanes and alkenes (Pignatello 1990) and trichloroethene in soil (Pavlostathis and Jaglal 1991). These experiments often display two-stage kinetic processes, of which the second is associated with irreversible binding (Karickoff 1984; Pavliostathis and Mathavan 1992).
3. The mobilization of sorbed xenobiotics is of serious concern in areas subjected to historical pollution, and this has motivated extensive investigations on desorption. For example, studies with sediment from New Bedford Harbor, MA, revealed both the significant role of organic carbon and that increased desorption of PCB congeners occurred in distilled water rather than in saline water; such data clearly support the concept of three phases in partitioning models (Brannon et al. 1991).
4. Probably most laboratory studies on sorption/desorption have used single substrates although this is almost certainly an oversimplification of most natural situations. An example of the significance of interactions is afforded by a study with poly(N,N-dimethylaminoethyl methacrylate) (Figure 3.1) of which a substantial fraction was irreversibly adsorbed on a sediment with high ion-exchange capacity; in addition, presorption of the polymer to the sediment significantly increased the subsequent sorption of naphthalene (Podoll and Irwin 1988).

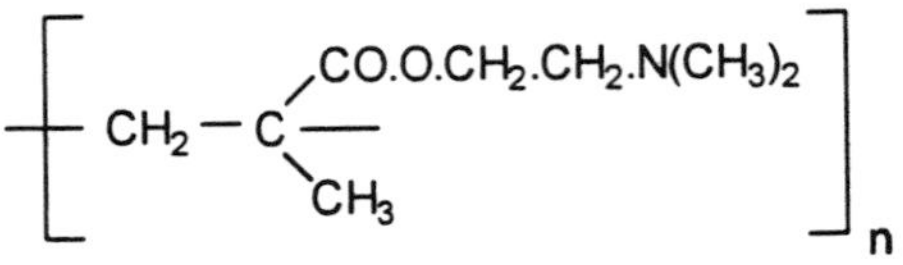

$$\left[\quad CH_2 - C \begin{array}{c} \nearrow CO.O.CH_2.CH_2.N(CH_3)_2 \\ \searrow CH_3 \end{array} \quad \right]_n$$

Figure 3.1   Poly(*N,N*-dimethylaminoethylmethylacrylate).

Aging and bioavailability — Results which indicate decreasing recoverability of xenobiotics from the sediment phase with increasing time from deposition may be accommodated under the general description of "aging"; this is the result of interactions between the xenobiotics and organic and inorganic components of the soil or sediment matrix. The mechanisms of some of these associations are discussed in Section 3.2.3. The aging of soils and sediments introduces a potentially serious indeterminacy into estimates of both the chemical concentration and the degree of bioavailability of xenobiotics and may, therefore, result in a serious ambiguity in correlating the exposure of biota to xenobiotics and the effects that are observed. Its significance has been illustrated, for example, with PAHs and with 2,4,5, 2′,4′,5′-hexachlorobiphenyl in the amphipod *Pontoporeia hoyi* (Landrum 1989) and may be even greater for compounds such as chloroguaiacols since a substantial fraction of these is chemically inaccessible in naturally aged sediments (Remberger et al. 1988). Aging has been demonstrated conclusively in the terrestrial environment, and the critical effect of residence time on the degree of recovery has been evaluated (Capriel et al. 1985; McCall and Agin 1985; Winkelmann and Klaine 1991). It seems unlikely, however, that these "bound" residues are merely static reserves, so that the critical issue is the extent to which they are desorbed and, thereby, become directly available to biota (Knezovich et al. 1987).

The degree of bioavailability of organic compounds depends critically on their chemical structure which determines the kinds of interaction that may take place within the solid phase. For example, linear alkylbenzenesulfonates are readily desorbed from sediments, so that their biodegradability and potential toxicity is largely unaffected by the presence of sediments (Hand and Williams 1987). On the other hand, benzo[a]pyrene, even though accessible to chemical extraction, appears to be available to biota only to a limited extent (Varanasi et al. 1985). It is possible to distinguish two apparently opposing environmental effects:

1. Sorption of toxicants may result in diminished exposure of biota to deleterious concentrations of xenobiotics. This has been clearly demonstrated in experiments under clearly defined laboratory systems involving dissolved organic carbon in the aquatic phase. These experiments showed reduced bioavailability and, hence, diminished toxicity. A number of organisms and a range of toxicants have been examined: for example, *Salmo salar* and a range of organochlorine compounds including both chlorophenolic and neutral compounds (Carlberg et al. 1986), *Oncorhynchus mykiss* (syn. *Salmo gairdneri*) and benzo[a]pyrene and 2,2′,5,5′-tetrachlorobiphenyl (Black and McCarthy 1988), and *Diporeia* sp. (syn. *Pontoporeia hoyi*) and PAHs and 2,2′,4,4′-tetrachlorobiphenyl (Landrum et al. 1987).

2. On the other hand, the persistence of xenobiotics may be increased if they are not accessible to the relevant degradative microorganisms. This is discussed more extensively in Chapter 4 (Section 4.5.3), so that only a few illustrative examples will be given here. Experiments on the biodegradation of compounds as diverse as *iso*propylphenyl diphenyl phosphate (Heitkamp et al. 1984), aliphatic esters of 4-aminobenzoate (Flenner et al. 1991) in spiked sediments, or substituted phenols in the presence of naturally occurring humic acids (Shimp and Pfaender 1985) support the view that *increased* persistence of xenobiotics is to be expected when the substrates are not freely available to microorganisms in the aquatic phase. Conclusions concerning the influence of sediment redox potential and pH on the degradation of pentachlorophenol (DeLaune et al. 1983), therefore, appear to be equally consistent with the role of binding of pentachlorophenol to the sediment phase. In all cases, the key issue is probably the rate of transport into the microbial cells, since rates of chemical hydrolysis of phosphorothioate esters under neutral conditions were apparently unaffected by sediment sorption (Macalady and Wolfe 1985); on the other hand, the presence of humic material reduced the rate of alkaline hydrolysis of the octyl ester of 2,4-dichlorophenoxyacetic acid (Perdue and Wolfe 1982). A similar apparent contradiction emerges from the results of studies in which the role of dissolved organic carbon either facilitated or retarded the transport of xenobiotics through porous material such as sand (Magee et al. 1991). These differences may, on the other hand, merely reflect significant differences in the structure of the humic material used in these experiments.

Regardless of details, it may be concluded that the mechanisms by which xenobiotics become associated with particulate matter and the degree to which these interactions are reversible are of cardinal importance in assessing the environment impact of xenobiotics.

3.2.3 Mechanisms of Interaction Between Xenobiotics and Components of Solid Matrices

Introduction — It is now appropriate to consider in more detail the mechanisms of interaction between xenobiotics and the components of soils and sediments. It may plausibly be assumed that the principles are applicable equally to both of these matrices — with one important exception: interactions in the terrestrial environment catalyzed by fungal enzymes will probably play, at most, a minimal role in most aquatic systems. Details of the mechanisms by which xenobiotics are "bound" to components of the sediment phase have seldom been established, although several plausible hypotheses have been put forward; proposed mechanisms of interaction include ionic binding, long-range (van der Waals) forces, or sorption by undefined mechanisms, although these are pragmatic and probably conceal the complexity of the molecular processes. Three major components of the sediment matrix have generally been considered:

1. Inorganic minerals dominated by the most abundant elements, such as aluminium, iron, and silicon.

2. Organic constituents, such as lignin-derived compounds and undefined compounds similar to humic and fulvic acids originating from the terrestrial environment.
3. Detrital material resulting from the decomposition of aquatic and sediment biota comprising both lipid and proteinaceous material.

Essentially three mechanisms of interaction may be discerned: (1) sorption involving interaction with inorganic components, (2) covalent reactions involving the organic constituents, and (3) physical entrapment.

1. Interactions with inorganic components — Extensive studies (Hayes et al. 1978a,b) have been directed to the sorption onto clay minerals of pyridinium and bipyridinium compounds which are valuable agrochemicals. The mechanisms were clearly different in a number of respects from those noted below for other types of compounds. Adsorption correlated with the cation exchange capacity of the clays, and when this was saturated, sorption was attributed to van der Waals interactions involving the pyridinium rings. As might be expected for these compounds, quaternization with methyl groups reduced the degree of adsorption. The sorption of a wide range of nitroaromatic compounds to mineral surfaces has been examined (Haderlein and Schwarzenbach 1993), and it has been proposed that interaction involves the formation of electron donor-acceptor complexes that are particularly strong for compounds containing several electron-withdrawing substituents such as nitro. The results suggested the possibly significant role of such interactions in the transport of such compounds in aquifers. It has also been hypothesized that partitioning of chlorinated catechols from the aquatic phase into the sediment phase took place through formation of complexes with Fe and Al components, and this has been correlated with simultaneous desorption of chlorocatechols, Fe and Al (Remberger et al. 1993). Collectively, these observations clearly demonstrate the importance of interactions between organic compounds and mineral surfaces.

2. Interactions involving organic components — The precise chemical structures of the organic components of soil and sediments are largely unknown, and terms such as "humic acid" and "fulvic" acid are primarily descriptive rather than representing chemically defined entities. The most detailed studies on the mechanisms of interaction have been directed to interactions between hypothetical structures of humic material and xenobiotics with phenolic and aromatic amino groups. Two essentially different mechanisms for binding of xenobiotics to organic components of solid matrices have been considered. They are (1) the chemical reaction between the xenobiotic and functional groups on the humus structure and (2) enzymatically mediated reactions of incorporation which merit attention as representing plausible models, particularly in the terrestrial environment.

Two examples of different types of chemical reactions may be used as illustration.

1. It has been hypothesized that carbonyl and quinone groups occur in humic material, and the presence of these has now been confirmed by a study in which diverse fulvic and humic acid samples were derivatized with ^{15}N hydroxylamine and the products examined by ^{15}N and ^{13}C-NMR (Thorn et al. 1992) (Figure 3.2). These observations underscore the relevance of the results from an earlier investigation (Parris 1980) in which the interaction of aromatic amines with

carbonyl and quinone groups was studied. It was shown that after a rapid reversible reaction, a slow irreversible reaction took place, probably involving addition of the amines to quinones followed by tautomerism and oxidation (Figure 3.3).

2. Structures in which phenols are covalently linked to C_3-guaiacyl residues have been examined as models for interaction between chlorophenols and lignin residues in humic acids (Zitzelsberger et al. 1987).

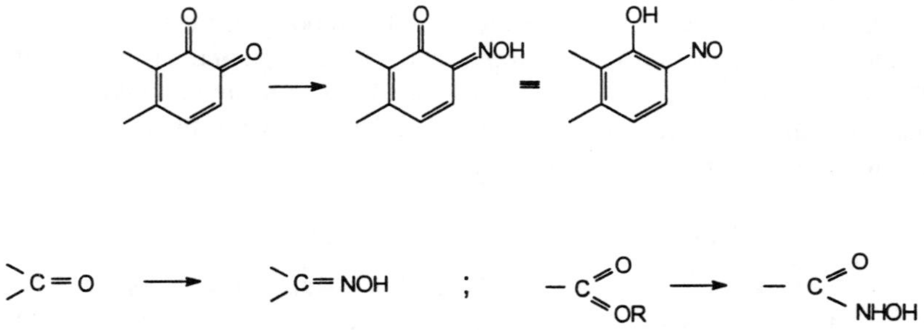

Figure 3.2 Structural entities identified in humic and fulvic acids by ^{15}N NMR after reaction with [^{15}N] hydroxylamine.

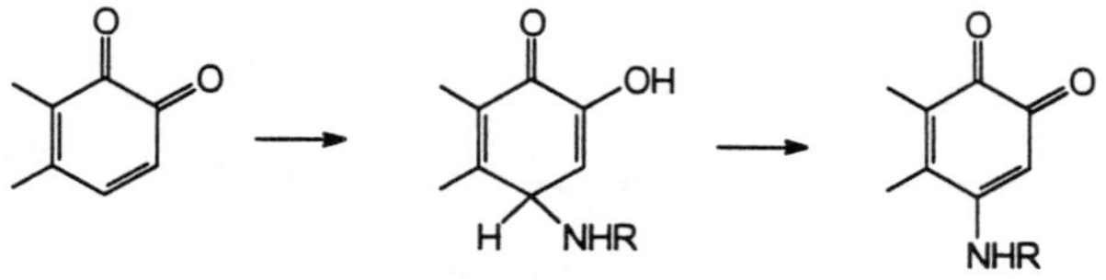

Figure 3.3 Reaction of amines with quinone groups.

Considerable attention has also been directed to enzymatic reactions mediated by fungal oxidoreductase enzymes such as phenol oxidase, peroxidase, and laccase; it should be appreciated that the conclusions from these studies are relevant primarily to the terrestrial environment where fungi are an important part of the microflora. These systems have been used to copolymerize structurally diverse xenobiotics including substituted anilines (Bollag et al. 1983) and benzo[a]pyrene quinone (Trenck and Sandermann 1981) to lignin-like structures. One great advantage of the use of these model systems is that it has been possible to isolate the products of the reactions and determine their chemical structure. Three examples may be given to illustrate the different substrates involved and the types of products that may be produced.

1. Reaction between halogenated phenols and syringic acid in the presence of laccase from the fungus *Rhizoctonia praticola* resulted in the formation of a series of diphenyl ethers containing one ring originating from the chlorophenol together with 1,2-quinonoid products resulting from partial O-demethylation and oxidation (Bollag and Liu 1985) (Figure 3.4). Comparable reactions have

also been postulated to occur between 2,4-dichlorophenol and fulvic acid (Sarkar et al. 1988).

2. A study using guaiacol and 4-chloroaniline and a number of oxidoreductases has demonstrated the synthesis of oligomeric quinonimines together with compounds resulting from the reaction of the aniline with diphenoquinones produced from guaiacol (Simmons et al. 1989) (Figure 3.5).

3. Direct evidence for the existence of covalent bonding between 2,4-dichlorophenol and peat humic acid in the presence of horseradish peroxidase has been provided by the results of an NMR study using 2,6-[^{13}C]-2,4-dichlorophenol (Hatcher et al. 1993). In the absence of suitable model compounds, interpretation of the results was based on estimated chemical shifts for a range of plausible structures. The most important contributions came from those with an ester linkage with the phenol group and covalent bonds between carbon atoms of the humic acid and C_4 (with loss of chlorine) and C_6 of the chlorophenol.

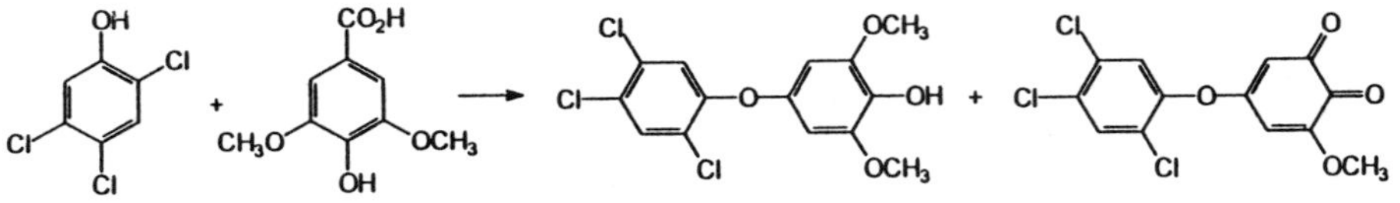

Figure 3.4 Reaction between 2,4,5-trichlorophenol and syringic acid catalyzed by laccase.

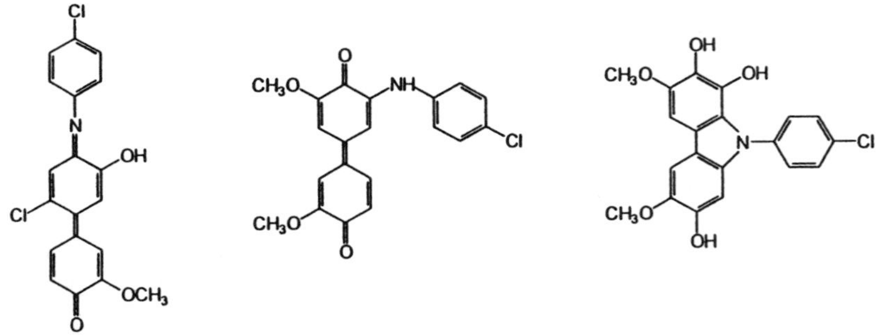

Figure 3.5 Products from the enzymatic copolymerization of guaiacol and 4-chloroaniline.

There is, therefore, extensive evidence that may be used to rationalize the occurrence of "bound" residues in soils, and this phenomenon is of particular significance for agrochemicals. Such processes influence not only their recovery by chemical procedures but also their biological effect and their biodegradability (Calderbank 1989): the latter is discussed further in Chapter 4 (Section 4.5.3). The extent to which these principles are applicable to aquatic systems appears not to have been established, though it is clearly possible that comparable mechanisms do exist.

3. Physical entrapment — Physical entrapment may also be significant for some molecules, and attention is drawn to the occurrence of micropores in soils and the role of these in retaining xenobiotics. This has been demonstrated for 1,2-dibromoethane, which is notoriously persistent in agricultural soils (Steinberg et al. 1987). Although this mechanism has seldom been considered as a quantitatively significant phenomenon in a wider context, the chemical stability of a wide range of clathrate compounds (Chapter 2, Section 2.2.2) may be worth examining (Hagan 1962), and the existence of complexes between the pyrazole phenyl ether herbicide and cyclodextrin (Garbow and Gaede 1992) could provide a viable model for such essentially physical interactions.

An important conclusion from all these studies is that the mechanisms of interaction are specifically related to the structure of the xenobiotic and that exclusive concentration of effort on neutral hydrophobic compounds may divert attention from important principles that are relevant to groups of structurally different compounds.

3.3 PHASE HETEROGENEITY: DISSOLVED ORGANIC CARBON, INTERSTITIAL WATER, AND PARTICULATE MATTER

Partitioning of a xenobiotic from the aquatic phase into biota or into the sediment phase is not a terminal process; indeed, from an environmental point of view, these represent merely the introduction of the xenobiotic into a complex network of interactions. At least three important factors should be evaluated (1) direct desorption mechanisms from the sediment phase into the aquatic phase, (2) resuspension of particulate matter from the sediment phase into the water column, and (3) the role of interstitial water in the sediment phase.

3.3.1 The Inhomogeneity of the Water Column

The preceding discussion has taken into account only partitioning between two single phases. In natural ecosystems, however, the situation is almost invariably more complex, and significant complications are introduced as a result of the inhomogeneity of the water column. This may often contain organic carbon in various states of aggregation, and the distinction between these is empirical rather than theoretical. A useful pragmatic procedure has used SepPak C-18 columns from which humus-bound xenobiotics are eluted directly, whereas the dissolved components are retained and may then be eluted with methanol (Landrum et al. 1984). There are, therefore, at least the following three states — sometimes referred to as phases — in which xenobiotics may be found in the water column (1) truly dissolved in the aqueous phase, (2) associated with dissolved (including high molecular weight) organic carbon, and (3) bound to particulate material.

It should be emphasized that these divisions are purely empirical, that dynamic interactions between them occur continuously, and that the relative concentrations in the various fractions will depend on the specific nature and the geographical location of the water mass. For example, in samples of water from sites in Lake Michigan, although the freely dissolved concentration of organochlorine compounds was dominant and the contribution from compounds bound to dissolved organic carbon was generally < 5% (Eadie et al. 1990), this may not necessarily be the case for other kinds of lakes or for brackish or marine systems. The exposure of biota may not, however, always be unequivocally defined. For example, the burden of freshwater mussels exposed to natural contaminants in Lake St. Clair in the presence of contaminated or uncontaminated sediments clearly showed that uptake was mediated principally via the water column — although the possible role of uptake via compounds sorbed to dissolved or particulate material was not unambiguously resolved (Muncaster et al. 1990). Attention has already been directed to the fact that the uptake of highly hydrophobic compounds into biota may take place via food rather than by direct uptake from the water mass (Section 3.1.1), and such compounds may be associated with organic compounds in any of the three states distinguished above. Reference has already been made (Section 3.2.2) to the possible detoxification which may be exerted by dissolved organic carbon.

3.3.2 The Role of Interstitial Water

There has been extensive interest in interstitial water in view of its significance in determining the exposure of sediment-dwelling biota to xenobiotics and its role in the dissemination of xenobiotics. It should be appreciated, however, that the term "interstitial" is operational rather than absolute. Although various methods have been advocated including pressure filtration, centrifugation at moderate centrifugal force (ca. $1000 \times$ g) is probably the preferred procedure. Interstitial water is not generally a homogeneous phase, and xenobiotics may occur in any of the states described in Section 3.3.1. At least for hydrophobic xenobiotics such as PCBs in coastal marine sediments, these compounds are associated in the interstitial water with organic carbon that is probably colloidal though of undetermined structure (Brownawell and Farrington 1986). The role of interstitial water in mediating the exposure of biota to xenobiotics is discussed in this section, while its significance in the dissemination of xenobiotics is discussed briefly in Section 3.5.1.

For sediment-dwelling organisms, one important factor which determines the degree of exposure to xenobiotics in the sediment phase is the partitioning from the true sediment phase into interstitial water from which the xenobiotic may then be accumulated by biota. Exposure of sediment biota to xenobiotics is, however, a complex process, since uptake may proceed either via particulate material or via interstitial water or by both routes. In the equilibrium partition model, the concentration of a xenobiotic in the interstitial water (C_{iw}) is given by the following relation:

$$C_{iw} = C_s/K_{oc}.f_{oc}$$

where C_s is the concentration in the sediment, K_{oc} the partition coefficient between water and "generic" organic carbon, and f_{oc} the fraction of organic carbon in the sediment. As an example, this relation has been verified for fluoranthene at interstitial concentrations less than 50 µg/L, though it was not valid at higher concentrations (Swartz et al. 1990).

It has been shown that the uptake of a number of halogenated xenobiotics into biota is mediated by interstitial water. Examples include the transport of chlorobenzenes into larvae of the midge *Chironomus decorus* (Knezovich and Harrison 1988) and of a range of chlorinated compounds into oligochaete worms (Oliver 1987). In addition, its significance may be inferred from the results of experiments on the uptake of PCB by the polychaete *Nereis diversicola* (Fowler et al. 1978). Interstitial water has, therefore, been widely used to assess the toxicity of sediments using a number of aquatic organisms (Carr et al. 1989; Ankley et al. 1991), and the equilibrium partition model justifies this application even for organisms such as the marine amphipods *Rhepoxynius abronius* and *Corophium spinicorne* which have potentially different routes of exposure to xenobiotics in the sediment phase (Swartz et al. 1990). A number of important facts have, however, emerged which illustrate the caution that should be exercised in the application and interpretation of the results of assays using interstitial water:

1. Care should be exercised to ensure that the test organisms realistically represent the situation near the water-sediment interface (Ankley et al. 1991).
2. The rate of accumulation of PAHs from spiked sediments by *Diporeia* sp. could not be predicted from measurements of partitioning between interstitial water and sediment particles (Landrum et al. 1991).
3. Although the toxicity of DDT to the amphipod *Hyalella azteca* decreased with increasing carbon content of the sediment, this was not the case for endrin (Nebeker et al. 1989), so that specific mechanisms of interaction even between neutral xenobiotics and the organic carbon in the sediment phase may be of determinative significance. The results with DDT are, in fact, consistent with evidence from equilibrium dialysis experiments of its association with dissolved humic material (Carter and Suffet 1982).

There are, therefore, unresolved factors which restrict the extrapolation of laboratory-determined partitioning parameters to field situations, and the results underscore the limitation of models for partitioning which encompass compounds with significantly different physico-chemical properties. There, therefore, remains a need for the application of direct assays for toxicity using true sediment organisms such as oligochaetes (Wiederholm et al. 1987), although the results even from such tests should be interpreted with caution in view of the complication of aging which has been briefly noted above.

3.3.3 The Role of Sediment and Particulate Matter in the Aquatic Phase

The dynamics of transport from the aqueous phase into biota, into sediments, and into the atmosphere are cardinal determinants of the dissemination of xenobiotics after discharge. For the aquatic phase, the complexity of the situation that may prevail in natural ecosystems may be illustrated by two apparently conflicting results from laboratory experiments using the same compound. The results of experiments in which fathead minnows (*Pimephales promelas*), the worm *Lumbriculus variegatus,* and two amphipods were exposed to hexachlorobenzene in water and in spiked sediment suggested that the sediment was a more efficient sink than the biota (Schuytema et al. 1990). On the other hand, the results from another study suggested that hexachlorobenzene was apparently desorbed from contaminated sediments and accumulated in algae (Autenrieth and DePinto 1991). The critical issues are, therefore, the *relative rates* of the various partition processes, and these may be determined specifically by the biota present in the system and by the lipid content of the relevant biota (Section 3.1.1); attention has already been directed (Section 3.2.2) to the degree of reversibility of these xenobiotic associations, and these factors underscore that care should be exercised in extrapolating to natural systems the results of laboratory experiments on partition.

Association between xenobiotics and particulate — though suspended — material should also be viewed in a broader perspective. On the one hand, stratification of the water mass of lakes may result in xenobiotic-associated particulate material from the upper layers of the lakes entering the superficial layers of the sediment surface; on the other hand, resuspension of sediment material at the sediment/water interface may bring about re-entry of such material into the water column. The extent of the various processes depends critically on the structures of the compounds so that, for example, more highly chlorinated PCBs in Lake Superior were lost from the water column, whereas the less highly chlorinated congeners entered the water phase from the sediments (Baker et al. 1985). Simplistic views that sediments function exclusively as "sinks" for highly hydrophobic organic compounds must, therefore, be viewed with caution, and the details of the dynamics of the water masses in lakes should be taken into consideration.

It may, therefore, be concluded that our understanding of the basic mechanisms — and in particular, the dynamics of the partitioning of xenobiotics between all phases, including truly dissolved organic matter, colloidal material, and particulate matter — and their relative importance in determining the exposure of biota to xenobiotics is strictly limited and that refinement of many unresolved details is highly desirable.

3.4 PARTITIONING BETWEEN THE AQUATIC PHASE AND THE ATMOSPHERE

It is obvious that volatile organic compounds such as low molecular weight hydrocarbons and chlorinated hydrocarbons may readily be partitioned into the atmosphere from the aquatic phase, and, indeed, this may be the major sink for such compounds (Smith et al. 1980). Attention has been directed especially to volatile chlorinated compounds in light of their role in atmospheric chemical reactions, in particular the destruction of ozone. Increasing evidence of the long-distance transport of apparently non-volatile compounds such as PCBs (Swackhamer and Armstrong 1986) which is discussed in greater detail in Section 3.5.3 illustrates, however, the need for assessing the magnitude of partitioning between the aquatic phase and the atmosphere even of quite large molecules that are presumably considerably less volatile than the chlorinated alkanes.

The rate of transfer from the aquatic phase to the atmosphere is a complex function of several mass-transfer parameters including Henry's law constant (H): this is defined as the ratio of the vapor pressure of a compound to its solubility in water, and the value of H is of particular significance in many natural situations. Because of experimental difficulties, values of H are frequently calculated rather than measured experimentally, though this may result in substantial errors in the estimated values. An extensive compilation of values of H has been provided by Mackay and Shiu (1981). Three experimental procedures have been used:

1. A direct method using a wetted-wall column has been used for several pesticides with values of H between 10^{-5} and 10^{-7} [dimensionless units] (Fendinger and Glotfelty 1988) and for selected pesticides, PAHs and PCBs (Fendinger and Glotfelty 1990).

2. A dynamic headspace gas-partitioning method has been used to assess the effect of dissolved organic carbon on the value of H for mirex (Yin and Hassett 1986), and with a gas chromatographic detection system for analysis, this is applicable to native water samples containing mirex.
3. A gas-purge system has been used for a number of chlorinated aromatic hydrocarbons and PAHs (ten Hulscher et al. 1992).

Particular attention should be directed to the units used for reporting values of H and those for both vapor pressure and solubility. Dimensionless values of H may be reported as exemplified above and are the least ambiguous. To convert them into conventional $kPa.m^3.mol^{-1}$, they should be multiplied by RT which has a value of 0.246 in this case. Vapor pressures may be expressed in atmospheres or torr (1 kPa = 7.5 torr = 9.87×10^{-3} atmos) and solubilities in $g.m^{-3}$ or $mol.m^{-3}$.

Estimates of the transfer rates for a number of aquatic environments have been made (Smith et al. 1981), from which it emerges clearly that the half-lives of compounds in the aqueous phase will be less than 10 d for compounds with H values > ca. 10^{-1} $kPa.m^3.mol^{-1}$ in lakes and > ca. 10^{-2} $kPa.m^3.mol^{-1}$ in rivers. This is consistent with the prediction (Mackay and Yuen 1980) that significant rates of volatilization occur for compounds having values of H > 10^{-1} $kPa.m^3.mol^{-1}$, and may still be significant at values of 10^{-3} $kPa.m^3.mol^{-1}$: a lower limit may be set for water which has a value of 3×10^{-5} $kPa.m^3.mol^{-1}$ The values of H (ten Hulscher et al. 1992) suggest appreciable rates of volatilization for all the chlorinated aromatic compounds (ranging from 0.192 $kPa.m^3.mol^{-1}$ for 1,3,5-trichlorobenzene to 0.016 $kPa.m^3.mol^{-1}$ for 2,2',5,5'-tetrachlorobiphenyl) but not for the PAHs that had values < 5×10^{-5} $kPa.m^3.mol^{-1}$ Partition from the aqueous phase into the atmosphere is particularly significant for compounds with appreciable vapor pressure and low water solubility (Mackay and Yuen 1980).

As for the other partition processes discussed above, empirical relations have been sought to relate the vapor pressure (P) to simple molecular parameters. Correlation between P and the boiling point (T_b) and melting point (T_m) has resulted in the proposal of the following equation (Mackay and Yuen 1980) which holds for a wide range of neutral compounds:

$$\ln P = 10.6 (1 - T_b/T) + 6.8 (1 - T_m/T)$$

where T is the ambient temperature. Correlations have also been used to estimate P for a few compounds including chlorinated guaiacols from gas-chromatographic retention times (Bidleman and Renberg 1985). Since values of P_{ow} have been correlated with aqueous solubility (S), it is simple to express H_c as a function of T_b and P_{ow}:

$$\ln H_c = 10.6 (1 - T_b/T) + \ln P_{ow} - 12.1$$

It is thus clear that there is a network of relations between the functions determining the following partitions: octanol/water (P_{ow}), water/biota (BCF), water/generic organic carbon in sediment (K_{oc}), water/atmosphere (H_c) — and between each of them and thermodynamic properties such as aqueous solubility (S), vapor pressure (P), and melting point (T_m).

3.5 DISSEMINATION OF XENOBIOTICS

Introduction — It is now appreciated that environmental hazards cannot be evaluated solely on the basis of the effects observed in the immediate neighborhood of point discharges. Xenobiotics may be transported over long distances and may then be recovered from samples remote from the source of initial discharge. This is particularly significant for the most persistent compounds such as organochlorines — in particular PCBs, PCCs, PCDDs and PCDFs — and some agrochemicals, and, indeed, the recovery of these from environmental matrices confirms the view that these compounds are not readily degraded and are, therefore, persistent in the environment. The effects of these compounds may, therefore, be exerted in pristine areas remote from the initial discharge that may be particularly sensitive to such perturbations. The important role of metabolites produced by microorganisms and of conjugates of metabolites excreted by higher biota should also be taken into consideration.

In the discussions presented in earlier sections, attention has been directed to the partition between pairs of phases: between the aqueous phase and biota including fish, algae, and higher plants; between the aqueous phase and sediments including particulate matter; and between the aqueous phase and the atmosphere. In most natural ecosystems, all of these are simultaneously present, and most of the partition processes are reversible to a greater or lesser extent. The distribution of a xenobiotic is, therefore, determined by the dynamics of sorption/desorption, bioconcentration/elimination after metabolism, and atmospheric deposition/evaporation from the aquatic phase. All of these are interconnected, and the transport of xenobiotics associated with both particulate matter and with algae may be mediated by processes as different as uptake by biota and deposition onto and incorporation into the sediment phase. Some examples of these partitioning processes that have been revealed in studies using mesocosms are discussed in Chapter 7 (Section 7.4.3). It is also important to take into account the geographical location, the morphometry, and the topographical surroundings of the water mass. For example, although the range of organochlorine compounds identified in samples of water and biota from Lake Baikal, Siberia was similar to that found in the Great Lakes of North America, the atmosphere-water partitioning in the lake and the input from snow-melt from the mountains that surround Lake Baikal will be very different (Kucklick et al. 1994). The spectrum of biota that may be exposed to the toxic effects of xenobiotics is, therefore, wide and is the subject of Chapter 7. It is important also to appreciate the progress of a xenobiotic through the food chain. Although few of these are so truncated as that in Antarctic waters involving phytoplankton-krill-baleen whales and crabeater seals (Knox 1970), all involve a primary producer (algae) and a succession through crustaceans and fish to sea mammals such as seals, dolphins, and whales — and in many cases ultimately — man. For example, in Arctic waters the succession: phytoplankton-copepod-cod-ringed seal-polar bear is important (Welch et al. 1992). At each stage, the partitions noted above will occur, even though true equilibrium will seldom be achieved. In the following sections, some general remarks are presented on the factors that determine the dissemination of xenobiotics in natural systems, and an attempt is made in Section 3.5.5 to present the role of models in providing an overview of all of these processes.

The potential ambiguity in experiments on bioconcentration resulting from metabolism by biota has already been noted (Section 3.1.3), but it is important to appreciate other significant consequences of metabolism by fish and higher biota:

1. Metabolism may serve as an effective mechanism for the elimination of the xenobiotic; examples of a number of important reactions are given in Chapter 7 (Section 7.5.1).
2. The metabolites may themselves be toxicants. A good example is provided by the diol epoxides of some PAHs that are putatively the causative agents of liver carcinomas in fish (Chapter 8, Section 8.2).
3. The metabolites may be excreted from the organism, and thereby the transformation products of the xenobiotic are introduced into the environment. An example of this is given in Section 3.5.2.

3.5.1 Transport Within Aquatic Systems: The Role of Water and Sediment

The physical transport of dissolved xenobiotics within aquatic systems may obviously take place under the influence of currents and tides. It has also been shown, however, that particulate matter and river sediment may be important vehicles and that the dissemination of xenobiotics may also take place by subsequent diffusion from these transported sediments into the water column. An attempt will be made to illustrate the operation of these mechanisms.

A study conducted in the St. Clair and Detroit rivers illustrated that, for hexachlorobenzene, octachlorostyrene, and PCBs, transport by the water mass and on suspended solids were of equal importance, although at least in these systems, river-bed sediments were of lesser significance (Lau et al. 1989). The extensive evidence for binding of lipophilic xenobiotics to particulate material in lakes (Baker et al. 1991) indicates that this may play an important role in the transport of xenobiotics; compounds sorbed to particulate matter with a low density may be effectively transported over long distances before deposition onto sediment surfaces. The dynamics of resuspension should also be considered, since it has been shown by the use of sediment traps that resuspension of superficial sediments may be quantitatively important and result in the relatively uniform distribution of xenobiotics in large sedimentation basins (Oliver et al. 1989). The possible geographical extent of such dissemination mechanisms may be illustrated from the results of a study which identified PCB congeners in sediment samples from Great Slave Lake, Northwest Territories, Canada (Mudroch et al. 1992). These compounds probably enter the lake by atmospheric deposition and resuspension of sediments into the MacKenzie River drainage system could then result in their transport into the Arctic Ocean in the Beaufort Sea. Contamination of rivers by leaching of agrochemicals from terrestrial systems may also be significant, and attention has been drawn to the significance of overland flow during heavy spring rainfall in addition to transport of these compounds by contaminated groundwater (Squillace and Thurman 1992).

Xenobiotics may have become associated with particulate matter including algae by processes that have been discussed in Section 3.2. This detrital material may then be transported within the water mass, and at any stage this may be deposited onto the sediments. The partition of the xenobiotic from the bulk sediments into interstitial water may then result in re-exposure of biota to the xenobiotic; this has been discussed in Section 3.3.2. Another important process is diffusion within the sediment phase and from interstitial water into the water column, and both of these are important in the dissemination of xenobiotics. For example, diffusion of PCBs within sediment cores in Lake Ontario has been established (Eisenreich et al. 1989), and there is good evidence to support the re-entry of PCBs into the water column by diffusion from interstitial water (Baker et al. 1985). The state of these xenobiotics is not unequivocally established, though they are probably associated with colloidal material (Brownawell and Farrngton 1986); the degree of their bioavailability, therefore, remains unresolved. It may also be noted that, for reactive compounds, their chemical stability may be increased by the association. For example, chlorocatechols that are associated with particulate matter in interstitial water are thereby protected from atmospheric oxidation that would result in rapid transformation of the freely dissolved compounds (Remberger et al. 1993).

The main factors that operate may be summarized as (1) transport of dissolved compounds within the aqueous phase, (2) transport of sediment- and particle-associated compounds within the water mass, (3) diagenesis within the sediment phase, and (4) diffusion via the interstitial water from the sediment phase into the water mass. The relative quantitative significance of these will depend on a number of factors including the hydrological conditions, the organic components in the system, and the structure of the xenobiotic.

3.5.2 Transport Within Aquatic Systems: The Role of Biota

Extensive discussion has already been devoted to the concentration of xenobiotics from the aquatic phase into biota (Section 3.1), and this may contribute significantly to the dissemination of the compounds. The illustrative examples include the following, some of which have already been noted in the context of bioconcentration:

1. The accumulation of xenobiotics into planktonic algae or plants (Section 3.1.1) may introduce the compound into higher organisms in the food chain such as fish, while the detrital material may eventually enter the sediment phase and be dispersed by any of the mechanisms noted in Section 3.5.1.
2. After deposition into sediments, particulate-associated xenobiotics may be desorbed into sediment interstitial water and thereby mediate exposure of the xenobiotic to sediment-dwelling organisms such as oligochaetes, amphipods, or chironomids, and thence into higher organisms in the food chain; this has been discussed in Section 3.3.2. It is important to appreciate that such processes may occur at sites remote from those at which the initial sorption to the particulate matter took place and that their extent will depend both on hydrological conditions in the water mass as well as on the specific nature of the association.
3. Xenobiotics may be accumulated from the aquatic phase into fish, many of which move over substantial distances, and some of which serve as prey to larger fish (Thomann and Connolly 1984); this may result in biomagnification that is discussed in Section 3.5.4. In addition, fish may metabolize the xenobiotic and excrete the conjugates into the aquatic system; this is discussed in greater detail in Chapter 7 (Section 7.5.1) and will often function as an effective mechanism

for the elimination of the xenobiotic. A single example will suffice to illustrate its role in the dissemination of the transformation products of compounds accumulated by fish from the aquatic phase: 3,4,5-trichloroveratrole (a bacterial metabolite of 3,4,5- and 4,5,6-trichloroguaiacol) is accumulated in zebra fish (*Brachydanio rerio*) where it can be metabolized with the formation of 3,4,5- and 4,5,6-trichloroguaiacol and 3,4,5-trichlorocatechol that are conjugated to sulfate and glucuronate. These water-soluble metabolites are then excreted into the aquatic phase resulting in dispersal not of the original xenobiotic but of its metabolites (Neilson et al. 1989).

4. More complex interacting systems that mediate dissemination deserve brief mention. One interesting possibility is illustrated by the demonstration that in a mescocosm system, 2,3,7,8-tetrachlorodibenzofuran may be accumulated from spiked sediments into the larval stages of insects. After emergence as flying insects, these may then be consumed by a range of aquatic and terrestrial species including birds (Fairchild et al 1992). Even though the magnitude of the transfer may be small, it should not be neglected in mass balance studies of lakes, and in a wider perspective, this suggests an additional process for the widespread dissemination of xenobiotics.

5. It has been suggested that bacteria may play a role in the transport of hydrophobic compounds in soils (Lindqvist and Enfield 1992), and this mechanism could potentially apply also to aquatic systems where such processes could reasonably be included under particulate transport. There are, however, obvious unresolved issues concerning the subsequent desorption and bioavailability of these sorbed compounds.

There are, therefore, a number of important roles that biota play in the dissemination of xenobiotics including (1) deposition of algal-associated xenobiotics onto the sediment phase, (2) progressive dissemination of xenobiotics via predators through the food chain, (3) elimination of xenobiotics as metabolites, and all of these may take place at any point in aquatic systems.

3.5.3 The Role of Atmospheric Transport

Introduction — It is clear from the preceding discussion that many xenobiotics will be recovered from matrices differing from those into which they were initially discharged. An important vector over long distances is the atmosphere, and the compounds may then reach remote areas both in the form of precipitation and as particulate deposition. The following example may be used to illustrate the magnitudes that may be involved in the atmospheric transport of particulate matter. Transport via dust particles that were subsequently deposited onto the surface of snow in the Canadian Arctic amounted to some thousands of tons and was attributed to atmospheric transport of dust from agricultural land in China (Welch et al. 1991). Increasing concern has been expressed over the contamination of hitherto pristine regions such as the Arctic and Antarctic. A general review has examined in detail the situation in the Arctic for a number of compounds including organochlorines and PAHs (Barrie et al. 1992), and others have been devoted specifically to the Arctic marine ecosystem (Muir et al. 1992), the Arctic terrestrial ecosystem (Thomas et al. 1992), and freshwaters in the Canadian Arctic (Lockhart et al. 1992). Airborne xenobiotics may also be deposited onto or accumulated in higher plants that may be consumed by herbivores and thus eventually transmitted to man (Riederer 1990).

The processes by which xenobiotics enter the atmosphere in the first place have received increasing attention: the principles have already been discussed in Section 3.4, and a model describing the various processes has been developed (Mackay et al. 1986). It has emerged that it is particularly compounds with low water solubility and appreciable vapor pressure that are effectively retransported from the aquatic phase into the atmosphere. This accounts for the recovery of compounds such as PCBs, PCCs (polychlorinated camphene consisting of chlorinated bornanes and bornenes), DDT, and some of the higher PAHs from widely distributed environmental samples. On the other hand, volatilization of α-hexachlorocyclohexane in the Great Lakes is appreciable only during the summer months, and the overall flux is from the atmosphere to the aquatic phase (McConnell et al. 1993).The complex dynamics of the equilibria between the phases are clearly demonstrated by extensive data from studies of PCBs at sites remote from possible direct discharge (Swackhamer and Hites 1988; Swackhamer et al. 1988). Some of the most significant conclusions from these elegant studies are of general significance and are worth noting. First, the distribution between the aqueous phase, the biota, and the sediment was complex and depended critically on the PCB congener involved, and second, although the sole source of PCB input was from the atmosphere and originated from remote sites of discharge, a substantial part of the burden from the atmosphere was returned from the aquatic phase after deposition, and this re-entry was accompanied by significant changes in the distribution of the various PCB congeners.

The latter conclusion is clearly supported by the results of an investigation that measured the concentrations of PCBs and PAHs in samples of air and water in a number of samples collected in Lake Superior. These were used to calculate fugacity gradients across the atmosphere-water interface (Baker and Eisenreich 1990), and the results clearly showed that during the summer months these xenobiotics were transferred from the aquatic phase into the atmosphere. Additional investigations in Lake Michigan revealed the significant roles of wind velocity and water concentration (Achman et al. 1993). In addition, all of these results clearly underscore the necessity of considering the different behavior of individual components of commercial products such as PCBs or of complex mixtures such as PAHs and chlorinated dibenzo-1,4-dioxins.

It is important to appreciate that partition processes also take place in the atmosphere; scavenging by rain and sorption to particulate matter are of particular importance. Their significance has been demonstrated for the chlorinated dibenzo-1,4-dioxins (Koester and Hites 1992) and accounts for the enhancement of octachlorodibenzo-1,4-dioxin which has consistently been observed to dominate the other congeners in environmental samples (Czuczwa and Hites 1986) except fish (Zacharewski et al. 1989) possibly due to its limited uptake (Bruggeman et al. 1984).

The long-distance dissemination of xenobiotics — Although complex chemical transformations — mainly photochemical — take place in the atmosphere, many chemically stable compounds may be transported intact via the atmosphere and subsequently enter the aquatic and terrestrial environments in the form of precipitation; once again, complications from binding of these compounds to particulate matter including, for example, soot particles must be taken into consideration, and this is particularly important for PAHs that are bound to particulate matter before emission. The virtual global distribution of some xenobiotics attests to the probable importance of atmosphertic transport and provides evidence that this is by no means restricted to highly volatile compounds. Nonetheless, some evidence suggests that for compounds including PCBs, hexachlorobenzene, and hexachlorocyclohexanes, although transport within the troposphere is important, its role in transport between the hemispheres is marginal (Wittlinger and Ballschmiter 1990). A few examples of long-range transport will be used to illustrate both the geographical range covered and the diversity of the compounds involved.

1. Persistent organochlorine compounds have been identified in rainfall (Strachan 1988), and compounds such as hexachlorocyclohexanes, α-endosulfan, and dieldrin have been detected in samples of snow even in remote areas such as the Canadian Arctic (Gregor and Gummer 1989).

2. Analysis of pine (*Pinus sylvestris*) needles collected over large areas of Europe showed the presence of a number of halogenated compounds including DDT, pentachlorophenol, and PCBs (Jensen et al. 1992). Whereas the origin of the DDT which has been banned for many years in Europe could be attributed to its application in an area of southern Germany, the source of pentachlorophenol which was highest in samples from Northen Sweden could not be unequivocally established.

3. Pine needles from North America have also been used for monitoring polychlorinated dibenzodioxins (Safe et al. 1992): the spectrum of the congeners which was dominated by octachlorodibenzodioxin and the most highly chlorinated dibenzofurans suggested their origin from impurities in commercial pentachlorophenol used for wood impregnation rather as a result of the combustion of chlorinated aromatic compounds. This is consistent with a similar conclusion drawn from previous results of the analysis of air particulates and sediments (Czuczwa and Hites 1986).

4. Analysis of lichens from the Great Lakes region has been used to assess the dissemination of a number of organochlorine compounds and has drawn attention to the significance of atmosphere-plant partitioning (Muir et al. 1993). The pattern of compounds that were found differed from those found in atmospheric or rainfall samples, and this fact illustrates clearly the specific influence of the chemical structure both of plant surfaces and of their surface area.

5. Samples of seal blubber from the high-Arctic archipelago Svalbard contained levels of polychlorinated camphenes comparable with those found in biota in the Baltic (Andersson et al. 1988), and the same group of compounds was dominant in samples of narwhal (*Monodon monoceros*) blubber from Pond Inlet, Baffin Island, in the Canadian high Arctic (Muir et al. 1992).

6. Residues of organochlorine compounds including such ubiquitous representatives as hexachlorobenzene, hexachlorocyclohexanes, DDT, and its metabolites have been found in samples of lichens and moss from several localities in the Antarctic (Focardi et al. 1991).

7. A variety of samples from an ice island in the Canadian Arctic contained a range of organochlorine compounds, and the role of atmospheric delivery was supported by their occurrence in samples of air, water, snow, zooplankton, and benthic amphipods (Bidleman et al. 1989). The significance of Arctic haze was inferred from comparable data collected in May and September from the same ice island (Hargrave et al. 1988).

8. The occurrence of PCBs in samples of cod, seal, and polar bear in the Canadian Arctic (Muir et al. 1988) attests to the ubiquitous distribution of these compounds, while the existence of biomagnification within this food chain is discussed further in Section 3.5.4.

9. The carcinogenic amino-α-carbolines, 2-amino-9H-pyrido[2,3-b]indole and its 3-methyl derivative (Figure 3.6) that are pyrolysis products of tryptophan have been identified in environmental samples including airborne particles and rainwater as well as cigarette-polluted indoor air (Manabe et al. 1992). If generally confirmed, this finding might suggest the existence of a potentially serious threat to human health.

10. Tri(4-chlorophenyl)methanol has been recovered from samples of marine mammals and bird eggs from the Arctic, the Antarctic, Australia, and North America and, although its source has been conjectured, this has not been conclusively identified (Jarman et al. 1992).

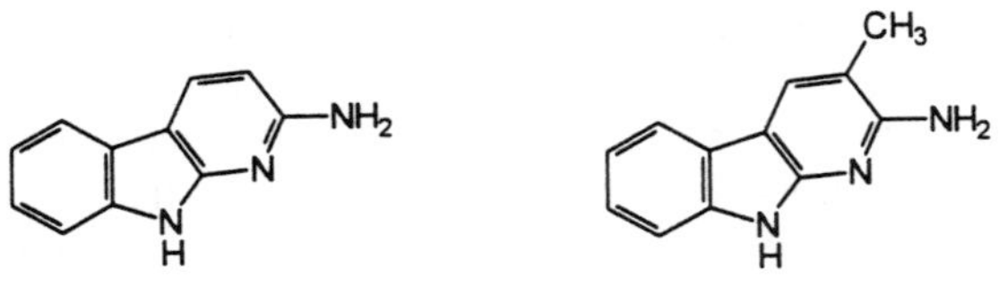

Figure 3.6 2-Amino- and 2-amino-3-methyl-9H-pyrido[2,3b]indole.

It is clear from this summary that polar regions may be substantially exposed to certain groups of organochlorine compounds, and it has been suggested on the basis of an analysis of their physico-chemical properties that polar regions may function as an effective condensation trap for xenobiotics of intermediate volatility (Wania and Mackay 1993).

3.5.4 Biomagnification

Biomagnification is the process whereby a compound enters a food chain at a lower trophic level and, via a series of predators, is increasingly concentrated in higher levels; biomagnification may be considered as sequential bioaccumulation, and some general comments have been given in the introduction to Section 3.5. Simplified fugacity models may not predict the existence of biomagnification, although bioenergetic models allow for biomagnification which may be especially significant for compounds with values of log $P_{ow} > 4$ (Connolly and Pedersen 1988). Considerable attention has been directed to theoretical considerations, and the data appear to suggest only moderate levels of biomagnification (Thomann 1989; Bierman 1990). It is probably fair to state, however, that there are very substantial approximations involved in such estimates since they inherit all the existing ambiguities of those for predicting bioconcentration potential. In addition, few results are reported for the concentrations of xenobiotics normalized to the same base — of which lipid weight is probably the most relevant and widely applicable. An attempt will be made, however, to illustrate the operation of this important translocation process.

Attention has already been drawn to a laboratory study (Sijm et al. 1992) in which successive generations of guppy (*Poecilia reticulata*) were exposed to PCB-contaminated food. A study with guppy (*P. reticulata*) and goldfish (*Carassius auratus*) and food contaminated with a mixture of 1,2,4,5-tetra-, penta- and hexachlorobenzene, 2,2′,4,4′,6,6′-hexachlorobiphenyl and mirex has illustrated the importance in biomagnification of food digestion and food absorption in the gastrointestinal tract (Gobas et al. 1993a). Biomagnification has been examined extensively in field samples from the marine food chain. The following examples involve not only different organisms but also different experimental methodologies.

1. Biomagnification using biota at different trophic levels has been carefully examined in the Arctic marine food chain — Arctic cod (*Boreogadus saida*), ringed seal (*Phoca hispida*), and polar bear (*Ursus maritimus*) (Muir et al. 1988), and biomagnification factors (BMFs) were calculated. The existence of biomagnification was clearly demonstrated, and a number of other important facts emerged:

   • BMFs for fish/seal and seal/bear were dependent on the sex of the seals.
   • In the fish/seal system, BMFs for DDE were greater than for total DDT, though for seal/bear values were < 1 for both compounds.
   • The highest BCFs were observed for the most highly chlorinated PCB congeners with 7 to 9 chlorine atoms; this is consistent with the unique pattern of PCB congeners in polar bears, which have relatively high concentrations of these congeners, but apparently successfully metabolize those with lower degrees of chlorination (Norstrom et al. 1988).

   Levels of biomagnification of highly chlorinated PCBs, of DDT, DDE and chlordane-related compounds were disturbingly high, and all of these compounds together with α-hexachlorocyclohexane have been identified in the diet of Inuit who consume and are essentially dependent on the harvest of ring seal, beluga, eider, and Arctic char (Cameron and Weis 1993). The more controversial issue of the risks as opposed to the benefits of such consumption has been addressed (Kinloch et al. 1992), and this is discussed again in a different context in Chapter 8 (Section 8.4.3). Dietary exposure to organochlorine compounds probably exists for all circumpolar inhabitants that pursue a traditional way of life, and attention has already been drawn (Section 3.5.3) to the suggestion that polar regions may be an effective condensation trap for such compounds.

2. Biomagnification has been investigated in a marine food chain in the western Pacific Ocean (Tanabe et al. 1984), and this study examined samples of zooplankton, myctophids (lanternfish) (*Diaphus suborbitalis*), squid (*Todarodes pacificus*), and striped dolphin (*Stenella coeruleoalba*) that were analyzed for a number of organochlorine compounds. In contrast to the results from the fish/seal/bear study, the spectrum of PCB congeners in myctophid, squid, and dolphin was essentially similar, though in comparison with the seawater, it was markedly enriched in the more highly chlorinated congeners. On the other hand, compared with squid, myctophid, and zooplankton, hexachlorocyclohexane isomers were completely different in striped dolphin with a dominance of the β-isomer and almost none of the α-isomer. Bioconcentration factors were presented in terms of total body weight, although concentrations were also given in terms of whole body lipid which ranged from ca. 3% in mycophid and squid to over 13% for striped dolphin. Once again, there is no doubt that biomagnification

existed — even though detailed kinetic analysis is not possible — and that the level of bioconcentration depended on the trophic level of the biota.

3. Other systems that have been examined include fish/cormorant (*Phalacrocorax carbo sinensis*) in Schleswig-Holstein on the German North Sea coast. These data were not presented on a comparable basis for the two components, however, so that numerical comparison with other systems cannot be made. PCB concentrations in the subcutaneous fat of cormorants were from 10 to 100 times those in muscle tissue of marine fish with tenfold values for freshwater fish (Scharenberg 1991). There is evidence that birds, and especially those consuming fish, have only low levels of microsomal monooxygenase activity (Walker 1983), although it is important to distinguish the differential activity of 3-methylcholanthrene-induced and phenobarbital-induced monooxygenase systems to the metabolism of different xenobiotics (Braune and Norstrom 1989), and this could have an important bearing on these observations on cormorants.

An example from a single freshwater system — the Great Lakes — nicely illustrates the differences that may be encountered at different trophic levels and for different organochlorine compounds. On the basis of a limited number of analyses, there appeared to be little biomagnification between zooplankton and forage fish for 2,3,7,8-tetrachlorodibenzo-1,4-dioxin, (Whittle et al. 1992), whereas values of 32 between herring gulls (*Larus argentatus*) and alewife (*Alosa pseudoharengus*) have been reported for 2,4,3′,4′-tetrachlorobiphenyl (Braune and Norstrom 1989). For PCBs in this system, the BMFs depended critically on both the position and the number of chlorine atoms with a maximum value of 219 for 2,3,4,5,2′,3′,4′,5′-octachlorobiphenyl, but were consistently higher than the values for biomagnification from alewife to salmonids calculated on the basis of data from a previous study (Oliver and Niimi 1988). This was attributed to the different energy demands on fish and birds which are much greater in the latter.

In summary, whereas there is no reason for doubting the existence and significance of biomagnification, four general comments may be worth inserting:

1. Experiments using goldfish (*Carassius auratus*) and a range of neutral organochlorine compounds including chlorinated benzenes, biphenyls, and octachlorostyrene showed that uptake was not dependent on the lipid content of the diet and was, therefore, determined primarily by passive diffusion into the gastrointestinal tract rather than by cotransport with lipids (Gobas et al. 1993b). Essentially similar conclusions may be drawn from experiments (Larsen et al. 1992) on the bioavailability of PCB congeners to the earthworm *Lumbricus rubellus*.

2. The investigations cited above demonstrated the selective accumulation in various organs, and that for the PCBs, this was congener specific. For example, decachlorobiphenyl was found only in the brain tissue of cormorants (Scharenberg 1991). This provides further support for the critical comments that have been raised against sum parameters.

3. Although the transport of PAHs from particulate matter in the aquatic phase through blue mussels (*Mytilus edulis*) to the common eider (*Somateria mollissima*) has been demonstrated in the Baltic Sea, there was a successive change in the relative concentrations of the various individual hydrocarbons (Broman et al.

1990). This may be attributed to increasing metabolic activity at the higher trophic levels that could result in the production of putatively mutagenic compounds. This is an issue of general importance that merits detailed evaluation (Braune and Norstrom 1989).
4. Investigations have been largely limited to neutral organochlorine compounds with a high potential for bioconcentration, and it would clearly be desirable to extend this to other groups of compounds including those containing polar substituents such as carboxylic acid, phenolic, or amino groups.

In summary, the following should be critically evaluated in discussion of biomagnification:

1. Normalization to a relevant base of the concentration of the xenobiotic at the various trophic levels.
2. Physiological differences in the organisms at the various trophic levels and their bioenergetics.
3. The metabolism of the xenobiotic as it is transported through the various trophic levels.

3.5.5 The Role of Models in Evaluating the Distribution of Xenobiotics

It will have become apparent from the preceding discussions that xenobiotics after discharge from a point source may enter any of the various environmental compartments: aquatic systems including biota and sediment, the atmosphere, terrestrial systems including soils, biota, and in the long run possibly the ultimate predator — man. Considerable effort has, therefore, been devoted to the development and application of models to evaluate this dissemination in quantitiative terms. These involve the concept of fugacity, and it seems appropriate at the beginning to examine this concept briefly.

In classical thermodynamics — that deals with the behavior of ideal reversible systems — the term is used to preserve the relationship between the free energy and the temperature and pressure by replacing the pressure with the fugacity (Partington 1950), and it may be shown that the concept can be applied also to other phases and to mixtures (Tolman 1950). It is important to realize, however, that the additional terms included in the van der Waals equation for real gases derive from interaction mechanisms between molecules that lie beyond the valid application of reversible thermodynamics (Jeans 1952). The application of the term to natural systems that are — by definition — seldom or never in equilibrium should, therefore, be constantly kept in mind and the reasons for departure from predictions sought in molecular mechanisms of interaction. Models are generally classified as belonging to Levels I and II which depict equilibrium conditions, whereas Level III takes into account non-equilibrium steady-state conditions. A clear presentation of the relevant assumptions as well as models for Levels IV and V has been given (Paterson and Mackay 1985). All of the models require the input of physico-chemical parameters for the compound(s) and, for Level III estimated rates of degradation; estimation of these probably presents the greatest uncertainty. Inevitably, the predictive value of all the models depends critically on the accuracy of the input data as well as on the environmental realism of the assumptions that are incorporated.

As pointed out in Chapter 1, modeling lies far beyond the expertise of the author, but in view of its importance, a few inevitably superficial remarks are offered by way of illustration.

1. A Level III study using data for 2,3,7,8-tetrachlorodibenzo-1,4-dioxin showed that the low water solubility, low vapor pressure, and high P_{ow} resulted in partitioning of the compound mainly into soil (69%) and sediment (29%) and that the major removal process is air advection. On the basis of reasonable assumptions, it was concluded that the major exposure route for man is via food, mainly meat and fruit and vegetables (Mackay et al. 1985).

2. A comprehensive application of Level III was made to the accumulation and elimination of a range of neutral halogenated xenobiotics by fish from food and water and took into account transport across the gills, activity in the gastrointestinal tract including metabolism, and the bioavailability in the water mass (Clark et al. 1990). The results illustrated the importance of the uptake of xenobiotics via the food that results in biomagnification and that this process is most significant for hydrophobic compounds that are poorly metabolized.

3. Concern has been expressed over human exposure to xenobiotics transported in the atmosphere and accumulated in higher plants that are consumed both directly or indirectly, for example, as milk from herbivores. A simple model for assessing the transport of xenobiotics from the atmosphere to vegetation has been developed (Riederer 1990) and was used to calculate equilibrium concentrations in plant tissues and bioconcentration factors from air to vegetation. The important issue of revolatilization was addressed since this determines the persistence of xenobiotics in plants.

4. Two different models were used to examine the dispersion of a limited number of chlorophenolic compounds in a segment of the Gulf of Bothnia (Kolset and Heiberg 1988); one of these, EXAMS takes into account ionized and non-ionized forms separately, whereas FEQUM does not. The application of models to such complex situations is extremely difficult in the absence of the relevant hydrological data, and the use of chloroform in this case to calibrate the model is somewhat questionable, since this compound differs so significantly in its physico-chemical properties from the chloroguaiacols and chlorocatechols that were the subject of study. In spite of this, generally reasonable distributions were obtained for the chloroguaiacols — though less so for tetrachlorocatechol.

The further successful application of such models would appear to depend on more explicit understanding of the mechanism of transport processes and metabolism, as well as more precise numerical definition of cardinal environmental parameters.

3.5.6 Leaching and Recovery from Other Solid Phases

The preceding discussion has focused attention on aquatic systems including sediments. Some xenobiotics are, however, deliberately applied to terrestrial systems; for example, modern agriculture is dependent on the use of substantial amounts of agrochemicals for control of weeds, insects, and pathogenic microorganisms. Concern has been expressed both about the residues of agrochemicals in food for human consumption and about the persistence and mobilization of these compounds in the soil. Extensive investigations have, therefore, been directed to the analysis of these residues. What is, of course, of equal significance is the extent to which non-target organisms are exposed to these potential toxicants, and this was the concern to which Rachel Carson originally drew attention.

It has already been pointed out that the exposure of biota to xenobiotics is critically determined by the dynamics of desorption processes and that these depend on the nature of the solid matrix: agricultural soils throughout the world are extremely variable both in texture, organic content, and mineral composition. Standard extraction procedures, some of which have been outlined in Chapter 2 (Section 2.2.5), have been widely adopted even though the level of recovery may be quite low, and this is plausibly attributable to association between the xenobiotic and humic components of the soil. The significance of physical entrapment has been illustrated by the extreme persistence of low molecular weight and normally volatile compounds such as 1,2-dibromoethane (Steinberg et al. 1987; Sawhney et al. 1988).

The whole issue of recoverability of xenobiotics from solid phases, of their desorption, and of their bioavailability has emerged increasingly in a wider context: the problem of solid waste disposal (Suflita et al. 1992) and the effectiveness of bioremediation of contaminated sediments (Harkness et al. 1993). Leaching from what are euphemistically termed landfills could result in the pollution of both ground and surface waters — and both sediment and biota — by a wide range of organic chemicals, most of which are probably recalcitrant and many of which are toxic. An illustrative example is given in Chapter 2, Section 2.5. This will almost certainly absorb substantial activity in the future and will claim increasing attention from chemists, microbiologists, and toxicologists to provide a basis for the rational application of control and recovery strategies.

3.6 MONITORING

Introduction — Many xenobiotics initially discharged into restricted geographical areas now have a world-wide distribution as a result of the various transport processes that have been discussed in this chapter. This knowledge has accumulated from extensive monitoring studies that have achieved increasing prominence and encompassed an ever-widening range of compounds. Monitoring may be used to provide evidence for evaluating the persistence of a xenobiotic in the environment and, at the same time, determining the extent to which it has entered environments geographically distant from the initial point of discharge. If, of course, a compound is continuously introduced into the environment, its recovery does not necessarily provide evidence of its persistence; an example of this is the phthalates whose ubiquity is clearly established even though they are degradable and cannot, therefore, be regarded as recalcitrant. On the other hand, an ingenious application of monitoring to demonstrate the microbial degradation of PCBs in contaminated natural sediments may be given. The sediments were analyzed for the presence of chlorobenzoates that had previously been established in laboratory experiments as microbial metabolites (Flanagan and May 1993; Chapter 5, Section 5.3.3). These results demonstrate both the necessity of laboratory experiments on biodegradation for interpreting the results of an environmental monitoring program and the value of carefully selected analytes for inclusion.

Even a cursory examination of the literature shows that analysis of virtually every environmental sample reveals contamination from polycyclic hydrocarbons — resulting from incomplete combustion processes — and a range of the more recalcitrant organohalogen compounds such as DDT (together with its degradation product DDE), PCBs, hexachlorocyclohexanes, compounds related to aldrin, and mixtures present in commercial toxaphene preparations. Possibly the most disturbing fact which has already been noted is the occurrence of these compounds in samples from remote and largely isolated locations in the Arctic and Antarctic.

Choice of samples — Although monitoring has traditionally been used to determine whether, in the final analysis, xenobiotics are persistent in the environment, the question of what kinds of samples to monitor and what compounds to analyze cannot easily or universally be answered. Samples from virtually all possible matrices have been used including the atmosphere, snow, water, sediments, and a diverse range of biota. Among biota, the U.S. National Contaminant Biomonitoring Program uses fish collected from a network of stations (Schmitt et al. 1990), and the Mussel Watch Program uses bivalves (Goldberg 1986; Sericano et al. 1993). Organisms as diverse as aquatic leeches (Metcalfe et al. 1984, 1988) and herring gulls (Norstrom et al. 1986) in the Great Lakes, cormorants in the North Sea coast of Germany (Scharenberg 1991), polar bears in the high Arctic (Muir et al. 1988), and insect larvae in the Hudson River (Novak et al. 1988) have been used less extensively. Monitoring programs using fish should take into account the season at which sampling is made, and their age, size, sex, and fat content, and analyze individuals rather than pooled samples to avoid serious errors in interpretation of the results (Bignert et al. 1993). Considerable care should be exercised in the choice of biota, and this should take into account their metabolic potential for the compounds that are being monitored. Indeed, advantage of this capability has been used for monitoring exposure to chlorophenolic compounds by analyzing fish bile for conjugates (Oikari and Kunnamo-Ojala 1987; Wachtmeister et al. 1991). Two apparently extreme situations may be used to illustrate potential problem areas. One apparent advantage of using sea-birds is the apparently low level of liver mixed-function-oxidase enzymes involved in the metabolism of most xenobiotics (Walker 1983); the xenobiotic may, therefore, be recovered metabolically intact, and this considerably simplifies the analysis of the samples. There is, however, evidence that illustrates the potential of birds to metabolize xenobiotics (Braune and Norstrom 1989). On the other hand, the unique pattern of PCB congeners found in polar bears attests to their ability to metabolize at least the less highly chlorinated congeners (Norstrom et al. 1988).

When sediments are employed, there are a number of critical experimental factors to which careful attention should be given; these include the size of the sample, the patchiness of natural sediments (Swartz et al. 1989), and the conditions for transport and preservation of the sample after collection. Concentrations of xenobiotics will generally be normalized to dry weight, amounts of organic carbon or to specific compound groups such as lipids, but even so, comparison between samples from different geographical regions or using different biota is often extremely difficult. It is, therefore, essential to define clearly at the outset the objective of the investigation; unfortunately, the importance of this is not always appreciated and may result in unnecessary cost and avoidable frustration.

Temporal record of input — Monitoring has also been carried out to provide a historical record of input to ecosystems and has generally been correlated in time with industrial production. This might be termed environmental archeology and has been most widely applied to organochlorine compounds, particularly PCBs and chlorinated dibenzo-1,4-dioxins. The results of such analyses generally provide, at the same time, convincing support for the widespread dissemination of many xenobiotics via atmospheric transport (Sections 3.4 and 3.5), but analysis for a specific compound may be invaluable in determining the details of dissemination process after discharge from a single point. An almost unique example is the use of 2,4-dipentylphenol to map its distribution in Lake Erie (Carter and Hites 1992); this was possible since the compound was produced virtually exclusively by a single plant and is an otherwise unusual xenobiotic. Over relatively short time spans (< 100 yr), peat bogs have been used to assess the magnitude of atmospheric input of xenobiotics including PCBs, DDT, hexchlorobenzene, and toxaphene (Rapaport and Eisenreich 1988); the advantage of these environments for such monitoring lies in the relatively low rates of degradation and minimal mobilizing of the xenobiotics in the organic matrix.

One of the cardinal issues is the dating of the samples. Ideally, advantage should be taken of archived samples collected at established dates, but this is possible only for samples of soils and some kinds of biota — and its unambiguous application depends critically on how adequately the samples have been preserved. A virtually unique example is provided by the analysis of soil samples from the Broadbalk Experiment at Rothamstead, England. This experiment began in 1843, and samples collected during more than a century have been used to investigate the deposition of PAHs which result from combustion processes of various kinds (Jones et al.1989). A survey of PCB levels carried out by the same groups using a wider range of samples that included other areas in England provided interesting details of temporal trends and the alterations in congener composition that have taken place probably mediated by volatilization (Alcock et al. 1993). More commonly, sediment samples have been investigated. In this case, cores have been collected and segments used simultaneously for radiodating using ^{210}Pb, ^{134}Cs, and for recent sediments ^{137}Cs or for older ones ^{14}C, and for chemical analysis. A striking example of its application is the finding of various chlorinated dibenzo-1,4-dioxins in sediment core samples from Japan (Hashimoto et al. 1990); especially notable is the finding of the 1,2,3,4,6,7,9-heptachloro and octachloro congeners in the deepest samples dated at ca. 8000 yr old. In this study, the possible complicating circumstances were apparently eliminated, but it may be valuable to summarize briefly the most important of them, since they are relevant to all monitoring studies that seek to provide an unequivocal temporal record of input. Analytical artifacts must, of course, be eliminated, and the levels of analytes in blanks must be essentially zero. Other less obvious factors may seriously compromise the interpretation of the results, so that their significance should be appreciated and taken into account:

1. Sediments may not remain undisturbed due to diffusion or bioturbation within the sediment phase; of these, bioturbation is probably dominant in many situations (Karickhoff and Morris 1985).
2. Sedimentation may not be uniform, so that some degree of inhomogeneity may exist within the sediment layers; this may introduce potentially serious errors, especially if pooled samples are analyzed (Sanders et al. 1992).
3. It cannot validly be assumed that even apparently recalcitrant compounds remain unaltered over extensive periods of time, and the possibility of their

transformation should be taken into consideration; this includes both abiotic reactions that are discussed in Chapter 4 (Sections 4.1.1 and 4.1.2) as well as those mediated by microorganisms that are the subject of detailed analysis in Chapter 6.

4. It has already been pointed out in Section 3.3.3 that resuspension of surficial sediments may result in re-entry of a part of the burden of the sediment phase into the aquatic phase.

Even data from the analysis of snow and ice samples from ice-caps in polar regions should be interpreted with caution in the light of possible compromising factors including ablation and percolation of melt water during the summer months.

Selection of analytes — The compounds chosen for analysis will generally belong to established groups of xenobiotics such as organochlorines, PAHs, or the structurally diverse range of agrochemicals, although halogenated anisoles that are transformation products of the corresponding phenols have begun to be included (Wittlinger and Ballschmiter 1990). In addition, attention should be drawn to the diversity of abiotic transformation products that have been noted in Chapter 2 (Section 2.5) and are discussed in greater detail in Chapter 4 (Sections 4.1.1 and 4.1.2), together with those mediated by microorganisms (Chapter 6) and higher organisms including fish (Chapter 7, Section 7.5). A good example of the use of established bacterial metabolites to provide evidence for the degradation of the lower congeners of PCBs in sediments is given in Chapter 5 (Section 5.3.3). All of these factors necessitate care both in the design and in the interpretation of the results of monitoring studies; particular care should be exercised in the matter of conclusive identification of analytes. It may, indeed, not be possible to identify even tentatively all the components in the samples; for example, a study of organic compounds in Florida sediments revealed three unidentified compounds in addition to PAHs, benzonaphthothiophene, and terpene-related compounds (Garcia et al. 1993). An example of the vigilance which may profitably be exercised in the analysis of environmental samples is illustrated by the very wide distribution of tris(4-chlorophenyl) methanol in marine mammals and bird eggs from widely separated geographical localities (Jarman et al. 1992). The results suggest a substantial input and although the source cannot be definitively determined, the compound has been used in the synthesis of polymers, agrochemicals, or synthetic dyes.

Commercial mixtures of compounds or effluents containing large numbers of compounds present particularly difficult choices in monitoring programs.

1. If a commercial product contains several components, it is clearly desirable to provide analytical data for each of them. For example, in the case of PCBs, most current studies provide data for specific congeners, and this is clearly motivated by several considerations.

   • The difference in volatility of the various congeners determines their redistribution among environmental compartments after deposition.
   • Some of the PCB congeners are more persistent than others, so that changes take place in the composition of the mixture after discharge. This may be due to microbial reactions such as anaerobic dechlorination, which is discussed more fully in Chapter 6 (Section 6.6), or the result of metabolism

by higher biota. An illustration in the cod-seal-polar bear food chain has been cited above.

- The toxicity of the congeners varies considerably so that attention has been directed particularly to the planar congeners that do not contain two or more chlorine atoms in positions adjacent to the phenyl-phenyl ring junction and which are significantly more toxic (De Voogt et al. 1990).

Since all 209 PCB congeners have been synthesized and GC analytical methods for their analysis have been developed (Mullin et al. 1984), there seems little justification for not incorporating these procedures into all investigations.

2. On the other hand, the complexity of some industrial effluents may make it difficult to propose a restricted number of specific compounds for analysis. An unusually difficult choice is presented by effluents from the production of bleached pulp with conventional technologies using molecular chlorine. Investigations on the distribution of organochlorine compounds in samples of sediment and biota from the Gulf of Bothnia and the Baltic Sea have made extensive use of measurements of the concentration of cyclohexane-extractable organically bound chlorine (Martinsen et al. 1988). The interpretative limitations of this procedure are, however, at least twofold. First, the origin of the source remains undefined in the absence of data for specific compounds related to the supposed industrial operation, and second, the biological effects cannot validly be correlated with such measurements, since there may exist other compounds inducing the observed effects. Analytical studies of extracts from sediments have substantiated the limitations of such parameters, since the greater part of the sample contained unchlorinated compounds, and a substantial fraction (>85%) of the organochlorine compounds remains unidentified (Remberger et al. 1990). In this case, additional complications have emerged from the newly established molecular weight distribution of the organochlorine components in the effluents (Jokela and Salkinoja-Salonen 1992) and the role of high molecular weight fractions in determining toxicity (Higashi et al. 1992).

Monitoring and ecoepidemiology — Monitoring may also be incorporated into ecoepidemiological investigations that are discussed in greater detail in Chapter 8 (Section 8.2). Some general comments may, however, be provided here. For example, above background concentrations of PCBs have been associated with impaired reproduction in several species of seals. Unless, however, it is shown that the compounds in question do, indeed, bring about the postulated biological effect, the correlation may be fortuitous. The complexity of the possible correlation of impaired reproduction in marine mammals and exposure to organochlorine compounds (Addison 1989) suggests a wider application of the *principle* of Koch's postulates for determining causal pathogenicity of bacteria to animals or plants. (1) Isolation of the putatively causative organism (substance), (2) demonstration of its pathogenicity (toxicity), and (3) recovery from biota infected with the organism isolated (affected by the putative toxicant).

The crux lies in fulfilment of the third criterion since, in natural ecosystems, biota are seldom exposed to only a single potential toxicant. This whole issue of association between biological effects and putative causes is discussed again in Chapter 8 (Section 8.2).

3.7 CONCLUSIONS

It has been shown that phase partitioning involves complex interactions, the details of which have not been completely resolved. A number of important conclusions may, however, be drawn from this discussion, and at least the following factors should be quantitatively evaluated, particularly in interpreting the results from a monitoring program:

1. The dynamics of partitioning between phases, and the role of interstitial water in sediments and soils. Specific attention should be directed to the quantitative significance of associations between xenobiotics and both organic and inorganic components of the soil and sediment phases and the extent to which these processes are reversible.
2. The importance of alterations in the structure of the xenobiotic after release should be taken into consideration; these changes may be mediated by both abiotic and biotic reactions and produce compounds with physical and chemical properties significantly different from the original xenobiotic.
3. It has been demonstrated that many organic compounds are effectively partitioned into the gas phase, and attention should be directed to this in light of its significance in the global distribution of xenobiotics.

REFERENCES

Achman, D.R., K.C. Hornbuckle, and S. J. Eisenreich. 1993. Volatilization of polychlorinated biphenyls from Green Bay, Lake Michigian (sic). *Environ. Sci. Technol.* 27: 75-87.

Addison, R.F. 1989. Organochlorines and marine mammal reproduction. *Can. J. Fish. Aquat. Sci.* 46: 360-368.

Alcock, R.E., A.E. Johnston, S.P. McGrath, M.L. Barrow, and K.C. Jones. 1993. Long-term changes in the polychlorinated biphenyl content of United Kingdom soils. *Environ. Sci. Technol.* 27: 1918-1923.

Andersson, Ö., C.-E. Linder, M. Olsson, L. Reutergårdh, U.B. Uvemo, and U. Wideqvist. 1988. Spatial differences and temporal trends of organochlorine compounds in biota from the northwestern hemisphere. *Arch. Environ. Contam. Toxicol.* 17: 755-765.

Ankley, G.T., M.K. Schubauer-Berigan, and J.R. Dierkes. 1991. Predicting the toxicity of bulk sediments to aquatic organisms with various test fractions: pore water vs. elutriate. *Environ.Toxicol. Chem.* 10: 1359-1366.

Autenrieth, R.L., and J.V. DePinto. 1991. Desorption of chlorinated hydrocarbons from phytoplankton. *Environ. Toxicol. Chem.* 10: 857-872.

Baker, J.E., and S.J. Eisenreich. 1990. Concentrations and fluxes of polycyclic aromatic hydrocarbons and polychlorinated biphenyls across the air-water interface of Lake Superior. *Environ. Sci. Technol.* 24: 342-352.

Baker, J.E., S.J. Eisenreich, and B.J. Eadie. 1991. Sediment trap fluxes and benthic recycling of organic carbon, polycyclic aromatic hydrocarbons, and polychlorobiphenyl congeners in Lake Superior. *Environ. Sci. Technol.* 25: 500-509.

Baker, J.E., S.J. Eisenreich, T.C. Johnson, and B.M. Halfman. 1985. Chlorinated hydrocarbon cycling in the benthic nepheloid layer of Lake Superior. *Environ. Sci. Technol.* 19: 854-861.

Barrie, L.A., D. Gregor, B. Hargrave, R. Lake, D. Muir, R. Shearer, B. Tracey, and T. Bidelman. 1992. Arctic contaminants: sources, occurrence and pathways. *Sci. Total Environ.* 122: 1-74.

Barron, M.G., S.M. Plakas, P.C. Wilga, and T. Ball. 1993. Absorption, tissue distribution and metabolism of chloropyrifos in channel catfish following waterborne exposure. *Environ. Toxicol. Chem.* 12: 1469-1476.

Bergen, B.J., W.G. Nelson, and R.J. Pruell. 1993. Partitioning of polychlorinated biphenyl congeners in the seawater of New Bedford Harbor, Massachusetts. *Environ. Sci. Technol.* 27: 938-942.

Bidleman, T.F., and L. Renberg. 1985. Determination of vapor pressures for chloroguaiacols, chloroveratroles and nonylphenol by gas chromatography. *Chemosphere* 14: 1475-1481.

Bidleman, T.F., G.W. Patton, M.D. Walla, B.T. Hargrave, W.P. Vass, P. Erickson, B. Fowler, V. Scott, and D.J. Gregor. 1989. Toxaphene and other organochlorines in Arctic Ocean fauna: evidence for atmospheric delivery. *Arctic* 42: 307-313.

Bierman, V.J. 1990. Equilibrium partitioning and biomagnification of organic chemicals in benthic animals. *Environ. Sci. Technol.* 24: 1407-1412.

Bignert, A., A Göthberg, S. Jensen, K. Litzén, T. Odsjö, M. Olsson, and L. Reutergårdh. 1993. The need for adequate biological sampling in ecotoxicological investigations: a retrospective study of twenty years pollution monitoring. *Sci. Tot. Environ.* 128: 121–139.

Black, M.C., and J.F. McCarthy. 1988. Dissolved organic macromolecules reduce the uptake of hydrophobic organic contaminants by the gills of rainbow trout (*Salmo gairdneri*). *Environ. Toxicol. Chem.* 7: 593-600.

Boese, B.L., H. Lee, D.T. Specht, and R.C. Randall. 1990. Comparison of aqueous and solid-phase uptake for hexachlorobenzene in the tellinid clam *Macoma nasuta* (Conrad): a mass balance approach. *Environ. Toxicol. Chem.* 9: 221-231.

Bollag, J.-M., R.D. Minard, and S.-Y. Liu. 1983. Cross-linkage between anilines and phenolic humus constituents. *Environ. Sci. Technol.* 17: 72-80.

Bollag, J.-M., and S.-Y. Liu. 1985. Copolymerization of halogenated phenols and syringic acid. *Pest. Biochem. Physiol.* 23: 261-272.

Brannon, J.M., T.E. Myers, D. Gunnison, and C.B. Price. 1991. Nonconstant polychlorinated biphenyl partitioning in New Bedford Harbor sediment during sequential batch leaching. *Environ. Sci. Technol.* 25: 1082-1087.

Braune, B.M., and R.J. Norstrom. 1989. Dynamics of organochlorine compounds in herring gulls: III. Tissue distribution and bioaccumulation in Lake Ontario gulls. *Environ. Toxicol. Chem.* 8: 957-968.

Broman, D., C. Näf, I. Lundbergh, and Y. Zebühr. 1990. An in situ study on the distribution, biotransformation and flux of polycyclic aromatic hydrocarbons (PAHs) in an aquatic food chain (seston-*Mytilus edulis* L. — *Somateria mollissima* L.) from the Baltic: an ecotoxicological perspective. *Environ. Toxicol. Chem.* 9: 429-442.

Brownawell, B.J., and J.W. Farrington. 1986. Biogeochemistry of PCBs in interstitial waters of a coastal marine sediment. *Geochim. Cosmochim. Acta* 50: 157-169.

Bruggeman, W.A., A. Opperhuizen, A. Wijbenga, and O. Hutzinger. 1984. Bioaccumulation of super-lipophilic chemicals in fish. *Toxicol. Environ. Chem.* 7: 173-189.

Bruggeman, W.A., J. Van der Stenen, and O. Hutzinger. 1982. Reversed-phase thin-layer chromatography of polynuclear aromatic hydrocarbons and chlorinated biphenyls. Relationship with hydrophobicity as measured by aqueous solubility and octanol-water partition coefficient. *J. Chromatogr.* 238: 335-346.

Calderbank, A. 1989. The occurrence and significance of bound pesticide residues in soil. *Revs. Environ. Contam. Toxicol.* 108: 1-1103.

Cameron, M., and I.M. Weis. 1993. Organochlorine contaminants in the country food diet of the Belcher Island Inuit, Northwest Territories, Canada. *Arctic* 46: 42-48.

Capriel, P., A. Haisch, and S.U. Khan.1985. Distribution and nature of bound (nonextractable) residues of atrazine in a mineral soil nine years after the herbicide application. *J. Agric. Food Chem.* 33: 567-569.

Carlberg, G.E., K. Martinsen, A. Kringstad, E. Gjessing, M. Grande, T. Källqvist, and J.U. Skåre. 1986. Influence of aquatic humus on the bioavailability of chlorinated micropollutants in Atlantic salmon. *Arch. Environ. Contam. Toxicol.* 15: 543-548.

Carr, R.S., J.W. Williams, and C.T.B. Fragata. 1989. Development and evaluation of a novel marine sediment pore water toxicity test with the polychaete Dinophilus gyrociliatus. *Environ. Toxicol. Chem.* 8: 533-543.

Carter, C.W., and I.H. Suffet. 1982. Binding of DDT to dissolved humic materials. *Environ. Sci. Technol.* 16: 735-740.

Carter, D.S., and R.A. Hites. 1992. Fate and transport of Detroit River derived pollutants throughout Lake Erie. *Environ. Sci. Technol.* 26: 1333-1341.

Chessels, M., D.W. Hawker, and D.W. Connel. 1992. Influence of solubility in lipid on bioconcentration of hydrophobic compounds. *Ecotoxicol. Environ. Saf.* 23: 260-273.

Chin, Y.-P., and P.M. Gschwend. 1992. Partitioning of polycyclic aromatic hydrocarbons to marine porewater organic colloids. *Environ. Sci. Technol.* 26: 1621-1626.

Chiou, C.T. 1985. Partition coefficients of organic compounds in lipid-water systems and correlations with fish bioconcentration factors. *Environ. Sci. Technol.* 19: 57-62.

Clark, K.E., F.A.P.C. Gobas, and D. Mackay. 1990. Model of organic chemical uptake and clearance by fish from food and water. *Environ. Sci. Technol.* 24: 1203-1213.

Clark, T.P., R.J. Norstrom, G.A. Fox, and H.T. Won. 1987. Dynamics of organochlorine compounds in herring gulls (*Larus argentatus*): II. a two-compartment model and data for ten compounds. *Environ. Toxicol. Chem.* 6: 547-559.

Connolly, J.P. 1991. Application of a food chain model to polychlorinated biphenyl contaminatiion of the lobster and winter flounder food chains in New Bedford Harbor. *Environ. Sci. Technol.* 25: 760-770.

Connolly, J.P., and C.J. Pedersen. 1988. A thermodynamic-based evaluation of organic chemical accumulation in aquatic organisms. *Environ. Sci. Technol.* 22: 99-103.

Czuczwa, J.M., and R.A. Hites. 1986. Airborne dioxins and dibenzofurans: sources and fates. *Environ. Sci. Technol.* 20: 195-200.

de Bruijn, J., and J. Hermens. 1991. Uptake and elimination kinetics of organophosphorous (sic) pesticides in the guppy (*Poecilia reticulata*): correlations with the octanol/water partition coefficient. *Environ. Toxicol. Chem.* 10: 791-804.

de Bruijn, J., F. Busser, W. Seinen, and J. Hermens. 1989. Determination of octanol/water partition coefficients for hydrophobic organic chemicals with the slow-stirring method. *Environ. Toxicol. Chem.* 8: 499-512.

DeLaune, R.D., R.P. Gambrell, and K.S. Reddy. 1983. Fate of pentachlorophenol in estuarine sediment. *Environ. Pollut. Ser. B.* 6: 297-308.

De Voe, H., M.M. Miller, and S.P. Wasik. 1981. Generator column and high pressure liquid chromatography for determining aqueous solubilities and octanol-water partition coefficients of hydrophobic substances. *J. Res. Natl. Bur. Stand. (U.S.)* 86: 361-366.

De Voogt, P., D.E. Wells, L. Reutergård, and U.A.T. Brinkman. 1990. Biological activity, determination and occurrence of planar, mono- and di-ortho PCBs. *Inter. J. Environ. Anal. Chem.* 40: 1-46.

de Wolf, W., J.H.M. de Bruijn, W. Seinen, and J.L.M. Hermens. 1992a. Influence of biotransformation on the relationship between bioconcentration factors and octanol-water partition coefficients. *Environ. Sci. Technol.* 26: 1197-1201.

de Wolf, W., W. Seinen, A. Opperhuizen, and J.L.M. Hermens.1992b. Bioconcentration and lethal body burden of 2,3,4,5-tetrachloroaniline in guppy, *Poecilia reticulata. Chemosphere* 25: 853-863.

de Wolf, W., W. Seinen, and J.L.M. Hermens. 1993. Biotransformation and toxicokinetics of trichloroanilines in fish in relation to their hydrophobicity. *Arch. Environ. Contam. Toxicol.* 25: 110-117.

Eadie, B.J., N.R. Morehead, and P.F. Landrum. 1990. Three-phase partitioning of hydrophobic organic compounds in Great Lakes waters. *Chemosphere.* 20: 161-178.

Eisenreich, S.J., P.D. Capel, J.A. Robbins, and R. Bourbonniere. 1989. Accumulation and diagenesis of chlorinated hydrocarbons in lake sediments. *Environ. Sci. Technol.* 23: 1116-1126.

Ernst, W. 1979. Factors affecting the evaluation of chemicals in laboratory experiments using marine organisms. *Ecotoxicol. Environ. Saf.* 3: 90.98.

Fairchild, W.L., D.C.G. Muir, R.S. Curri, and A.L. Yarechewski. 1992. Emerging insects as a biotic pathway for movement of 2,3,7,8-tetrachlorodibenzofuran from lake sediments. *Environ. Toxicol. Chem.* 11: 867-872.

Fendinger, N.J, and D.E. Glotfelty. 1988. A labratory method for the experimental determination of air-water Henry's Law constants for several pesticides. *Environ. Sci. Technol.* 22: 1289-1293.

Fendinger, N.J., and D.E. Glotfelty. 1990. Henry's Law constants for selected pesticides, PAHs and PCBs. *Environ. Toxicol. Chem.* 9: 731-735.

Fernandez, P. and J.M. Bayona. 1992. Use of off-line gel permeation chromatography-normal-phase liquid chromatography for the determination of polycyclic aromatic compounds in environmental samples and standard reference materials (air particulate matter and marine sediment). *J. Chromatogr.* 625: 141-149.

Flanagan, W.P., and R.J. May. 1993. Metabolite detection as evidence for naturally occurring aerobic PCB degradation in Hudson River sediments. *Environ. Sci. Technol.* 27: In press.

Flenner, C.K., J.R. Parsons, S.M. Schrap, and A. Opperhuizen. 1991. Influence of suspended sediment on the biodegradation of alkyl esters of *p*-aminobenzoic acid. *Bull. Environ. Contam. Toxicol.* 47: 555- 560.

Focardi, S., C. Gaggi, G. Chemello, and E. Bacci. 1991. Organochlorine residues in moss and lichen samples from two Antarctic areas. *Polar Rec.* 162: 241-244.

Fowler, S.W., G.G. Polikarpov, D.L. Elder, P. Parsi, and J.-P. Villeneuve.1978. Polychlorinated biphenyls: accumulation from contaminated sediments and water by the polychaete *Nereis diversicolor. Mar. Biol.* 48: 303-309.

Fox, G.A., S.W. Kennedy, R.J. Norstrom, and DC. Wigfield. 1988. Porphyria in herring gulls: a biochemical response to chemical contamination of Great Lakes food chains. *Environ. Toxicol. Chem.* 7: 831-839.

Garbarini, D.R., and L.W. Lion. 1986. Influence of the nature of soil organics on the sorption of toluene and trichloroethylene. *Environ. Sci. Technol.* 20: 1263-1269.

Garcia, K.L., J.J. Delfino, and D.H. Powell. 1993. Non-regulated organic compounds in Florida sediments. *Water Res.* 27: 1601-1613.

Garbow, J.R., and B.J. Gaede. 1992. Analysis of a phenyl ether herbicide-cyclodextrin inclusion complex by CPMAS^{13}C NMR. *J. Agric. Food Chem.* 40: 156-159.

Gauthier, T.D., W.R. Selz, and C.V.L. Grant. 1987. Effects of structural and compositional variations of dissolved humic materials on pyrene K_{oc} values. *Environ. Sci. Technol.* 21: 243-248.

Gobas, F.A.P.C., K.E. Clark, W.Y. Shiu, and D. Mackay. 1989. Bioconcentration of polybrominated benzenes and biphenyls and related superhydrophobic chemicals in fish: role of bioavailability and elimination into the feces. *Environ. Toxicol. Chem.* 8: 231-245.

Gobas, F.A.P.C., X. Zhang, and R. Wells, 1993a. Gastrointestinal magnification: the mechanism of biomagnification and food chain accumulation of organic chemicals. *Environ. Sci. Technol.* 27: 2855-2863.

Gobas, F.A.P.C., J.R. McCorquodale, and G.D. Haffner. 1993b. Intestinal absorption and biomagnification of organochlorines. *Environ. Toxicol. Chem.* 12: 567-576.

Goldberg, E.D. 1986. The mussel watch concept. *Environ. Monit. and Assessment* 7: 91-103.

Gregor, D.J., and W.D. Gummer. 1989. Evidence of atmospheric transport and deposition of organochlorine pesticides and polychlorinated biphenyls in Canadian Arctic snow. *Environ. Sci. Technol.* 23: 561-566.

Haderlein, S.B., and R.P. Schwarzenbach. 1993. Adsorption of substituted nitrobenzenes and nitrophenols to mineral surfaces. *Environ. Sci. Technol.* 27: 316-326.

Hagan, M. 1962. *Clathrate inclusion compounds.* Reinhold Publishing, New York.

Hand, V.C., and G.K. Williams. 1987. Structure-activity relationships for sorption of linear alkylbenzenesulfonates. *Environ. Sci. Technol.* 21: 370-373.

Hargrave, B.T., W.P. Vass, P.E. Erickson, and B.R. Fowler. 1988. Atmospheric transport of organochlorines to the Arctic Ocean. *Tellus* 40B: 480-493.

Harkness, M.R., J.B. McDermott, D.A. Abramowicz, J.J. Salvo, W.P. Flanagan, M.L. Stephens, F.J. Mondello, R.J. May, J.H. Lobos, K.M. Carroll, M.J. Brennan, A.A. Bracco, K.M. Fish, G.L. Warner, P.R. Wilson, D.K. Dietrich, D.T. Lin, C.B. Morgan, and W.L.Gately. 1993. In situ stimulation of aerobic PCB biodegradation in Hudson River sediments. *Science* 259: 503-507.

Hashimoto, S., T. Wakimoto, and R. Tatsukawa. 1990. PCDDs in the sediments accumulated about 8120 years ago from Japanese coastal areas. *Chemosphere* 21: 825-835.

Hatcher, P.G., J.M. Bortiatynski, R.D. Minard, J. Dec, and J.-M. Bollag. 1993. Use of high-resolution ^{13}C NMR to examine the enzymatic covalent binding of ^{13}C-labeled 2,4-dichlorophenol to humic substances. *Environ. Sci. Technol.* 27: 2098-2103.

Hawker, D.W., and D.W. Connell. 1986. Bioconcentration of lipophilic compounds by some aquatic organisms. *Ecotoxicol. Environ. Saf.* 11: 184-197.

Hayes, M.H.B., M.E. Pick, and B.A. Thoms. 1978a. The influence of organocation structure on the adsorption of mono- and of bipyridinium cations by expanding lattice clay minerals. I. Adsorption by Na^+-montmorillonite. *J. Colloid Interface Sc.* 65: 254-265.

Hayes, M.H.B., M.E. Pick, and B.A. Thoms. 1978b. The influence of organocation structure on the adsorption of mono- and of bipyridinium cations by expanding lattice clay minerals. II. Adsorption by Na^+-vermiculite. *J. Colloid Interface Sci.* 65: 266-275.

Heitkamp, M.A., J.N. Huckins, J. D. Pettie, and J.L. Johnson. 1984. Fate and metabolism of isopropylphenyl diphenyl phosphate in freshwater sediments. *Environ. Sci. Technol.* 18: 434-439.

Higashi, R.M., G.N. Cherr, J.M. Shenker, J.M. Macdonald, and D.G. Crosby. 1992. A polar high molecular mass constituent of bleached kraft mill effluent is toxic to marine organisms. *Environ. Sci. Technol.* 26: 2413-2420.

Hinman, M.L., and S.J. Klaine. 1992. Uptake and translocation of selected organic pesticides by the rooted aquatic plant *Hydrilla verticillata* Royle. *Environ. Sci. Technol.* 26: 609-613.

Ingebrigtsen, K., H. Hektoen, T. Andersson, E.K. Wehler, Å. Bergman, and I. Brandt. 1992. Enrichment of metabolites in the cerebrospinal fluid of cod (*Gadus morhua*) following oral administration of hexachlorobenzene and 2,4',5-trichlorobiphenyl. *Pharmacol. Toxicol.* 71: 420-425.

Isaacson, P.J., and C.R. Frink. 1984. Nonreversible sorption of phenolic compounds by sediment fractions: the role of sediment organic matter. *Environ. Sci. Technol.* 18: 43-48.

Jarman, W.M., M. Simon, R.J. Norstrom, S.A. Burns, C.A. Bacon, B.R.T. Simoneit, and R.W. Risebrough. 1992. Global distribution of tris(4-chlorophenyl)methanol in high trophic level birds and mammals. *Environ. Sci. Technol.* 26: 1770-1774.

Jeans, J. 1952. *An introduction to the kinetic theory of gases* pp. 63-68. Cambridge University Press, Cambridge.

Jensen, S., G. Eriksson, H. Kylin, and W.M.J. Strachan. 1992. Atmospheric pollution by persistent organic compounds: monitoring with pine needles. *Chemosphere* 24: 229-245.

Jokela, J,K., and M. Salkinoja-Salonen. 1992. Molecular weight distributions of organic halogens in bleached kraft pulp mill effluents. *Environ. Sci. Technol.* 26: 1190-1197.

Jones, K.C., J.A. Stratford, K.S. Waterhouse, E.T. Furlong, W. Giger, R.A. Hites, C. Schaffner, and A.E. Johnston. 1989. Increases in the polynuclear aromatic hydrocarbon content of an agricultural soil over the last century. *Environ. Sci. Technol.* 23: 95-101.

Karickhoff, S.W. 1981. Semi-empirical estimation of sorption of hydrophobic pollutants on natural sediments and soils. *Chemosphere* 10: 833-846.

Karickhoff, S.W. 1984. Organic pollutant sorption in aquatic systems. *J. Hydraul. Eng.* 100: 707-735.

Karickhoff, S.W., and K.R. Morris. 1985. Impact of tubificid oligochaetes on pollutant transport in bottom sediments. *Environ. Sci. Technol.* 19: 51-56.

Kinloch, D., H.V. Kuhnlein, and D. Muir. 1992. Inuit foods and diet. A preliminary assessment of benefits and risks. *Sci. Total Environ.* 122: 245-276.

Knezovich, J.P., and F.L. Harrison. 1988. The bioavailability of sediment-sorbed chloroben-zenes to larvae of the midge, *Chironomus decorus*. *Ecotoxicol. Environ. Saf.* 15: 226-241.

Knezovich, J.P., F.L. Harrison, and R.G. Wilhelm. 1987. The bioavailability of sediment-sorbed organic chemicals: a review. *Water Air Soil Pollut.* 32: 233-245.

Knox, G.A. 1970. Antarctic marine ecosystems pp. 69-96. In *Antarctic Ecology* (Ed. M.W. Holdgate). Vol. 1. Academic Press, London.

Koester, C.J., and R.A. Hites. 1992. Wet and dry deposition of chlorinated dioxins and furans. *Environ. Sci. Technol.* 26: 1375-1382.

Kolset, K., and A. Heiberg. 1988. Evaluation of the 'Fugacity' (FEQUM) and the 'EXAMS' chemical fate and transport models: a case study on the pollution of the Norrsundet Bay (Sweden). *Water Sci. Technol.* 20 [2]: 1-12.

Krone, C.A., D.G,. Burrows, D.W. Brown, P.S. Robisch, A.J. Friedman, and D.C. Malins. 1966. Nitrogen-containing aromatic compounds in sediments from a polluted harbor in Puget Sound. *Environ. Sci. Technol.* 20: 1144-1150.

Kucklick, J.R., T.F. Bidleman, L.L. McConnell, M.D. Walla, and G.P. Ivanov. 1994. Organo-chlorines in water and biota of Lake Baikal, Siberia. *Environ. Sci. Technol.* 28: 31-37.

Landrum, P.F. 1989. Bioavailability and toxicokinetics of polycyclic aromatic hydrocarbons sorbed to sediments for the amphipod *Pontoporeia hoyi*. *Environ. Sci. Technol.* 23: 588-595.

Landrum, P.F., B.J. Eadie, and W.R. Faust. 1991. Toxicokinetics and toxicity of a mixture of sediment-associated polycyclic aromatic hydrocarbons to the amphipod *Diporeia* sp. *Environ. Toxicol. Chem.* 10: 35-46.

Landrum, P.F., M.D. Reinhold, S.R. Nihart, and B.J. Eadie. 1985. Predicting the bioavailability of organic xenobiotics to *Pontoporeia hoyi* in the presence of humic and fulvic materials and natural dissolved organic matter. *Environ. Toxicol. Chem.* 4: 459-467.

Landrum, P.F., S.R. Nihart, B.J. Eadie, and L.R. Herche. 1987. Reduction in bioavailability of organic contaminants to the amphipod *Pontoporeia hoyi* by dissolved organic matter of sediment interstitial waters. *Environ. Toxicol. Chem.* 6: 11-20.

Landrum, P.F., S.R. Nihart, B.J. Eadie, and W.S. Gardner. 1984. Reverse-phase separation for determining pollutant binding to Aldrich humic acid and dissolved organic carbon of natural waters. *Environ. Sci. Technol.* 18: 187-192.

Larsen, B., F. Pelusio, H. Skejö, and A. Paya-Perez. 1992. Bioavailability of polychlorinated biphenyl congeners in the soil to earthworm (*L. rubellus*) system. *Internat. J. Environ. Anal. Chem.* 46: 149-162.

Lau, Y.L., B.G. Oliver, and B.G. Krishnappan. 1989. Transport of some chlorinated contami-nants by the water, suspended sediments, and bed sediments in the St. Clair and Detroit Rivers. *Environ. Toxicol. Chem.* 8: 293-301.

Lee, P.W. 1985. Fate of fenvalerate (pydin insecticide) in the soil environment. *J. Agric. Food Chem.* 33: 993-998.

Lindqvist, W.L., R., and C.G Enfield. 1992. Biosorption of dichlorodiphenyltrichloroethane and hexachlorobenzene in groundwater and its implications for facilitated transport. *Appl. Environ. Microbiol.* 58: 2211-2218.

Lockhart, W.L., R. Wagemann, B. Tracey, D. Sutherland, and D.J. Thomas. 1992. Presence and implications of chemical contaminants in the freshwaters of the Canadian Arctic. *Sci. Total Environ.* 122: 165-243.

Lydy, M.J., J.T. Oris, P.C. Baumann, and S.W. Fisher. 1992. Effects of sediment organic carbon content on the elimination rates of neutral lipophilic compounds in the midge (*Chironomus riparius*). *Environ. Toxicol. Chem.* 11: 347-356.

Macalady, D.L., and N.L. Wolfe. 1985. Effects of sorption and abiotic hydrolyses. 1. Organophosphorothioate esters. *J. Agric. Food Chem.* 33: 167-173.

Macdonald, R.W., W.J. Cretney, N. Crewe, and D. Paton. A history of octachlorodibenzo-*p*-dioxin, 2,3,7,8-tetrachlorodibenzofuran, and 3,3',4,4'-tetrachlorobiphenyl contamination in Howe Sound, British Columbia. *Environ. Sci. Technol.* 26: 1544-1550.

Mackay, D. 1982. Correlation of bioconcentration factors. *Environ. Sci. Technol.* 16: 274-278.

Mackay, D., and T.K. Yuen. 1980. Volatilization rates of organic contaminants from rivers. *Water Pollut. Res. J. Can.* 15: 83-98.

Mackay, D., and W.Y. Shiu. 1981. Vapor pressure of organic compounds. *J. Phys. Chem. Ref. Data* 10: 220-243.

Mackay, D., S. Paterson, and B. Cheung. 1985. Evaluating the environmental fate of chemicals. The fugacity — Level III approach as applied to 2,3,7,8, TCDD. *Chemosphere* 14: 859-863.

Mackay, D., S. Paterson, and W.H. Schroeder. 1986. Model describing the rates of transfer processes of organic chemicals between atmosphere and water. *Environ. Sci. Technol.* 20: 810-816.

Magee, B.R., L.W. Lion, and A.T. Lemley. 1991. Transport of dissolved organic macromolecules and their effect on the transport of phenanthrene in porous media. *Environ. Sci. Technol.* 25: 323-331.

Malins, D.C., B.B. McCain, D.W. Brown, S.-L. Chan, M.S. Meyers, J.T. Landahl, P.G. Prohaska, A.J. Friedman, L.D. Rhodes, D.G. Burrows, W.D. Gronlund, and H.O. Hodgins. 1984. Chemical pollutants in sediments and diseases of bottom-dwelling fish in Puget Sound. *Environ. Sci. Technol.* 18: 705-713.

Manabe, S., O. Wada, M. Morita, S. Izumikawa, K. Asakuno, and H. Suzuki. 1992. Occurrence of carcinogenic amino-a-carbolines in some environmental samples. *Environ. Pollut.* 75: 301-305.

Marple, L., B. Berridge, and L. Throop. 1986. Measurement of the water-octanol partition coefficient of 2,3,7,8-tetrachlorodibenzo-*p*-dioxin. *Environ. Sci. Technol.* 20: 397-399.

Martinsen, K., A. Kringstad, and G.E. Carlberg. 1988. Methods for determination of sum parameters and characterization of organochlorine compounds in spent bleach liquors from pulp mills and water, sediment and biological samples from receiving waters. *Water Sci. Technol.* 20[2]: 13-24.

McCall, P.J., and G.L. Agin. 1985. Desorption kinetics of picloram as affected by residence time in the soil. *Environ. Toxicol. Chem.* 4: 37-44.

McConnell, L.L., W.E. Cotham, and T.F. Bidleman. 1993. Gas exchange of hexachlorocyclohexane in the Great Lakes. *Environ. Sci. Technol.* 27: 1304-1311.

Mehrle, P.M., D.R. Buckler, E.E. Little, L.M. Smith, J.D. Petty, P.H. Peterman, D.L. Stalling, G.M. De Graeve, J.J. Coyle, and W.J. Adams. 1988. Toxicity and bioconcentratiion of 2,3,7,8-tetrachlorodibenzodioxin and 2,3,7,8-tetrachlorodibenzofuran in rainbow trout. *Environ. Toxicol. Chem.* 7: 47-62.

Metcalfe, J.L., M.E. Fox, and J.H. Carey. 1984. Aquatic leeches (*Hirudinea*) as bioindicators of organic chemical contaminants in freshwater ecosystems. *Chemosphere* 13: 143-150.

Metcalfe, J.L., M.E. Fox, and J.H. Carey. 1988. Freshwater leeches (*Hirudinea*) as a sceening tool for detecting organic contaminants in the environment. *Environ. Monit. Assess.* 11: 147-169.

Miller, M.M., S.P. Wasik, G.-L. Huang, W.-Y. Shiu, and D. Mackay. 1985. Relationships between octanol-water partition coefficients and aqueous solubility. *Environ. Sci. Technol.* 19: 522-529.

Mudroch, A., R.J. Allan, and S.R. Joshi. 1992. Geochemistry and organic contaminants in the sediments of Great Slave Lake, Northwest Territories, Canada. *Arctic* 45: 10-19.

Muir, D.C.G., and A.L. Yarechewski. 1988. Dietary accumulation of four chlorinated dioxin congeners by rainbow trout and fathead minnows. *Environ. Toxicol. Chem.* 7: 227-236.

Muir, D.C.G., A.L. Yarechewski, and A. Knoll. 1986. Bioconcentration and disposition of 1,3,6,8- tetrachlorodibenzo-*p*-dioxin and octachlorodibenzo-*p*-dioxin by rainbow trout and fathead minnows. *Environ. Toxicol. Chem.* 5: 261-272.

Muir, D.C.G., R.J. Norstrom, and M. Simon. 1988. Organochlorine contaminants in Arctic marine food chains: accumulation of specific polychlorinated biphenyls and chlordane-related compounds. *Environ. Sci. Technol.* 22: 1071-1079.

Muir, D.C.G., C.A. Ford, N.P. Grift, and R.E.A. Stewart. 1992. Organochlorine contaminants in narwhal (*Monodon monoceros*) from the Canadian Arctic. *Environ. Pollut.* 75: 307-316.

Muir, D.C.G., R. Wagemann, B.T. Hargrave, D.J. Thomas, D.B. Peakall, and R.J. Norstrom. 1992. Arctic marine ecosystem contamination. *Sci. Total Environ.* 122: 75-134.

Muir, D.C.G., M.D. Segstro, P.M. Welbourn, D. Toom, S.J. Eisenreich, C.R. Macdonald, and D.M. Whelpdale. 1993. Pattern of accumulation of airborne organochlorine contaminants in lichens from the upper Great Lakes region of Ontario. *Environ. Sci. Technol.* 27: 1201-1210.

Mullin, M., C.M. Pochini, S. McCrindle, M. Romkes, S.H. Safe, and L.M. Safe. 1984. High-resolution PCB analysis: synthesis and chromatographic properties of all 209 PCB congeners. *Environ. Sci. Technol.* 18: 468-476.

Muncaster, B.W., P.D.N. Herbert, and R. Lazar. 1990. Biological and physical factors affecting the body burden of organic contaminants in freshwater mussels. *Arch. Environ. Contam. Toxicol.* 19: 25-34.

Mäkelä, T.P., T. Petänen, J. Kukkonen, and A.O. Oikari. 1991. Accumulation and depuration of chlorinated phenolics in the freshwater mussel (*Anodonta anatina L.*). *Ecotoxicol. Environ. Saf.* 22: 153-163.

Nebeker, A.V., G.S. Schuytema, W.L. Griffis, J.A. Barbitta, and L.A. Carey. 1989. Effect of sediment organic carbon on survival of *Hyalella azteca* exposed to DDT and endrin. *Environ. Toxicol. Chem.* 705-718.

Neilson, A.H., H. Blanck, L. Förlin, L. Landner, P. Pärt, A. Rosemarin, and M. Söderström. 1989. Advanced hazard assessment of 4,5,6-trichloroguaiacol in the Swedish Environment. In *Chemicals in the Aquatic Environment*, pp. 329-374. Ed. L. Landner, Springer-Verlag, Berlin.

Neilson, A.H., A.-S. Allard, S. Fischer, M. Malmberg, and T. Viktor. 1990. Incorporation of a subacute test with zebra fish into a hieracrchical system for evaluating the effect of toxicants in the aquatic environment. *Ecotoxicol. Environ. Saf.* 20: 82-97.

Niimi, A.J., and G.P. Dookhran. 1989. Dietary absorption efficiencies and elimination rates of polycyclic aromatic hydrocarbons (PAHs) in rainbow trout (*Salmo gairdneri*). *Environ.Toxicol. Chem.* 8: 719-722.

Niimi, A.J., H.B. Lee, and G.P. Kissoon. 1989. Octanol/water partition coefficients and bioconcentration factors of chloronitrobenzenes in rainbow trout (*Salmo gairdneri*). *Environ. Toxicol. Chem.* 8: 817-823.

Norstrom, R.J., M. Simon, D.C.G Muir, and R.E. Schweinsberg. 1988. Organochlorine contaminants in Arctic marine food chains: identification, geographical distribution, and temporal trends in polar bears. *Environ. Sci. Technol.* 22: 1063-1071.

Norstrom, R.J., T.P. Clark, D.A. Jeffrey, H.T. Won, and A.P. Gilman. 1986. Dynamics of organochlorine compounds in herring gulls (*Larus argentatus*): 1. Distribution and clearance of [^{14}C]DDE in free-living herring gulls (*Larus argentatus*). *Environ. Toxicol. Chem.* 5: 41-48.

Novak, M.A., A.A. Reilly, and S.J. Jackling. 1988. Long-term monitoring of polychlorinated biphenyls in the Hudson River (New York) using caddisfly larvae and other macroinvertebrates. *Arch. Environ. Contam. Toxicol.* 17: 699-710.

Oikari, A., and T. Kunnamo-Ojala. 1987. Tracing of xenobiotic contamination in water with the aid of fish bile metabolites: a field study with caged rainbow trout (*Salmo gairdneri*). *Aquat. Pollut.* 9: 327-341.

Oliver, B.G. 1987. Biouptake of chlorinated hydrocarbons from laboratory-spiked and field sediments by oligochaete worms. *Environ. Sci. Technol.* 21: 785-790.

Oliver, B.G., and A.J. Niimi. 1985. Bioconcentration of some halogenated organics for rainbow trout: limitations in their use for prediction of environmental residues. *Environ. Sci. Technol.* 19: 842-849.

Oliver, B.G., and A.J. Niimi. 1988. Trophodynamic analysis of PCB congeners and other chlorinated hydrocarbons in the Lake Ontario ecosystem. *Environ. Sci. Technol.* 22: 388-397.

Oliver, B.G., M.N. Charlton, and R.W. Durham. 1989. Distribution, redistribution, and geo-chronology of polychlorinated biphenyl congeners and other chlorinated hydrocarbons in Lake Ontario sediments. *Environ. Sci. Technol.* 23: 200-208.

Opperhuizen, A., P. Serné, and J.M.D. van Steen. 1988. Thermodynamics of fish/water and octan-1-ol/water partitioning of some chlorinated benzenes. *Environ. Sci. Technol.* 22: 286-292.

Ou, L., K.S Edvardsson, and P.S.C. Rao. 1985. Aerobic and anaerobic degradation of Aldicarb in soils. *J. Agric. Food Chem.* 53: 72-78.

Parris, G.E. 1980. Covalent binding of aromatic amines to humates. 1. Reactions with carbonyls and quinones. *Environ. Sci. Technol.* 14: 1099-1106.

Partington, J.R. 1950. *Thermodynamics* pp. 112-114. Constrable & Company Ltd. London.

Paterson, S., and D. Mackay. 1985. The fugacity concept in environmental modeling pp. 121-140. In *Handbook of environmental chemistry* (Ed. O. Hutzinger), Vol. 2 Part C. Springer-Verlag, Berlin.

Pavliostathis, S.G., and G.N. Mathavan. 1992. Desorption kinetics of selcted volatile organic compounds from field contaminated soils. *Environ. Sci. Technol.* 26: 532-538.

Pavliostathis, S.G., and K. Jaglal. 1991. Desorptive behavior of trichloroethylene in contaminated soil. *Environ. Sci. Technol.* 25: 274-.279.

Perdue, E.M., and N.L. Wolfe. 1982. Modification of pollutant hydrolysis kinetics in the presence of humic substances. *Environ. Sci. Technol.* 16: 847-852.

Pereira, W.E., C.E. Rostad, C.T. Chiou, T.I. Brinron, L.B. Barber, D.K. Demcheck, and C.R. Demas. 1988. Contamination of estuarine water, biota, and sediment by halogenated organic compounds: a field study. *Environ. Sci. Technol.* 22: 772-778.

Peterman, P.H., and J.J. Delfino. 1990. Identification of isopropylbiphenyl, alkyl diphenylmethanes, diisopropylnaphthalene, linear alkyl benzenzes and other polychlorinated biphenyl replacement compounds in effluents, sediments and fish in the Fox River system, Wisconsin. *Biomed. Environ. Mass Spectrom.* 19: 755-770.

Pignatello, J.J. 1990. Slowly reversible sorption of aliphatic halocarbons in soils. I. Formation of residual fractions. *Environ. Toxicol. Chem.* 9: 1107-1115.

Podoll, R.T., and K.C. Irwin. 1988. Sorption of cationic oligomers on sediments. *Environ. Toxicol. Chem.* 7: 405-415.

Prahl, F.G., and R. Carpenter. 1983. Polycyclic aromatic hydrocarbon (PAH)-phase associations in Washington coastal sediment. *Geochim. Cosmochim. Acta* 47: 1013-1023.

Pärt, P. 1990. The perfused fish gill preparation in studies of the bioavailability of chemicals. *Ecotoxicol. Environ. Saf.* 19: 106-115.

Rapaport, R.A., and S.J. Eisenreich. 1988. Historical atmospheric inputs of high molecular weight chlorinated hydrocarbons to Eastern North America. *Environ. Sci. Technol.* 22: 931-941.

Remberger, M., P.-Å. Hynning, and A.H. Neilson. 1988. Comparison of procedures for recovering chloroguaiacols and chlorocatechols from contaminated sediments. *Environ. Toxicol. Chem.* 7: 795- 805.

Remberger, M., P.-Å. Hynning, and A.H. Neilson.1990. Gas chromatographic analysis and gas chromatographic-mass spectrometric identification of components in the cyclohexane-extractable fraction from contaminated sediment samples. *J. Chromatogr.* 508: 159-178.

Remberger, M., P.-Å. Hynning, and A.H. Neilson.1993. Release of chlorinated catechols from a contaminated sediment. *Environ. Sci. Technol.* 27: 158-164.

Renberg, L. O., S.G. Sundström, and A.-C. Rosén-Olofsson. 1985. The determination of partition coefficients of organic compounds in technical products and waste waters for the estimation of their bioaccumulation potential using reverse phase thin layer chromatography. *Toxicol. Environ Chem.* 10: 333-349.

Riederer, M. 1990. Estimating partitioning and transport of organic chemicals in the foliage/ atmosphere system: discussion of a fugacity-based model. *Environ. Sci. Technol.* 24: 829-837.

Safe, S., K.W. Brown, K.C. Donnelly, C.S. Anderson, K.V. Markiewicz, M.S. McLachlan, A. Reischi, and O. Hutzinger. 1992. Polychlorinated dibenzo-p-dioxins and dibenzofurans associated wirth wood-preserving chemical sites: biomonitoring with pine needles. *Environ. Sci. Technol.* 26: 394-396.

Sanders, G., K.C. Jones, J. Hamilton-Taylor, and H. Dörr. 1992. Historical inputs of polychlorinated biphenyls and other organochlorines to a dated lacustrine sediment core in rural England. *Environ. Sci. Technol.* 26: 1815-1821.

Sarkar, J.M., R.L. Malcolm, and J.-M. Bollag. 1988. Enzymatic coupling of 2,4-dichlorophenol to stream fulvic acid in the presence of oxidoreductases. *Soil Sci. Am. J.* 52: 688-694.

Sawhney, B.L., J.J. Pignatello, and A.M. Steinberg. 1988. Determination of 1,2-dibromoethane (EDB) in field soils: implications for volatile organic compounds. *J. Environ. Qual.* 17: 149-152.

Scharenberg, W. 1991. Cormorants (*Phalacrocorax carbo sinensis*) as bioindicators for polychlorinated biphenyls. *Arch. Environ. Contam. Toxicol.* 21: 536-540.

Scheunert, I., E. Topp, A. Attar, and F. Korte. 1994. Uptake pathways of chlorobenzenes in plants and their correlation with N-octanol/water partition coefficients. *Ecotoxicol. Environ. Saf.* 27: 90-104.

Schmitt, C.J., J.L. Zajick, and P.H. Peterman. 1990. National contaminants biomonitoring program: residues of organochlorine chemicals in U.S. freshwater fish, 1976-1984. *Arch. Environ. Contam. Toxicol.* 19: 748-781.

Schrap, S.M., and A. Opperhuizen. 1990. Relationship between bioavailability and hydrophobicity: reduction of the uptake of organic chemicals by fish due to sorption on particles. *Environ. Toxicol. Chem.* 9: 715-724.

Schuytema, G.S., D.F. Krawczyk, W.L. Griffis, A.V. Nebeker, and M.L. Robideaux. 1990. Hexachlorobenzene uptake by fathead minnows and macroinvertebrates in recirculating sediment/water systems. *Arch. Environ. Contam. Toxicol.* 19: 1-9.

Sericano, J.L., T.L. Wade, J.M. Brooks, E.L. Atlas, R.R. Fay, and D.L. Wilkinson. 1993. National status and trends Mussel Watch Program: chlordane-related compounds in Gulf of Mexico oysters, 1986-90. *Environ. Pollut.* 82: 23-32.

Servos, M., and D.C.G. Muir. 1989. Effect of suspended sediment concentration on the sediment to water partition coefficient for 1,3,6,8-tetrachlorodibenzo-p-dioxin. *Environ. Sci. Technol.* 23: 1302-1306.

Shimp, R., and F.P. Pfaender.1985. Influence of naturally occurring humic acids on biodeg-radation of monosubstituted phenols by aquatic bacteria. *Appl. Environ. Microbiol.* 49: 402-407.

Shiu, W.Y., W. Doucette, F.A.P.C. Gobas, A. Andren, and D. Mackay. 1988. Physical-chemical properties of chlorinated dibenzo-*p*-dioxins. *Environ. Sci. Technol.* 22: 651-658.

Sijm, D.T.H.M., W. Seinen, and A. Opperhuizen. 1992. Life-cycle biomagnification study in fish. *Environ. Sci. Technol.* 26: 2162-2174.

Simmons, K.E., R.D. Minard, and J.-M. Bollag. 1989. Oxidative co-oligomerization of guaiacol and 4-chloroaniline. *Environ. Sci. Technol.* 23: 115-121.

Skaar, D.R., B.T. Johnson, J.R. Jones, and J.N. Huckins. 1981. Fate of kepone and mirex in a model aquatic environment: sediment, fish, and diet. *Can. J. Fish. Aquat. Sci.* 38: 931-938.

Smith, A.E. 1985. Persistence and transformation of the herbicides [14]C Fenoxaprop-ethyl and [14]C Fenthiaprop-ethyl in two prairie soils under laboratory and field conditions. *J. Agric. Food Chem.* 33: 483-488.

Smith, J.H., D.C. Bomberger, and D.I. Haynes. 1980. Prediction of the volatilization rates of high-volatility chemicals from natural water bodies. *Environ. Sci.Technol.* 14: 1332-1337.

Smith, J.H., D.C. Bomberger, and D.I. Haynes. 1981. Volatilization rates of intermediate and low volatility chemicals from water. *Chemosphere* 10: 281-289.

Southworth, G.R., C.C. Keffer, and J.J. Beauchamp. 1980. Potential and realized bioconcentration. A comparison of observed and predicted bioconcentration of azaarenes in the fathead minnow (*Pimephales promelas*). *Environ. Sci. Technol.* 14: 1529-1531.

Southworth, G.R., C.C. Keffer, and J.J. Beauchamp. 1981. The accumulation and disposition of benz(a)acridine in the fathead minnow, *Pimephales promelas. Arch. Environ. Contam. Toxicol.* 10: 561-569.

Squillace, P.J., and E.M. Thurman. 1992. Herbicide transport in rivers: importance of hydrol-ogy and geochemistry in nonpoint-source contamination. *Environ. Sci. Technol.* 26: 538-545.

Stauffer, T.B., W.C. MacIntyre, and D.C. Wickman. 1989. Sorption of nonpolar organic chemicals on low-carbon-content aquifer materials. *Environ. Toxicol. Chem.* 8: 845-852.

Stephanou, E.G. 1992. α,ω-dicarboxylic acid salts and α,ω-dicarboxylic acids. Photooxidation products of unsaturated fatty acids, present in marine aerosols and marine sediments. *Naturwiss.* 79: 128-131.

Steinberg, S.M., J.J. Pignatello, and B.L. Sawhney. 1987. Persistence of 1,2-dibromoethane in soils: entrapment in intraparticle micropores. *Environ. Sci. Technol.* 21: 1201-1208.

Strachan, W.M.J. 1988. Toxic contaminants in rainfall in Canada: 1984. *Environ. Toxicol. Chem.* 7: 871-877.

Suflita, J.M., C.P. Gerba, R.A. Ham, A. C. Palmisano, W.L. Rathje, and J.A. Robinson. 1992. The world's largest landfill. A multidisciplinary investigation. *Environ. Sci. Technol.* 26: 1486-1495.

Swackhamer, D.L., and D.E. Armstrong. 1986. Estimation of the atmospheric and nonatmospheric contributions and losses of polychlorinated biphenyls for Lake Michigan on the basis of sediment records of remote lakes. *Environ. Sci. Technol.* 20: 879-883.

Swackhamer, D.L., and R.A. Hites. 1988. Occurrence and bioaccumulation of organochlorine compounds in fishes from Siskiwit Lake, Isle Royale, Lake Superior. *Environ. Sci. Technol.* 22: 543—548.

Swackhamer, D.L., and R.S. Skoglund. 1993. Bioaccumulation of PCBs by algae: kinetics versus equilibrium. *Environ. Toxicol. Chem.* 12: 831-838.

Swackhamer, D.L., B.D. McVeety, and R.A. Hites. 1988. Deposition and evaporation of polychlorobiphenyl congeners to and from Siskiwit Lake, Isle Royale, Lake Superior. *Environ. Sci. Technol.* 22: 664.-672.

Swain, W.R. 1991. Effects of organochlorine chemicals on the reproductive outcome of humans who consumed contamined Great Lakes fish: an epidemiologic consideration. *J. Toxicol. Environ. Health* 33: 587-639.

Swartz, R.C., D.W. Schults, T. H. Dewitt, G.R. Ditsworth, and J.O. Lamberson. 1990. Toxicity of fluoranthene in sediments to marine amphipods: a test of the eqiilibrium partitioning approach to sediment quality criteria. *Environ. Toxicol. Chem.* 9: 1071-1080.

Swartz, R.C., P.F. Kemp, D.W. Schults, G.R. Ditsworth, and R.J. Ozretich. 1989. Acute toxicity of sediment from Eagle Harbor, Washington, to the infaunal amphipod *Rhepoxynius abronius*. *Environ. Toxicol. Chem.* 8: 215-222.

Tadokoro, H., and Y. Tomita. 1987. The relationship between bioaccumulation and lipid content in fish. pp. 363-373. In *QSAR in Environmental Toxicology* — II. (Ed. K.L.E. Kaiser). Reidel Publishers, Dordrecht.

Tanabe, S., H. Tanaka, and R. Tatsukawa. 1984. Polychlorobiphenyls, ΣDDT,hexachlorocyclohexane isomers in the western North Pacific ecosystem. *Arch. Environ.Contam. Toxicol.* 731-738.

ten Hulscher, Th. E.M., L.E. van der Velde, and W.A. Bruggeman. 1992. Temperature dependence of Henry's Law constants for selected chlorobenzenes, polychlorinated biphenyls and polycyclic aromatic hydrocarbons. *Environ. Toxicol. Chem.* 11: 1595-1603.

Thomann, R.V. 1989. Bioaccumulation model of organic chemical distributiion in aquatic food chains. *Environ. Sci. Technol.* 23: 699-707.

Thomann, R.V., and J.P. Connolly. 1984. Model of PCB in the Lake Michan lake trout food chain. *Environ. Sci. Technol.* 18: 65-71.

Thomas, D.J., B. Tracey, H. Marshall, and R.J. Norstrom. 1992. Arctic terrestrial ecosystem contamination. *Sci. Total Environ.* 122: 135-164.

Thorn, K.A., J.B. Arterburn, and M.A. Mikita. 1992. ^{15}N and ^{13}C NMR investigation of hydroxylamine-derivatized humic substances. *Environ. Sci. Technol.* 26: 107-116.

Tolman, R.C. 1938. *The principles of statistical mechanics* p. 602. Oxford University Press, Oxford.

Trenck, T.v.d., and H. Sandermann. 1981. Incorporation of benzo[a]pyrene quinones into lignin. *FEBS Lett.* 125: 72-76.

van der Oost, R., H. Heida, and A. Opperhuizen. 1988. Polychlorinated biphenyl congeners in sediments, plankton, molluscs, crustaceans, and eel in a freshwater lake: implications of using reference chemicals and indicator organisms in bioaccumulation studies. *Arch. Environ. Contam. Toxicol.* 17: 721-729.

Varanasi, U., W.L. Reichert, J.E. Stein, D.W. Brown, and H.R. Sanborn. 1985. Bioavailability and biotransformation of aromatic hydrocarbons in benthic organisms exposed to sediment from an urban estuary. *Environ. Sci. Technol.* 19: 836-841.

Vassilaros, D.L., P.W. Stoker, G.M. Booth, and M.L. Lee. 1982. Capillary gas chromatographic determination of polycyclic aromatic compounds in vertebrate fish tissue. *Anal. Chem.* 54: 106-112.

Veith, G.D., D.L. DeFoe, and B.V. Bergstedt. 1979a. Measuring and estimating the bioconcentration factor of chemicals in fish. *J. Fish. Res. Board Can.* 36: 1040-1048.

Veith, G.D., N.M. Austin, and R.T. Morris. 1979b. A rapid method for estimating log P for organic chemicals. *Water Res.* 13: 43-47.

Wachtmeister, C.A., L. Förlin, K.C. Arnoldsson, and J. Larsson. 1991. Fish bile as a tool for monitoring aquatic pollutants: studies with radioactively labeled 4,5,6-trichloroguaiacol. *Chemosphere* 22: 39-46.

Walker, C.H. 1983. Pesticides and birds-mechanisms of selective toxicity. *Agric. Ecosystems Environ.* 9: 211-226.

Wania, F., and D. Mackay. 1993. Global fractionation and cold condensation of low volatility organochlorine compounds in polar regions. *Ambio* 22: 10-18.

Welch, H.E., D.C.G. Muir, B.N. Billeck, W.L. Lockhart, G.J. Brunskill, H.J. Kling, M.P. Olson, and R.M. Lemoine. 1991. Brown snow: a long-range transport event in the Canadian Arctic. *Environ. Sci. Technol.* 25: 280-286.

Welch, H.E., M.A. Bergmann, T.D. Siferd, K.A. Martin, M.F. Curtis, R.E. Crawford, R.J. Conover, and H. Hop. 1992. Energy flow through a marine ecosystem of the Lancaster Sound Region, Arctic Canada. *Arctic* 45: 343-357.

Welling, W., and J.W. de Vries. 1992. Bioconcentration kinetics of the organophosphate insecticide chlorpyrifos in guppies (*Poecilia reticulata*). *Ecotoxicol. Environ. Saf.* 23: 64-75.

Westcott, N.D., and B.L. Worobey. 1985. Novel solvent extraction of lindane from soil. *J. Agric. Food Chem.* 33: 58-60.

Whittle, D.M., D.B. Sargent, S.Y. Huestis, and W.H. Hyatt. 1992. Foodchain accumulation of PCDD and PCDF isomers in the Great Lake aquatic community. *Chemosphere* 25: 181-184.

Wiederholm, T., A.-M. Wiederholm, and G. Milbrink. 1987. Bulk sediment bioassays with five species of fresh-water oligochaetes. *Water Air Soil Pollut.* 36: 131-154.

Winkelmann, D.A., and S.J. Klaine.1991. Degradation and bound residue formation of four atrazine metabolites, deethylatrazine, deisopropylatrazine, dealkylatrazine and hydroxyatrazine, in a western Tennessee soil. *Environ. Toxicol. Chem.*10: 347-354.

Wittlinger, R., and K. Ballschmiter. 1990. Studies of the global baseline pollution XIII. C6-C14 organohalogens (a and g [sic]-HCH, HCB, PCB, 4,4'-DDT, 4,4-DDE, *cis*- and *trans*-chlordane, *trans*-nonachlor, anisols) in the lower troposphere of the southern Indian Ocean. *Fresenius J. Anal. Chem.* 336: 193-200.

Woodburn, K.B., W.J. Doucette, and A.W. Andren. 1984. Generator column determination of octanol/water partition coefficients for selected polychlorinated biphenyl congeners. *Environ. Sci. Technol.* 18: 457-459.

Yin, C., and J.P. Hassett. 1986. Gas-partitioning approach for laboratory and field studies of mirex fugacity in water. *Environ. Sci. Technol.* 20: 1213-1217.

Zacharewski, T., L. Safe, S. Safe, B. Chittim, D. DeVault, K. Wiberg, P.-A. Bergqvist, and C. Rappe. 1989. Comparative analysis of polychlorinated dibenzo-*p*-dioxin and dibenzofuran congeners in Great Lakes fish extracts by gas chromatography-mass spectrometry and in vitro enzyme induction activities. *Environ. Sci. Technol.* 23: 730-735.

Zitzelsberger, W., W. Ziegler, and P.R. Wallnöfer. 1987. Stereochemistry of the degradation of veratrylglycerol-β-2,4-dichlorophenyl-ether, a model compound for lignin-bound xenobiotic residues by *Phenerochaete chrysosporium, Corynebacterium equi* and photo-sensitized rioboflavin. *Chemosphere* 16: 1137-1142.

CHAPTER 4

Persistence: Introduction

SYNOPSIS

An overview is presented of the factors that determine the persistence of xenobiotics including the role of both abiotic and biotic reactions. Examples are given of photochemical and of chemical transformations including hydrolysis, dehalogenation, oxidation, and reduction and of the significance of these in determining the analytes that may have to be included in monitoring programs. It is pointed out that combinations of abiotic and biotic processes may be of determinative significance. Biotic reactions are discussed in detail, and the important distinction between biodegradation and biotransformation is emphasized. Attention is directed to the metabolic potential of groups of bacteria that have been less extensively examined; these include enteric organisms, ammonia-oxidizers, algae, and anaerobic phototrophic bacteria. The significance of electron acceptors other than oxygen is noted, and examples are illustrated with organisms using nitrate and related compounds and those growing anaerobically by reduction of $Fe(III)$, $Mn(IV)$, or $U(VI)$. Some of the important reactions mediated by yeasts, fungi, and algae are outlined. Attention is directed to metabolic interactions where several organisms and a single substrate are present or where several substrates and a single organism occur. Examples are given of metabolic limitations imposed by enzyme regulatory mechanisms and of metabolic situations where a single readily degraded substrate is present in addition to a more recalcitrant xenobiotic. Factors which may critically determine the biodegradability of xenobiotics in natural systems are summarized; these include the oxygen concentration, the substrate concentration, the synthesis of natural emulsifying agents, the nature of transport mechanisms, and the important question of the bioavailability of the xenobiotic. A number of incompletely resolved issues are discussed including biodegradation in pristine environments, estimation of the rates of metabolic reactions both in laboratory and natural systems, and the significance of toxic metabolites. Brief comments are added on the role of catabolic plasmids.

INTRODUCTION

Procedures for the analysis of environmental samples have been outlined in Chapter 2, and the processes that determine the dissemination of xenobiotics after discharge from point sources have been discussed in Chapter 3. In the next three chapters, the factors that determine the ultimate fate of xenobiotics will be discussed. This chapter attempts to present an overview of the factors that determine the persistence of xenobiotics, while Chapter 5 will be devoted to experimental procedures for carrying out the relevant investigations and Chapter 6 to a detailed examination of the pathways taken for the degradation of a wide range of structurally diverse xenobiotics. Attention will be focused on microorganisms, and, in particular, on bacteria which are the most important degradative organisms in virtually all natural aquatic ecosystems. A selection of references to books and review articles dealing with the topics covered in all three chapters is given separately at the end of this chapter; some of the classic reviews of Dagley are included since, although these are now dated in respect of detail, they provide a valuable perspective on problems of biodegradation and incorporate many valuable ideas together with references to the older literature. In addition, unlike many current reviews, they are not primarily directed to organochlorine compounds.

It was the persistence of DDT which raised the greatest alarm over its extensive use during the years 1940 to 1968. Although levels since its banning have decreased dramatically, those of its metabolite DDE may still be appreciable and serve to sustain the initial concern. Many organic compounds have become environmentally suspect, but it is especially the highly chlorinated ones such as the PCBs, PCCs, and mirex which have acquired the reputation of being unacceptable on account of their apparent persistence. As a result of these fears, there has emerged a general concern with all synthetic chlorinated organic compounds (Hileman 1993) which may possibly have deflected interest from other groups which merit comparable attention. It should, of course, be appreciated that, on the other hand, a number of compounds and products such as modern plastics have been valued for their stability under a variety of conditions — and are produced with this end in view.

For these reasons, studies on biodegradation began to occupy a central position in discussions on the environmental impact of organic chemicals, and the complexities have been clearly presented (Landner 1989). It should be appreciated at the outset that the terms "persistent" and "recalcitrant" are relative rather than absolute, since probably most chemical structures can be degraded or transformed by microorganisms. The crucial issue is the *rate* at which the reactions occur, and the area between slowly degradable compounds and truly persistent ones is often unresolved. For example, in spite of the fact that degradation of some PCB congeners has been demonstrated under aerobic conditions, and biotransformation (dechlorination) under anaerobic conditions, these compounds are still recoverable from many environmental samples; they must, therefore, be regarded as persistent. The critical questions are both what reactions take place and the rate at which they occur in the environment into which the compound is discharged. Both of these should be addressed in investigations aimed at incorporating environmental relevance.

Two essentially different processes determine the persistence of an organic compound in the aquatic environment. The first are abiotic reactions, and for some groups of compounds, these reactions may be dominant in determining their fate. The second are biotic reactions mediated by a wide range of organisms. Greatest attention will be devoted to microorganisms in the next three chapters, although a brief discussion of the metabolism of xenobiotics by higher organisms is given in Chapter 7 (Section 7.5).

4.1 ABIOTIC REACTIONS

Virtually any of the plethora of reactions known in organic chemistry may be exploited for the abiotic degradation of xenobiotics. Hydrolytic reactions may convert compounds such as esters, amides, or nitriles into the corresponding carboxylic acids, or ureas and carbamides into the amines. These abiotic reactions may, therefore, be the first step in the degradation of such compounds; the transformation products may, however, be resistant to further chemical transformation so that their ultimate fate is dependent upon subsequent microbial reactions. For example, for urea herbicides and related compounds the limiting factor is the rate of microbial degradation of the chlorinated anilines which are the initial hydrolysis products. The role of abiotic reactions should, therefore, always be taken into consideration and should be carefully evaluated in all laboratory experiments on biodegradation and biotransformation (Chapter 5, Section 5.3). It should be appreciated that the results of experiments directed to microbial degradation are probably discarded if they show substantial interference from abiotic reactions. A good illustration of the complementary roles of abiotic and biotic processes is offered by the degradation of tributyl tin compounds. Earlier experiments (Seligman et al. 1986) had demonstrated the degradation of tributyltin to dibutyltin primarily by microbial processes. It was subsequently shown, however, that an important abiotic reaction mediated by fine-grained sediments resulted in the formation also of monobutyltin and inorganic tin (Stang et al. 1992). It was therefore concluded that both processes were important in determining the fate of tributyl tin in the marine environment.

A study of the carbamate biocides, carbaryl and propham, illustrates the care that should be exercised in determining the relative importance of chemical hydrolysis, photolysis, and bacterial degradation (Figure 4.1) (Wolfe et al. 1978a). For carbaryl, the half-life for hydrolysis increased from 0.15 d at pH 9 to 1500 d at pH 5, while that for photolysis was 6.6 d; biodegradation was too slow to be significant. On the other hand, the half-lifes of propham for hydrolysis and photolysis were $>10^4$ d and 121 d — so greatly exceeding the half-life of 2.9 d for biodegradation that abiotic processes would be considered to be of subordinate significance. Close attention to structural features of xenobiotics is, therefore, clearly imperative before making generalizations on the relative signficance of alternative degradative pathways. Further comments on the cooperation of biotic and photochemical reactions are made in Section 4.1.1.

4.1.1 Photochemical Reactions

Photochemical reactions may be important especially in areas of high solar irradiation, or on the surface of soils, or in aquatic systems containing ultraviolet absorbing humic and fulvic acids (Zepp et al. 1981a,b). It has also been shown (Zepp and Schlotzhauer 1983) that, although the presence of algae may *enhance* photometabolism, this is subservient to direct photolysis at the cell densities likely to be encountered in rivers and lakes. It should be noted that different products may be produced in natural river water and in buffered medium; for example, photolysis of triclopyr (3,5,6-trichloro-2-pyridyloxyacetic acid) in sterile medium at pH 7 resulted in hydrolytic replacement of one chlorine atom, whereas in river water the

Figure 4.1 Carbaryl (**a**) and propham (**b**).

ring was degraded to form oxamic acid as the principal product (Woodburn et al. 1993). Particular attention has understandably, therefore, been directed to the photolytic degradation of biocides — including agrochemicals — that are applied to terrestrial systems. The suite of reactions involved may be quite complex and involve several steps including hydrolysis, hydroxylation, dechlorination, or the cleavage of aromatic rings. A few examples may be used to illustrate the most widely encountered types of reactions. It should be emphasized that photochemical reactions may often produce molecules structurally more complex and less susceptible to degradation than their precursors, even though the deep-seated rearrangements induced in complex compounds such as terpenes by UV irradiation are not likely to be encountered in environmental situations.

1. Atrazine is successively transformed to 2,4,6-trihydroxy-1,3,5-triazine (Pelizzetti et al. 1990) by dealkylation of the alkylamine side chains and hydrolytic displacement of the ring chlorine and amino groups (Figure 4.2).
2. Pentachlorophenol produces a wide variety of transformation products including chloranilic acid (2,5-dichloro-3,6-dihydroxybenzo-1,4-quinone) by hydrolysis and oxidation, a dichlorocyclopentanedione by ring contraction, and dichloromaleic acid by cleavage of the aromatic ring (Figure 4.3) (Wong and Crosby 1981).
3. The main products of photolysis of 3-trifluoromethyl-4-nitrophenol are 2,5-dihydroxybenzoate produced by hydrolysis of the nitro group and oxidation of the trifluoromethyl group, together with a compound identified as a condensation product of the original compound and the dihydroxybenzoate (Figure 4.4) (Carey and Cox 1981).
4. The potential insecticide that is a derivative of tetrahydro-1,3-thiazine undergoes a number of reactions resulting in some 43 products of which the dimeric azo compound is the principal one in aqueous solutions (Figure 4.5) (Kleier et al. 1985).
5. The herbicide trifluralin undergoes a photochemical reaction in which the *n*-propyl side-chain of the amine reacts with the vicinal nitro group to form the benzpyrazine (Figure 4.6) (Soderquist et al. 1975).
6. Heptachlor and *cis*-chlordane both of which are chiral, form caged or half-caged structures (Figure 4.7) on irradiation, and these products have been identified in biota from the Baltic, from the Arctic, and from the Antarctic (Buser and Müller 1993).
7. In the presence of both light and hydrogen peroxide, 2,4-dinitrotoluene is oxidized to the corresponding carboxylic acid; this is then decarboxylated to 1,3-dinitrobenzene which is hydroxylated to various nitrophenols and nitrocatechols before cleavage of the aromatic rings (Figure 4.8) (Ho 1986).

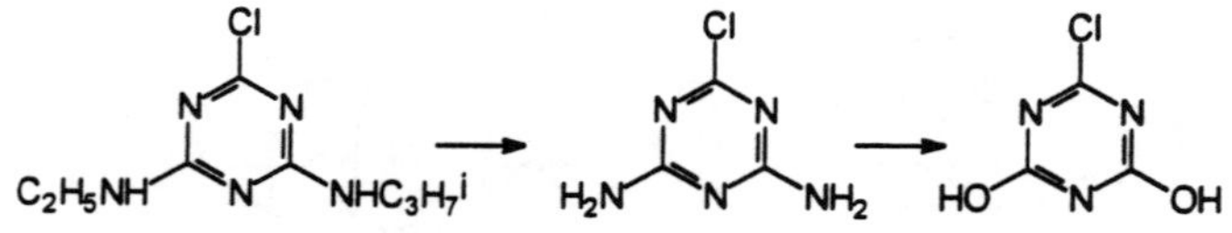

Figure 4.2 Photochemical transformation of atrazine.

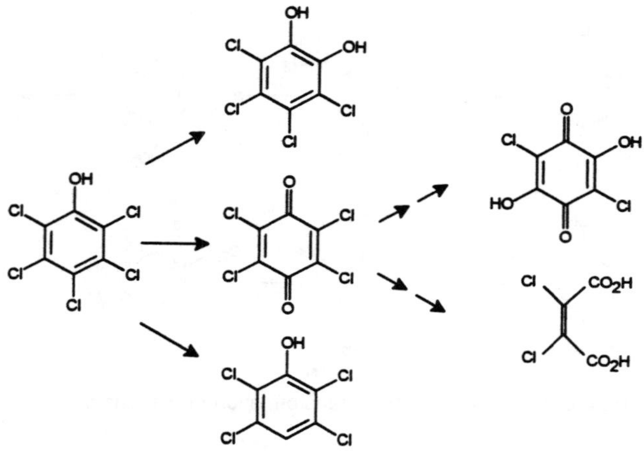

Figure 4.3 Photochemical transformation of pentachlorophenol.

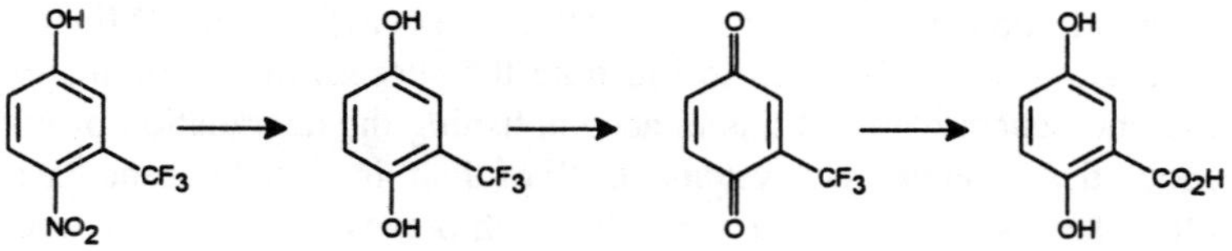

Figure 4.4 Photochemical transformation of 3-trifluoromethyl-4-nitrophenol.

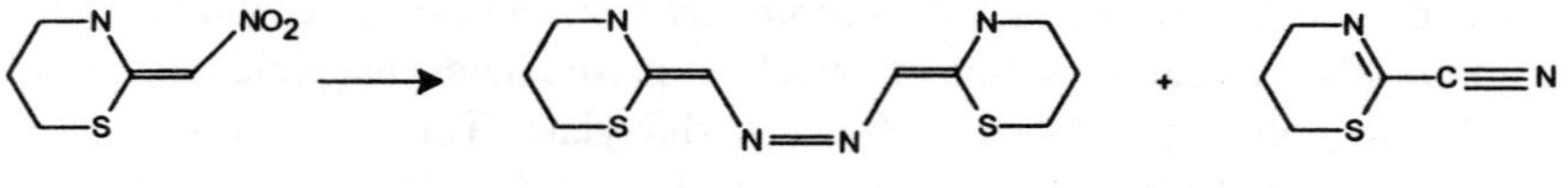

Figure 4.5 Photochemical transformation of a tetrahydro1,3-thiazine.

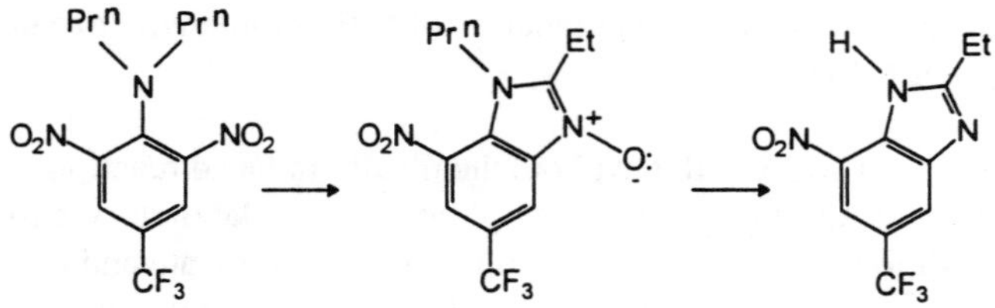

Figure 4.6 Photochemical transformation of trifluralin.

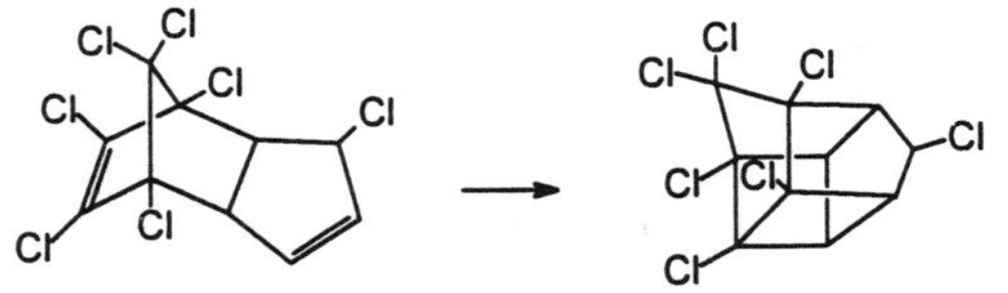

Figure 4.7   Photochemical transformation of chlordane.

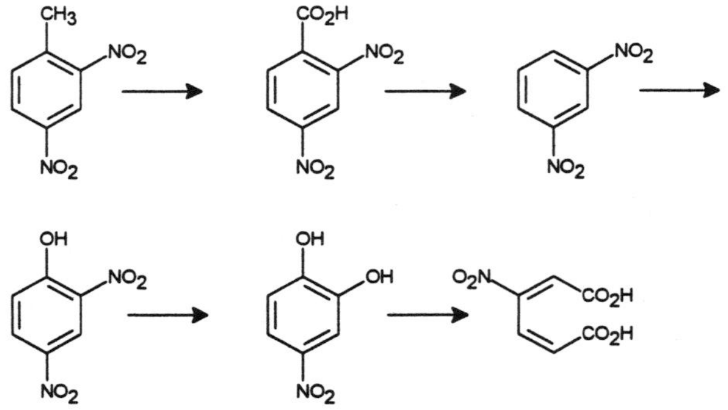

Figure 4.8   Photochemical transformation of 2,4-dinitrotoluene.

It has been shown that a combination of photolytic and biotic reactions may result in enhanced degradation of xenobiotics in municipal wastewater treatment systems, for example, of chlorophenols (Miller et al. 1988a) and benz[a]pyrene (Miller et al. 1988b). Two examples may be used to illustrate the success of a combination of microbial and photochemical reactions in accomplishing the degradation of widely different xenobiotics in natural ecosystems. Both of them involved marine bacteria, and it, therefore, seems plausible to assume that such processes might be especially important in warm-water marine environments.

1. The degradation of pyridine dicarboxylates (Amador and Taylor 1990).
2. The degradation of 3-and 4-trifluoromethylbenzoate: the microbial transformation resulted in the formation of catechol intermediates that were converted into 7,7,7-trifluoro-hepta-2,4-diene-6-one carboxylate. This was subsequently degraded photochemically with the loss of fluoride (Taylor et al. 1993). This degradation may be compared to the purely photochemical degradation of 3-trifluoromethyl-4-nitrophenol that has already been noted and contrasted with the resistance to microbial degradation of trifluoromethylbenzoates that is noted in Chapter 6 (Section 6.10).

It has also been suggested that photochemically induced reactions may take place between biocides and biomolecules of plant cuticles; laboratory experiments have examined addition reactions between DDT and methyl oleate and have been used to illustrate reactions which result in the production of "bound" DDT residues (Figure 4.9) (Schwack 1988).

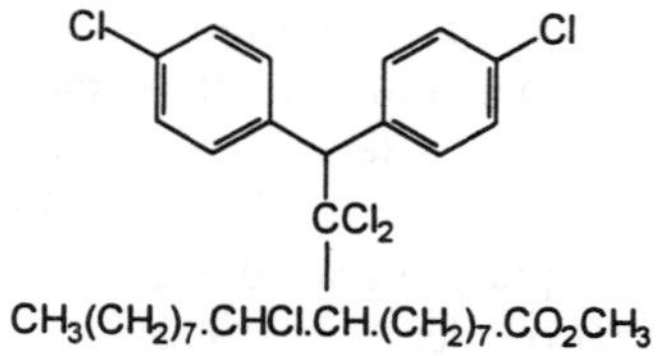

Figure 4.9   Product of reaction between DDT and methyl oleate.

These examples illustrate transformation reactions of xenobiotics, but it should be noted that naturally occurring compounds may also undergo comparable reactions. For example, the occurrence of C_8 and C_9 dicarboxylic acids in samples of atmospheric particles and in recent sediments (Stephanou 1992; Stephanou and Stratigakis 1993) has been attributed to photochemical degradation of unsaturated carboxylic acids that are widespread in almost all biota. Atmospheric reactions between unsaturated aliphatic and alicyclic hydrocarbons and hydroxyl and nitrate radicals have achieved increasing prominence. The formation of peroxyacetyl nitrate from isoprene (Grosjean et al. 1993a) and of peroxypropionyl nitrate (Grosjean et al. 1993b) from *cis*-3-hexen-1-ol that is derived from higher plants may be given as illustration of important contributions to atmospheric degradation.

4.1.2 Chemically Mediated Reactions

An attempt is made to present some examples of chemical degradation or transformation on the basis of a classification of the reactions that take place.

Hydrolytic reactions — Organic compounds containing carbonyl groups flanked by alkoxy groups (esters) or by amino- or substituted amino groups (amides, carbamates, and ureas) may be hydrolyzed by purely abiotic reactions under appropriate conditions of pH; the generally high pH of seawater (ca. 8.2) may be noted so that chemical hydrolysis may be quite important in this environment. On the other hand, although very few natural aquatic ecosystems have pH values sufficiently low for acidic hydrolysis to be of major importance, this may be important in terrestrial systems. It is, therefore, important to distinguish between alkaline or neutral and acidic hydrolytic mechanisms. It should also be appreciated that both hydrolytic and photolytic mechanisms may operate simultaneously and that the products may not necessarily be identical.

Substantial numbers of important agrochemicals contain the carbonyl groups noted above, so that abiotic hydrolysis may be the primary reaction in their transformation; the example of carbaryl has already been cited (Wolfe et al. 1978a). The same general principles may be extended to phosphate and thiophosphate esters, although in these cases, it is important to bear in mind the stability to hydrolysis of primary and secondary phosphate esters under neutral or alkaline conditions that prevail in most natural ecosystems. On the other hand, sulfate esters and sulfamides are generally quite resistant to chemical hydrolysis except under rather drastic conditions so that their hydrolysis is generally mediated by sulfatases and sulfamidases. A few examples will be given to illustrate the diversity of hydrolytic reactions; these involve structurally diverse agrochemicals that may enter aquatic systems by leaching.

1. The cyclic sulfite of α- and β-endosulfan (Singh et al. 1991).
2. The carbamate phenmedipham that results in the intermediate formation of *m*-tolyl *iso*cyanate (Figure 4.10) (Bergon et al. 1985).
3. 2-(thiocyanomethylthio)benzthiazole with initial formation of 2-thiobenzthiazole. This metabolite is then rapidly degraded photochemically to benzthiazole and 2-hydroxybenzthiazole (Brownlee et al. 1992).
4. Aldicarb undergoes simple hydrolysis at pH values above 7, whereas at pH values below 5 an elimination reaction intervenes (Figure 4.11) (Bank and Tyrrell 1984).
5. The sulfonyl urea sulfometuron methyl is stable at neutral or alkaline pH values but is hydrolyzed at pH 5 to methyl 2-aminosulfonylbenzoate that is cyclized to saccharin (Figure 4.12) (Harvey et al. 1985). The original compound is completely degraded to CO_2 by photolysis.
6. The pyrethroid insecticides fenvalerate and cypermethrin are hydrolyzed under alkaline conditions at low substrate concentrations, but at higher concentrations, the initially formed 3-phenoxybenzaldehyde reacts further with the substrate to form dimeric compounds (Figure 4.13) (Camilleri 1984).

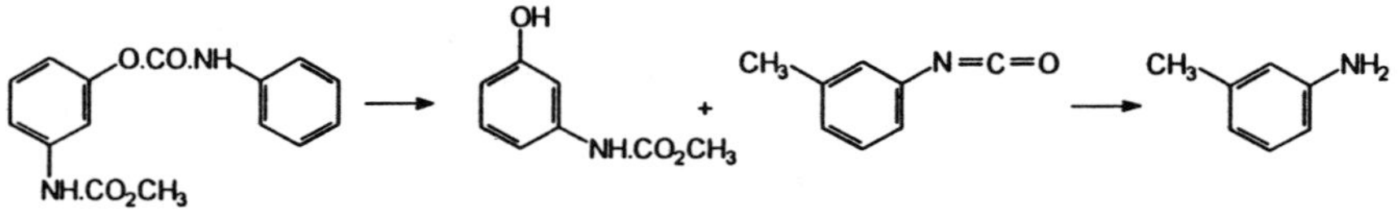

Figure 4.10 Hydrolysis of phenmedipham.

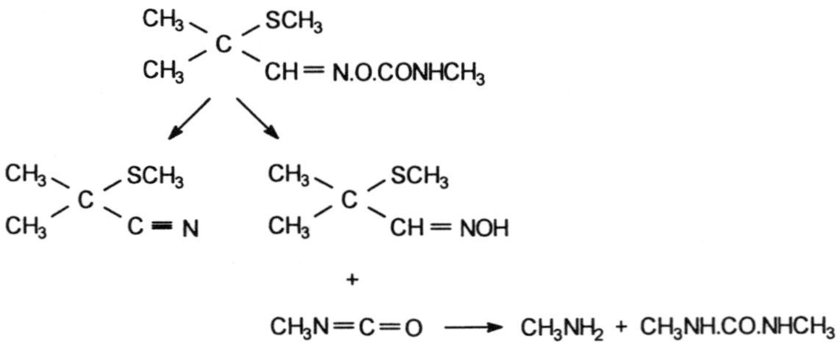

Figure 4.11 Hydrolysis of aldicarb.

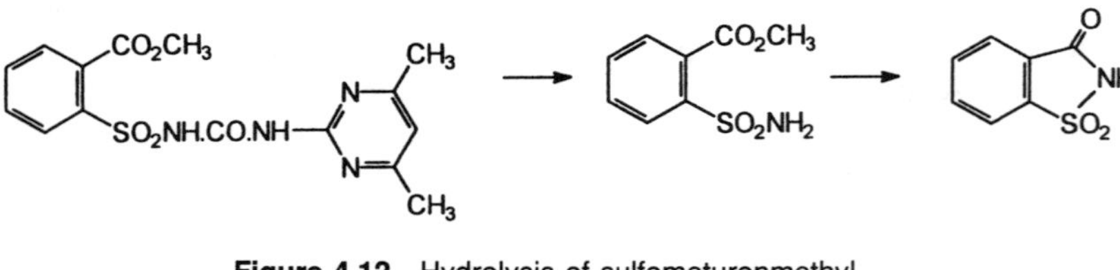

Figure 4.12 Hydrolysis of sulfometuronmethyl.

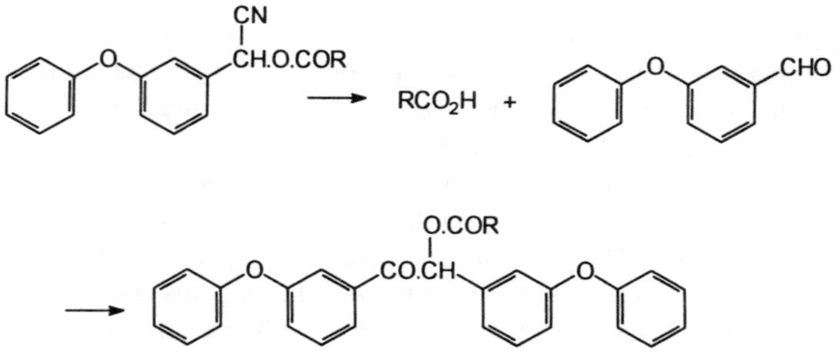

Figure 4.13 Hydrolysis of the pyrethroid insecticides fenvalerate and cypermethrin.

Three important comments are worth noting:

1. It should be emphasized that abiotic hydrolysis generally accomplishes only a single step in the ultimate degradation of the compounds that have been used for illustration. The intervention of subsequent biotic reactions is, therefore, almost invariably necessary for their complete mineralization. These reactions are discussed more fully in Chapter 6.

2. The operation of these hydrolytic reactions is independent of the oxygen concentration of the system so that — in contrast to biotic degradation and transformation — these reactions may occur effectively under both aerobic and anaerobic conditions.

3. Rates of hydrolysis may be influenced by the presence of dissolved organic carbon or sediment, and the effect is determined by the structure of the compound and by the kinetics of its association with these components. For example, whereas the neutral hydrolysis of chlorpyrifos was unaffected by sorption to sediments, the rate of alkaline hydrolysis was considerably slower (Macalady and Wolfe 1985); humic acid also reduced the rate of alkaline hydrolysis of 1-octyl 2,4-dichlorophenoxyacetate (Perdue and Wolfe 1982). Conversely, sediment sorption had no effect on the neutral hydrolysis of 4-chlorostilbene oxide, although the rate below pH 5 where acid hydrolysis dominates was reduced (Metwally and Wolfe 1990).

Dehalogenation reactions — The mechanism of chemical dechlorination of a range of organochlorine compounds has received increasing prominence. Attention has been directed to the role of corrins and porphyrins in the absence of biological systems, and a number of structurally diverse compounds have been shown to be dechlorinated including DDT (Zoro et al. 1974), lindane (Marks et al. 1989), mirex (Holmstead 1976), C_1 chloroalkanes (Krone et al. 1989), C_2 chloroalkenes (Gantzer and Wackett 1991), and C_2 chloroalkanes (Schanke and Wackett 1992). Detailed mechanistic examination of the dehydrochlorination of pentachloroethane to tetrachloroethene reveals, however, the potential complexity of this reaction and the possibly significant role of pentachloroethane in the abiotic transformation of hexachloroethane (Roberts and Gschwend 1991). The specificity of corrins and porphyrins is of particular interest, since it seems to be significantly less than that of the enzymes generally implicated in microbial dechlorination. At the same time, however, it should be appreciated that both porphyrins and corrins are constituents of many bacteria — the porphyrins as prosthetic groups of cytochromes and the corrins as the chromophore in vitamin B_{12} coenzyme and related compounds. The interesting — though possibly philosophical rather than scientific — question then arises whether reactions carried out by cells containing these pyrrolic compounds are biochemically or chemically mediated. The study of these reactions may, however, help to elucidate the mechanism of microbial dechlorination reactions, and this is illustrated further in Chapter 6 (Sections 6.4.4 and 6.6). Interest in the adverse environmental effects of chlorofluoroalkanes has stimulated interest in their anaerobic degradation which is probably mediated by abiotic reactions possibly involving porphyrins (Lovely and Woodward 1992; Lesage et al. 1992), although it should be noted that the C-F bond is apparently retained in the products (Lesage et al. 1992).

The foregoing reactions involve reductive dechlorination or elimination, but nucleophilic displacement of chloride may also be important in some circumstances. This has been examined with dihalomethanes using HS^- at concentrations that might be encountered in environments where active anaerobic sulfate-reduction is taking place. The rates of reaction with HS^- exceeded those for hydrolysis, and at pH values above 7 in systems in equilibrium with elementary sulfur, the rates with polysulfide exceeded those with HS^-. The principal product from dihalomethanes is the polythiomethylene $HS-(CH_2.S)_nH$ (Roberts et al. 1992).

Oxidation reactions — Considerable attention has been directed to waste destruction by Fenton's reagent — hydrogen peroxide in the presence of Fe^{2+} or Fe^{3+} — both in the presence of oxygen and under the influence of irradiation. The reaction involves hydroxyl radicals and has been studied particularly intensively for the destruction of chlorinated phenoxyacetic acid herbicides (Sun and Pignatello 1993); systematic investigations have been carried out on the effect of pH, the molar ratio of H_2O_2/substrate, and the possible complications resulting from the formation of iron complexes. Although this reaction may have limited environmental relevance except under rather special circumstances, attention is drawn to it since, under conditions where the concentration of oxidant is limiting, intermediates may be formed that are stable and that may possibly exert adverse environmental effects. Two examples may be used to illustrate the formation of intermediates, although it should be emphasized that total destruction of the relevant xenobiotics under optimal conditions can be successfully accomplished. The structure of the products that are produced by the action of Fenton's reagent on chlorobenzene are shown in Figure 4.14a (Sedlak and Andren 1991) and those from 2,4-dichlorophenoxyacetate in Figure 4.14b (Sun and Pignatello 1993).

Reduction reactions — Reduction of monocyclic aromatic nitro compounds with reduced sulfur compounds mediated by a naphthoquinone or an iron porphyrin has been demonstrated (Schwarzenbach et al. 1990), and it has been suggested that this reaction may be significant in determining the fate of these compounds in the environment. A comparable reaction has also been observed in the filtrate from a strain of *Streptomyces* sp. that is known to synthesize cinnaquinone (2-amino-3-carboxy-5-hydroxybenzo-1,4-quinone) and the 6,6'diquinone (dicinnaquinone) as secondary metabolites (Glaus et al. 1992). The reduction of nitrobenzenes in the

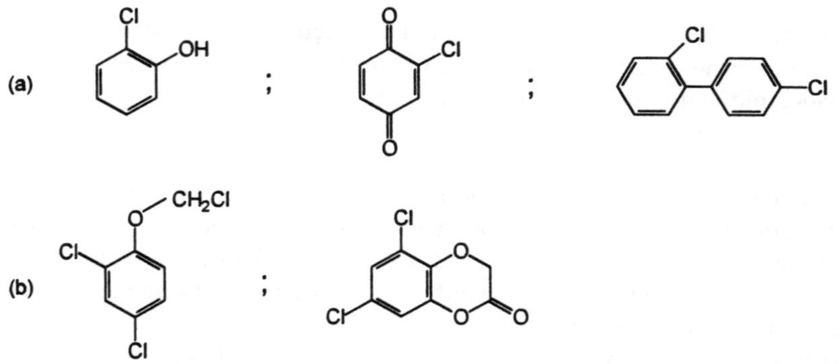

Figure 4.14 Transformation products from **(a)** chlorobenzene and **(b)** 2,4-dichlorophen-oxyacetate.

presence of sulfide and natural organic matter from a variety of sources has also been demonstrated and could be expected to be a significant abiotic process in some natural systems (Dunnivant et al. 1992).

The biodegradation of N-heterocylic aromatic compounds frequently involves a reductive step (Chapter 6), but purely chemical reduction may take place under highly anaerobic conditions and has, for example, been encountered with the substituted 1,2,4-triazolo[1,5a]pyrimidine Flumetsulam (Wolt et al. 1992).

Reaction of amines with nitrite and related compounds — Two examples may be given to illustrate the intrusion of chemical reactions between xenobiotics and nitrite or related compounds that are produced microbiologically from nitrate; this is the major — and possibly, generally, the only — metabolic function of the bacteria.

1. A strain of *Escherichia coli* produces a naphthotriazole from 2,3-diaminonaphthalene and nitrite that is formed from nitrate by the action of nitrate reductase. The initial product is NO which is converted into the active nitrosylating agent by reactions with oxygen; this then reacts chemically with the amine (Ji and Hollocher 1988). A comparable reaction may plausibly account for the formation of dimethylnitrosamine by *Pseudomonas stutzeri* during growth with dimethylamine in the presence of nitrite (Mills and Alexander 1976). On the other hand, for other organisms including *E. coli*, enzymatic reactions may be involved though their nature has not been clearly delineated.
2. The formation of 3,3′,4,4′-tetrachloroazobenzene, 1,3-*bis*(3,4-dichlorophenyl)-triazine and 3,3′,4,4′-tetrachlorobiphenyl from 3,4-dichloroaniline and nitrate by *E. coli* presumably involves intermediate formation of the diazonium compound by reaction of the amine with nitrite (Corke et al. 1979).

In view of concern over the presence of nitrate in groundwater, the possible environmental significance of these or analogous reactions should not be overlooked.

Thermal reactions during combustion — The products of incomplete combustion may be associated with particulate matter before their discharge into the atmosphere, and these may ultimately enter the aquatic and terrestrial environments in the form of precipitation and dry deposition. The spectrum of compounds involved is quite extensive and a number of them are formed by reactions between hydrocarbons and inorganic sulfur or nitrogen constituents of air. Some illustrative examples include the following.

1. The pyrolysis of vinylidene chloride produces a range of chlorinated aromatic compounds including polychlorinated benzenes, styrenes, and naphthalenes (Yasahura and Morita 1988), and a series of chlorinated acids including chlorobenzoic acids has been identified in emissions from a municipal incinerator (Mowrer and Nordin 1987).
2. Nitroaromatic compounds have been identified in diesel engine emissions (Salmeen et al. 1984), and attention has been directed particularly to 1,8- and 1,6-dinitropyrene (Figure 4.15) since these compounds are mutagenic and possibly carcinogenic (Nakagawa et al. 1983).
3. A wide range of azaarenes including acridines and benzoacridines, 4-azafluorene and 10-azabenzo[a]pyrene (Figure 4.16) has been identified in particulate samples of urban air, and some of these have been recovered from contaminated sediments (Yamauchi and Handa 1987).
4. Ketonic and quinonoid derivatives of aromatic hydrocarbons have been identified in automobile (Alsberg et al. 1985) and diesel exhaust particulates (Levsen 1988), and have been recovered from samples of marine sediments (Fernandéz et al. 1992).
5. Halogenated phenols, particularly 2-bromo-, 2,4-dibromo-, and 2,4,6-tribromophenol, have been identified in automotive emissions and are the products of thermal reactions involving dibromoethane fuel additive (Müller and Buser 1986). It can, therefore, no longer be assumed that such compounds are exclusively the products of biosynthesis by marine algae.

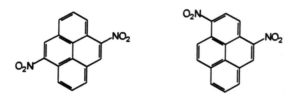

Figure 4.15   1,6- and 1,8-dinitropyrene.

Figure 4.16   Azaarenes identified in particulate samples of urban air.

Reactions in the atmosphere — Although chemical transformations in the atmosphere are peripheral to this discussion, these reactions should be kept in mind since their products may subsequently enter the aquatic and terrestrial environments. The persistence and the toxicity of these secondary products are, therefore, directly relevant to this discussion. Atmospheric transformations may be mediated both by photochemical and by free radical reactions involving reactive entities such as nitrate radicals and hydroxyl radicals. Most attention has hitherto been directed to the kinetics of degradation and possibly less to the identification of the products of the reactions. Under simulated atmospheric conditions, the kinetics of the reactions between hydroxyl and nitrate radicals and a range of anthropogenic compounds including aliphatic and aromatic hydrocarbons (Tuazon et al. 1986), terpenes (Atkinson et al. 1985a), amines (Atkinson et al. 1987), and heterocyclic compounds (Atkinson et al. 1985b) have been examined and the products identified in some cases. The formation of phosgene in the atmosphere from geminal dichloroethenes has been rationalized on the basis of reactions with hydroxyl radicals (Grosjean 1991), though the lifetime of such a reactive compound will be limited both in the atmosphere and after deposition on the aquatic and terrestrial environments. Another group of reactive compounds that has been identified in the atmosphere are dimethyl and monomethyl hydrogen sulfates (Eatough et al. 1986), and although the occurrence of these compounds is clearly correlated with the presence of SO_x, the mechanism of their formation has not been resolved. Attention is particularly directed to the possibility of further reactions mediated by both of these highly reactive — and toxic — compounds.

The products from the atmospheric oxidation of monoterpenes in the presence of NO have been rationalized in terms of reactions between the terpenes with hydroxyl radicals and ozone and between carbonyl transformation products with hydroxyl radicals and ozone (Grosjean et al. 1992). A wide range of aldehydes and ketones may be produced, and the possible occurrence of analogous transformation products in environmental samples should be considered. An interesting example is the formation of the mutagenic 2-nitro- and 6-nitro-6H-dibenzo[b,d]pyran-6-ones (Figure 4.17) by the oxidation of phenanthrene in the presence of NO_x and methyl nitrite as a source of hydroxyl radicals (Helmig et al. 1992a). These nitro compounds have been identified in samples of ambient air (Helmig et al. 1992b) and add further examples to the list of mononitroarenes that already include 2-nitropyrene and 2-nitrofluoranthene. It appears plausible to suggest that comparable reactions are involved in the formation of the dinitroarenes identified in diesel exhaust.

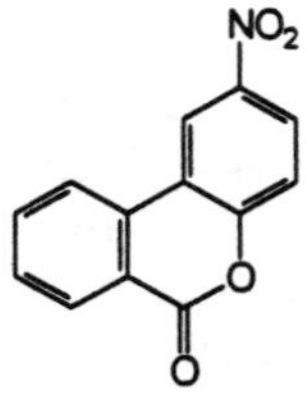

Figure 4.17   Product from the photochemical reaction of phenanthrene and NO_x.

4.2 BIOTIC REACTIONS

It is generally conceded that biotic reactions are of primary significance in determining the fate and persistence of organic compounds in most natural aquatic ecosystems. By way of providing continuity with the abiotic reactions that have been discussed above, one example may be given of the long chain of events which may bring about environmental effects through the subtle interaction of biotic and abiotic reactions: 2-dimethylthiopropionic acid is produced by algae and by the marsh grass *Spartina alternifolia* and may then be metabolized in sediment slurries under anoxic conditions to dimethyl sulfide (Kiene and Taylor 1988) and by aerobic bacteria to methyl sulfide (Figure 4.18) (Taylor and Gilchrist 1991). Dimethyl sulfide — and possibly also methyl sulfide — are oxidized in the troposphere to sulfuric and methanesulfonic acids, and it has been suggested that these compounds may play a critical role in promoting cloud formation (Charleson et al. 1987). It should be added that sulfide itself can be biologically methylated to methyl sulfide, and this is noted in Chapter 6 (Section 6.11.4). The long-term effect of the biosynthesis of methyl sulfides on climate alteration may be considerable — and yet at first glance, this seems far removed from the production of an osmolyte by higher plants, its metabolism in aquatic systems, or microbial methylation.

$$(CH_3)_2S.CH_2.CO_2H \longrightarrow CH_3.S.CH_2.CH_2.CO_2H \longrightarrow HS.CH_2.CH_2.CO_2H + CH_3SH$$

Figure 4.18   Aerobic metabolism of 2-dimethylthiopropionic acid.

It is essential at the start to make a clear distinction between biodegradation and biotransformation.

Biodegradation under aerobic conditions results in the mineralization of an organic compound to carbon dioxide and water and — if the compound contains nitrogen, sulphur, phosphorus, or chlorine — with the release of ammonium (or nitrite), sulfate, phosphate, or chloride; these inorganic products may then enter well-established geochemical cycles. Under anaerobic conditions, methane may be formed in addition to carbon dioxide, and sulfate may be reduced to sulfide.

During *biotransformation*, on the other hand, only a restricted number of metabolic reactions is accomplished, and the basic framework of the molecule remains essentially intact. Some illustrative examples of biotransformation reactions include the following:

1. The hydroxylation of dehydroabietic acid by fungi (Figure 4.19) (Kutney et al. 1982).
2. The epoxidation of alkenes by bacteria (Patel et al. 1982; van Ginkel et al. 1987); this is discussed again in Chapter 6 (Section 6.1.3).
3. The formation of 16-chlorohexadecyl-16-chlorohexadecanoate from hexadecyl chloride by *Micrococcus cerificans* (Figure 4.20) (Kolattukudy and Hankin 1968).
4. The *O*-methylation of chlorophenols to anisoles by fungi (Gee and Peel 1974; Cserjesi and Johnson 1972) and by bacteria (Suzuki 1978; Rott et al. 1979; Neilson et al. 1983; Häggblom et al. 1988).
5. The formation of glyceryl-2-nitrate from glyceryl trinitrate by *Phanerochaete chrysosporium* (Servent et al. 1991) (Figure 4.21).

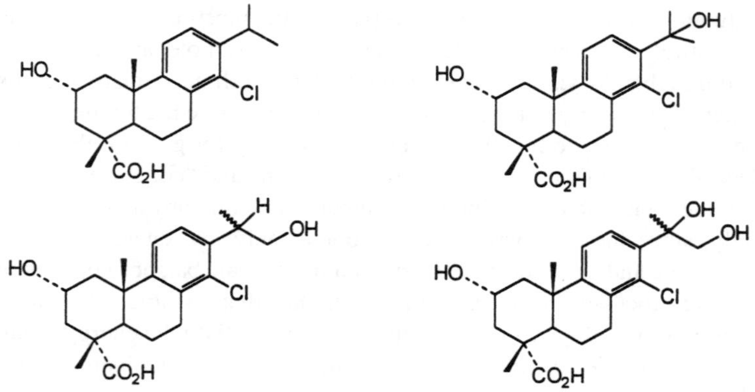

Figure 4.19 Metabolites produced from dehydroabietic acid by *Mortierella isabellina*.

$$CH_3.[CH_2]_{14}.CH_2Cl \longrightarrow ClCH_2[CH_2]_{14}.CH_2.O.CO.[CH_2]_{14}.CH_2Cl$$

Figure 4.20 Formation of an ester during the metabolism of a chloroalkane.

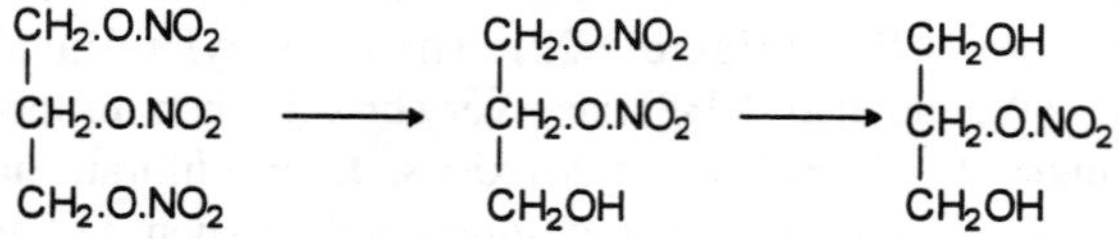

Figure 4.21 Metabolism of glyceryl trinitrate by *Phanaerochaete chrysosporium*.

The initial biotransformation products may, in some cases, be incorporated into cellular material. For example, the carboxylic acids formed by the oxidation of long-chain *n*-alkyl chlorides were incorporated into cellular fatty acids by strains of *Mycobacterium* sp.(Murphy and Perry 1983), and metabolites of metolachlor that could only be extracted from the cells with acetone were apparently chemically bound to unidentified sulfur-containing cellular components (Liu et al. 1989).

More extensive details of a wider range of microbial transformation reactions will be found throughout Chapter 6.

Biodegradation and biotransformations are, of course, alternatives, but they are not mutually exclusive. For example, it has been suggested that for chlorophenolic compounds the *O*-methylation reaction may be an important alternative to reactions that bring about their degradation (Allard et al. 1987). Apart from the environmental significance of biotransformation reactions, many of them have enormous importance in biotechnology: for example, in the synthesis of sterol derivatives and in reactions that take advantage of the oxidative potential of methylotrophic and methanotrophic bacteria (Lidstrom and Stirling 1990) and of rhodococci (Finnerty 1992). A few of these and related reactions are discussed further in Chapter 6 (Section 6.11).

It is important also to consider the degradation of xenobiotics in the wider context of metabolic reactions carried out by the cell. The cell must obtain energy to carry out essential biosynthetic (anabolic) reactions for its continued existence and to enable growth and cell division to take place. The substrate cannot, therefore, be degraded entirely to carbon dioxide or methane, for example, and a portion must be channeled into the biosynthesis of essential molecules. Indeed, many organisms will degrade xenobiotics only in the presence of a suitable, more readily degraded growth substrate that supplies both cell carbon and the energy for growth; this is discussed later (Section 4.4.2) in the context of "cometabolism" and "concurrent metabolism". Growth under anaerobic conditions is demanding both physiologically and bio-chemically, since the cells will generally obtain only low energy yields from the growth substrate and must additionally maintain a delicate balance between oxidative and reductive processes. True fermentation implies that a single substrate is able to provide carbon for cell growth and, at the same time, satisfy the energy requirements of the cell. A simple example of fermentation is the catabolism of glucose by facultatively anaerobic bacteria to pyruvate, which is further transformed into a variety of products including acetate, butyrate, propionate, or ethanol by different organisms. On the other hand, a range of electron acceptors other than oxygen may be used under anaerobic conditions to mediate oxidative degradation of the carbon substrate at the expense of the reduction of the electron acceptors. For example, the following reductions may be coupled to oxidative degradation: nitrate to nitrogen (or nitrous oxide), sulfate to sulfide, carbonate to methane, fumarate to succinate, trimethylamine-N-oxide to trimethylamine, or dimethylsulfoxide to dimethyl sulfide (Styrvold and Strøm 1984) (Figure 4.22). The dimethyl sulfoxide reductase from *Escherichia coli* (Weiner et al. 1988) has a very broad substrate specificity and is able to reduce a range of sulfoxides and N-oxides. Even chlorate has been shown to support the growth of an anaerobic community at the expense of acetate (Malmqvist et al. 1991), and a facultatively anaerobic organism *Thauera selenatis* has been shown to be able to use selenate as an electron acceptor for anaerobic growth (Macy et al. 1993). This should be carefully distinguished from the situation in which selenate is gratuitously reduced during aerobic growth (Maiers et al. 1988) and from anaerobic sulfate reduction that is accomplished only under strictly anaerobic con-ditions. The environments required by the relevant organisms are directly related to the redox potential of the relevant reactions, so that increasingly reducing conditions

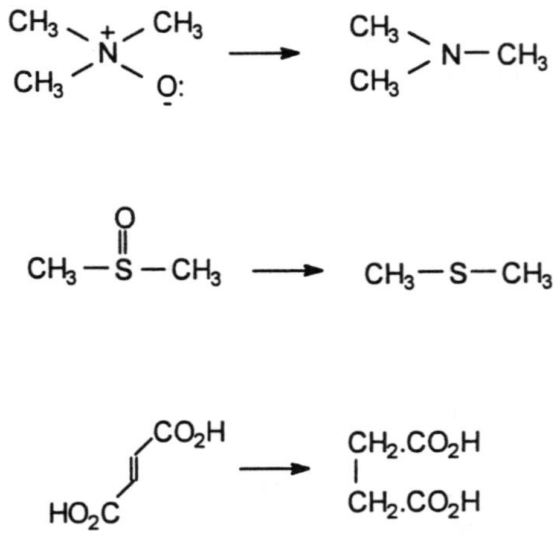

Figure 4.22 Examples of alternative electron acceptors and the reduction products.

are required for reduction of nitrate, sulfate, and carbonate. Examples of the degradation of xenobiotics under these various conditions are discussed further in Section 4.3.2.

It is important to underscore the fact that carbon dioxide is required not only for the growth of strictly phototrophic and lithotrophic organisms; many organisms which are heterotrophic may require carbon dioxide for their growth. Some illustrative examples are anaerobic bacteria, such as the acetogens, methanogens, and the propionic bacteria, and aerobic bacteria that degrade propane (MacMichael and Brown 1987), the branched hydrocarbon 2,6-dimethyloct-2-ene (Fall et al. 1979), or oxidize carbon monoxide (Meyer and Schlegel 1983). The lag after diluting glucose-grown cultures of *Escherichia coli* into fresh medium may, indeed, be eliminated by the addition of $NaHCO_3$, and this is consistent with the requirement of this organism for low concentrations of CO_2 for growth (Neidhardt et al. 1974). The role of CO_2 in the anaerobic biotransformation of aromatic compounds is discussed in greater detail in Chapter 6 (Section 6.7.3).

4.3 THE SPECTRUM OF ORGANISMS

It is probably no exaggeration to state most xenobiotics can be metabolized — albeit, to varying degrees — under the appropriate conditions, providing, of course, that the compound is not lethally toxic. Nonetheless, even such normally toxic compounds as carbon monoxide (Meyer and Schlegel 1983), cyanide (Harris and Knowles 1983), toluene (Claus and Walker 1964), and fluoroacetate (Meyer et al. 1990) can be metabolized by bacteria.

Although the cardinal importance of fungi in the terrestrial environment is unquestioned, most attention in the aquatic environment has been directed to bacteria. It is worth pointing out, however, that because of the similarity of the metabolic systems of fungi to those of mammalian systems, it has been suggested that fungi could be used as models for screening purposes, and several interesting examples have been provided (Smith and Rosazza 1983; Griffiths et al. 1992). The fungus *Cunninghamella elegans* has attracted particular attention (Section 4.3.5), and it may be noted that the reactions involved in the transformation of alachlor by this organism (Section 4.3.5) (Pothuluri et al. 1993) are similar to those carried out by mammalian systems and to the biotransformation of the analogous metalochlor by bluegill sunfish (*Lepomis macrochirus*) and by a soil actinomycete that are noted in Chapter 7 (Section 7.5.1). A wide range of taxonomically diverse fungi have also been used for the synthesis of less readily available compounds, and a few examples are given in Chapter 6 (Section 6.11.2).

Illustrations of the plethora of pathways used by bacteria for the degradation and biotransformation of xenobiotics will be provided in Chapter 6, but it is appropriate here to say something of the spectrum of metabolic potential of specific organisms, especially those that have hitherto achieved less prominence in discussions of biodegradation and biotransformation. It should be appreciated that, in natural situations, bacteria may be subjected to severe nutrient limitation so that they are compelled to reproduce at extremely low rates in order to conserve their metabolic energy (Kjellberg et al.1987; Siegele and Kolter 1992). This does not necessarily mean, however, that these organisms have negligible metabolic potential towards xenobiotics.

Other slow-growing organisms may be well adapted to the natural environment (Poindexter 1981), although they may not be numerically dominant among organisms isolated by normal procedures.

4.3.1 Aerobic and Facultatively Anaerobic Bacteria

The well-established metabolic versatility of groups such as the pseudomonads and their relatives and the methylotrophs and methanotrophs has possibly deflected attention from other groups which may be present in aquatic systems and which may play an important role in determining the fate of xenobiotics. Although the potential of other gram-negative groups such as the acinetobacters, moraxellas, and species of *Alcaligenes* is well established, Gram-positive groups seem to have generally achieved less prominence in aquatic systems. In the succeeding paragraphs, some examples of the metabolic importance of a few of these less prominent groups of organisms are presented.

Gram-positive aerobic bacteria — The metabolic versatility of organisms belonging to the genera *Mycobacterium* and *Rhodococcus* is becoming well established, and some examples will be given as illustration.

Mycobacteria can oxidize short-chain alkenes (DeBont et al. 1980), and organisms known under the invalid specific name "*Mycobacterium paraffinicum*" (Wayne et al. 1991) are able to degrade a number of alkanes. It has also been shown that strains of mycobacteria growth with propane are able to oxidize the apparently unrelated substrate trichloroethylene (Wackett et al. 1989). There has also been a revival of interest in the role of mycobacteria in the degradation of polycyclic hydrocarbons including pyrene (Heitkamp et al. 1988), naphthalene (Kelley et al. 1990), and phenanthrene (Guerin and Jones 1988). A strain of *Mycobacterium* sp. that is able to use all of these as a sole source of carbon and energy has also been isolated (Boldrin et al. 1993). This seems appropriate in view of the historical importance of similar organisms even though those designated as mycobacteria (Gray and Thornton 1928) were not acid-fast, and some at least would be currently assigned to the genus *Rhodococcus*. This illustrates a potentially serious taxonomic pitfall which must be avoided since the distinction between the genera *Mycobacterium* and *Rhodococcus* has not always been unequivocal in the older literature (Finnerty 1992). There has been increasing interest in rhodococci, and this has been sustained by their potential application in biotechnology (Finnerty 1992). Rhodococci have a wide metabolic potential, and, for example, an organism capable of degrading acetylene and assigned to the genus *Mycobacterium* almost certainly belongs to the genus *Rhodococcus* (DeBont et al. 1980). Considerable interest has been expressed in the chlorophenol-degrading organism *R. chlorophenolicus* (Apajalahti et al. 1986), partly motivated by its potential for application to bioremediation of chlorophenol-contaminated industrial sites and a strain of *Rhodococcus* sp. is capable of degrading a number of chlorinated aliphatic hydrocarbons including vinyl chloride and trichloroethene as well as the aromatic hydrocarbons benzene, naphthalene and biphenyl (Malachowsky et al. 1994). A number of biotransformations have also been accomplished using rhodococci, and these include, for example, the hydrolysis of nitriles, which is discussed in Chapter 6 (Section 6.11.1), and the reduction of the conjugated C=C double bond in 2-nitro-1-phenylprop-1-ene (Sakai et al. 1985).

A satisfying evaluation of the metabolic potential of microorganisms in natural ecosystems should not, therefore, fail to consider these organisms, which are certainly widespread, and to distinguish between rates of degradation and metabolic potential; slow-growing organisms may be extremely important in degrading xenobiotics in natural ecosystems.

Facultative anaerobes

1. *Enterobacteriaceae* — Facultatively anaerobic bacteria, and especially those belonging to the family *Enterobacteriaceae,* have a long history as agents of disease in man although there is extensive evidence for their occurrence in a wide variety of environmental samples; methods for their identification and classification have, therefore, been extensively developed and have traditionally used the ability to ferment a wide range of carbohydrates as taxonomic characters. This has had the possibly unfortunate effect of deflecting attention from the capability of these organisms for the degradation of other classes of substrates, although their ability to utilize substrates such as 3- and 4-hydroxybenzoates (Véron and Le Minor 1975) and nicotinate (Grimond et al. 1977) under aerobic conditions has been quite extensively used for taxonomic classification. A few examples illustrating their ability to degrade and transform diverse xenobiotics may be used to draw attention to the metabolic capabilities of these somewhat neglected organisms:

   • The degradation of DDT by organisms designated *Aerobacter aerogenes* (possibly currently *Klebsiella aerogenes*) (Wedemeyer 1967) (Figure 4.23) and the partial reductive dechlorination of methoxychlor by *K. pneumoniae* (Baarschers et al. 1982).

   • The biotransformation of methyl phenyl phosphonate to benzene by *K. pneumoniae* (Cook et al. 1979) (Figure 4.24a).

   • The biotransformation of γ-hexachlorocyclohexane to tetrachloro-cyclohexene by *Citrobacter freundii* (Figure 4.24b) (Jagnow et al 1977).

   • The biotransformation of 2,4,6-trihydroxy-1,3,5-triazine and atrazine under anaerobic conditions by an unidentified facultative anaerobe (Jessee et al. 1983).

   • The biotransformation of 3,5-dibromo-4-hydroxybenzonitrile to the corresponding acid by a strain of *K. pneumoniae* ssp. *ozaenae* which uses the substrate as sole source of nitrogen (Figure 4.24c) (McBride et al. 1986).

   • The decarboxylation of 4-hydroxycinnamic acid to 4-hydroxystyrene and of ferulic acid (3-methoxy-4-hydroxycinnamic acid) to 4-vinylguaiacol by several strains of *Hafnia alvei* and *H. protea* and by single strains of *Enterobacter cloacae* and *K. aerogenes* (Figure 4.24d) (Lindsay and Priest 1975). Several taxa of *Enterobacteriaceae* including *K. pneumoniae, Ent. aerogenes,* and *Proteus mirabilis* are able to decarboxylate the amino acid histidine that is abundant in the muscle tissue of scombroid fish (Yoshinaga and Frank 1982) and the histamine produced has been associated with an incident of scombroid fish poisoning (Taylor et al. 1979).

   • The metabolism of ferulic acid (3-methoxy-4-hydroxycinnamic acid) by *Ent. cloacae* to a number of products including phenylpropionate and benzoate (Figure 4.25) (Grbic-Galic 1986).

   • Utilization of uric acid as a nitrogen source by strains of *Aer. aerogenes, K. pneumoniae,* and *Serratia kiliensis* (Rouf and Lomprey 1968).

- The sequential reduction of the nitro groups of 2,6-dinitrotoluene by *Salmonella typhimurium* (Sayama et al. 1992) — a taxon that is not generally noted for its degradative capability — is noted again in Chapter 6, Section 6.8.2.
- The taxonomic application of the ability of enteric organisms to grow with hydroxylated aromatic carboxylic acids has been noted above. Details of the degradation of 3- and 4-hydroxyphenylacetate (Cooper and Skinner 1980) and of 3-hydroxyphenylpropionic acid (Burlingame and Chapman 1983) have been established, and it has been demonstrated that the enzyme that carries out the hydroxylation has a wide substrate range extending to 4-methylphenol and even to 4-chlorophenol (Prieto et al. 1993).

2. Gram-positive organisms — The genus *Staphylococcus* is traditionally associated with disease in man, but the demonstration (Monna et al. 1993) that a strain of *Staphylococcus auriculans* — isolated by enrichment with dibenzofuran and with no obvious clinical association — could degrade this substrate and carry out limited biotransformation of fluorene and dibenzo-1,4-dioxin may serve to illustrate the unsuspected metabolic potential of facultatively anaerobic Gram-positive organisms.

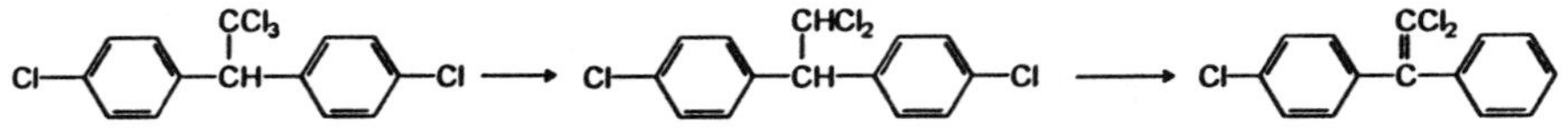

Figure 4.23 Degradation of DDT by *Aerobacter aerogenes*.

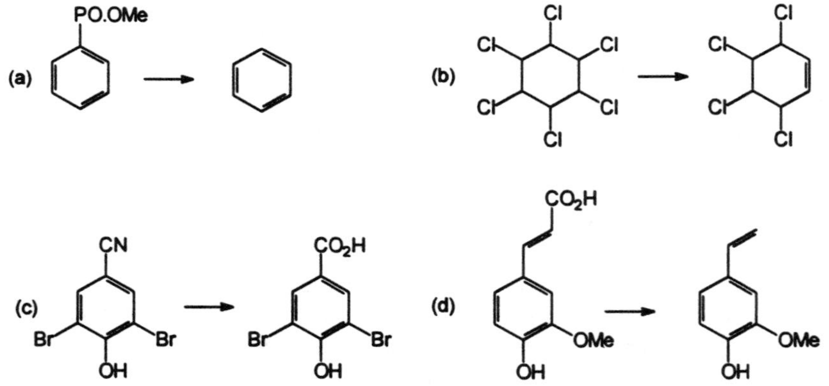

Figure 4.24 Examples of biotransformations carried out by *Enterobacteriaceae*: **(a)** methylphenyl phosphonate, **(b)** γ-hexachlorocyclohexane, **(c)** 3,5-dibromo-4-hydroxybenzonitrile, **(d)** decarboxylation of ferulic acid.

The range of substrates for monooxygenase systems — Extensive evidence has revealed that bacterial monooxygenase systems have a remarkable metabolic versatility. A few of the reactions carried out by the monooxygenase system of methylotrophic bacteria are summarized in Figure 4.26, and it is on account of this that these organisms have received attention for their technological potential (Lidstrom and Stirling 1990). An equally wide metabolic potential has also been demonstrated for cyclohexane monooxygenase which has been shown to accomplish two broad types of reaction: one in which formally nucleophilic oxygen reacts with the substrate and the other in which formally electrophilic oxygen is involved (Figure 4.27) (Branchaud and Walsh 1985). In addition, it has emerged that the monooxygenase system in methylotrophs is similar to that in the nitrite-oxidizing bacteria and that the spectrum of biotransformations is carried out by the latter equally wide.

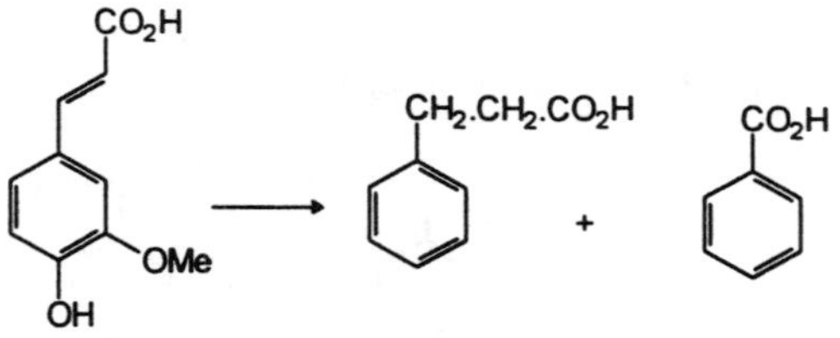

Figure 4.25 Metabolism of ferulic acid by *Enterobacter cloacae*.

The following illustrates some of the biotransformations which have been observed with *Nitrosomonas europaea*, and these are particularly interesting since this organism has an obligate dependency on CO_2 as carbon source and has traditionally been considered to be extremely limited in its ability to use organic carbon for growth:

- Oxidation of benzene to phenol and 1,4-dihydroxybenzene (Figure 4.28) (Hyman et al. 1985).
- Oxidation of alkanes (C_1 to C_8) to alkanols and alkenes (C_2-C_5) to epoxides (Hyman et al. 1988).
- Oxidation of chloroalkanes at carbon atoms substituted with a single chlorine atom to the corresponding aldehyde (Rasche et al. 1991).
- Oxidation of the trichloromethyl group in 2-chloro-6-(trichloromethyl)-pyridine to the corresponding carboxylic acid (Vannelli and Hooper 1992) at high oxygen concentrations during cooxidation of ammonia or hydrazine. At low oxygen concentrations in the presence of hydrazine reductive dechlorination to 2-chloro-6-dichloromethylpyridine occurs (Vannelli and Hooper 1993) is an example of a rather unusual hydrolytic reaction.
- A range of sulfides including methylsulfide, tetrahydrothiophene, and phenylmethylsulfide are oxidized to the corresponding sulfoxides (Juliette et al. 1993).

As is the case for methylotrophic bacteria, such transformations are probably confined neither to a single organism nor to strains of specific taxa within the group. For example, both *Nitrosococcus oceanus* and *Nitrosomonas europaea* are able to oxidize methane to CO_2 (Jones and Morita 1983; Ward 1987). The versatility of this group of organisms clearly motivates a reassessment of their ecological significance, particularly in the marine environment where they are widely distributed.

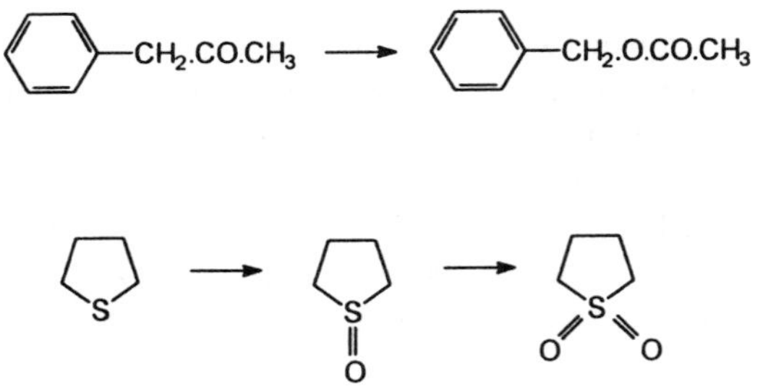

Figure 4.27 Examples of the reactions mediated by cyclohexane monooxygenase.

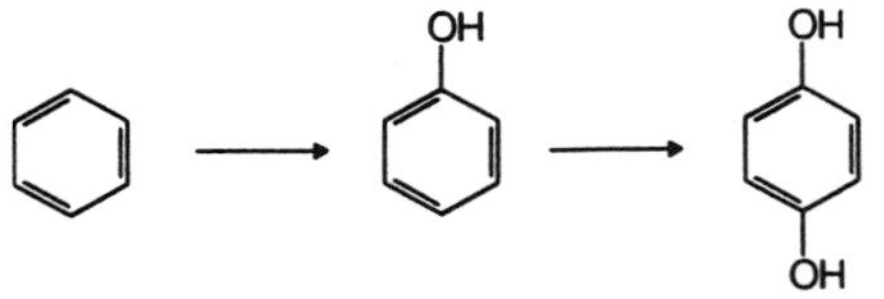

Figure 4.28 Metabolism of benzene by *Nitrosomonas europaea*.

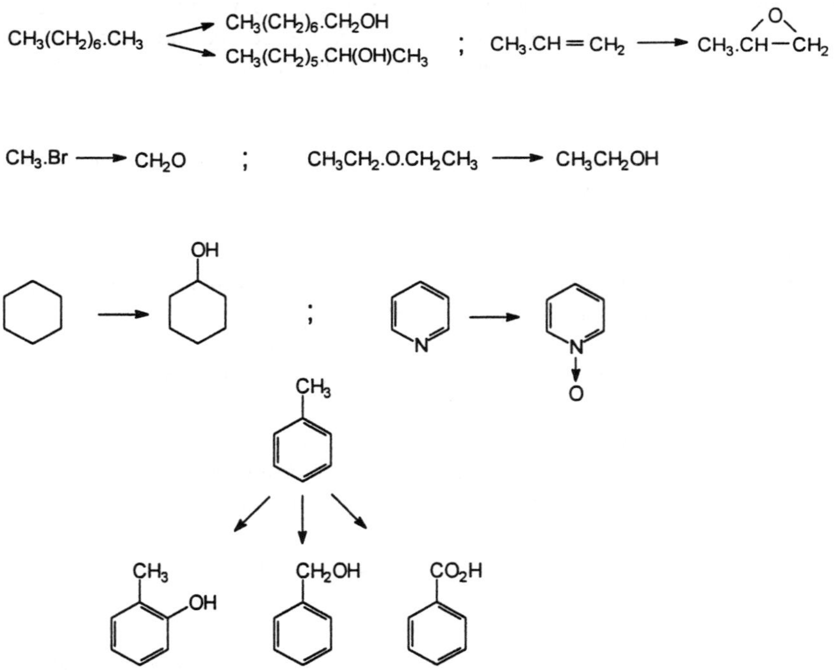

Figure 4.26 Examples of reactions mediated by the monooxygenase system of methylotrophic bacteria.

4.3.2 Organisms Using Electron Acceptors Other Than Oxygen

A number of facultatively anaerobic bacteria are able to carry out a respiratory metabolism in the absence of oxygen using alternative inorganic electron acceptors.

These include nitrate, fumarate, trimethylamine-*N*-oxide, or dimethylsulfoxide (Section 4.2), and an attempt will be made here to illustrate the metabolic potential of organisms under these conditions. It should be noted that the conditions under which these function and their regulation depend on the organism. For example, whereas in *Escherichia coli*, oxygen represses the synthesis of the other reductases, and under anaerobic conditions the reductases for fumarate, DMSO and TMAO, are repressed by nitrate, this is not the case for *Wolinella succinogenes* in which sulfur represses the synthesis of the more positive electron acceptors nitrate and fumarate (Lorenzen et al. 1993).

Nitrate and related compounds — The degradation of organic compounds with nitrate in the absence of oxygen — denitrification or nitrate dissimilation — has been known for a long time and has been used as a valuable character in bacterial classification; the products are either dinitrogen or nitrous oxide, and the reaction is generally inhibited by oxygen so that it occurs to a significant extent only under anoxic conditions. Renewed interest has been focused on degradation of xenobiotics under anaerobic conditions in the presence of nitrate — possibly motivated by the extent of leaching of nitrate fertilizer from agricultural land into groundwater. In studies with such organisms, a clear distinction should be made between degradation of the substrate under three conditions which may or may not be biochemically equivalent (1) aerobic conditions, (2) anaerobic conditions in the presence of nitrate, and (3) strictly anaerobic conditions in the absence of any electron acceptor where fermentation is predominant. The last of these is briefly discussed in Section 4.3.3. A few examples are given as illustration of the degradations that have been observed with facultatively anaerobic organisms using nitrate as electron acceptor.

- The degradation of carbon tetrachloride to CO_2 by a *Pseudomonas* sp. (Criddle et al. 1990).
- The degradation of benzoate (Taylor and Heeb 1972; Williams and Evans 1975; Ziegler et al. 1987) and phthalate (Nozawa and Maruyama 1988; Afring and Taylor 1981).
- The degradation of alkyl benzenes (Hutchins 1991; Evans et al. 1991a,b; Altenschmidt and Fuchs 1991).

The mechanism of many of these reactions remains incompletely resolved, and some of them are discussed further in Chapter 6 (Section 6.7.3); the potential complexities of the pathways for the aromatic compounds may be illustrated by the demonstration that phenol undergoes a reversible exchange reaction with CO_2 to produce 4-hydroxybenzoate before dehydroxylation to benzoate (Figure 4.29) (Tschech and Fuchs 1989).

It is also important to emphasize that an organism that can degrade a given substrate under conditions of nitrate dissimilation may not necessarily display this potential under aerobic conditions. For example, a strain of *Pseudomonas* sp. could be grown with vanillate under anaerobic conditions in the presence of nitrate but could not be grown under aerobic conditions with the same substrate. On the other hand, cells grown anaerobically with nitrate and vanillate were able to oxidize vanillate under both aerobic and anaerobic conditions; the cells were also able to

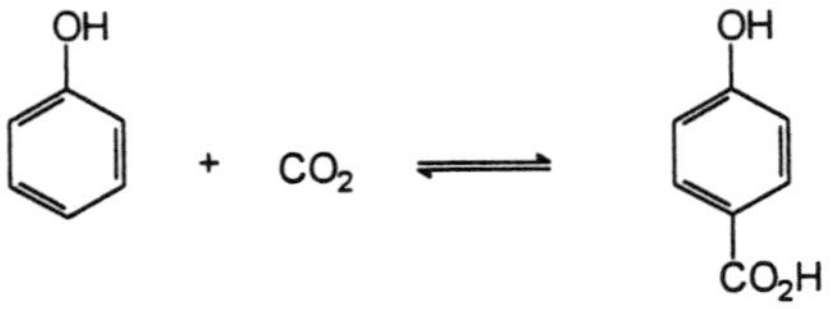

Figure 4.29 Exchange reaction between phenol and CO_2.

demethylate a much wider spectrum of aromatic methoxy compounds under anaerobic conditions than under aerobic conditions (Taylor 1983). Such subtleties should be clearly appreciated and taken into consideration in evaluating the degradative potential of comparable organisms under different physiological conditions.

Iron (III), manganese (IV) and U(VI) — Many bacteria that ferment organic substrates are able to reduce Fe(III) to Fe(II) gratuitously, but these organisms are apparently unable to use the energy of this reduction for growth and couple this to the oxidation of organic substrates. There are, however, bacteria that can accomplish this, and some of them can also effectively use Mn(IV) and U(VI). There are two broadly different groups of organisms with this metabolic potential: one is facultatively anaerobic; the other, strictly anaerobic.

1. Strains of *Shewanella putrefaciens* (*Alteromonas putrefaciens*) are widely distributed in environmental samples and are generally considered as aerobic organisms with the capability of reducing thiosulphate in complex media to sulfide. They are also able, however, to grow anaerobically using Fe(III) as electron acceptor and to oxidize formate, lactate, or pyruvate. These substrates cannot, however, be completely oxidized to CO_2 since the acetate produced from the C_2 compounds is not further metabolized; Mn(IV) may also function in a similar way (Lovley et al. 1989). The bioenergetics of the system has been examined in cells of another strain of this organism grown anaerobically on lactate with either fumarate or nitrate as electron acceptors, and respiration-linked proton translocation in response to Mn(IV), fumarate, or oxygen was clearly demonstrated (Myers and Nealson 1990). Levels of Fe(III) reductase, nitrate reductase, and nitrite reductase are elevated by growth under microaerophilic conditions, and the organism probably possesses three reductase systems, each of which apparently consists of low-rate and high-rate components (DiChristina 1992). Another organism with hitherto unknown taxonomic affinity has been isolated (Caccavo et al. 1992) and is able to couple the oxidation of lactate to the reduction of Fe(III), Mn(IV) and U(VI).
2. A strictly anaerobic organism designated GS-15 and assigned to the newly described taxon *Geobacter metallireducens* (Lovley et al. 1993) is able to use Fe(III) as electron acceptor for the oxidation of a number of compounds including acetate (Lovley and Lonergan 1990) and toluene and phenols under anaerobic conditions (Champine and Goodwin 1991). This organism is also able to oxidize acetate by reduction of Mn(IV) and U(VI) (Lovley et al. 1993).

These transformations illustrate important processes for the cycling of organic carbon in sediments where Fe(III) may have been precipitated, and it seems likely that comparable geochemical cycles involving manganese (Lovley and Phillips 1988) will also achieve greater prominence (Lovley 1991; Nealson and Myers 1992). It has also been suggested that such organisms could be used for the immobilization of soluble U(vi) in wastewater containing both U(vi) and organic compounds by conversion to insoluble U(iv) (Lovley et al. 1991).

4.3.3. Anaerobic Bacteria

There are a number of reasons for the increased interest in transformations carried out by anaerobic bacteria. One is that many xenobiotics are partitioned from the aquatic phase into the sediment phase after discharge into the aquatic environment (Chapter 3, Section 3.2). Another is that sediments in the vicinity of industrial discharge often contain, in addition, readily degraded organic matter; the activity of aerobic and facultatively anaerobic bacteria then renders these sediments effectively anaerobic.

The fate of xenobiotics in the environment is, therefore, significantly determined by the degradative activity of anaerobic bacteria.

The terms anaerobic and anoxic are purely operational and imply merely the absence of air or oxygen, and the absolute distinction between aerobic and anaerobic organisms is becoming increasingly blurred; the problem of defining anaerobic bacteria is, therefore, best left to philosophers. Possibly, the critical issue is the degree to which low concentrations of oxygen are either necessary for growth or toxic. During growth of bacteria in the absence of externally added electron acceptors, the term fermentation implies that a redox balance is achieved between the substrate (which may include CO_2) and its metabolites. A few examples are given to illustrate the apparently conflicting situations that may be encountered and the gradients of response that may exist.

1. Although strictly anaerobic bacteria do not generally grow in the presence of high potential electron acceptors such as oxygen or nitrate, an intriguing exception is provided by an obligately anaerobic organism that uses nitrate as electron acceptor during the degradation of resorcinol (Gorny et al. 1992). This isolation draws attention to the unknown extent to which such organisms exist in natural systems, since strictly anaerobic conditions are seldom used for the isolation of organisms using nitrate as electron acceptor; other formally similar organisms are either facultatively anaerobic or have a fermentative metabolism. On the other hand, the normally strictly anaerobic sulfate-reducing organisms *Desulfobulbus propionicus* and *Desulfovibrio desulfuricans* may grow by reducing nitrate or nitrite to ammonia using hydrogen as electron donor (Seitz and Cypinka 1986). In the presence of nitrate, the acetogen *Clostridium thermoaceticum* oxidizes the *O*-methyl groups of vanillin or vanillate to CO_2 without production of acetate that is the normal product in the absence of nitrate (Seifritz et al. 1993).

2. Some anaerobic organisms such as clostridia are appreciably tolerant of exposure to oxygen, and indeed others such as *Wolinella succinogenes* that have hitherto been classified as anaerobes are, in fact, microaerophilic (Han et al.

1991). On the other hand, organisms such as *Nitrosomonas europaea* that normally obtain energy for growth by oxidation of ammonia to nitrite may apparently bring about denitrification of nitrite under conditions of oxygen stress (Poth and Focht 1985) or under anaerobic conditions in the presence of pyruvate (Abeliovich and Vonshak 1992).

3. Attention is drawn in Section 4.5.1 to the role of oxygen concentration so that it is appropriate to note the existence of microaerotolerant or microaerophilic organisms. The example of *W. succinogenes* has been given above, and another is provided by *Malonomonas rubra* that uses malonate as sole source of carbon and energy (Dehning and Schink 1989a). It should be noted, however, that *Propionigenium modestum* that obtains its energy by decarboxylating succinate to propionate is a strictly anaerobic organism (Schink and Pfennig 1982).

Attention may be briefly directed to two groups of anaerobic bacteria that display metabolic versatility to structurally diverse compounds — clostridia and sulfate-reducers. The classical studies on the anaerobic metabolism of amino acids, purines, and pyrimidines by clostridia not only set out the relevant experimental procedures, and thereby laid the foundations for virtually all future investigations, but also brought to light the importance and range of coenzyme-B_{12}-mediated rearrangements. These reactions were mainly carried out by species of clostridia, and, indeed, some of these degradations belong to the classical age of microbiology in Delft (Liebert 1909). The number of different clostridia investigated may be gained from the following selected examples: the classic purine-fermenting organisms *Clostridium acidurici* and *Cl. cylindrosporum* ferment several purines including uric acid, xanthine, and guanine, while the degradation of pyrimidines such as orotic acid is accomplished by *Cl. oroticum*. A large number of clostridia including *Cl. perfringens, Cl. saccharobutyricum, Cl. propionicum, Cl. tetani, Cl. sporogenes,* and *Cl. tetanomorphum* ferment a range of single amino acids, while some participate in the Stickland reaction involving two amino acids. In Chapter 6 the pathways employed for the degradation of aminoacids are discussed in Section 6.7.1 and those for the degradation of N-containing heterocyclic compounds in Section 6.7.4. The spectrum of compounds degraded by anaerobic sulfate-reducing bacteria is continuously widening and now includes, for example, alkanes (Aeckersberg et al. 1991), nicotinic acid (Imhoff-Stuckle and Pfennig 1983), indoles (Bak and Widdel 1986), benzoate (Tasaki et al. 1991), and catechol (Szewzyk and Pfennig 1987). Further details of some of the pathways involved are provided in Chapter 6 (Section 6.7).

It is appropriate to mention briefly a few other aspects of anaerobic organisms.

1. Syntrophic associations are briefly discussed in Section 4.4.1
2. Compounds such as oxalate (Dehning and Schink 1989b) and malonate (Dehning et al. 1989; Janssen and Harfoot 1992) that are degraded by decarboxylation with only a modest energy contribution are, nevertheless, able to support the growth of the appropriate organisms, and the bioenergetics of the microaerophilic *Malonomonas rubra* has been investigated (Hilbi et al. 1993).
3. Extensive effort has been directed to the anaerobic dechlorination of aromatic compounds, though hitherto only two pure cultures seem to have been isolated:

one is a sulfate reducing organism *Desulfomonile tiedjei* (DeWeerd et al 1990); the other is a spore-forming organism (Madsen and Licht 1992). Possibly, the experimental difficulty in isolating pure cultures of such organisms with a defined metabolic capacity for xenobiotics is still a limiting factor in investigating the enzymology of putative metabolic pathways.

4.3.4 Phototrophic Organisms

The metabolic significance of algae and cyanobacteria has received relatively limited attention in spite of the fact that they are important components of many ecosystems and may, for example, in the marine environment, be of particular significance. Whereas the heterotrophic growth of algae at the expense of simple carbohydrates, amino acids, lower aliphatic carboxylic acids, and simple polyols is well documented (Neilson and Lewin 1974), the potential of algae for metabolism of xenobiotics has been much less extensively explored. Among these metabolic possibilities which have received less attention than they possibly deserve are the following:

1. The transformation — though not apparently the degradation — of naphthalene has been examined in cyanobacteria and microalgae including representatives of green, red, and brown algae, and diatoms (Cerniglia et al. 1980a, 1982) and the transformation of biphenyl (Cerniglia et al. 1980b) and aniline (Cerniglia et al. 1981) and methyl naphthalenes (Cerniglia et al. 1983) by cyanobacteria (Figure 4.30). Phenanthrene is metabolized by *Agmenellum quadruplicatum* to the trans-9,10-dihydrodiol by a monooxygenase system, and the transiently formed 1-hydroxyphenanthrene is *O*-methylated (Narro et al. 1992).
2. A number of green algae are able to use aromatic sulfonic acids (Figure 4.31a) (Soeder et al. 1987) and aliphatic sulfonic acids (Figure 4.31b) (Biedlingmeier and Schmidt 1983) as a source of sulfur.
3. Representatives of the major groups of algae are able to use a range of amino acids as nitrogen sources (Neilson and Larsson 1980).
4. The transformation of DDT to DDE — albeit, in rather low yield — by elimination of one molecule of HCl has been observed in several marine algae (Rice and Sikka 1973).

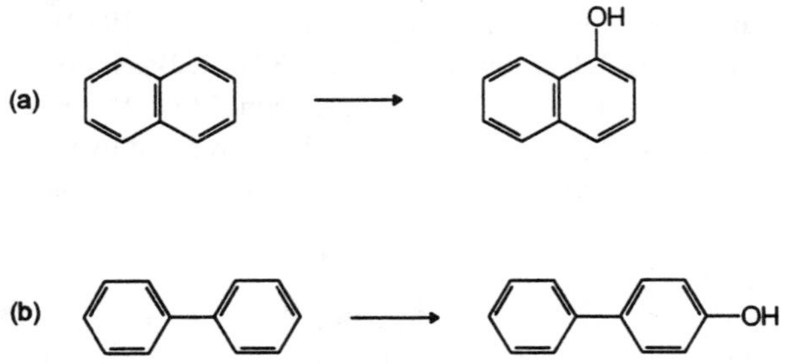

Figure 4.30 The biotransformation of **(a)** naphthalene and **(b)** biphenyl by algae.

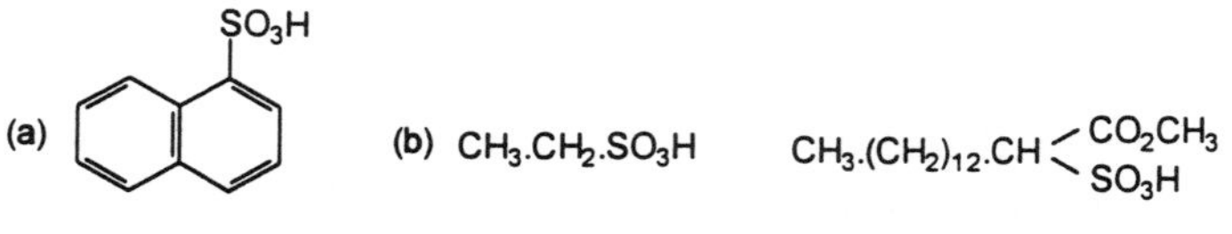

Figure 4.31   Organosulfur compounds used as S-sources by algae.

There has been a revival of interest in the metabolic potential of anaerobic phototrophic bacteria, particularly the purple non-sulfur organisms that can degrade aromatic compounds (Khanna et al. 1992). Such organisms are widely distributed in appropriate ecosystems and may, therefore, play a significant role in the degradation of xenobiotics. Less appears to be known of the potential of other anaerobic phototrophs, such as the purple and green sulphur bacteria, to degrade xenobiotics.

4.3.5 Eukaryotic Organisms: Fungi and Yeasts

In the terrestrial environment — and possibly also in a few specialized aquatic ecosystems — fungi and yeasts play a cardinal role in biodegradation and biotransformation. The role of yeasts in the coastal marine environment is illustrated from results of their frequency and metabolic potential for transformation of phenanthrene and benz[a]anthracene (MacGillivray and Shiaris 1993).

Biotransformation (hydroxylation) of a wide range of PAHs and related compounds including biphenyl, naphthalene, anthracene, phenanthrene, 4-, and 7-methylbenz[a]anthracene, 7,12-dimethylbenz[a]anthracene (Figure 4.32) has been examined in a number of fungi, most extensively in species of *Cunninghamella* especially in *C. elegans* (McMillan et al. 1987). Hydroxylation of benzimidazole (Seigle-Murandi et al. 1986) by *Absidia spinosa* and of biphenyl ether by *C. echinulata* (Seigle-Murandi et al. 1991) has also been studied on account of the industrial interest in the metabolites. The biotransformation of alachlor (2-chloro-N-methoxymethyl-N-[2,6-diethylphenyl]acetamide) by *C. elegans* has already been mentioned (Section 4.3) and involves primarily hydroxylation at the benzylic carbon atom and loss of the methoxymethyl group (Pothuluri et al. 1993). In all of these cases, the reactions are purely biotransformations since the aromatic rings of these compounds are not destroyed.

Particular attention has been focused on the white-rot fungus *Phanerochaete chrysosporium* on account of its ability to degrade lignin and to metabolize a wide range of unrelated compounds including PAHs (Bumpus 1989) and organochlorine compounds such as DDT (Bumpus and Aust 1987), PCBs (Eaton 1985), lindane and chlordane (Kennedy et al. 1990), pentachlorophenol (Mileski et al. 1988), and 2,7-dichlorobenzo-1,4-dioxin (Valli et al. 1992a). The novel pathways for the degradation of 2,4-dichlorophenol (Valli and Gold 1991) and 2,4-dinitrotoluene (Valli et al.

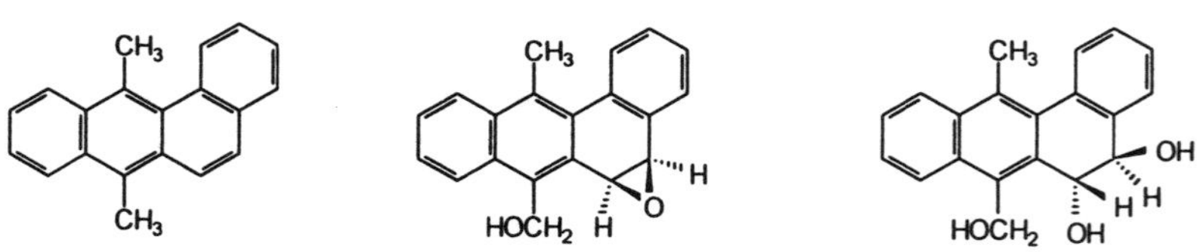

Figure 4.32   Biotransformation of 7,12-dimethylbenz[a]anthracene by fungi.

1992b) are discussed in a wider context in Chapter 6 (Section 6.5.2 and Section 6.8.2). Degradation of all these compounds is apparently mediated by two peroxidase systems — lignin peroxidases and manganese-dependent peroxidases — and a laccase that is produced by several white-rot fungi though not by *P. chrysopsporium*. A serious interpretative ambiguity has, however, emerged from the observation that lignin peroxidase is able to *polymerize* a range of putative aromatic precursors to lignin, but that it is not the functional enzyme in the *depolymerization* of lignin (Sarkanen et al. 1991). The regulation of the synthesis of these oxidative enzymes is complex and is influenced by nitrogen limitation, the growth status of the cells, and the concentration of manganese in the medium (Perez and Jeffries 1990). In addition, it seems clear that monooxyganase and epoxide hydrolase activities are also involved, since the biotransformation of phenanthrene takes place even in the absence of peroxidase systems (Sutherland et al. 1991). Further comments on the synthesis of these enzyme systems is given in Chapter 5 (Section 5.2.2). Clearly, therefore, a number of important unresolved issues remain. In addition, attention has already been drawn in Chapter 3 (Section 3.2.3) to the role of fungal redox systems in the covalent linking of xenobiotics to aromatic components of humus in soils.

The metabolic capabilities of yeasts have attracted attention in different contexts, and it has emerged that, in contrast to many fungi, they are able to bring about disruption of aromatic rings. A few examples that illustrate the various possibilities are given.

1. Ring cleavage clearly occurs during the metabolism of phenol (Walker 1973) by the yeast *Rhodotorula glutinis* and of aromatic acids by various fungi (Cain et al. 1968; Durham et al. 1984; Gupta et al. 1986).
2. Comparable ring cleavage reactions have also been found in studies on the metabolism of aromatic compounds by the yeast *Trichosporon cutaneum* whose metabolic versatility is indeed quite comparable with that of bacteria; examples of this may be found in the degradation of phenol and resorcinol (Gaal and Neujahr 1979), of tryptophan and anthranilate (Anderson and Dagley 1981a), and of aromatic acids (Anderson and Dagley 1981b).
3. The ability to grow at the expense of 4-hydroxy- and 3,4-dihydroxybenzoate has been used for classification of medically important yeasts including *Candida parapsilosis* (Cooper and Land 1979).
4. The phenol-assimilating yeast *Candida maltosa* degraded a number of phenols even though these were unable to support growth; hydroxylation of 3-chloro- and 4-chlorophenol produced initially 4-chlorocatechol and then 5-chloropyrogallol (Polnisch et al. 1992).

Clearly then, yeasts possess metabolic potential for the degradation of xenobiotics little inferior to that of many bacterial groups, so that their role in natural ecosystems justifies the greater attention that has been directed to them (MacGillivray and Shiaris 1993).

4.3.6 Other Organisms

Limited investigations have revealed the metabolic potential of other eukaryotic microorganisms including that of the apochlorotic alga (protozoan) *Prototheca zopfii* that is able to degrade aliphatic hydrocarbons (Walker and Pore 1978; Koenig and

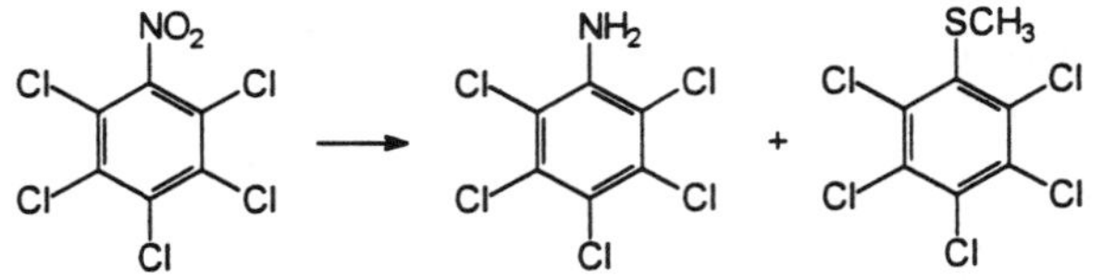

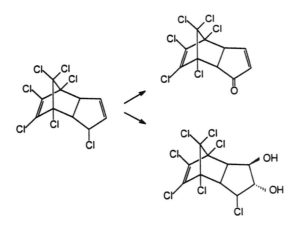

Figure 4.33 Biotransformation of pentachloronitrobenzene by *Tetrahymena thermophila*.

Ward 1983) and of *Tetrahymena thermophila* that transforms pentachloronitrobenzene to the corresponding aniline and pentachlorothioanisole (Figure 4.33) (Murphy et al. 1982). *Daphnia magna* has been shown to bring about dechlorination and limited oxidation of heptachlor (Figure 4.34) (Feroz et al. 1990).

The metabolic potential of fish has already been briefly noted in the context of bioconcentration (Chapter 3, Section 3.1.3) and is discussed more extensively in Chapter 7 (Section 7.5.1) in the context of toxicology. It is, therefore, sufficient here merely to state that fish may effectively metabolize a wide range of xenobiotics even though complete degradation is seldom achieved; instead, simple biotransformations are carried out, such as introduction of hydroxyl groups into aromatic rings followed by conjugation with sulfate or glucuronic acid.

Figure 4.34 Biotransformation of heptachlor by *Daphnia magna*.

4.4 INTERACTIONS

4.4.1 Single Substrates: Several Organisms

Cultures of a single microorganism will occur naturally only in circumstances where extreme selection pressure operates, for example, in hydrothermal vents or in the sediments of Antarctic lakes. Generally, many different organisms with diverse metabolic potential will exist side by side, so that metabolic interactions are probably the rule rather than the exception in natural ecosystems. Both the nature and the tightness of the association may vary widely, and in some cases, degradation of a single compound may necessitate the cooperation of two (or more) organisms. Some well-defined interactions and the different mechanisms underlying the cooperation are illustrated by the following examples:

1. *One of the organisms fulfills the need for a growth requirement by the other.* Examples are provided by vitamin requirements of one organism that is provided by the other: biotin in co-cultures of *Methylocystis* sp. and *Xanthobacter* sp. (Lidstrom-O'Connor et al. 1983) and thiamin in cocultures of *Pseudomonas aeruginosa* and an undefined *Pseudomonas* sp. that degrade the phosphonate herbicide glyphosate (Moore et al. 1983).

2. *One organism may be able to carry out only a single step in the biodegradation.* Many examples among aerobic organisms have been provided (Reanney et al. 1983; Slater and Lovatt 1984) so that this is probably a widespread situation; three examples will, therefore, suffice as illustration:

   - The degradation of parathion is carried out by a mixed culture of *Pseudomonas stutzeri* and *P. aeruginosa* (Daughton and Hsieh 1977) in which the 4-nitrophenol initially formed by the former is metabolized by the latter (Figure 4.35).
   - The degradation of 4-chloroacetophenone is accomplished by a mixed culture of an *Arthrobacter* sp. and a *Micrococcus* sp. The first organism is able to carry out all the degradative steps except the conversion to 4-chlorocatechol of the intermediate 4-chlorophenol that is toxic to the first organism (Havel and Reineke 1993). The details of the pathway are discussed again in Chapter 6 (Section 6.2.1).
   - The aerobic degradation of polyethylene glycol is carried out by a consortium of a *Flavobacterium* sp. and a *Pseudomonas* sp. in which the latter is required for the degradation of the glycollate produced by the former (Figure 4.36) (Kawai and Yamanaka 1986).

3. *Two organisms are required to maintain the redox balance.* Among anaerobic bacteria, hydrogen transfer is important, since the redox balance must be maintained and the hydrogen concentration in the mixed cultures may be critical. Interspecies hydrogen transfer has been demonstrated especially among populations of rumen bacteria containing methanogens where the concentration of hydrogen must be limited for effective functioning of the consortia. Illustrative examples have been summarized (Wolin 1982), and a few additional comments on degradative reactions that are dependent on hydrogen transfer mediated by one of the cooperating organisms are added here.

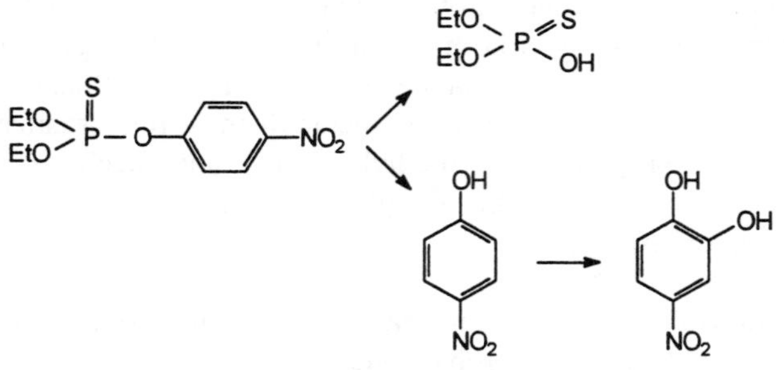

Figure 4.35 Degradation of parathion by a mixed culture of two pseudomonads.

$$HO.(CH_2.CH_2.O)_n.CH_2.CH_2.O.CH_2.CH_2OH \longrightarrow HO.(CH_2.CH_2.O)_n.CH_2.CH_2.O.CH_2.CO_2H$$

$$\longrightarrow \quad HO.(CH_2.CH_2.O)_n.CH_2.CH_2OH \quad + \quad HO.CH_2.CO_2H$$

Figure 4.36 Degradation of polyethylene glycol by a mixed culture involving successive depolymerization.

Stable metabolic associations generally between pairs of anaerobic bacteria have been termed syntrophs, and these are effective in degrading a number of aliphatic carboxylic acids and benzoate under anaerobic conditions; these reactions have been discussed in reviews (Schink 1992; Lowe et al. 1993) that provide lucid accounts of the role of syntrophs in the degradation of complex organic matter, while a number of important degradations and transformations of aromatic compounds by mixed cultures are discussed in Chapter 6 (Section 6.7.3). Two examples may be used here to illustrate the experimental intricacy of the problems besetting the study of syntrophic metabolism under anaerobic conditions.

1. Oxidation under anaerobic conditions of long-chain aliphatic carboxylic acids was established in syntrophic cultures of *Clostridium bryantii/Desulfovibrio* sp. (Stieb and Schink 1985), *Syntrophomonas sapovorans/Methanospirillum hungatei* (Roy et al. 1986), and *Syntrophomonas wolfei* in coculture with H_2-utilizing anaerobic bacteria (McInerney et al. 1981). The role of the second syntroph was to metabolize the reducing equivalents produced by oxidation of the carboxylic acids. *S. wolfei* was subsequently, however, adapted to grow with crotonate in pure culture (Beaty and McInerney 1987), and this procedure was also used for *Cl. bryantii*. 16S rRNA sequence analysis was then used to show the close relationship of these two organisms and to assign them to a new genus *Syntrophospora* (Zhao et al. 1990). On the other hand, anaerobic oxidation of carboxylic acids with chain lengths of up to ten has been demonstrated in pure cultures of species of *Desulfonema* (Widdel et al. 1983), and even aliphatic hydrocarbons may be completely oxidized to CO_2 by a sulfate-reducing bacterium (Aeckersberg et al. 1991).

2. There has been considerable interest in the anaerobic degradation of propionate that is a fermentation product of many complex substrates, and syntrophic associations of acetogenic and methanogenic bacteria have been obtained. As in the example noted above, during metabolism of propionate in a syntrophic culture (Houwen et al. 1991), the methanogens serve to remove hydrogen produced during the oxidation of propionate to acetate. Growth of syntrophic propionate-oxidizing bacteria in the absence of methanogens has, however, been accomplished using fumarate as sole substrate (Plugge et al. 1993). Fumarate plays a central role in the metabolism of this organism, since it is produced from propionate via methylmalonate and succinate, and fumarate itself is metabolized by the acetyl-CoA cleavage pathway via malate, oxalacetate, and pyruvate.

The interaction of two organisms is clearly not obligatory for the ability to degrade these carboxylic acids under anaerobic conditions.

4.4.2 Cometabolism and Related Phenomena

In natural ecosystems, it is seldom indeed that either a pure culture or a single substrate exists; in general, several substrates will be present, and these will include compounds of widely varying susceptibility to microbial degradation. The phenomenon where degradation occurs in the presence of two substrates has been termed "cooxidation" or less specifically "cometabolism" or "concurrent metabolism". Unfortunately, the term cometabolism (Horvath 1972) has been used in different and conflicting ways; since the prefix "co" implies "together", it should not, therefore, legitimately be extended to situations where only a single substrate is present and for which the term "biotransformation" is unambiguous and seems entirely adequate. In its application to reactions carried out by methanotrophic and methylotrophic bacteria, it is not always evident that both substrates are obligately required. Detailed discussions have been presented (Dalton and Stirling 1982), and some of the conflicting aspects have been briefly summarized (Neilson et al. 1985).

The phenomenon merits careful analysis, however, since important metabolic principles lie behind most of the experiments, even though confusion may have arisen as a result of ambiguous terminology. An attempt is therefore made to ignore semantic implications and to adopt a wide perspective in discussing this environmentally important issue. A pragmatic point of view has been adopted, and the following examples attempt to illustrate the kinds of experiments which have been carried out under various conditions.

Experiments in which only a single substrate is present — Organisms may be obtained after elective enrichment with a given substrate but are subsequently shown to be unable to use the substrate for growth, although they are able to accomplish its partial metabolism; such situations may be adequately classified as biotransformation. A typical example is the partial oxidation of 2,3,6-trichlorobenzoate to 3,5-dichlorocatechol (Figure 4.37) (Horvath 1971). In some cases, strains unselected by enrichment have been used, for example, oxidation of 2-chlorophenol or 3-chlorobenzoate to 3-chlorocatechol (Figure 4.38) by several strains of bacteria grown with non-chlorinated substrates, such as phenol, naphth-2-ol, or naphthalene, and that

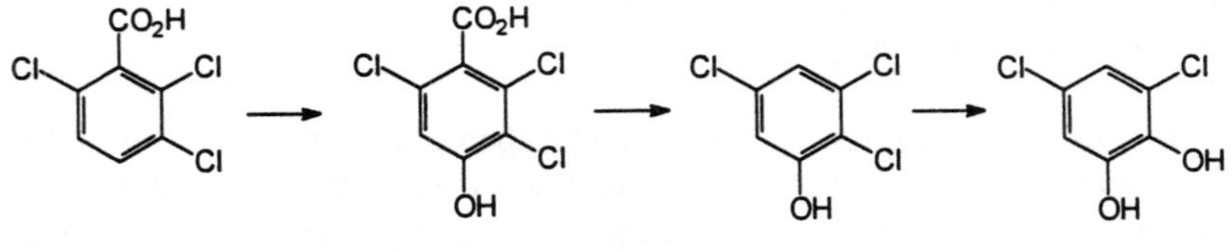

Figure 4.37   Biotransformation of 2,3,6-trichlorobenzoate.

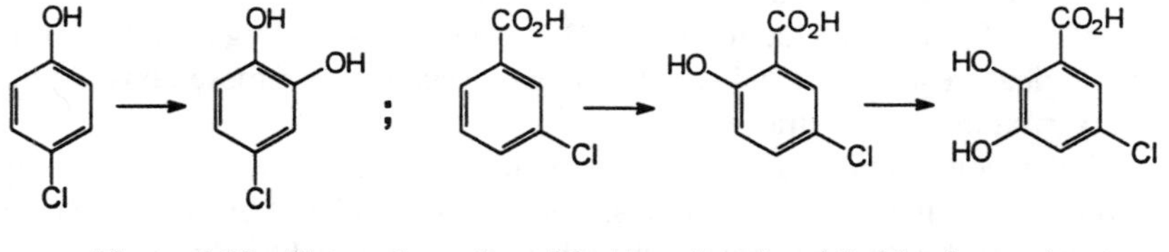

Figure 4.38   Biotransformation of 2-chlorophenol and 3-chlorobenzoate.

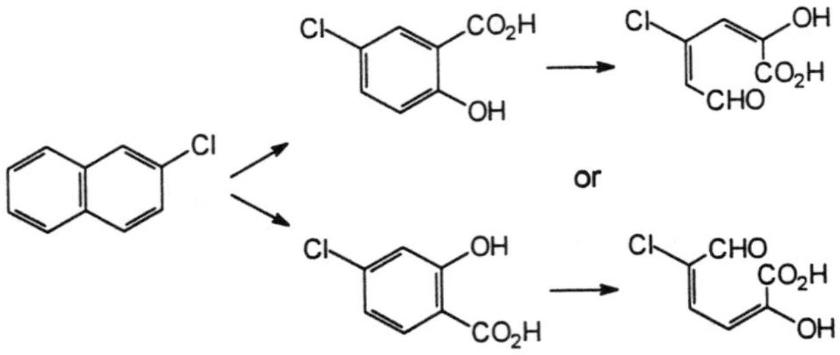

Figure 4.39 Biotransformation of 2-chloronaphthalene.

were unable to use the chlorinated compounds for growth (Spokes and Walker 1974), and oxidation of nitrobenzene to 3-nitrocatechol by strains of *Pseudomonas* sp. grown with toluene or chlorobenzene (Haigler and Spain 1991).

The induction of catabolic enzymes by pre-exposure to an analogue substrate — Cells may be grown before exposure to the xenobiotic on an analogue substrate; although the xenobiotic is extensively degraded, it cannot be used alone to support growth of the cells. Examples include the oxidation of 2-chloronaphthalene to chloro-2-hydroxy-6-ketohexa-2,4-dienoic acids (Figure 4.39) by cells in which biotransformation of the substrate was induced by growth with naphthalene (Morris and Barnsley 1982) or the oxidation of methylbenzothiophenes by cells grown with 1-methylnaphthalene (Saflic et al. 1992). Comparable situations could be encountered during experiments on bioremediation in which it may be experimentally expedient to grow cells on a suitable analogue (Klecka and Maier 1988) or to introduce the organisms into the contaminated site (Harkness et al. 1993). There are, however, inherent dangers in this procedure since, for example, cells able to degrade 5-chlorosalicylate (Crawford et al. 1979), 2,6-dichlorotoluene (Vandenbergh et al. 1981), and pentachlorophenol (Stanlake and Finn 1982) were unable to degrade the non-halogenated analogues, and the degradation of 4-nitrobenzoate was inhibited by benzoate even though the strain could use either substrate separately (Haller and Finn 1978).

Sophisticated metabolic interactions — Considerable attention has been directed to situations in which more sophisticated interactions have been uncovered, and some illustrative examples are provided.

1. Ring fission during degradation of chloroaromatic compounds is generally catalyzed by catechol-1,2-dioxygenases, whereas for the corresponding methyl compounds, a catechol-2,3-dioxygenase is involved. These pathways are generally incompatible due to the inactivation of catechol-2,3-oxygenase by 3-chlorocatechol, although simultaneous degradation of chloro- and methyl-substituted aromatic compounds has been shown to occur in some strains (Taeger et al. 1988; Pettigrew et al. 1991). This phenomenon is highly relevant to the biological treatment of industrial effluents since most of these consist of complex mixtures of substrates.

2. Simultaneous metabolism of closely related substrates may be restricted by the synthesis of inhibitory metabolites. For example, cells of an *Acinetobacter* sp. grown with 4-chlorobenzoate could dehalogenate 3,4-dichlorobenzoate, although

the organism cannot use this as sole growth substrate, and the dichlorobenzoate inhibited both its own metabolism and that of the growth substrate 4-chlorobenzoate (Adriaens and Focht 1991a,b).

3. A strain of *Pseudomonas* sp. grew with a wide range of aromatic compounds including phenol, benzoate, benzene, toluene, naphthalene, and chlorobenzene (Haigler et al. 1992), and mixtures of these substrates were degraded in continuous cultures without evidence for accumulation of metabolites. In this case, degradation depended on the presence of a non-specific toluene dioxygenase and of induction of enzymes for the *ortho* (1:2), *meta* (2:3), and the modified *ortho* pathways for degradation of the catechols. The results implied the existence of a metabolic situation additional to that normally encompassed by the terms biodegradation and biotransformation; enzymes induced by the presence of one substrate facilitated the degradation of another substrate that is not normally used as sole source of carbon and energy. These results have obvious implications for implementing bioremediation programs.

Enzyme induction by growth on unrelated compounds — Enzymes necessary for the metabolism of a substrate may be induced by growth on structurally unrelated compounds. A striking example is the degradation of trichloroethene by different strains of *Pseudomonas* sp. grown with phenol (Folsom et al. 1990) or with toluene. This result may be consistent with the widely different pathways that may be followed in toluene degradation (Figure 4.40): *P. putida* F1 by the classical toluene dioxygenase system (Zylstra et al. 1989), *P. cepacia* G4 by monooxygenation to 2-methylphenol (Shields et al. 1989), and *P. mendocina* KR by monooxidation to 4-methylphenol (Whited and Gibson 1991). A plausibly comparable situation exists for strains of *Pseudomonas* sp. and *Rhodococcus erythropolis* that were obtained by enrichment with *iso*propylbenzene and that could be shown to oxidize trichloroethene (Dabrock et al. 1992); in addition, one of the pseudomonads could oxidize 1,1-dichloroethene, vinyl chloride, trichloroethane, and 1,2-dichloroethane. This situation may be of widespread occurrence, and further examples of its existence will be facilitated by insight into the mechanisms of pathways for biodegradation.

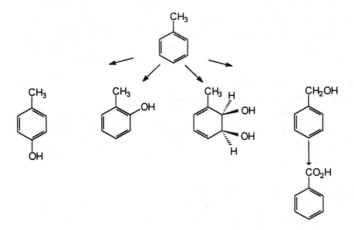

Figure 4.40 Pathways for the biotransformation of toluene.

The role of readily degraded substrates — Whereas the results from experiments on biodegradation in which readily degraded substrates such as glucose are added have probably restricted relevance to natural ecosystems in which such substrates exist in negligible concentration, situations in which readily degraded substrates are present in addition to those less readily degraded undoubtedly occur in biological waste treatment systems. In these circumstances, at least three broadly different metabolic situations may exist:

1. The presence of glucose may suppress degradation of a recalcitrant compound. Examples are: (a) strains of *Pseudomonas pickettii* degrade 2,4,6-trichlorophenol, and its degradation is induced by several other chlorophenols but is repressed in the presence of glucose or succinate (Kiyohara et al. 1992) and (b) the presence of glucose decreased the rate of degradation of phenol in a natural lake-water community, though the rate was increased by arginine (Rubin and Alexander 1983).

2. The presence of glucose may, on the other hand, enhance the degradation of a recalcitrant compound. Several different metabolic situations may be discerned, each representing a different mechanism for the stimulatory effect.

   • Experiments in which degradation of fluorobenzoates by a mixed bacterial flora was enhanced by the presence of glucose might plausibly be attributed to an increase in the cell density of the appropriate organism(s) (Horvath and Flathman 1976); a comparable conclusion could also be drawn from the data for the degradation of 2,4-dichlorophenoxyacetate and O,O-dimethyl-O-[3-methyl-4-nitrophenyl] phosphorothioate in cyclone fermentors (Liu et al. 1981).

   • The presence of readily degraded substrates such as glutamate, succinate, or glucose had a stimulatory effect on the degradation of pentachlorophenol by a *Flavobacterium* sp., possibly by ameliorating the toxic effects of pentachlorophenol (Topp et al. 1988), and enhanced the ability of natural communities to degrade a number of xenobiotics (Shimp and Pfaender 1985a).

   • The presence of glucose facilitated the anaerobic dechlorination of pentachlorophenol and may plausibly be attributed to the increased level of reducing equivalents (Henriksen et al. 1991). A comparable phenomenon is the enhancement of the dechlorination of tetrachloroethene in anaerobic microcosms by the addition of carboxylic acids including lactate, propionate, butyrate, and crotonate (Gibson and Sewell 1992).

3. The xenobiotic may be degraded in preference to glucose. This is encountered in a phenol-utilizing strain of the yeast *Trichosporon cutaneum* that possesses a partially constitutive catechol 1,2-dioxygenase (Shoda and Udaka 1980) and illustrates the importance of regulatory mechanisms in determining the degradation of xenobiotics. Constitutive synthesis of the appropriate enzyme systems may indeed be of determinative significance in many natural ecosystems.

In some circumstances, therefore, the presence of readily degraded substrates may clearly facilitate degradation of more recalcitrant xenobiotics, though this is neither universally the case nor has a generally valid mechanism for these positive effects emerged. Whereas the addition of metabolizable analogues may increase the overall

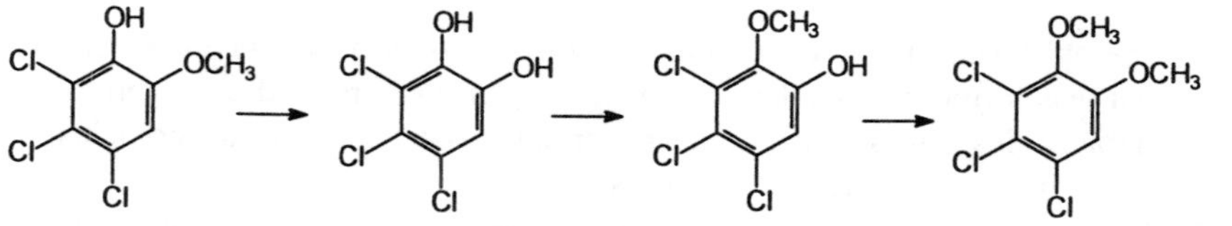

Figure 4.41 Biotransformation of 4,5,6-trichloroguaiacol by *Rhodococcus* sp. during growth with vanillate.

rates of degradation (Klecka and Maier 1988), it should be emphasized that the presence of readily degraded substrates in enrichments would generally be expected to be counterselective to the development of organisms degrading a given xenobiotic and that the observed enhancements summarized above were generally observed during relatively short time intervals.

Simultaneous metabolism of two structurally related substrates in which only one of them serves as growth substrate during biotransformation of the other may exist. A simple example is the *O*-demethylation of 4,5,6-trichloroguaiacol to 3,4,5-trichlorocatechol followed by successive *O*-methylation during growth of a strain of *Rhodococcus* sp. at the expense of vanillate (Figure 4.41) (Allard et al. 1985).

4.5 DETERMINATIVE PARAMETERS

4.5.1 Physical Parameters

Physical parameters such as temperature, salinity, pH, and oxygen concentration may critically determine the persistence or otherwise of a xenobiotic under natural conditions, and these should, therefore, be critically evaluated. Experiments can be carried out under any of the conditions that simulate the natural environment, and these can be imposed both during isolation of the organisms by enrichment and incorporated into the design of subsequent experiments on biodegradation and biotransformation. In practice, most experiments are carried out with mesophiles and at pH values in the vicinity of pH 7 — presumably motivated by the fact that these are — or are assumed to be — prevalent in natural ecosystems. It should also be emphasized that water temperatures during the winter in high latitudes in both the northern and southern hemispheres are low — probably under 10°C — so that greater advantage should be taken of psychrophiles, particularly in investigations aimed at providing realistic estimates of rates for microbial transformation under natural conditions. With the notable exception of the anaerobic sulfate-reducing bacteria, surprisingly few investigations have used truly marine bacteria in spite of the fact that substantial quantities of xenobiotics are discharged into the sea either directly or via the input from rivers. An example of the use of undefined populations of sulfate-reducing bacteria is given in Chapter 5 (Section 5.3.2) and Chapter 8 (Section 8.5.2).

The question of oxygen concentration is a good deal more complicated: whereas the extreme conditions in which oxygen concentration is high, for example, in well-mixed surface waters, or essentially absent, for example, in organic-rich deep sediments, are readily visualized and realized in laboratory experiments, it is worth drawing attention to a number of less obvious situations. The question of the oxygen tolerance of bacteria has already been discussed in Section 4.3.3.

1. The existence of microaerophilic organisms, such as *Wolinella succinogenes* that was formerly considered an anaerobe, or conversely the oxygen tolerance of many clostridia suggests that such organisms may occupy an ecological niche between the two extremes noted above.

2. Attention has already been drawn to facultatively anaerobic organisms (Section 4.3.1) that can employ either fermentative or oxidative modes for the metabolism of appropriate substrates, such as carbohydrates, and to the existence of others (Section 4.3.2) that can use nitrate as an alternative electron acceptor in the absence of oxygen; such organisms have, therefore, two metabolic options, although they are generally mutually exclusive. An illustrative example of the metabolic flexibility of facultatively anaerobic organisms is provided by the type strains of all the species of the enteric genus *Citrobacter*. These organisms are able to degrade a number of amino acids including glutamate using either respiratory or fermentative metabolism, and they can rapidly switch between these alternatives (Gerritse and Gottschal 1993). Under anaerobic conditions, the initial steps of glutamate degradation involve formation of 3-methylaspartate, mesaconate, and citramalate that are typical of clostridial fermentations (Chapter 6, Section 6.7.1). Although the oxygen concentration in natural environments may be highly variable, organisms with such a high degree of metabolic flexibility may reasonably be presumed to be at an advantage.

3. There is substantial evidence that organisms that are strictly dependent on the aerobic metabolism of substrates for growth and replication may nonetheless accomplish biodegradations and biotransformations under conditions of low oxygen concentration. Indeed, such conditions may inadvertently prevail in laboratory experiments using dense cell suspensions. It is, of course, important to remember that the *growth* of these aerobic organisms is strictly dependent on the availability of oxygen. A good illustration of the environmental role of putatively aerobic organisms under conditions of low oxygen concentration is provided by the demonstration that the rate of biodegradation of hexadecane in a marine enrichment culture was unaffected until oxygen concentrations were lower than 1% saturation (Michaelsen et al. 1992). It should also be appreciated that the oxygen concentration required for accomplishing different stages of a biodegradation pathway may differ significantly, and two examples may be used as illustration.

   • The degradation of pentachlorophenol by *Rhodococcus chlorophenolicus* proceeds by initial formation of tetrachlorohydroquinone; whereas its formation is oxygen-dependent, its subsequent degradation can be accomplished in the absence of oxygen (Apajalahti and Salkinoja-Salonen 1987).

   • Biotransformation as opposed to biodegradation may, in fact, be favored by limited oxygen concentration; a good example is provided by the synthesis of $7\alpha,12\beta$-dihydroxy-1,4-androstadien-3,17-dione from cholic acid by a strain of *Pseudomonas* sp. (Smith and Park 1984), in which oxygen limitation restricts the rate of C-9α hydroxylation that precedes degradation.

Increasing attention has been directed to the degradation of xenobiotics in aquifers, and it has been shown that most of the relevant bacteria are associated with fine particles rather than existing as free entities; it is, therefore, important to include such material in laboratory experiments using unenriched communities that attempt to simulate natural conditions (Holm et al. 1992). These bacterial associations should also be evaluated in the context of the bioavailability of the substrate that is discussed in Section 4.5.3.

4.5.2 The Influence of Substrate Concentration

Probably most laboratory experiments on biodegradation and biotransformation have been carried out using relatively high concentrations of the appropriate substrates, even though these may be far in excess of those that are likely to be encountered in natural ecosystems (Subba-Rao et al. 1982; Alexander 1985). This limitation is particularly severe during conventional tests for biodegradability and seriously restricts the degree to which the results of these experiments are environmentally relevant. Except in the immediate neighborhood of industrial discharge, low concentrations of xenobiotics are almost certainly the rule rather than the exception in natural situations, and a number of significant experimental observations should be taken into consideration.

Utilization of low substrate concentrations — Investigations into the flora of natural waters have revealed the presence of bacteria able to grow with extremely low substrate concentrations, and these have been termed oligotrophs (Poindexter 1981). It is, of course, well established that organisms such as *Aeromonas hydrophila* (Van der Kooij et al. 1980), *Pseudomonas aeruginosa* (Van der Klooij et al. 1982), and a species of *Spirillum* (Van der Kooij and Hijnen 1984) may proliferate in natural waters supplemented with low concentrations of additional organic carbon, though the significance of such organisms in natural systems burdened with xenobiotics is less clearly resolved. The critical issue would seem to be the effectiveness of substrate transport into the cells. Although it has been suggested that there exist bacteria specially adapted to such conditions, it seems doubtful on the basis of existing evidence whether it is really justifiable to make a distinction between eutrophic and oligotrophic bacteria (Martin and MacLeod 1984).

Possible existence of threshold concentrations — An issue which has received some — though almost certainly insufficient — attention exists when bacteria are exposed to extremely low concentrations of a xenobiotic — of the order of ng/L or less. For example, although the rates of biodegradation of phenol, benzoate, benzylamine, 4-nitrophenol, and di(2-ethylhexyl)phthalate in natural lake water were linear over a wide range of substrate concentrations between ng/L and µg/L (Rubin et al. 1982), it has been shown that the rates of degradation of 2,4-dichlorophenoxyacetate at concentrations of the order of µg/L were extremely low (Boethling and Alexander 1979). This observation has subsequently been extended to a greater range of compounds (Hoover et al. 1986). These and other data could be interpreted as supporting the concept of a threshold concentration below which growth and degradation does not take place — or occurs at insignificant rates. Although the reasons for the existence of such threshold concentrations have not been clearly resolved, a number of plausible hypotheses may be suggested: (1) the substrate concentrations may be too low for effective transport into the cells, or (2) there may be a limiting substrate concentration required for induction of the appropriate catabolic enzymes. At low substrate concentrations, the necessary enzymes would simply not be synthesized, and this could be the determining factor in some circumstances (Janke 1987). Experiments with chlorinated benzenes in which the effect of substrate concentration was examined in batch cultures and in recirculating fermentors showed that although substrates could be degraded completely in the former, a residual concentration of the substrate persisted in the latter (van der Meer et al. 1992).

All these observations emphasize that tests for biodegradability carried out at high substrate concentrations may not adequately predict the rates of degradation occurring in natural ecosystems where only low concentrations of xenobiotics are encountered (Alexander 1985). This phenomenon is, therefore, of enormous environmental importance, since it would imply the possibility of extreme persistence of low concentrations in natural ecosystems. The further exploration of this phenomenon is probably only limited in practice by the access to analytical methods for measuring sufficiently accurately substrate concentrations at the level of ng/L or lower. Most studies that have been carried out have, therefore, used ^{14}C-labeled substrates, which necessarily limits the range of compounds accessible and restricts the elucidation of metabolic pathways in which only biotransformation or partial mineralization has occurred.

Strategies used by cells for substrates with low or negligible water solubility — In Chapter 5 this problem will be discussed in the context of the design of laboratory experiments on biodegradation and biotransformation, but a more formal approach will be adopted here. Organisms have developed their own strategies to circumvent the low water solubility of their substrates, and a few examples may be given as illustration:

1. They may produce extracellular enzymes which attack the substrate without the need for transport into the cell; examples are cellulase, DNAse, or gelatinase.
2. They may synthesize surface-active emulsifying compounds during growth. This problem has been extensively investigated because of its commercial application (Gerson and Jajic 1979), and some of the key conclusions from a wide range of studies may be briefly summarized.

   • Glycolipids consisting of long-chain carboxylic acids and rhamnose (Itoh and Suzuki 1972; Rendell et al. 1990) or trehalose (Suzuki et al. 1969; Singer et al. 1990) have been isolated during growth of a number of different bacteria on *n*-alkanes. The rhamnolipid surfactant produced by a strain of *Pseudomonas* sp. was effective in enhancing degradation of octadecane (Zhang and Miller 1992), though the concentration used in these experiments was rather high (300 mg/L) to encourage its practical application.

   • A polyanionic heteropolysaccharide (emulsan) is produced during growth of a strain of *Acinetobacter calcoaceticus* with hydrocarbon mixtures, and the high molecular weight polymer is necessary for emulsifying activity (Shoham and Rosenberg 1983). The value of emulsan for treating oil spills seems, however, equivocal in the light of results that demonstrate reduced biodegradation in its presence (Foght et al. 1989).

   • A *Rhodococcus* sp. synthesizes a glycolipid during growth with *n*-alkanes and *n*-alkanols but not with carboxylic acids, triglycerides, or carbohydrates, and its formation was favored by nitrogen limitation (Singer and Finnerty

1990). These conditions may, of course, be counterproductive to optimal growth of the organisms.

- The synthesis of an emulsifying agent produced by *Candida lipolytica* is inducible during growth with a number of *n*-alkanes but is not synthesized during growth with glucose (Cirigliano and Carman 1984).

A number of different situations have, therefore, clearly emerged, and it seems premature to draw general conclusions, especially in respect of the application of these natural surfactants to the bioremediation of oil spills.

Transport mechanisms — The mechanisms whereby carbohydrates, carboxylic acids, and glycerol are transported across the bacterial cell membrane prior to metabolism have been elucidated in great detail. By contrast, very little effort has apparently been directed to the transport of xenobiotics. It has been suggested that active transport systems for benzoate (Thayer and Wheelis 1982) and for mandelate (Higgins and Mendelstam 1972) are involved, though the mechanisms have not been defined. A detailed investigation (Groenewegen et al. 1990) has examined the uptake of 4-chlorobenzoate by a coryneform bacterium which degraded this compound, and the following observations were made:

1. The uptake was inducible and occurred in cells grown with 4-chlorobenzoate but not with glucose
2. A proton motive force (Δp) driven mechanism was almost certainly involved,
3. Uptake could not take place under anaerobic conditions unless an electron acceptor such as nitrate was present.

The transport of toluene-4-sulfonate into *Comamonas testosteroni* has also been examined (Locher et al. 1993), and again rapid uptake required growth of the cells with toluene-4-sulfonate or 4-methylbenzoate. From the results of experiments with various inhibitors, it was concluded that a toluenesulfonate anion/proton symport system operates rather than transport driven by a difference in electrical potential ($\Delta\psi$).

These interesting results illustrate the potential complexity of this hitherto neglected aspect of the degradation of xenobiotics and suggest that extension to other organisms and to a wider range of xenobiotics is merited.

4.5.3 Bioavailability: "Free" and "Bound" Substrates

In the discussion of partition in Chapter 3 (Sections 3.2.2 and 3.2.3), it has been pointed out that xenobiotics exist not only in the "free" state but also in association with organic and mineral components of the particles in the water mass and in the sediment phase. This association is a central determinant of the persistence of xenobiotics in the environment, since the extent to which such residues are accessible to microbial attack is largely unknown, and its magnitude is critical also to the effectiveness of bioremediation (Harkness et al. 1993). Attention is also drawn to the critical role of complexes between xenobiotics and Fe in determining biodegradability, and this is discussed in Chapter 5 (Section 5.2.2). Only a brief attempt is made to place this important subject in perspective and to illustrate some of the major unresolved issues. At least two important facts should be borne in mind in evaluating the results that are used as illustration:

1. Although the most persuasive evidence for the significance of bioavailability comes from experiments in terrestrial systems, there seems no reason to doubt the validity of translating the principles to aquatic sediments that contain organic matter structurally resembling that of soils.
2. Most experiments have used spiked samples and, therefore, do not take into account the cardinal issue of aging after deposition that has been discussed in Chapter 3 (Section 3.2.2) and that should be critically evaluated.

Examples illustrating the generality of the concept include the following:

1. Montmorillonite complexes with benzylamine at concentrations below 200 µg/L decreased the extent of mineralization in lake water samples, though a similar effect was not noted with benzoate (Subba-Rao and Alexander 1982). Even in apparently simple systems, general conclusions cannot, therefore, be drawn even for two not entirely dissimilar aromatic compounds, both of which are readily degradable under normal circumstances in the dissolved state.
2. Suspensions of 2,4-dichlorophenoxyacetate sorbed to sterile soils were completely protected from degradation by either free or sorbed bacteria, and degradation of the substrate required access by the bacteria to the free compound in solution (Ogram et al. 1985). Rates of degradation in a high organic soil were lower than for a low organic soil (Greer and Shelton 1992), and this further supports the significance of desorption of the xenobiotic in determining its biodegradability.
3. ^{14}C-labeled 2,4-dichlorophenol bound to synthetic or natural humic acids or polymerized by H_2O_2 and peroxidase was mineralized to CO_2 only to a limited extent (<10%), and the greater part remained bound to the polymers (Dec et al. 1990).
4. ^{14}C-labeled 3,4-dichloroaniline-lignin conjugates were degraded to $^{14}CO_2$ by *Phanerochaete chrysosporium* as effectively as the free compound (Arjmand and Sandermann 1985), and it was therefore concluded that these "bound" residues were not persistent in the environment. This may, however, represent a special case for several reasons (1) although this organism is able to degrade lignin, the relevance of such organisms in most aquatic environments is possibly marginal; (2) the lignin peroxidases implicated in lignin degradation are generally extracellular so that soluble substrates are probably not necessary.
5. The presence of humic acids had a detrimental effect on the degradation of substituted phenols by a microbial community after lengthy adaption to the humic acids and was not alleviated by the addition of inorganic nutrients (Shimp and Pfaender 1985b). The diminished number of organisms with degradative capability was responsible for the reduced degree of degradation, so that the predominant effect was probably the toxicity of the humic acids even towards adapted microorganisms.
6. In short-term experiments with carbofuran (2,3-dihydro-2,2-dimethyl-7-benzofuranyl-*N*-methylcarbamate), degradation was accomplished by organisms in an enrichment culture obtained from soils with a low carbon content where sorption of the substrate is low but was essentially absent in cultures obtained from soils with a high organic matter (Singh and Sethunathan 1992).

All of the preceding investigations have been concerned with polar compounds. It may, therefore, be valuable to provide a perspective by giving some examples that examined neutral compounds.

1. The aerobic mineralization of α-hexachlorocyclohexane by endemic bacteria in the soil is limited by the rate of its desorption and by intraparticle mass transfer (Rijnaarts et al. 1990).
2. Whereas degradation of the readily extractable toluene in spiked soil by *Pseudomonas putida* was rapidly accomplished, there was a residue that was degraded much more slowly at a rate that was apparently dependent on its desorption (Robinson et al. 1990).
3. The extent of bioremediation of sediments contaminated with PCBs appears to be limited by the association of a significant fraction with organic components of the sediment phase (Harkness et al. 1993).

The results of these experiments in both aquatic systems and terrestrial systems may profitably be viewed against the extensive evidence for the persistence of agrochemicals in the terrestrial environment. Considerable effort has been directed to the issue of bound residues of agrochemicals (Calderbank 1989) and to its significance in determining both their biological effects and their persistence. At the same time, it should be appreciated that from an economic point of view, enhanced rates of degradation of agrochemicals in the terrestrial environment may be highly undesirable (Racke and Coats 1990). Nonetheless, care must be taken in making generalizations. For example, whereas it has been established that soil microorganisms may significantly increase the evolution of $^{14}CO_2$ from ^{14}C parathion (Racke and Lichtenstein 1985), this was not observed with chlorpyrifos (Racke et al. 1990).

Analysis of these diverse observations clearly demonstrates that the persistence of a xenobiotic in the aquatic or in the terrestrial environment may be significantly increased if the xenobiotic is bound either to inorganic minerals or to any of a range of complex natural polymers such as humic and fulvic acids. One of the key issues is the rate of desorption of the xenobiotic from the matrices; this has already been discussed in Chapter 3 (Section 3.2.2), and this may depend critically on the mechanism of the association (Section 3.2.3). The degree of bioavailability may also depend on the nature of the relevant organisms, since, for example, soil-sorbed naphthalene was degraded at markedly different rates by two naphthalene-degrading organisms (Guerin and Boyd 1992), and at low substrate concentrations, 2,4-dichlorophenoxyacetate was degraded at different rates by the two strains that were examined (Greer and Shelton 1992). Although a number of important unresolved issues remain, it is clear that the degree of bioavailability of xenobiotics in natural systems introduces an important additional uncertainty in extrapolating the results of studies on biodegradation and biotransformation of "free" xenobiotics in laboratory experiments to processes and rates in natural ecosystems.

4.5.4 Pre-Exposure: Pristine and Contaminated Environments

Experimental aspects of the elective enrichment procedure are discussed in some detail in Chapter 5, but the question of its existence and significance in natural populations already exposed to xenobiotics is conveniently addressed here. It is important to distinguish between induction (or derepression) of catabolic enzymes and selection for a specific phenotype. The former is a relatively rapid response, so that exposure of samples from uncontaminated areas to xenobiotics for a period of weeks or months would be expected to result in the selection of organisms with degradative potential rather than merely be the result of low rates of enzyme induction. The results which have been obtained generally support the view that pre-exposure increases the number of organisms capable of degrading a given xenobiotic, though only a few attempts have been made to quantify the number of organisms involved. There is convincing evidence that exposure to unusual substrates in laboratory experiments elicits the synthesis of genes for their degradation (Mortlock 1982) and increasing support for the view that exposure to xenobiotics increases the probability of mutations that are favorable to the degradation of these substrates (Hall 1990; Thomas et al. 1992). The following examples attempt to illustrate the spectrum of responses that have been observed.

1. Rates of mineralization of the more readily degraded PAHs, such as naphthalene and phenanthrene, were greater in samples from PAH-contaminated areas than in those from pristine sediments, although even in the former, the rates for benza[a]anthracene and benz[a]pyrene were extremely low (Herbes and Schwall 1978).

2. Experiments using marine sediment slurries have examined the effect of pre-exposure to various aromatic hydrocarbons on the rate of subsequent degradation of the same or other hydrocarbons. The results clearly illustrated the complexity of the selection process; for example, whereas pre-exposure to benzene, naphthalene, anthracene, or phenanthrene enhanced the rate of mineralization of naphthalene, similar pre-exposure to naphthalene stimulated the degradation of phenanthrene but had no effect on that of anthracene (Bauer and Capone 1988).

3. In experiments using soil samples from a pristine aquifer exposed in the laboratory to a range of compounds, widely diverse responses were observed (Aelion et al. 1987):

   - The bacterial population was apparently already adapted to some of the compounds, such as phenol, 4-chlorophenol, and 1,2-dibromoethane, at the start of the experiment, and these substrates were therefore rapidly degraded.
   - No adaptation was found with chlorobenzene or 1,2,4-trichlorobenzene, and only slight mineralization was observed.
   - A linear increase in the rate of degradation with increasing length of exposure was noted for some generally readily degraded compounds such as aniline.
   - True adaptation by selection of the appropriate organisms was observed only for 4-nitrophenol.

   It, therefore, seems premature to draw general conclusions on the influence of pre-exposure on the biodegradability of structurally different substrates.

4. Systematic studies on the degradation of 4-nitrophenol (Spain et al. 1984) showed that the rates of adaptation in a natural system were comparable to those observed in a laboratory test system and were associated with an increase in the number of degrading organisms by up to one thousand-fold.

5. It has been consistently observed that a wide range of agrochemicals applied successively to the same plots are increasingly readily degraded, presumably due to enrichment of the appropriate degradative microorganisms (Racke and Coats 1990); examples include compounds as diverse as naphthalene (Gray and Thornton 1928), γ-hexachlorocyclohexane (Wada et al. 1989) and triazines (Cook and Hütter 1981).

6. Dehalogenation of polychlorinated or polybrominated biphenyls was more rapid in cultures using inocula prepared from sediments contaminated with the chlorinated or brominated biphenyl, respectively (Morris et al. 1992).

These results are particularly relevant to bioremediation and suggest that organisms originally isolated from areas either contaminated naturally (Fredrickson et al. 1991) or as a result of industrial activity (Grosser et al. 1991) may be particularly attractive. The discussion hitherto has been devoted to the issue of selection, but for the sake of completeness, it should be noted that the alternative approach of deliberately adding an inducer has also been examined. Salicylate is an inducer of the enzymes for degradation of naphthalene, and its addition to soil has been shown to result in a modest increase in the number of organisms degrading naphthalene (Ogunseitan et al. 1991). Since, however, naphthalene is such a readily degraded compound with a low level of persistence, the wider evaluation of this interesting idea would have to be explored before its application to practical situations could be justified.

This discussion may also be viewed against the background of studies on genetic transfer of catabolic activity towards a given xenobiotic. Considerable care must be exercised in the interpretation even of data which seem supportive of this process, since selection and enrichment from a small population of organisms initially present must be excluded. One interesting investigation on the biodegradation of aniline revealed the existence of two genotypes differing in their tolerance to the substrate. The dominant organism, which was originally present, assimilated aniline at micromolar concentrations but was inhibited at higher concentrations; a mutant could, however, be isolated from a population of several hundred cells or by continuous culture, and this organism tolerated millimolar concentrations. Populations of the two organisms in a natural system were apparently regulated by the prevailing concentration of aniline (Wyndham 1986).

In attempting to synthesize the results of all these experiments, it is hardly possible to escape the conclusion that our understanding of the processes regulating the population dynamics of microorganisms in natural systems is limited.

4.6 RATES OF METABOLIC REACTIONS

4.6.1 Kinetic Aspects

The question of the rates at which xenobiotics are degraded or transformed is of cardinal importance, since it is upon their quantification that a given compound can in the final analysis be designated persistent or otherwise (Battersby 1990). It should be appreciated at the outset that, even if acceptable rates of degradation are observed in laboratory experiments, the final assessment of persistence depends upon the demonstration that the compound is indeed degraded under natural conditions. And in the long run, it is the latter fact that is of primary environmental significance. As will have been appreciated from earlier discussions, however, persistence is determined not only by rates of biotic and abiotic degradation but also by the accessibility of the substrate which may have been concentrated into biota or associated with organic or inorganic components in the water mass or in the sediment phase. There are a number of important issues which must be addressed in discussions of rates, and some of them are briefly discussed in the following paragraphs.

1. Even if rate constants can be measured in laboratory experiments, these must be normalized to the number of microbial cells. This may pose only minor problems with an axenic culture in the laboratory and has been consistently used in some investigations where well-defined kinetics prevail (Allard et al. 1987), but this becomes a major problem in natural situations: how many of the organisms are metabolically active in accomplishing the given reaction? Specific DNA probes have been used for detection of genes coding for heavy metal resistance (Diels and Mergeay 1990) and for the detection of pathogens (Samadpour et al. 1990). This procedure has, however, been less extensively applied to organisms of catabolic significance (Sayler et al. 1985; Holben et al. 1992), although the values obtained for 2,4-dichlorophenoxyacetic-degrading populations agreed well with those using conventional most-probable-number methods (Holben et al. 1992). This is an attractive approach even if the availability of the relevant probes presents a practical limitation to its more widespread application.

2. Although some possible kinetic models have been assembled (Simkins and Alexander 1984) and their application has been evaluated (Simkins et al. 1986), microbial reactions may not follow well-defined kinetics. They may, for example, exhibit multi-phase kinetics which have been illustrated during the transformation of methyl parathion by a *Flavobacterium* sp. (Lewis et al. 1985); system I was a high-affinity, low-capacity system, whereas system II was the opposite. The results of such studies underscore the potential errors which may invalidate predictions which neglect multi-phase kinetics.

3. Attempts have been made to apply the structure-activity concept (Hansch and Leo 1979) to environmental problems, and this has been successfully applied to the rates of hydrolysis of carbamate pesticides (Wolfe et al. 1978b) and of esters of chlorinated carboxylic acids (Paris et al. 1984). This has been extended to correlating rates of biotransformation with the structure of the substrates and has been illustrated with a number of single-stage reactions. Some examples are sufficient to illustrate the application of this procedure.

   • Rates of bacterial hydroxylation of substituted phenols to catechols by *Pseudomonas putida* correlated well with the van der Waals radii of the substituents (Paris et al. 1982), and this was also demonstrated for the biotransformation of anilines to catechols both by this strain and by a natural population of bacteria (Paris and Wolfe 1987).
   • Rates of hydrolysis of substituted aromatic amides by the bacterial population of pond water correlated well with the infrared C=O stretching frequencies of the substrates (Steen and Collette 1989).

- Rates of anaerobic dechlorination of aromatic hydrocarbons (Peijnenburg et al. 1992) and of the hydrolysis of aromatic nitriles under anaerobic conditions (Peijnenburg et al. 1993) have been correlated with a number of parameters including Hammett σ-constants, inductive parameters and evaluations of the soil/water partitioning.

There are, however, some important limitations to a general extension of this principle for a number of reasons: biodegradation — as opposed to biotransformation — of complex molecules necessarily involves a number of sequential reactions, each of whose rates may be determined by complex regulatory mechanisms. For novel compounds containing structural entities that have not been previously investigated, the level of prediction must necessarily be limited by the lack of relevant data.

4. The significant and incompletely resolved issue of the possible occurrence of threshold substrate concentrations below which rates of biodegradation may become extremely low or even non-existent has already been noted in Section 4.5.2.

Too olympian a view of the problem of rates should not, however, be adopted; an overly critical attitude should not be allowed to pervade the discussions — provided that the limitations of the procedures that are used are clearly appreciated and set forth. In view of the great practical importance of quantitative estimates of persistence to microbial attack, any procedure — even if it provides merely orders of magnitude — should not be neglected.

4.6.2 Metabolic Aspects: Nutrients

In many natural ecosystems, microbial growth and metabolism are limited by the concentrations of inorganic nutrients such as nitrogen, phosphorus, or even iron. Systematic investigation of such limitation on biodegradation has seldom been carried out except when the xenobiotic contains N or P that is used for growth of the cells. Two important examples of the importance of nutrient limitation in determining the realization of biodegradation may, however, be noted: (1) expression of lignin peroxidases in *Phanerochaete chrysosporium* is induced by N-limitation and also by the concentration of Mn (II) in the medium (Perez and Jeffries 1990); this is discussed further in Chapter 5 (Section 5.2.2) and (2) under conditions of selenium starvation, *Clostridium purinilyticum* degrades uric acid by an unusual pathway involving cleavage of the iminazole ring to produce 5,6-diaminouracil which is then degraded to formate, acetate, glycine, and CO_2 (Dürre and Andreesen 1982).

In experiments using field samples containing natural assemblies of microorganisms, two effects of supplementation with nitrogen or phosphorus have been encountered: (1) a decreased lag phase was observed before transformation of 4-methylphenol (Lewis et al. 1986), and (2) the bacterial population increased, though there was no effect on the second-order rates of transformation of a number of compounds including phenol and the agrochemicals propyl 2,4-dichlorophenoxyacetate, methyl parathion, and methoxychlor (Paris and Rogers 1986).

Limitation in the concentrations of these inorganic nutrients does not, therefore, appear to have a dramatic effect on the persistence of the relatively few compounds that have been examined systematically.

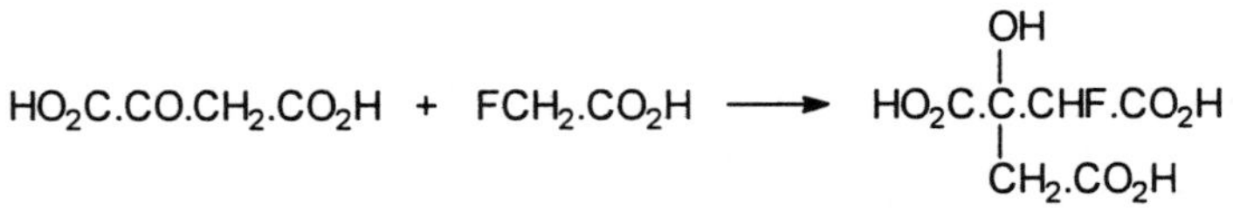

$$\text{HO}_2\text{C.CO.CH}_2\text{.CO}_2\text{H} \;+\; \text{FCH}_2\text{.CO}_2\text{H} \;\longrightarrow\; \underset{\overset{|}{\text{CH}_2\text{.CO}_2\text{H}}}{\overset{\overset{\text{OH}}{|}}{\text{HO}_2\text{C.C.CHF.CO}_2\text{H}}}$$

Figure 4.42   Incorporation of fluoracetate into fluorocitrate.

Consistent with the preceding comments on the metabolism of xenobiotics in the presence of additional carbon substrates (Section 4.4.2), the situation with carbon additions may be quite complex and need not be addressed in detail again. Two simple examples may be given. In these, addition of glucose apparently elicited two differing responses, though it should be emphasized that since the concentration of such readily degradable substrates in natural aquatic systems will be extremely low, the environmental relevance of such observations will inevitably be restricted: (1) the rate of mineralization of phenol by the flora of natural lake water decreased (Rubin and Alexander 1983) and (2) the rate of mineralization of 4-nitrophenol was enhanced in lake water inoculated with a *Corynebacterium* sp. when rates of mineralization were low (Zaidi et al. 1988).

4.7 REGULATION AND TOXIC METABOLITES

There are a number of examples in which metabolites are produced which themselves toxify the organism responsible for their synthesis. The classic example is fluoroacetate (Peters 1952), which enters the TCA cycle and is thereby converted into fluorocitrate (Figure 4.42); this compound effectively inhibits aconitase — the enzyme involved in the next metabolic step — so that cell metabolism itself is inhibited with the resulting death of the cell. It should be noted, however, that bacteria able to degrade fluoroacetate to fluoride exist, so that some organisms have developed the capability for overcoming this toxicity (Meyer et al. 1990). The issue of toxic metabolites merits attention in a wider context (Chapter 7, Section 7.5), and it has also been documented among microorganisms in a number of diverse metabolic situations.

1. Biotransformation of chlorobenzene by *Pseudomonas putida* grown on toluene or benzene resulted in the formation of 3-chlorocatechol; this inhibited further metabolism by catechol 2,3-dioxygenase, so that its presence resulted in the formation of catechols even from benzene and toluene (Gibson et al. 1968; Klecka and Gibson 1981; Bartels et al. 1984). The same situation has emerged in the degradation of 3-chlorobenzoate; the inhibition that would result from the inhibitory effect of 3-chlorocatechol on catechol 2,3-dioxygenase is lifted by the synthesis of catechol 1,2-dioxygenase (Reinecke et al. 1982). This is, indeed, a general strategy that is discussed again in Chapter 6 (Section 6.5.1).
2. The converse situation occurs during the degradation of 4-chlorobenzoate in which the synthesis of catechol 1,2-dioxygenase circumvents the production of chloroacetaldehyde by the action of catechol 2,3-dioxygenase (Reinecke et al. 1982).

3. The degradation of trichloroethene by methylotrophic bacteria involves the epoxide as intermediate (Little et al. 1988); further transformation of this results in the production of CO which may toxify the bacterium, both by competition for reductant and by enzyme inhibition (Henry and Grbic-Galic 1991). The inhibitory effect of CO may, however, be effectively overcome by adding a reductant such as formate.

These observations may be considered in the wider perspective of metabolic regulation, of which only a few examples have been parenthetically discussed. Very few studies have been directed primarily to the genetics and the regulation of the enzymes for the degradation of xenobiotics; probably most of the enzymes are inducible, and this is consistent with the fact that most strains have been isolated after specific enrichment with the xenobiotic. In the case of biotransformation, however, there are sporadic examples of the constitutive synthesis of enzymes. The case of the partially constitutive synthesis of catechol 1,2-dioxygenase in the yeast *Trichosporon cutaneum* (Shoda and Udaka 1980) has already been noted (Section 4.4.2). Another example is the system carrying out the *O*-methylation of halogenated phenolic compounds that was apparently constitutive (Neilson et al. 1988); this observation is consistent with the isolation of the strains by enrichment with C_1 compounds structurally unrelated to the halogenated substrates. The *O*-methylation reaction may function primarily as a detoxification system, so that constitutive synthesis of the enzyme would clearly be advantageous to the survival of the cells.

4.8 CATABOLIC PLASMIDS

Plasmids may be defined as fragments of DNA which replicate outside the bacterial chromosome, and they are important in a number of different contexts:

- As carriers of antibiotic resistance: the emergence of antibiotic-resistant strains has had serious repercussions in the application of antibiotic therapy and has seriously increased the danger of nosicomial infections.
- The presence of unusual carbohydrate fermentation patterns — particularly for lactose — and the ability to use citrate among *Enterobacteriaceae* has hindered — and in some cases, jeopardized — the identification of pathogenic strains including *Salmonella typhimurium*.
- Resistance to heavy metals such as Hg may be mediated by plasmid-borne genes.
- Genes coding for the catabolism of a large number of diverse xenobiotics are carried on plasmids; this is more germane to the present discussion, and some examples are given in the table.

Probably the greatest interest in catabolic plasmids stems from the possibility of constructing strains having increased metabolic potential towards xenobiotics and from the potential application of such strains to waste treatment systems. Considerable effort has been devoted to aromatic chlorinated compounds, and conspicuous success has been achieved in overcoming problems with, for example, synthesis of

Plasmids Carrying Genes Coding for the Biodegradation or Biotransformation of Xenobiotics

Hydrocarbons and related compounds	
Octane	Chakrabarty et al. 1973
Camphor	Rheinwald et al. 1973
Citronellol, geraniol	Vandenbergh and Wright 1983
Linalool	Vandenbergh and Cole 1986
Toluene and xylene	Williams and Worsey 1976
Naphthalene	Connors and Barnsley 1982
Dibenzothiophene	Monticello et al. 1985
Halogenated compounds	
Chloroalkanoates	Kawasaki et al. 1981; Hardman et al. 1986
Dichloromethane	Gälli and Leisinger 1988
Chlorobenzoates	Chaterjee et al. 1981
Chlorophenoxyacetates	Don and Pemberton 1985; Chaudry and Huang 1988
Chlorobiphenyls	Furukawa and Chakrabarty 1982; Shields et al.1985
Chloridazon	Kreiss et al. 1981
Bromoxynil	Stalker and McBride 1987
Chlorobenzenes	van der Meer et al. 1991
Other structural groups	
6-aminohexanoate cyclic dimer	Negoro et al. 1980; Kanagawa et al. 1989
Parathion	Serdar et al. 1982; Mulbry et al. 1986
Nicotine	Brandsch et al. 1982
Cinnamic acid	Andreoni and Bestetti 1986
Ferulic acid	Andreoni and Bestetti 1986
S-ethyl-N,N-dipropylthiocarbamate	Tam et al. 1987
Aniline	Anson and Mackinnon 1984
2-aminobenzenesulfonate	Jahnke at al. 1990
1,1′-dimethyl-4,4′-bipyridinium dichloride	Salleh and Pemberton 1993

toxic metabolites such as 3-chlorocatechol during the degradation of chlorobenzenes by natural strains. There are several important issues which should, however, be addressed before considering the application of such artificially constructed strains to biological treatment systems:

1. Competition with endemic strains which may eventually outnumber and eliminate the introduced strains.
2. Genetic instability if selection pressure is removed through fluctuations in the loading of the xenobiotic.
3. Concern over the discharge of such strains into the environment: attempts have, therefore, been made to incorporate safeguards so that the strains are unlikely to survive in competition with natural strains in the ecosystem.

The questions of plasmid transmission and the stability of plasmids in natural ecosystems have received somewhat less attention, so that great caution should be exercised in drawing general conclusions on the basis of the somewhat fragmentary evidence from laboratory experiments. Some important principles may be illustrated with the following observations:

1. A study with a strain of plasmid-borne antibiotic-resistant *Escherichia coli* indicated that the strain did not transmit these plasmids to indigenous strains after introduction into the terrestrial environment (Devanas et al. 1986).

2. In enteric bacteria carrying thermosensitive plasmids coding for the utilization of citrate and for resistance to antibiotics, it has been shown (Smith et al. 1978) that rates of transmission are negligible at 37°C but appreciable at 23°C — a temperature more nearly approaching that which prevails in most natural ecosystems.

3. There is evidence that, even in the absence of selective pressure exerted by the presence of a xenobiotic, bacterial populations may retain a small number of organisms carrying the relevant degradative plasmids. For example, strains of *Pseudomonas putida,* in which the degradation of toluene is mediated by genes on the non-conjugative TOL plasmid, maintain a small population of cells carrying the plasmid even in the absence of toluene (Keshavarz et al. 1985). It has also been shown (Duetz et al. 1994) that cultures of *Pseudomonas putida* grown with growth-limiting concentrations of succinate express TOL catabolic genes in both the upper and lower pathways in response to *o*-xylene (that is not metabolized) but fail to do so during non-limited growth with succinate.

4. Strains of indigenous groundwater bacteria, and a strain of *Pseudomonas putida* carrying a TOL plasmid, were introduced into microcosms prepared from a putatively pristine aquifer. DNA-specific probes were used to monitor the numbers of organisms carrying the genotypes, and it was found that the stability of genotypes for the degradation of toluene was maintained in the absence of selective pressure over the eight-week period of the experiment (Jain et al. 1987).

5. A strain of *Alcaligenes* sp. carrying a plasmid bearing the genes for the degradation of 3-chlorobenzoate was introduced into a freshwater microcosm system, and a specific DNA probe was used to enumerate the organisms bearing this gene (Fulthorpe and Wyndham 1989). A number of generally important results were obtained:

   • The presence of 3-chlorobenzoate was needed to maintain the catabolic genotype.
   • The number of probe-positive organisms often greatly outnumbered that of the original organisms determined by plate counts.
   • The nature of these "additional" probe-positive organisms is unknown.

Clearly, therefore, a great deal of investigation is required before even a tentative generalization can be hazarded on the cardinal issues of the stability of plasmids in natural ecosystems, the extent to which these plasmids are transmissible, and the stability of the genotypes in the absence of selective pressure. At least some of the apparently conflicting views may be attributed to the different organisms that have been used, including their nutritional demands — compounded by the widely varying environments in which their stability has been examined (Sobecky et al. 1992). Currently, the greatest volume of research is apparently being devoted to purely genetic aspects of these problems.

4.9 CONCLUSIONS

It may be valuable to summarize the broader conclusions that may be drawn from the extensive literature.

1. It is imperative to distinguish clearly between the degradation and the transformation of xenobiotics; this applies equally to biotic and abiotic processes. Failure to do so may result in erroneous estimates of the fate of a compound and of its toxicity that may be mediated by its metabolites.

2. Conventional assays for determining the biodegradability of a xenobiotic do not provide a rational basis for extrapolating the results to conditions that occur in the natural environment.

3. Abiotic reactions, including those induced photochemically, may be important in degradation and, in addition, may act in concert with microbial processes.

4. A wide range of taxonomically distinct organisms is able to degrade or transform xenobiotics, and reactions carried out under anaerobic conditions merit continued attention. Sequential transformation reactions carried out by different organisms may be necessary to accomplish degradation of the initial substrate, and attempts to mimic such situations in laboratory experiments have been illuminating.

5. The extent to which a xenobiotic is susceptible to microbial attack depends on a number of environmental factors, including the concentration of the substrate, its bioavailability, and the presence of other substrates as well as on physical parameters such as temperature, salinity, and pH.

6. Attention should be directed to the complex reactions that take place in the natural environment and, in particular, to the situation that exists when microorganisms are exposed to several substrates at different concentrations and to the complex kinetics that may result.

REFERENCES: REVIEWS AND BOOKS

Anthony, C. 1982. *The biochemistry of methylotrophs*. Academic Press, London.

Berry, D.F., A.J. Francis, and J.-M. Bollag. 1987. Microbial metabolism of homocyclic and heterocyclic aromatic compounds under anaerobic conditions. *Microbiol. Revs.* 51: 43-59.

Chakrabarty, A.M. (Ed.). 1982. *Biodegradation and detoxification of environmental pollutants*. CRC Press, Boca Raton.

Cook, A.M. 1987. Biodegradation of *s*-triazine xenobiotics. *FEMS Microbiol. Revs.* 46: 93-116.

Dagley, S. 1975. A biochemical approach to some problems of environmental pollution pp.81-138. In *Essays in biochemistry* Vol. 11 (Eds. P.N. Campbell and W.N. Aldridge). Academic Press, London.

Dagley, S. 1977. Microbial degradation of organic compounds in the biosphere pp.121-170. In *Survey of progress in chemistry* Vol. 8 (Ed. A.F. Scott). Academic Press, New York.

Dagley, S. 1978. Determinants of biodegradability. *Quart. Revs. Biophys.* 11: 577-602.

Dalton, H., and D.I. Stirling. 1982. Co-metabolism. *Phil. Trans. R. Soc. London.* B 297: 481-496.

Ensley, B.D. 1991. Biochemical diversity of trichloroethylene metabolism. *Ann. Rev. Microbiol.* 45: 283-299.

Evans, W.C., and G. Fuchs. 1988. Anaerobic degradation of aromatic compounds. *Ann. Rev. Microbiol.* 42: 289-317.

Gibson, D.T. (Ed.). 1984. *Microbial degradation of organic compounds*. Marcel Dekker, New York.

Gibson, G.R. 1990. Physiology and ecology of the sulfate-reducing bacteria. *J. Appl. Bacteriol.* 69: 769-797.

Hagedorn, S.R., R.S. Hanson, and D.A. Kunz (Eds.).1988. *Microbial metabolism and the carbon cycle.* Harwood Academic Publishers, Chur, Switzerland.

Hardman, D.J. 1991. Biotransformation of halogenated compounds. *Crit. Revs. Biotechnol.* 11: 1-40.

Häggblom, M. 1990. Mechanisms of bacterial degradation and transformation of chlorinated monoaromatic compounds. *J. Basic Microbiol.* 30: 115-141.

Häggblom, M. 1992. Microbial breakdown of halogenated aromatic pesticides and related compounds. *FEMS Microbiol. Revs.* 103: 29-72.

Kamely, D., A. Chakrabarty, and G.S. Omenn (Eds.). 1989. *Biotechnology and Biodegradation. Advances in Applied Biotechnology* volume 4. Gulf Publishing Company, Houston.

Leisinger, T., A.M. Cook, R. Hütter, and J. Nüesch (Eds.). 1981. *Microbial degradation of xenobiotics and recalcitrant compounds.* Academic Press, London.

Lowe, S.E., M.K. Jain, and J.G. Zeikus. 1993. Biology, ecology, and biotechnological applications of anaerobic bacteria adapted to environmental stresses in temperature, pH, salinity, or substrates. *Microbiol. Revs.* 57: 451-509.

Mandelstam, J., K. McQuillen, and I. Dawes.1982. *Biochemistry of bacterial growth.* Blackwell Scientific Publications, Oxford.

Mohn, W.W., and J.M. Tiedje. 1992. Microbial reductive dehalogenation. *Microbiol. Rev.* 56: 482-507.

Neilson, A.H. 1990. The biodegradation of halogenated organic compounds. *J. Appl. Bacteriol.* 69: 445-470.

Neilson, A.H., A.-S. Allard, and M. Remberger. 1985. Biodegradation and transformation of recalcitrant compounds pp. 29-86. In *Handbook of environmental chemistry* (Ed. O. Hutzinger), Vol. 2 Part C. Springer-Verlag, Berlin.

Racke, D., and J.R. Coats (Eds.). 1990. *Enhanced biodegradation of pesticides in the environment.* ACS Symposium Series 426. American Chemical Society, Washington, D.C.

Reineke, W., and H.-J. Knackmuss. 1988. Microbial degradation of haloaromatics. *Ann. Rev. Microbiol.* 42: 263-287.

Sims, G.K., and E.J. O'Loughlin. 1989. Degradation of pyridines in the environment. *Crit. Rev. Environ. Control.* 19: 309-340.

Slater, J..H., R. Whittenbury, and J.W.T. Wimpenny (Eds.). 1983. *Microbes in their natural environments.* Society for General Microbiology Symposium 34. Cambridge University Press, Cambridge.

Vogels, G.D., and C. Van Der Drift. 1976. Degradation of purines and pyrimidines by microorganisms. *Bacteriol. Rev.* 40: 403-468.

Wackett, L.P., M.S.P. Logan, F.A. Blocki, and C. Bao-li. 1992. A mechanistic perspective on bacterial metabolism of chlorinated methanes. *Biodegradation* 3: 19-36.

Watkinson, R.J. (Ed.). 1978. *Developments in biodegradation of hydrocarbons — 1.* Applied Science Publishers, London.

Wolin, M.J. 1982. Hydrogen transfer in microbial communities. In *Microbial interactions and communities.* (Eds. A.T. Bull and J.H. Slater) Vol.1 pp. 323-356. Academic Press, London.

Zehnder, A.J.B. (Ed.). 1988. *Biology of anaerobic microorganisms.* John Wiley and Sons, New York.

REFERENCES: OTHERS

Abeliovich, A., and A. Vonshak. 1992. Anaerobic metabolism of *Nitrosomonas europaea.* *Arch. Microbiol.* 158: 267-270.

Adriaens, P., and D.D. Focht. 1991a. Evidence for inhibitory substrate interactions during cometabolism of 3,4-dichlorobenzoate by *Acinetobacter* sp. strain 4-CB1. *FEMS Microbiol. Ecol.* 85: 293-300.

Adriaens, P., and D.D. Focht. 1991b. Cometabolism of 3,4-dichlorobenzoate by *Acinetobacter* sp. strain 4-CB1. *Appl. Environ. Microbiol.* 57: 173-179.

Aeckersberg, F., F. Bak, and F. Widdel. 1991. Anaerobic oxidation of saturated hydrocarbons to CO_2 by a new type of sulfate-reducing bacterium. *Arch. Microbiol.* 156: 5-14.

Aelion, C.M., C.M. Swindoll, and F.K. Pfaender. 1987. Adaptation to and biodegradation of xenobiotic compounds by microbial communities from a pristine aquifer. *Appl. Environ. Microbiol.* 53: 2212-2217.

Afring, R.P., and B.F. Taylor. 1981. Aerobic and anaerobic catabolism of phthalic acid by a nitrate-respiring bacterium. *Arch. Microbiol.* 130: 101-104.

Alexander, M. 1985. Biodegradation of organic chemicals. *Environ. Sci. Technol.* 18: 106-111.

Allard, A.-S., M. Remberger, and A.H. Neilson. 1985. Bacterial *O*-methylation of chloroguaiacols: effect of substrate concentration, cell density and growth conditions. *Appl. Environ. Microbiol.* 49: 279-288.

Allard, A.-S., M. Remberger, and A.H. Neilson. 1987. Bacterial *O*-methylation of halogen-substituted phenols. *Appl. Environ. Microbiol.* 53: 839-845.

Alsberg, T., U. Stenberg, R. Westerholm, M. Strandell, U. Rannug, A. Sundvall, L. Romert, V. Bernson, B. Petterson, R. Toftgård, M. Franzén, M. Jansson, J.Å. Gustafsson, K.E. Egebäck, and G. Tejle. 1985. Chemical and biological characterization of organic material from gasoline exhaust particles. *Environ. Sci. Technol.* 19: 43-50.

Altenschmidt, U., and G. Fuchs. 1991. Anaerobic degradation of toluene in denitrifying *Pseudomonas* sp.: indication for toluene methylhydroxylation and benzoyl-CoA as central aromatic intermediate. *Arch. Microbiol.* 156: 152-158.

Amador, J.A., and B.F. Taylor. 1990. Coupled metabolic and photolytic pathway for degradation of pyridinedicarboxylic acids, especially dipicolinic acid. *Appl. Environ. Microbiol.* 56: 1352-1356.

Anderson, J.J., and S. Dagley. 1981b. Catabolism of aromatic acids in *Trichosporon cutaneum*. *J. Bacteriol.* 141: 534-543.

Anderson, J.J., and S. Dagley. 1981a. Catabolism of tryptophan, anthranilate, and 2,3-dihydroxybenzoate in *Trichosporon cutaneum*. *J. Bacteriol.* 146: 291-297.

Andreoni, V., and G. Bestetti. 1986. Comparative analysis of different *Pseudomonas* strains that degrade cinnamic acid. *Appl. Environ. Microbiol.* 52: 930-934.

Anson, J.G., and G. Mackinnon. 1984. Novel plasmid involved in aniline degradation. *Appl. Environ. Microbiol.* 48: 868-869.

Apajalahti, J.H., P. Kärpänoja, and M.S. Salkinoja-Salonen. 1986. *Rhodococcus chlorophenolicus* sp. nov., a chlorophenol-mineralizing actinomycete. *Int. J. Syst. Bacteriol.* 36: 246-251.

Apajalahti, J.H., and M.S. Salkinoja-Salonen. 1987. Dechlorination and *para*-hydroxylation of polychlorinated phenols by *Rhodococcus chlorophenolicus*. *J. Bacteriol.* 169: 675-681.

Arjmand, M., and H. Sandermann. 1985. Mineralization of chloroaniline/lignin conjugates and of free chloroanilines by the white rot fungus *Phanerochaete chrysosporium*. *J. Agric. Food Chem.* 33: 1055-1060.

Atkinson, R., E.C. Tuazon, T.J. Wallington, S.M. Aschmann, J. Arey, A.M. Winer, and J.N. Pitts. 1987. Atmospheric chemistry of aniline, *N,N*-dimethylaniline, pyridine, 1,3,5-triazine and nitrobenzene. *Environ. Sci. Technol.* 21: 64-72.

Atkinson, R., S.M. Aschmann, A.M. Winer, and J.N. Pitts. 1985a. Kinetics and atmospheric implications of the gas-phase reactions of NO_3 radicals with a series of monoterpenes and related organics at $294 \pm 2K$. *Environ. Sci. Technol.* 19: 159-163.

Atkinson, R., S.M. Aschmann, A.M. Winer, and W.P.L. Carter. 1985b. Rate constants for the gas-phase reactions of NO_3 radicals with furan, thiophene, and pyrrole at $295 + 1$ K and atmospheric pressure. *Environ. Sci. Technol.* 19: 87-90.

Baarschers, W.H., A.I. Bharaty, and J. Elvish. 1982. The biodegradation of methoxychlor by *Klebsiella pneumoniae*. *Can. J. Microbiol.* 28: 176-179.

Bak, F., and F. Widdel. 1986. Anaerobic degradation of indolic compounds by sulfate-reducing enrichment cultures, and description of *Desulfobacterium indolicum* gen. nov., sp. nov. *Arch. Microbiol.* 146: 170-176.

Bank, S., and R.J. Tyrrell. 1984. Kinetics and mechanism of alkaline and acidic hydrolysis of aldicarb. *J. Agric. Food Chem.* 32: 1223-1232.

Bartels, I., H.-J. Knackmuss, and W. Reineke. 1984. Suicide inactivation of catechol 2,3-dioxygenase from *Pseudomonas putida* mt-2 by 3-halocatechols. *Appl. Environ. Microbiol.* 47: 500-505.

Battersby, N.S. 1990. A review of biodegradation kinetics in the aquatic environment. *Chemosphere* 21: 1243-1284.

Bauer, J.E., and D.G. Capone. 1988. Effects of co-occurring aromatic hydrocarbons on degradation of individual polycyclic aromatic hydrocarbons in marine sediment slurries. *Appl. Environ. Microbiol.* 54: 1649-1655.

Beaty, P.S., and M.J. McInerney. 1987. Growth of *Syntrophomonas wolfei* in pure culture on crotonate. *Arch. Microbiol.* 147: 389-393.

Bergon, M., N.B. Hamida, and J.-P. Calmon. 1985. Isocyanate formation in the decomposition of phenmedipham in aqueous media. *J. Agric. Food Chem.* 33: 577-583.

Biedlingmeier, J.J., and A. Schmidt. 1983. Arylsulfonic acids and some S-containing detergents as sulfur sources for growth of *Chlorella fusca*. *Arch. Microbiol.* 136: 124-130.

Boethling, R.S., and M. Alexander. 1979. Effect of concentration of organic chemicals on their biodegradation by natural microbial communities. *Appl. Environ. Microbiol.* 37: 1211-1216.

Boldrin, B., A. Tiehm, and C. Fritzsche. 1993. Degradation of phenanthrene, fluorene, fluoranthene, and pyrene by a *Mycobacterium* sp. *Appl. Environ. Microbiol.* 59: 1927-1930.

Branchaud, B.P., and C.T. Walsh. 1985. Functional group diversity in enzymatic oxygenation reactions catalyzed by bacterial flavin-containing cyclohexanone oxygenase. *J. Amer. Chem. Soc.* 107: 2153-2161.

Brandsch, R., A. E. Hinkkanen, and K. Decker. 1982. Plasmid-mediated nicotine degradation in *Arthrobacter oxidans*. *Arch. Microbiol.* 132: 26-30.

Brownlee, B.G., J.H. Carey, G.A. MacInnes, and I.T. Pellizzari. 1992. Aquatic environmental chemistry of 2-(thiocyanomethylthio)benzothiazole and related benzothiazoles. *Environ. Toxicol. Chem.* 11: 1153-1168.

Bumpus, J.A. 1989. Biodegradation of polycyclic aromatic hydrocarbons by *Phanerochaete chrysosporium*. *Appl. Environ. Microbiol.* 55: 154-158.

Bumpus, J.A., and S.D. Aust. 1987. Biodegradation of DDT [1,1,1-trichloro-2,2-*bis*(4-chlorophenyl)ethane] by the white rot fungus *Phanerochaete chrysosporium*. *Appl. Environ. Microbiol.* 53: 2001-2008.

Burlingame, R., and P.J. Chapman. 1983. Catabolism of phenylpropionic acid and its 3-hydroxy derivative by *Escherichia coli*. *J. Bacteriol.* 155: 113-121.

Buser, H.-R., and M.D. Müller. 1993. Enantioselective determination of chlordane components, metabolites, and photoconversion products in environmental samples using chiral high-resolution gas chromatography and mass spectrometry. *Environ. Sci. Technol.* 27: 1211-1220.

Caccavo, F., R.P. Blakemore, and D.R. Lovley. 1992. A hydrogen-oxidizing, Fe(III)-reducing microorganism from the Great Bay estuary, New Hampshire. *Appl. Environ. Microbiol.* 58: 3211-3216.

Cain, R.B., R.F. Bilton, and J.A. Darrah. 1968. The metabolism of aromatic acids by microorganisms. *Biochem. J.* 108: 797-832.

Calderbank, A. 1989. The occurrence and significance of bound pesticide residues in soil. *Revs. Environ. Contam. Toxicol.* 108: 1-1103.

Camilleri, P. 1984. Alkaline hydrolysis of some pyrethroid insecticides. *J. Agric. Food Chem.* 32: 1122-1124.

Carey, J.H., and M.E. Cox. 1981. Photodegradation of the lampricide 3-trifluoromethyl-4-nitrophenol (TFM) 1. Pathway of the direct photolysis in solution. *J. Great Lakes Res.* 7: 234-241.

Cerniglia, C. E., J.P. Freeman, J.R. Althaus, and C. van Baalen. 1983. Metabolism and toxicity of 1- and 2-methylnaphthalene and their derivatives in cyanobacteria. *Arch. Microbiol.* 136: 177-183.

Cerniglia, C.E., C. van Baalen, and D.T. Gibson. 1980b. Oxidation of biphenyl by the cyanobacterium, *Oscillatoria* sp., strain JCM. *Arch. Microbiol.* 125: 203-207.

Cerniglia, C.E., D. T. Gibson, and C. van Baalen. 1980a. Oxidation of naphthalene by cyanobacteria and microalgae. *J. Gen. Microbiol.* 116: 495-500.

Cerniglia, C.E., D.T. Gibson, and C. van Baalen. 1982. Naphthalene metabolism by diatoms isolated from the Kachemak Bay region of Alaska. *J. Gen. Microbiol.* 128: 987-990.

Cerniglia, C.E., J.P. Freeman, and C. van Baalen. 1981. Biotransformation and toxicity of aniline and aniline derivatives in cyanobacteria. *Arch. Microbiol.* 130: 272-275.

Chakrabarty, A.M., G. Chou, and I.C. Gunsalus. 1973. Genetic regulation of octane dissimilation plasmid in *Pseudomonas*. *Proc. Natl. Acad. Sci. U.S.A.* 70: 1137-1140.

Champine, J.E., and S. Goodwin. 1991. Acetate catabolism in the dissimilatory iron-reducing isolate GS-15. *J. Bacteriol.* 173: 2704-2706.

Charleson, R.J., J.E. Lovelock, M.O. Andreae, and S.G. Warren. 1987. Oceanic phytoplankton, atmospheric sulfur, cloud albedo and climate. *Nature* (London) 326: 655-661.

Chatterjee, D.K., S.T. Kellog, S. Hamada, and A.M. Chakrabarty. 1981. Plasmid specifying total degradation of 3-chlorobenzoate by a modified *ortho* pathway. *J. Bacteriol.* 146: 639-646.

Chaudry, G.S., and G.H. Huang. 1988. Isolation and characterization of a new plasmid from a *Flavobacterium* sp. which carries the genes for degradation of 2,4-dichlorophenoxyacetate. *J. Bacteriol.* 170: 7897-3902.

Cirigliano, M.C., and G.M. Carman. 1984. Isolation of a bioemulsifier from *Candida lipolytica*. *Appl. Environ. Microbiol.* 48: 747-750.

Claus, D., and N. Walker. 1964. The decomposition of toluene by soil bacteria. *J. Gen. Microbiol.* 36: 107-122.

Cook, A.M., and R. Htter. 1981. Degradation of s-triazines: a critical view of biodegradation pp. 237-249. In *Microbial degradation of xenobiotics and recalcitrant compounds*. T. Leisinger, A.M. Cook, R. Hütter, and J. Nüesch (Eds.). Academic Press, London.

Cook, A.M., C.G. Daughton, and M. Alexander. 1979. Benzene from bacterial cleavage of the carbon-phosphorus bond of phenylphosphonates. *Biochem. J.* 184: 453-455.

Connors, M.A., and E.A. Barnsley. 1982. Naphthalene plasmids in pseudomonads. *J. Bacteriol.* 149: 1096-1101.

Cooper, B.H., and G.A. Land. 1979. Assimilation of protocatechuate acid and *p*-hydroxybenzoic acid as an aid to laboratory identification of *Candida parapsilosis* and other medically important yeasts. *J. Clin. Microbiol.* 10: 343-345.

Cooper, R.A., and M. A. Skinner. 1980. Catabolism of 3- and 4-hydroxyphenylacetate by the 3,4-dihydroxyphenylacetate parthway in *Escherichia coli*. *J. Bacteriol.* 143: 302-306.

Corke, C.T., N.J. Bunce, A.-L. Beaumont, and R.L. Merrick. 1979. Diazonium cations as intermediates in the microbial transformations of chloroanilines to chlorinated biphenyls, azo compounds and triazenes. *J. Agric. Food Chem.* 27: 644-646.

Crawford, R.L., P.E. Olson, and T.D. Frick. 1979. Catabolism of 5-chlorosalicylate by a *Bacillus* isolated from the Mississippi River. *Appl. Environ. Microbiol.* 38: 379-384.

Criddle, C.S., J.T. DeWitt, D. Grbic-Galic, and P.L. McCarty. 1990. Transformation of carbon tetrachloride by *Pseudomonas* sp. strain KC under denitrifying conditions. *Appl. Environ. Microbiol.* 56: 3240-3246.

Cserjesi, A.J., and E.L. Johnson. 1972. Methylation of pentachlorophenol by *Trichoderma virginatum*. *Can. J. Microbiol.* 18: 45-49.

Dabrock, B., J. Riedel, J. Bertram, and G. Gottschalk. 1992. Isopropylbenzene (cumene) — a new substrate for the isolation of trichloroethene-degrading bacteria. *Arch. Microbiol.* 158: 9-13.

Dalton, H., and D.I. Stirling. 1982. Co-metabolism. *Phil. Trans. R. Soc. London.* B 297: 481-496.

Daughton, C.G., and D.P.H. Hsieh. 1977. Parathion utilization by bacterial symbionts in a chemostat. *Appl. Environ. Microbiol.* 34: 175-184.

DeBont, J.A.M., S.B. Prtimrose, MD. Collins, and D. Jones. 1980. Chemical studies on some bacteria which utilize gaseous unsaturated hydrocarbons. *J. Gen. Microbiol.* 117: 97-102.

Dec, J., K.L. Shuttleworth, and J.-M. Bollag. 1990. Microbial release of 2,4-dichlorophenol bound to humic acid or incorporated during humification. *J. Environ. Qual.* 19: 546-551.

Dehning, I., and B. Schink.1989a. *Malonomonas rubra* gen. nov. sp. nov., a microaerotolerant anaerobic bacterium growing by decarboxylation of malonate. *Arch. Microbiol.* 151: 427-433.

Dehning, I., and B. Schink.1989b. Two new species of anaerobic oxalate-fermenting bacteria, *Oxalobacter vibrioformis* sp. nov. and *Clostridium oxalicum* sp. nov, from sediment samples. *Arch. Microbiol.* 153: 79-84.

Dehning, I., M. Stieb, and B. Schink.1989. Sporomusa malonica sp. nov., a homoacetogenic bacterium growing by decarboxylation of malonate or succinate. *Arch. Microbiol.* 151: 421-426.

Devanas, M.A., D. Rafaeli-Eshkol, and G. Stotsky. 1986. Survival of plasmid-containing strains of *Escherichia coli* in soil: effect of plasmid size and nutrients on survival of hosts and maintenance of plasmids. *Curr. Microbiol.* 13: 269-277.

DeWeerd, K.A., L. Mandelco, R.S. Tanner, C.R. Woese, and J.M. Suflita. 1990. *Desulfomonile tiedjei* gen. nov. and sp. nov., a novel anaerobic, dehalogenating, sulfate-reducing bacterium. *Arch. Microbiol.* 154: 23-30.

DiChristina, T.J. 1992. Effects of nitrate and nitrite on dissimilatory iron reduction by *Shewanella putrefaciens* 200. *J. Bacteriol.* 174: 1891-1896.

Diels, L., and M. Mergeay. 1990. DNA probe-mediated detection of resistant bacteria from soils highly polluted by heavy metals. *Appl. Environ. Microbiol.* 56: 1485-1491.

Don, R.H., and J.M. Pemberton. 1985. Genetic and physical map of the 2,4-dichlorophenoxyacetic acid-degradative plasmid pJP4. *J. Bacteriol.* 161: 466-468.

Duetz, W.A., S. Marqués, C. de Jong, J.L. Ramos, and J.G. van Andel. 1994. Inducibility of the TOL catabolic pathway in *Pseudomonas putida* (pWWO) growing on succinate in continuous culture: evidence of carbon catabolite repression control. *J. Bacteriol.* 176: 2354-2361.

Dunnivant, F.M., R.P. Schwarzenbach, and D.L. Macalady. 1992. Reduction of substituted nitrobenzenes in aqueous solutions containing natural organic matter. *Environ. Sci. Technol.* 26: 2133-2141.

Durham, D.R., C.G. McNamee, and D.P. Stewart. 1984. Dissimilation of aromatic compounds in *Rhodotorula graminis*: biochemical characterization of pleiotrophically negative mutants. *J. Bacteriol.* 160: 771-777.

Dürre, P., and J.R. Andreesen. 1982. Anaerobic degradation of uric acid via pyrimidine derivatives by selenium-starved cells of *Clostridium purinolyticum*. *Arch. Microbiol.* 131: 255-260.

Eaton, D.C. 1985. Mineralization of polychlorinated biphenyls by *Phanerochaete chrysosporium*, a lignolytic fungus. *Enzyme Microbiol. Technol.* 7: 194-196.

Eatough, D.J., V.F. White, L.D. Hansen, N.L. Eatough, and J.L. Cheney. 1986. Identification of gas-phase dimethyl sulfate and monomethyl hydrogen sulfate in the Los Angeles atmosphere. *Environ. Sci. Technol.* 20: 867-872.

Evans, P. J., D.T. Mang, K.S. Kim, and L.Y. Young. 1991a. Anaerobic degradation of toluene by a denitrifying bacterium. *Appl. Environ. Microbiol.* 57: 1139-1145.

Evans, P.J., D.T. Mang, and L.Y. Young. 1991b. Degradation of toluene and m-xylene and transformation of o-xylene by denitrifying enrichment cultures. *Appl. Environ. Microbiol.* 57: 450-454.

Fall, R.R., J.I. Brown, and T.L. Schaeffer. 1979. Enzyme recruitment allows the biodegradation of recalcitrant branched hydrocarbons by *Pseudomonas citronellolis. Appl. Environ. Microbiol.* 38: 715-722.

Fernández, P., M. Grifoll, A.M. Solanas, J. M. Bayiona, and J. Albalgs. 1992. Bioassay-directed chemical analysis of genotoxic compounds in coastal sediments. *Environ. Sci. Technol.* 26: 817-829.

Feroz, M., A.A. Podowski, and M.A.Q. Khan. 1990. Oxidative dehydrochlorination of heptachlor by *Daphnia magna. Pesticide Biochem. Physiol.* 36: 101-105.

Finnerty, W.R. 1992. The biology and genetics of the genus *Rhodococcus. Ann. Rev. Microbiol.* 46: 193-218.

Foght, J.M., D.L. Gutnick, and D.W.S. Westlake. 1989. Effect of emulsan on biodegradation of crude oil by pure and mixed bacterial cultures. *Appl. Environ. Microbiol.* 55: 36-42.

Folsom, B.R., P.J. Chapman, and P.H. Pritchard. 1990. Phenol and trichloroethylene degradation by *Pseudomonas cepacia* G4: kinetics and interactions between substrates. *Acinetobacter* sp. strain 4-CB1. *Appl. Environ. Microbiol.* 56: 1279-1285.

Fredrickson, J.K., F.J. Brockman, D.J. Workman, S.W. Li, and T.O. Stevens. 1991. Isolation and characterization of a subsurface bacterium capable of growth on toluene, naphthalene, and other aromatic compounds. *Appl. Environ. Microbiol.* 57: 796-803.

Fulthorpe, R.R., and R. C. Wyndham. 1989. Survival and activity of a 3-chlorobenzoate-catabolic genotype in a natural system. *Appl. Environ. Microbiol.* 55: 1584-1590.

Furukawa, K., and A. M. Chakrabarty. 1982. Involvement of plasmids in total degradation of chlorinated biphenyls. *Appl. Environ. Microbiol.* 44: 619-626.

Gaal, A., and H.Y. Neujahr. 1979. Metabolism of phenol and resorcinol in *Trichosporon cutaneum. J. Bacteriol.* 137: 13-21.

Gälli, R., and T. Leisinger. 1988. Plasmid analysis and cloning of the dichloromethane-utilizing genes of *Methylobacterium* sp. DM4. *J. Gen. Microbiol.* 134: 943-952.

Gantzer, C.J., and L.P. Wackett. 1991. Reductive dechlorination catalyzed by bacterial transition-metal coenzymes. *Environ. Sci. Technol.* 25: 715-722.

Gee, J.M., and J.L. Peel. 1974. Metabolism of 2,3,4,6-tetrachlorophenol by microorganisms from broiler house litter. *J. Gen. Microbiol.* 85: 237-243.

Gerritse, J., and J.C. Gottschal. 1993. Oxic and anoxic growth of a new *Citrobacter* species on amino acids. *Arch. Microbiol.* 160: 51-61.

Gerson, D.F., and J.E. Jajic. 1979. Comparison of surfactant production from kerosene by four species of *Corynebacterium. Antonie van Leeuwenhoek* 45: 81-94.

Gibson, S.A., and G.W. Sewell. 1992. Stimulation of reductive dechlorination of tetrachloroethene in anaerobic aquifer microcosms by addition of short-chain organic acids or alcohols. *Appl. Environ. Microbiol.* 58: 1392-1393.

Gibson, D.T., J.R. Koch, C.L. Schuld, and R.E. Kallio. 1968. Oxidative degradation of aromatic hydrocarbons by microorganisms. II. Metabolism of halogenated aromatic hydrocarbons. *Biochem.* 7: 3795-3802.

Glaus, M.A., C.G. Heijman, R.P. Schwarzenbach, and J. Zeyer. 1992. Reduction of nitroaromatic compounds mediated by *Streptomyces* sp. exudates. *Appl. Environ. Microbiol.* 58: 1945-1951.

Gorny, N., G. Wahl, A. Brune, and B. Schink. 1992. A strictly anaerobic nitrate-reducing bacterium growing with resorcinol and other aromatic compounds. *Arch. Microobiol.* 158: 48-53.

Gray, P.H.H., and H.G. Thornton. 1928. Soil bacteria that decompose certain aromatic compounds. *Centbl. Bakt.* (2 Abt.) 73: 74-96.

Grbic-Galic, D. 1986. O-demethylation, dehydroxylation, ring-reduction and cleavage of aromatic substrates by Enterobacteriaceae under anaerobic conditions. *J. Appl. Bacteriol.* 61: 491-497.

Greer, L.E., and D.R. Shelton. 1992. Effect of inoculant strain and organic matter content on kinetics of 2,4-dichlorophenoxyacetic acid degradation in soil. *Appl. Environ. Microbiol.* 58: 1459-1465.

Griffiths, D.A., D.E. Brown, and S.G. Jezequel. 1992. Biotransformation of warfarin by the fungus *Beauveria bassiana. Appl. Microbiol. Biotechnol.* 37: 169-175.

Grimond, P.A.D., F. Grimond, H.L.C. Dulong de Rosnay, and P.H.A. Sneath. 1977. Taxonomy of the genus *Serratia. J. Gen. Microbiol.* 98: 39-66.

Groenewegen, P.E.J., A.J.M. Diessen, W.M. Konigs, and J.A.M. de Bont. 1990. Energy-dependent uptake of 4-chlorobenzoate in the *coryneform bacterium* NTM-1. *J. Bacteriol.* 172: 419-423.

Grosjean, D. 1991. Atmospheric chemistry of toxic contaminants. 4. Saturated halogenated aliphatics: methyl bromide, epichlorhydrin, phosgene. *J. Air Waste Manage. Assoc.* 41: 56-61.

Grosjean, D., E.L. Williams II, and J.H. Seinfeld. 1992. Atmospheric oxidation of selected terpenes and related carbonyls: gas-phase carbonyl products. *Environ. Sci. Technol.* 26: 1526-1533.

Grosjean, D., E.L. Williams II, and E. Grosjean. 1993a. A biogenic precursor of peroxypropionyl nitrate: atmospheric oxidation of *cis*-3-hexen-1-ol. *Environ. Sci. Technol.* 27: 979-981.

Grosjean, D., E.L. Williams II, and E. Grosjean. 1993b. Atmospheric chemistry of isoprene and its carbonyl products. *Environ. Sci. Technol.* 27: 830-840.

Grosser, R.J., D. Warshawsky, and J.R. Vestal. 1991. Indigenous and enhanced mineralization of pyrene, benzo[a]pyrene, and carbazole in soils. *Appl. Environ. Microbiol.* 57: 3462-3469.

Guerin, W.F., and G.E. Jones. 1988. Mineralization of phenanthrene by a *Mycobacterium* sp. *Appl. Environ. Microbiol.* 54: 937-944.

Guerin, W.F., and S.A. Boyd. 1992. Differential bioavailability of soil-sorbed naphthalene by two bacterial species. *Appl. Environ. Microbiol.* 58: 1142-1152

Gupta, J.K., C. Jebsen, and H. Kneifel. 1986. Sinapic acid degradation by the yeast *Rhodotorula graminis. J. Gen. Microbiol.* 132: 2793-2799.

Häggblom, M., D. Janke, P.J.M. Middeldorp, and M. Salkinoja-Salonen. 1988. O-methylation of chlorinated phenols in the genus *Rhodococcus. Arch. Microbiol.* 152: 6-9.

Haigler, B.E., and J.C. Spain. 1991. Biotransformation of nitrobenzene by bacteria containing toluene degradative pathways. *Appl. Environ. Microbiol.* 57: 3156-3162.

Haigler, B.E., C.A. Pettigrew, and J.C. Spain. 1992. Biodegradation of mixtures of substituted benzenes by *Pseudomonas* sp. strain JS150. *Appl. Environ. Microbiol.* 58: 2237-2244.

Hall, B.G. 1990. Spontaneous point mutations that occur more often when advantageous than when neutral. *Genetics* 126: 5-16.

Haller, H.D., and R.K. Finn. 1978. Kinetics of biodegradation of *p*-nitrobenzoate and inhibition by benzoate in a pseudomonad. *Appl. Environ. Microbiol.* 35: 890-896.

Han, Y.-H., R.M. Smibert, and N.R. Krieg. 1991. *Wolinella recta, Wolinella curva, Bacteroides ureolyticus,* and *Bacteroides gracilis* are microaerophiles, not anaerobes. *Int. J. System. Bacteriol.* 41: 218-222.

Hansch, C., and A. Leo. 1979. *Substituent constants for correlation analysis in chemistry and biology.* John Wiley & Sons, Inc., New York.

Hardman, D.J., P.C. Gowland, and J.H. Slater. 1986. Large plasmids from soil bacteria enriched on halogenated alkanoic acids. *Appl. Environ. Microbiol.* 51: 44-51.

Harkness, M.R., J.B. McDermott, D.A. Abramowicz, J.J. Salvo, W.P. Flanagan, M.L. Stephens, F.J. Mondello, R.J. May, J.H. Lobos, K.M. Carroll, M.J. Brennan, A.A. Bracco, K.M. Fish, G.L. Warner, P.R. Wilson, D.K. Dietrich, D.T. Lin, C.B. Morgan, and W.L.Gately. 1993. In situ stimulation of aerobic PCB biodegradation in Hudson River sediments. *Science* 259: 503-507.

Harris, R., and C.J. Knowles. 1983. Isolation and growth of a *Pseudomonas* species that utilizes cyanide as a source of nitrogen. *J. Gen. Microbiol.* 129: 1005-1011.

Harvey, J., J.J. Dulka, and J.J. Anderson. 1985. Properties of sulfometuroin methyl affecting its environmental fate: aqueous hydrolysis and photolysis, mobility and adsorption on soils, and bioaccumulation potential. *J. Agric. Food Chem.* 33: 590-596.

Havel, J., and W. Reineke. 1993. Microbial degradation of chlorinated acetophenones. *Appl. Environ. Microbiol.* 59: 2706-2712.

Heitkamp, M.A., J.P. Freeman, D.W. Miller, and C.E. Cerniglia. 1988. Pyrene degradation by a *Mycobacterium* sp.: identification of ring oxidation and ring fission products. *Appl. Environ. Micobiol.* 54: 2556-2565.

Helmig, D., J. Arey, W.P. Harger, R. Atkinson, and J. López-Cancio. 1992a. Formation of mutagenic nitrodibenzopyranones and their occurrence in ambient air. *Environ. Sci. Technol.* 26: 622-624

Helmig, D., J. L pez-Cancio, J. Arey, W.P. Harger, and R. Atkinson. 1992b. Quantification of ambient nitrodibenzopyranones: further evidence for atmospheric mutagen formation. *Environ. Sci. Technol.* 26: 2207-2213.

Henriksen, H.V., S. Larsen, and B.K. Ahring. 1991. Anaerobic degradation of PCP and phenol in fixed-film reactors: the influence of an additional substrate. *Water Sci. Technol.* 24: 431-436.

Henry, S.M., and D. Grbic-Galic. 1991. Inhibition of trichloroethylene oxidation by the transformation intermediate carbon monoxide. *Appl. Environ. Microbiol.* 57: 1770-1776

Herbes, S.E., and L.R. Schwall. 1978. Microbial transformation of polycyclic aromatic hydrocarbons in pristine and petroleum-contaminated sediments. *Appl. Environ. Microbiol.* 35: 306-316.

Higgins, S.J., and J. Mandelstam. 1972. Evidence for induced synthesis of an active transport factor for mandelate in *Pseudomonas putida. Biochem. J.* 126: 917-922.

Hilbi, H., R. Hermann, and P. Dimroth. 1993. The malonate decarboxylase enzyme system of *Malonomonas rubra:* evidence for the cytoplasmic location of the biotin-containing component. *Arch. Microbiol.* 160: 126-131.

Hileman, B. 1993. Concerns broaden over chlorine and chlorinated hydrocarbons. *Chem. Eng. News* 71 (16): 11-20.

Ho, P.C. 1986. Photooxidation of 2,4-dinitrotoluene in aqueous solution in the presence of hydrogen peroxide. *Environ. Sci. Technol.* 20: 260-267.

Holben, W.E., J.K. Jansson, B.K. Chelm, and J.M. Tiedje. 1988. DNA probe method for the detection of specific microorganisms in the soil bacterial community. *Appl. Environ. Microbiol.* 54: 703-711.

Holben, W.E., B.M. Schroeter, V.G.M. Calabrese, R.H. Olsen, J.K. Kukor, V.O. Biederbeck, A.E. Smith, and J.M. Tiedje. 1992. Gene probe analysis of soil microbial populations selected by amendment with 2,4-dichlorophenoxyacetic acid. *Appl. Environ. Microbiol.* 58: 3941-3948.

Holm, P.E., P.H. Nielsen, H.-J. Albrechtsen, and T.H. Christensen. 1992. Importance of unattached bacteria and bacteria attached to sediment in determining potentials for degradation of xenobiotic organic contaminants in an aerobic aquifer. *Appl. Environ. Microbiol.* 58: 3020-3026.

Holmstead, R.L. 1976. Studies of the degradation of mirex with an iron (II) porphyrin model system. *J. Agric. Food Chem.* 24: 620-624.

Hoover, D.G., G.E. Borgonovi, S.H. Jones, and M. Alexander. 1986. Anomalies in mineralization of low concentrations of organic compounds in lake water and sewage. *Appl. Environ. Microbiol.* 51: 226-232.

Horvath, R.S. 1971. Cometabolism of the herbicide 2,3,6-trichlorobenzoate. *J. Agric. Food Chem.* 19: 291-293.

Horvath, R.S. 1972. Microbial co-metabolism and the degradation of organic compounds in nature. *Bacteriol. Revs.* 36: 146-155.

Horvath, R.S., and P. Flathman. 1976. Co-metabolism of fluorobenzoates by natural microbial populations. *Appl. Environ. Microbiol.* 31: 889-891.

Houwen, F.P., C. Dijkema, A.J.M. Stams, and A.J.B. Zehnder. 1991. Propionate metabolism in anaerobic bacteria; determination of carboxylation reactions with ^{13}C-NMR spectroscopy. *Biochim. Biophys. Acta* 1056: 126-132.

Hutchins, S.R. 1991. Biodegradation of monoaromatic hydrocarbons by aquifer microorganisms using oxygen, nitrate or nitrous oxide as the terminal electron acceptors. *Appl. Environ. Microbiol.* 57: 2403-2407.

Hyman, M.R., A.W. Sansome-Smith, J.H. Shears, and P.M. Wood. 1985. A kinetic study of benzene oxidation to phenol by whole cells of *Nitrosomonas europaea* and evidence for the further oxidation of phenol to hydroquinone. *Arch. Microbiol.* 143: 302-306.

Hyman, M.R., I.B. Murton, and D.J. Arp. 1988. Interaction of ammonia monooxygenase from Nitrosomonas europaea with alkanes, alkenes, and alkynes. *Appl. Environ. Microbiol.* 54: 3187-3190.

Imhoff-Stuckle, D., and N. Pfennig. 1983. Isolation and characterization of a nicotinic acid-degrading sulfate-reducing bacterium, *Desulfococcus niacini* sp. nov. *Arch. Microbiol.* 136: 194-198.

Itoh, S., and T. Suzuki. 1972. Effect of rhamnolipids on growth of *Pseudomonas aeruginosa* mutant deficient in n-paraffin-utilizing ability. *Agr. Biol. Chem.* 36: 2233-2235.

Jagnow, G., K. Haider, and P.-C. Ellwardt. 1977. Anaerobic dechlorination and degradation of hexachlorocyclohexane by anaerobic and facultatively anaerobic bacteria. *Arch. Microbiol.* 115: 285-292.

Jahnke, M., T. El-Banna, R. Klintworth, and G. Auling. 1990. Mineralization of orthanilic acid is a plasmid-associated trait in *Alcaligenes* sp. O-1. *J. Gen. Microbiol.* 136: 2241-2249.

Jain, R.K., G.S. Sayler, J.T. Wilson, L. Houston, and D. Pacia. 1987. Maintenance and stability of introduced genotypes in groundwater aquifer material. *Appl. Environ. Microbiol.* 53: 996-1002.

Janke, D. 1987. Use of salicylate to estimate the threshold inducer level for de novo synthesis of the phenol-degrading enzymes in *Pseudomonas putida* strain H. *J. Basic Microbiol.* 27: 83-89.

Janssen, P.H., and C.G. Harfoot. 1992. Anaerobic malonate decarboxylation by *Citrobacter diversus. Arch. Microbiol.* 157: 471-474.

Jessee, J.A., R.E. Benoit, A.C. Hendricks, G.C. Allen, and J.L. Neal. 1983. Anaerobic degradation of cyanuric acid, cysteine, and atrazine by a facultative anaerobic bacterium. *Appl. Environ. Microbiol.* 45: 97-102.

Ji, X.-B., and T.C. Hollocher. 1988. Mechanism for nitrosation of 2,3-diaminonaphthalene by *Escherichi coli:* enzymatic production of NO followed by O_2-dependent chemical nitrosation. *Appl. Environ. Microbiol.* 54: 1791-1794.

Jones, R.D., and R.Y. Morita. 1983. Methane oxidation by *Nitrosococcus oceanus* and *Nitrosomonas europaea. Appl. Environ. Microbiol.* 45: 401-410.

Juliette, L.Y., M.R. Hyman, and D.J. Arp. 1993. Inhibition of ammonia oxidation in *Nitrosomonas europaea* by sulfur compounds: thioethers are oxidized to sulfoxides by ammonia monooxygenase. *Appl. Environ. Microbiol.* 59: 3718-3727.

Kanagawa, K., S. Negoro, N. Takada, and H. Okada. 1989. Plasmid dependence of *Pseudomonas* sp. strain NK87 enzymes that degrade 6-aminohexanoate-cyclic dimer. *J. Bacteriol.* 171: 3181-3186.

Kawai, F., and H. Yamanaka. 1986. Biodegradation of polyethyelene glycol by symbiotic mixed culture (obligate mutualism). *Arch. Microbiol.* 146: 125-129.

Kawasaki, H., H. Yahara, and K. Tonomura. 1981. Isolation and characterization of plasmid pUO1 mediating dehalogenation of haloacetate and mercury resistance in *Moraxella* sp. B. *Agric. Biol. Chem.* 45: 1477-1481.

Kelley, I., J.P. Freeman, and C.E. Cerniglia. 1990. Identification of metabolites from degradation of naphthalene by a *Mycobacterium* sp. *Biodegradation* 1: 283-290.

Kennedy, D.W., S.D. Aust, and J.A. Bumpus. 1990. Comparative biodegradation of alkyl halide insecticides by the white rot fungus, *Phanerochaete chrysosporium* (BKM-F-1767). *Appl. Environ. Microbiol.* 56: 2347-2353.

Keshavarz, T., M.D. Lilly, and P.C. Clarke. 1985. Stability of a catabolic plasmid in continuous culture. *J. Gen. Microbiol.* 131: 1193-1203.

Khanna, P., B. Rajkumar, and N. Jothikumar. 1992. Anoxygenic degradation of aromatic substances by *Rhodopseudomonas palustris. Curr. Microbiol.* 25: 63-67.

Kiene, R.P., and B.F. Taylor. 1988. Demethylation of dimethylsulfoniopropionate and production of thiols in anoxic marine sediments. *Appl. Environ. Microbiol.* 54: 2208-2212.

Kiyohara, H., T. Hatta, Y. Ogawa, T. Kakuda, H. Yokoyama, and N. Takizawa. 1992. Isolation of *Pseudomonas pickettii* strains that degrade 2,4,6-trichlorophenol and their dechlorination of chlorophenols. *Appl. Environ. Microbiol.* 58: 1276-1283.

Kjellberg, S., M. Hermansson, P. Mårdén, and G.W. Jones. 1987. The transient phase between growth and nongrowth of heterotrophic bacteria with emphasis on the marine environment. *Ann. Rev. Microbiol.* 41: 25-49.

Klecka, G.M., and D.T. Gibson. 1981. Inhibition of catechol 2,3-dioxygenase from *Pseudomonas putida* by 3-chlorocatechol. *Appl. Environ. Microbiol.* 41: 1159-1165.

Klecka, G.M., and W.J. Maier. 1988. Kinetics of microbial growth on mixtures of pentachlorophenol and chlorinated aromatic compounds. *Biotechnol. Bioeng.* 31: 328-335.

Kleier, D., I. Holden, J.E. Casida, and L.O. Ruzo. 1985. Novel photoreactions of an insecticidal nitromethylene heterocycle. *J. Agric. Food Chem.* 33: 998-1000.

Koenig, D.W., and H.B. Ward. 1983. *Prototheca zopfii* Krüger strain UMK-13 growth on acetate or *n*-alkanes. *Appl. Environ. Microbiol.* 45: 333-336.

Kolattukudy, P.E., and L. Hankin. 1968. Production of omega-haloesters from alkyl halides by *Micrococcus cerificans. J. Gen. Microbiol.* 54: 145-153.

Kreiss, M., J. Eberspächer, and F. Linmgens. 1981. Detection and characterization of plasmids in Chloridazon and Antipyrin degrading bacteria. *Zbl. Bakteriol. Hyg., I. Abt. Orig.* C2: 45-60.

Krone, U.E., R.K. Thauer, and H.P.C. Hogenkamp. 1989. Reductive dehalogenation of chlorinated C1-hydrocarbons mediated by corrinoids. *Biochemistry* 28: 4908-4914.

Kunisaki, N., and M. Hayashi. 1979. Formation of *N*-nitrosamines from secondary amines and nitrite by resting cells of *Escherichia coli* B. *Appl. Environ. Microbiol.* 37: 279-282.

Kutney, J.P., E. Dimitriadis, G.M. Hewitt, P.J. Salisbury, and M. Singh. 1982. Studies related to biological detoxification of kraft mill effluent. IV: The biodegradation of 14-chlorodehydroabietic acid wirth *Mortierella isabellina. Helv. Chem. Acta* 65: 1343-1350.

Landner, L. (Ed.) 1989. *Chemicals in the aquatic environment,* Springer-Verlag, Berlin.

Lesage, S., S. Brown, and K.R. Hosler. 1992. Degradation of chlorofluorocarbon-113 under anaerobic conditions. *Chemosphere* 24: 1225-1243.

Levsen, K. 1988. The analysis of diesel particulate. *Fresenius Z. Anal. Chem.* 331: 467-478.

Lewis, D.L., H.P. Kollig, and R.E. Hodson. 1986. Nutrient limitation and adaptation of microbial populations to chemical transformations. *Appl. Environ. Microbiol.* 51: 598-603.

Lewis, D.L., R.E. Hodson, and L.F. Freeman. 1985. Multiphase kinetics for transformation of methyl parathion by *Flavobacterium* species. *Appl. Environ. Microbiol.* 50: 553-557.

Lidstrom, M.E., and D.I. Stirling. 1990. Methylotrophs: genetics and commercial applications. *Ann. Rev. Microbiol.* 44: 27-58.

Lidstrom-O'Connor, M.E., G.L. Fulton, and A.E. Wopat. 1983. 'Methylobacterium ethanolicum': a syntrophic association of two methylotrophic bacteria. *J. Gen. Microbiol.* 129: 3139-3148.

Liebert, F. 1909. The decomposition of uric acid by bacteria. *Proc. K. Acad. Ned. Wetensch.* 12: 54-64.

Lindsay, R.F., and F.G. Priest. 1975. Decarboxylation of substituted cinnamic acids by enterobacteria: the influence on beer flavor. *J. Appl. Bacteriol.* 39: 181-187.

Little, C.D., A.V. Palumbo, S.E. Herbes, M.E. Lidström, R.L. Tyndall, and P.J. Gilmour. 1988. Trichloroethylene biodegradation by a methane-oxidizing bacterium. *Appl. Environ. Microbiol.* 54: 951-956.

Liu, D., W.M.J. Strachan, K. Thomson, and K. Kwasniewska. 1981. Determination of the biodegradability of organic compounds. *Environ. Sci. Technol.* 15: 788-793.

Liu, S.-Y., Z. Zheng, R. Zhang, and J.-M. Bollag. 1989. Sorption and metabolism of metolachlor by a bacterial community. *Appl. Environ. Microbiol.* 55: 733-740.

Locher, H.H., B. Poolman, A.M. Cook, and W.N. Konings. 1993. Uptake of 4-toluene sulfonate by *Comamonas testosteroni* T-2. *J. Bacteriol.* 175: 1075-1080.

Lorenzen, J.P., A. Kröger, and G. Unden. 1993. Regulation of anaerobic respiratory pathways in *Wolinellla succinogenes* by the presence of electron acceptors. *Arch. Microbiol.* 159: 477-483.

Lovley, D.R. 1991. Dissimilatory Fe(III) and Mn(IV) reduction. *Microbiol. Rev.* 55: 259-287.

Lovley, D.R., and D.J. Lonergan. 1990. Anaerobic oxidation of toluene, phenol, and *p*-cresol by the dissimilatory iron-reducing organism, GS-15. *Appl. Environ. Microbiol.* 56: 1858-1864.

Lovley, D.R., and E.J.P. Phillips. 1988. Novel mode of microbial energy metabolism: organic carbon oxidation coupled to dissimilatory reduction of iron or manganese. *Appl. Environ. Microbiol.* 54: 1472-1480.

Lovley, D.R., and J.C. Woodward. 1992. Consumption of freons CFC-11 and CFC-12 by anaerobic sediments and soils. *Environ. Sci. Technol.* 26: 925-929.

Lovley, D.R., E.J.P. Phillips, and D.J. Lonergan. 1989. Hydrogen and formate oxidation coupled to dissimilatory reduction of iron or mangenese by *Alteromonas putrefaciens*. *Appl. Environ. Microbiol.* 55: 700-706.

Lovley, D.R., E.J.P. Phillips, Y. A. Gorby, and E.R. Landa. 1991. Microbial reduction of uranium. *Nature* 350: 413-416.

Lovley, D.R., S.J. Giovannoni, D.C. White, J.E. Champine, E.J.P. Phillips, Y.A. Gorby, and S. Goodwin. 1993. *Geobacter metallireducens* gen nov. sp. nov., a microorganism capable of coupling the complete oxidation of organic compounds to the reduction of iron and other metals. *Arch. Microbiol.* 159: 336-344.

Lowe, S.E., M.K. Jain, and J.G. Zeikus. 1993. Biology, ecology, and biotechnological applications of anaerobic bacteria adapted to environmental stresses in temperature, pH, salinity, or substrates. *Microbiol. Revs.* 57: 451-509.

Macalady, D.L. and N.L. Wolfe. 1985. Effects of sediment sorption and abiotic hydrolyses.1. Organophosphorothioate esters. *J. Agric. Food Chem.* 33: 167-173.

MacGillivray, A.R., and M.P. Shiaris. 1993. Biotransformation of polycyclic aromatic hydrocarbons by yeasts isolated from coastal sediments. *Appl. Environ. Microbiol.* 59: 1613-1618.

MacMichael, G.J., and L.R. Brown. 1987. Role of carbon dioxide in catabolism of propane by "Nocardia paraffinicum" (*Rhodococcus rhodochrous*). *Appl. Environ. Microbiol.* 53: 65-69.

Macy, J.M., S. Rech, G. Auling, M. Dorsch, E. Stackebrandt, and L.I. Sly. 1993. *Thauera selenatis* gen. nov., sp. nov., a member of the beta subclass of *Proteobacteria* with a novel type of anaerobic respiration. *Int. J. Syst. Bacteriol.* 43: 135-142.

Madsen, T., and D. Licht. 1992. Isolation and characterization of an anaerobic chlorophenol-transforming bacterium. *Appl. Environ. Microbiol.* 58: 2874-2878.

Maiers, D.T., P.L. Wichlacz, D.L. Thompson, and D.F. Bruhn. 1988. Selenate reduction by bacteria from a selenium-rich environment. *Appl. Environ. Microbiol.* 54: 2591-2593.

Malachowsky, K.J., T.J. Phelps, A.B. Teboli, D.E. Minnikin, and D.C. White. 1994. Aerobic mineralization of trichloroethylene, vinyl chloride, and aromatic compounds by *Rhodococcus species. Appl. Environ. Microbiol.* 60: 542-548.

Malmqvist, Å., T. Welander, and L. Gunnarsson. 1991. Anaerobic growth of microorganisms with chlorate as an electron acceptor. *Appl. Environ. Microbiol.* 57: 2229-2232.

Marks, T.S., J.D. Allpress, and A. Maule. 1989. Dehalogenation of lindane by a variety of porphyrins and corrins. *Appl. Environ. Microbiol.* 55: 1258-1261.

Martin, P., and R.A. MacLeod. 1984. Observations on the distinction between oligotrophic and eutrophic marine bacteria. *Appl. Environ. Microbiol.* 47: 1017-1022.

McBride, K.E., J.W. Kenny, and D.M. Stalker. 1986. Metabolism of the herbicide bromoxynil by *Klebsiella pneumoniae* subspecies *ozaenae. Appl. Environ. Microbiol.* 52: 325-330.

McInerney, M.J., M.P. Bryant, R.B. Hespell, and J.W. Costerton. 1981. *Syntrophomonas wolfei* gen. nov., sp. nov., an anaerobic, syntrophic, fatty acid-oxidizing bacterium. *Appl. Environ. Microbiol.* 41: 1029-1039.

McMillan, D. C., P.P. Fu, and C.E. Cerniglia. 1987. Stereoselective fungal metabolism of 7,12-dimethylbenz[a]anthracene: identification and enantiomeric resolution of a K-region dihydrodiol. *Appl. Environ. Microbiol.* 53: 2560-2566.

Metwally, M. E.-S., and N.L. Wolfe. 1990. Hydrolysis of chlorostilbene oxide. II. Modelling of hydrolysis in aquifer samples and in sediment-water systems. *Environ. Toxicol. Chem.* 9: 963-973.

Meyer, J.J.M., N. Grobbelaar, and P.L. Steyn. 1990. Fluoroacetate-metabolizing pseudomonad isolated from *Dichapetalum cymosum. Appl. Environ. Microbiol.* 56: 2152-2155.

Meyer, O., and H.G. Schlegel. 1983. Biology of aerobic carbon monoxide-oxidizing bacteria. *Ann. Rev. Microbiol.* 37: 277-310.

Michaelsen, M., R. Hulsch, T. Höpner, and L. Berthe-Corti. 1992. Hexadecane mineralization in oxygen-controlled sediment-seawater cultivations with autochthonous microorganisms. *Appl. Environ. Microbiol.* 58: 3072-3077.

Mileski, G., J.A. Bumpus, M.A. Jurek, and S.D. Aust. 1988. Biodegradation of pentachlorophenol by the white rot fungus, *Phanerochaete chrysosporium. Appl. Environ. Microbiol.* 54 .: 2885-2889.

Miller, R.M., G.M. Singer, J.D. Rosen, and R. Bartha. 1988a. Sequential degradation of chlorophenols by photolytic and microbial treatment. *Environ. Sci. Technol.* 22: 1215-1219.

Miller, R.M., G.M. Singer, J.D. Rosen, and R. Bartha. 1988b. Photolysis primes biodegradation of benzo[a]pyrene. *Appl. Environ. Microbiol.* 54: 1724-1730.

Mills, A.L., and M. Alexander. 1976. *N*-nitrosamine formation by cultures of several microorganisms. *Appl. Environ. Microbiol.* 31: 892-895.

Monna, L., T. Omori, and T. Kodama. 1993. Microbial degradation of dibenzofuran, fluorene, and dibenzo-*p*-dioxin by *Staphylococcus auriculans* DBF63. *Appl. Environ. Microbiol.* 59: 285-289.

Monticello, D.J., D. Bakker, and W.R. Finnerty. 1985. Plasmid-mediated degradation of dibenzothiophene by *Pseudomonas* species. *Appl. Environ. Microbiol.* 49: 756-760.

Moore, J.K., H.D. Braymer, and A.D. Larson. 1983. Isolation of a *Pseudomonas* sp. which utilizes the phosphonate herbicide glyphosate. *Appl. Environ. Microbiol.* 46: 316-320.

Morris, C.M., and E. A. Barnsley. 1982. The cometabolism of 1- and 2-chloronaphthalene by pseudomonads. *Can. J. Microbiol.* 28: 73-79.

Morris, P.J., J.F. Quensen III, J.M. Tiedje, and S.A. Boyd. 1992. Reductive debromination of the commercial polybrominated biphenyl mixture Firemaster BP6 by anaerobic microorganisms from sediments. *Appl. Environ. Microbiol.* 58: 3249-3256.

Mortlock, R.P. 1982. Metabolic acquisition through laboratory selection. *Ann. Rev. Microbiol.* 34: 37-66.

Mowrer, J., and J. Nordin. 1987. Characterization of halogenated organic acids in flue gases from municipal waste incinerators. *Chemosphere* 16: 1181-1192.

Mulbry, W.W., J.S. Karns, P.C. Kearney, J.O. Nelson, C.S. McDaniel, and J.R. Wild. 1986. Identification of a plasmid-borne Parathion hydrolase gene from *Flavobacterium* sp. by Southern hybridization with *opd* from *Pseudomonas diminuta. Appl. Environ. Microbiol.* 51: 926-930.

Müller, M.D., and H.-R. Buser. 1986. Halogenated aromatic compounds in automotive emissions from leaded gasoline additives. *Environ. Sci. Technol.* 20: 1151-1157.

Murphy, G.l., and J.J. Perry. 1983. Incorporation of chlorinated alkanes into fatty acids of hydrocarbon-utilizing mycobacteria. *J. Bacteriol.* 156: 1158-1164.

Murphy, S.E., A. Drotar, and R. Fall. 1982. Biotransformation of the fungicide pentachloronitrobenzene by *Tetrahymena thermophila. Chemosphere* 11: 33-39.

Myers, C.R., and K.H. Nealson. 1990. Respiration-linked proton translocation coupled to anaerobic reduction of manganese(IV) and iron (III) in *Shewanella putrefaciens. J. Bacteriol.* 172: 6232-6238.

Nakagawa, R., S. Kitamori, K. Horikawa, K. Nakashima, and H. Tokiwa. 1983. Identification of dinitropyrenes in diesel-exhaust particles. Their probable presence as the major mutagens. *Mutation Res.* 124: 201-211.

Narro, M.L., C.E. Cerniglia, C. Van Baalen, and D.T. Gibson. 1992. Metabolism of phenanthrene by the marine cyanobacterium *Agmenellum quadruplicatum* PR-6. *Appl. Environ. Microbiol.* 58: 1351-1359.

Nealson, K.H., and C.R. Myers. 1992. Microbial reduction of manganese and iron: new approaches to carbon cycling. *Appl. Environ. Microbiol.* 58: 439-443.

Negoro, S., H. Shinagawa, A. Naata, S. Kinoshita, T. Hatozaki, and H. Okada. 1980. Plasmid control of 6-aminohexanoic acid cyclic dimer degradation enzymes of *Flavobacterium* sp. KI72. *J. Bacteriol.* 143. 238-345.

Neidhardt, F.C., P.L. Bloch, and D.F. Smith. 1974. Culture medium for enterobacteria. *J. Bacteriol.* 119: 736-747.

Neilson, A.H., and T. Larsson. 1980. The utilization of organic nitrogen for growth of algae: physiological aspects. *Physiol. Plant.* 48: 542-553.

Neilson, A.H., and R.A. Lewin. 1974. The uptake and utilization of organic carbon by algae: an essay in comparative biochemistry. *Phycologia* 13: 227-264.

Neilson, A.H., A.-S. Allard, P.-Å. Hynning, M. Remberger, and L. Landner. 1983. Bacterial methylation of chlorinated phenols and guiaiacols: formation of veratroles from guaiacols and high molecular weight chlorinated lignin. *Appl. Environ. Microbiol.* 45: 774-783.

Neilson, A.H., A.-S. Allard, and M. Remberger. 1985. Biodegradation and transformation of recalcitrant compounds pp. 29-86. In *Handbook of environmental chemistry* (Ed. O. Hutzinger), Vol. 2 Part C. Springer-Verlag, Berlin.

Neilson, A.H., C. Lindgren, P.-Å. Hynning, and M. Remberger. 1988. Methylation of halogenated phenols and thiophenols by cell extracts of gram-positive and gram-negative bacteria. *Appl. Environ. Microbiol.* 54: 524-530.

Nozawa, T., and Y. Maruyama. 1988. Anaerobic metabolism of phthalate and other aromatic compounds by a denitrifying bacterium. *J. Bacteriol.* 170: 5778-5784.

Ogram, A.V., R.E. Jessup, L.T. Ou, and P.S.C. Rao. 1985. Effects of sorption on biological degradation rates of (2,4-dichlorophenoxy)acetic acid in soils. *Appl. Environ. Microbiol.* 49: 582-587.

Ogunseitan, O.A., I.L. Deklgado, Y.-L. Tsai, and B.H. Olson, 1991. Effect of 2-hydroxybenzoate on the maintenance of naphthalene-degrading pseudomonads in seeded and unseeded soils. *Appl. Environ. Microbiol.* 57: 2873-2879.

Paris, D.F., and J. E. Rogers. 1986. Kinetic concepts for measuring microbial rate constants: effects of nutrients on rate constants. *Appl. Environ. Microbiol.* 51: 221-225.

Paris, D.F., and N.L. Wolfe. 1987. Relationship between properties of a series of anilines and their transformation by bacteria. *Appl. Environ. Microbiol.* 53: 911-916.

Paris, D.F., N.L. Wolfe, and W.C. Steen. 1982. Structure-activity relationships in microbial transformation of phenols. *Appl. Environ. Microbiol.* 44: 153-158.

Paris, D.F., N.L. Wolfe, and W.C. Steen. 1984. Microbial transformation of esters of chlorinated carboxylic acids. *Appl. Environ. Microbiol.* 47: 7-11.

Patel, R.N., C.T. Hou, A.I. Laskin, and A. Felix. 1982. Microbial oxidation of hydrocarbons: properties of a soluble monooxygenase from a facultative methane-utilizing organisms *Methylobacterium* sp. strain CRL-26. *Appl. Environ. Microbiol.* 44: 1130-1137.

Peijnenburg, W.J.G.M., M.J. 't Hart, H.A. den Hollander, D van de Meent, H.H. Verboom, and N.L. Wolfe. 1992. QSARs for predicting reductive transformation constants of halogenated aromatic hydrocarbons in anoxic sediment systems. *Environ. Toxicol. Chem.* 11: 301-314.

Peijnenburg, W.J.G.M., K.G.M. de Beer, H.A. den Hollander, M.H.L. Stegeman, and H. Verboom. 1993. Kinetics, products, mechanisms and QSARs for the hydrolytic transformation of aromatic nitriles in anaereobic sediment slurries. *Environ. Toxicol. Chem.* 12: 1149-1161.

Pelizzetti, E., V. Maurino, C. Minero, V. Carlin, E. Pramauro, O. Zerbinati, and M.L. Tosata. 1990. Photocatalytics degradation of atrazine and other *s*-triazine herbicides. *Environ. Sci. Technol.* 24: 1559-1565.

Perdue, E.M., and N.L. Wolfe. 1982. Modification of pollutant hydrolysis kinetics in the presence of humic substances. *Environ. Sci. Technol.* 16: 847-852.

Perez, J., and T.W. Jeffries. 1990. Mineralization of ^{14}C-ring-labeled synthetic lignin correlates with the production of lignin peroxidase, not of manganese peroxidase or laccase. *Appl. Environ. Microbiol.* 56: 1806-1812.

Peters, R. 1952. Lethal synthesis. *Proc. Roy. Soc. (London)* B 139: 143-170.

Pettigrew, C.A., B.E. Haigler, and J.C. Spain. 1991. Simultaneous biodegradation of chlorobenzene and toluene by a *Pseudomonas* strain. *Appl. Environ. Microbiol.* 57: 157-162.

Plugge, C.M., C. Dijkema, and A.J.M. Stams. 1993. Acetyl-CoA cleavage patyhways in a syntrophic propionate oxidizing bacterium growing on fumarate in the absence of methanogens. *FEMS Microbiol. Lett.* 110: 71-76.

Poindexter, J.S. 1981. Oligotrophy. Fast and famine existence. *Adv. Microb. Ecol.* 5: 63-89.

Polnisch, E., H. Kneifel, H. Franzke, and K.L. Hofmann. 1992. Degradation and dehalogenation of monochlorophenols by the phenol-assimilating yeast *Candida maltosa*. *Biodegradation* 2: 193-199.

Poth, M., and D.D. Focht. 1985. [15]N kinetic analysis of N_2O production by *Nitrosomonas europaea*: an examination of nitrifier denitrification. *Appl. Environ. Microbiol.* 49: 1134-1141.

Pothuluri, J.V., J.P. Freeman, F.E. Evans, T.B. Moorman, and C.E. Cerniglia. 1993. Metabolism of alachlor by the fungus *Cunninghamella elegans*. *J. Agric. Food Chem.* 41: 483-488.

Prieto, M.A., A. Perez-Aranda, and J.L. Garcia. 1993. Characterization of an *Escherichia coli* aromatic hydroxylase with a broad substrate range. *J. Bacteriol.* 175: 2162-2167.

Racke, K.D, and J.R. Coats. 1990. *Enhanced biodegradation of pesticides in the environment.* American Chemical Society Symposium Series 426. American Chemical Society, Washington, D.C.

Racke, K.D., and E.P. Lichtenstein. 1985. Effects of soil microorganisms on the release of bound ^{14}C residues from soils previously treated with [^{14}C]parathion. *J. Agric. Food Chem.* 33: 938-943.

Racke, K.D., D.A. Laskowski, and M.R. Schultz. 1990. Resistance of chloropyrifos to enhanced biodegradation in soil. *J. Agric. Food Chem.* 38: 1430-1436.

Rasche, M.E., M.R. Hyman, and D.J. Arp. 1991. Factors limiting aliphatic chlorocarbon degradation by *Nitrosomonas europaea*: cometabolic inactivation of ammonia monooxygenase and substrate specificity. *Appl. Environ. Microbiol.* 57: 2986-2994.

Reanney, D.C., P.C. Gowland, and J.H. Slater. 1983. Genetic interactions among microbial communities. *Symp. Soc. Gen. Microbiol.* 34: 379-421.

Reinecke, W., D.J. Jeenes, P.A. Williams, and H.-J. Knackmuss. 1982. TOL plasmid pWWO in constructed halobenzoate-degrading *Pseudomonas* strains: prevention of meta pathway. *J. Bacteriol.* 150: 195-201.

Rendell, N.B., G.W. Taylor, M. Somerville, H. Todd, R. Wilson, and P.J. Cole. 1990. Characterization of *Pseudomonas* rhamnolipids. *Biochim. Biophys. Acta* 1045: 189-193.

Rheinwald, J.G., A.M. Chakrabarty, and I.C. Gunsalus. 1973. A transmissible plasmid controlling camphor oxidation in *Pseudomonas putida. Proc. Natl. Acad. Sci. U.S.A.* 70: 885-889.

Rice, C.P., and H.C. Sikka, 1973. Uptake and metabolism of DDT by six species of marine algae. *J. Agric. Food Chem.* 21: 148-152.

Rijnaarts, H.H.M., A. Bachmann, J.C. Jumelet, and A.J.B. Zehnder. 1990. Effect of desorption and intraparticle mass transfer on the aerobic biomineralization of alpha-hexachlorocyclohexane in a contaminated calcareous soil. *Environ. Sci. Technol.* 24: 1349-1354.

Roberts, A.L., and P.M. Gschwend. 1991. Mechanism of pentachloroethane dehydrochlorination to tetrachloroethylene. *Environ. Sci. Technol.* 25: 76-86.

Roberts, A.L., P.N. Sanborn, and P.M. Gschwend. 1992. Nucleophilic substitution of dihalomethanes with hydrogen sulfide species. *Environ. Sci. Technol.* 26: 2263-2274.

Robinson, K.G., W.S. Farmer, and J.T. Novak. 1990. Availability of sorbed toluene in soils for biodegradation by acclimated bacteria. *Water Res.* 24: 345-350.

Rott, B., S. Nitz, and F. Korte. 1979. Microbial decomposition of sodium pentachlorophenolate. *J. Agric. Food Chem.* 27: 306-310.

Rouf, M.A., and R.F. Lomprey. 1968. Degradation of uric acid by certain aerobic bacteria. *J. Bacteriol.* 96: 617-622.

Roy, F., E. Samain, H.C. Dubourguier, and G. Albagnac. 1986. *Synthrophomonas* (sic) *sapovorans* sp. nov., a new obligately proton reducing anaerobe oxidizing saturated and unsaturated long chain fatty acids. *Arch. Microbiol.* 145: 142-147.

Rubin, H.E., and M. Alexander. 1983. Effect of nutrients on the rates of mineralization of trace concentrations of phenol and *p*-nitrophenol. *Environ. Sci. Technol.* 17: 104-107.

Rubin, H.E., R.V. Subba-Rao, and M. Alexander. 1982. Rates of mineralization of trace concentrations of aromatic compounds in lake water and sewage samples. *Appl. Environ. Microbiol.* 43: 1133-1138.

Saflic, S., P.M. Fedorak, and J.T. Andersson. 1992. Diones, sulfoxides, and sulfones from the aerobic cometabolism of methylbenzothiophenes by *Pseudomonas* strain BT1. *Environ. Sci. Technol.* 26: 1759-1764.

Sakai, K., A. Nakazawa, K. Kondo, and H. Ohta. 1985. Microbial hydrogenation of nitroolefins. *Agric. Biol. Chen.* 49: 2231-2236.

Salleh, M.A., and J.M. Pemberton. 1993. Cloning of a DNA region of a *Pseudomonas* plasmid that codes for detoxification of the herbicide paraquat. *Curr. Microbiol.* 27: 63-67.

Salmeen, I.T., A.M. Pero, R. Zator, D. Schuetzle, and T.L. Riley. 1984. Ames assay chromatograms and the identification of mutagens in diesel particle extracts. *Environ. Sci. Technol.* 18: 375-382.

Samadpour, M., J. Liston, J.E. Ongerth, and P.I. Tarr. 1990. Evaluation of DNA probes for detection of Shiga-like-toxin-producing *Escherichia coli* in food and calf fecal samples. *Appl. Environ. Microbiol.* 56: 1212-1215.

Sarkanen, S., R.A. Razal, T. Piccariello, E. Yamamoto, and N.G. Lewis. 1991. Lignin peroxidase: toward a clarification of its role in vivo. *J. Biol. Chem.* 266: 3636-3643.

Sayama, M., M. Inoue, M.-A. Mori, Y. Maruyama, and H. Kozuka. 1992. Bacterial metabolism of 2,6-dinitrotoluene with *Salmonella typhimurium* and mutagenicity of the metabolites of 2,6- dinitrotoluene and related compounds. *Xenobiotica* 22: 633-640.

Sayler, G.S., M.S. Shields, E.T. Tedford, A. Breen, S.W. Hooper, K.M. Sirotkin, and J.W. Davis. 1985. Application of DNA-DNA colony hybridization to the detection of catabolic genotypes in environmental samples. *Appl. Environ. Microbiol.* 49: 1295-1303.

Schanke, C.A., and L.P. Wackett. 1992. Environmental reductive elimination reactions of polychlorinated ethanes mimicked by transition-metal coenzymes. *Environ. Sci. Technol.* 26: 830-833.

Schink, B. 1992. Syntrophism among prokaryotes pp. 276-299. In *The Prokaryotes* (Eds. A. Balows, H.G. Trüper, M. Dworkin, W. Harder, and K.-H. Schleifer). Springer-Verlag, Heidelberg.

Schink, B., and N. Pfennig. 1982. *Propionigenium modestum* gen. nov. sp. nov. a new strictly anaerobic, nonsporing bacterium growing on succinate. *Arch. Microbiol.* 133: 209-216.

Schwack, W. 1988. Photoinduced additions of pesticides to biomolecules. 2. Model reactions of DDT and methoxychlor with methyl oleate. *J. Agric. Food Chem.* 36: 645-648.

Schwarzenbach, R.P., R. Stierliu, K. Lanz, and J. Zeyer. 1990. Quinone and iron porphyrin mediated reduction of nitroaromatic compounds in homogeneous aqueous solution. *Environ. Sci. Technol.* 24: 1566-1574.

Sedlak, D.L., and A.W. Andren. 1991. Oxidation of chlorobenzene with Fenton's reagent. *Environ. Sci. Technol.* 25: 777-782.

Seifritz, C., S.L. Daniel, A. Gößner, and H. L. Drake. 1993. Nitrate as a preferred electron sink for the acetogen *Clostridium thermoaceticum*. *J. Bacteriol.* 175: 8008-8013.

Seigle-Murandi, F., R. Steiman, F. Chapella, and C. Luu Duc.1986. 5-hydroxylation of benzimidazole by Micromycetes. II. Optimization of production with *Absidia spinosa*. *Appl. Microbiol. Biotechnol.* 25: 8-13.

Seigle-Murandi, F.M., S.M.A. Krivobok, R.L. Steiman, J.-L. A. Benoit-Guyod, and G.-A. Thiault. 1991. Biphenyl oxide hydroxylation by *Cunninghamella echinulata*. *J. Agric. Food Chem.* 39: 428-430.

Seitz, H.-J., and H. Cypinka. 1986. Chemolithotrophic growth of *Desulfovibrio desulfuricans* with hydrogen coupled to ammonification with nitrate or nitrite. *Arch. Microbiol.* 146: 63-67.

Seligman, P.F., A.O. Valkirs, and R.F. Lee. 1986. Degradation of tributytin in San Diego Bay, California, waters. *Environ. Sci. Technol.* 20: 1229-1235.

Serdar, C.M., D.T. Gibson, D. M. Munnecke, and J.H. Lancaster. 1982. Plasmid involvement in Parathion hydrolysis in *Pseudomonas diminuta*. *Appl. Environ. Microbiol.* 44: 246-249.

Servent, D., C. Ducrorq, Y. Henry, A. Guissani, and M. Lenfant. 1991. Nitroglycerin metabolism by *Phanerochaete chrysosporium*: evidence for nitric oxide and nitrite formation. *Biochim. Biophys. Acta* 1074: 320-325.

Shields, M.S., S.W. Hooper, and G.S. Sayler. 1985. Plasmid-mediated mineralization of 4-chlorobiophenyl. *J. Bacteriol.* 163: 882-889.

Shields, M.S., S.O. Montgomery, P.J. Chapman, S.M. Cuskey and P.H. Pritchard.1989. Novel pathway of toluene catabolism in the trichloroethylene-degrading bacterium G4. *Appl. Environ. Microbiol.* 55: 1624-1629.

Shimp, R.J., and F.K. Pfaender. 1985a. Influence of easily degradable naturally occurring carbon substrates on biodegradation of monosubstituted phenols by aquatic bacteria. *Appl. Environ. Microbiol.* 49: 394-401.

Shimp, R., and F.K. Pfaender. 1985b. Influence of naturally occurring humic acids on biodegradation of monosubstituted phenols by aquatic bacteria. *Appl. Environ. Microbiol.* 49: 402-407.

Shoda, M., and S. Udaka. 1980. Preferential utilizatiion of phenol rather than glucose by *Trichosporon cutaneum* possessing a partiallty constitutive catechol 1,2-oxygenase. *Appl. Environ. Microbiol.* 39: 1129-1133.

Shoham, Y., and E. Rosenberg. 1983. Enzymatic depolymerization of emulsan. *J. Bacteriol.* 156: 161-167.

Siegele, D.A., and R. Kolter. 1992. Life after log. *J. Bacteriol.* 174: 345-348.

Simkins, S., and M. Alexander. 1984. Models for mineralization kinetics with the variables of substrate concentration and population density. *Appl. Environ. Microbiol.* 47: 1299-1306.

Simkins, S., R. Mukherjee, and M. Alexander. 1986. Two approaches to modeling kinetics of biodegradation by growing cells and application of a two-compartment model for mineralization kinetics in sewage. *Appl. Environ. Microbiol.* 51: 1153-1160.

Singer, M.E.V., and W.R. Finnerty. 1990. Physiology of biosurfactanat synthesis by *Rhodococcus species* H13-A. *Can. J. Microbiol.* 36: 741-745.

Singer, M.E.V., W.R. Finnerty, and A. Tunelid. 1990. Physical and chemical properties of a biosurfactant synthesized by *Rhodococcus* sp. H13-A. *Can. J. Microbiol.* 36: 746-750.

Singh, N., and N. Sethunathan. 1992. Degradation of soil-sorbed carbofuran by an enrichment culture from carbofuran-retreated Azolla plot. *J. Agric. Food Chem.* 40: 1062-1065.

Singh, N.C., T.P. Dasgupta, E.V. Roberts, and A. Mansingh. 1991. Dynamics of pesticides in tropical conditions. 1. Kinetic studies of volatilization, hydrolysis, and photolysis of dieldrin and α- and β-endosulfan. *J. Agric. Food Chem.* 39: 575-579.

Slater, J.H., and D. Lovatt. 1984. Biodegradation and the significance of microbial communties pp. 439-485. In *Microbial degradation of organic compounds* (Ed. D.T. Gibson). Marcel Dekker, Inc., New York.

Smith, H.W., Z. Parsell, and P. Green. 1978. Thermosensitive antibiotic resistance plasmids in enterobacteria. *J. Gen. Microbiol.* 109: 37-47.

Smith, M.G., and R.J. Park. 1984. Effect of restricted aeration on catabolism of cholic acid by two *Pseudomonas* species. *Appl. Environ. Microbiol.* 48: 108-113.

Smith, R.V., and J.P. Rosazza. 1983. Microbial models of mammalian metabolism. *J. Nat. Prod.* 46: 79-91.

Sobecky, P.A., M.A. Schell, M.A. Moran, and R.E. Hodson. 1992. Adaptation of model genetically engineered microorganisms to lake water: growth rate enhancements and plasmid loss. *Appl. Environ. Microbiol.* 58: 3630-3637.

Soderquist, C.J., D.G. Crosby, K.W. Moilanen, J.N. Seiber, and J.E. Woodrow. 1975. Occurrence of trifluralin and its photoproducts in air. *J. Agric. Food Chem.* 23: 304-309.

Soeder, C.J., E. Hegewald, and H. Kneifel. 1987. Green microalgae can use naphthalenesulfonic acids as sources of sulfur. *Arch. Microbiol.* 148: 260-263.

Spain, J.C., P.A. van Veld, C.A. Monti, P.H. Pritchard, and C.R. Cripe. 1984. Comparison of *p*-nitrophenol biodegradation in field and laboratory test systems. *Appl. Environ. Microbiol.* 48: 944-950.

Spokes, J.R., and N. Walker. 1974. Chlorophenol and chlorobenzoic acid co-metabolism by different genera of soil bacteria. *Arch. Microbiol.* 96: 125-134.

Stalker, D.M., and K. E. McBride. 1987. Cloning and expresssion in *Escherichia coli* of a *Klebsiella ozaenae* plasmid-borne gene encoding a nitrilase specific for the herbicide Bromoxynil. *J. Bacteriol.* 169: 955-960.

Stang, P.M., R.F. Lee, and P.F. Seligman. 1992. Evidence for rapid, nonbiological degradation of tributyltin compounds in autoclaved and heat-treated fine-grained sediments. *Environ. Sci. Technol.* 26: 1382-1387.

Stanlake, G.J., and R.K. Finn. 1982. Isolation and characterization of a pentachlorophenol-degrading bacterium. *Appl. Environ. Microbiol.* 44: 1421-1427.

Steen, W.C., and T.W. Collette. 1989. Microbial degradation of seven amides by suspended bacterial populations. *Appl. Environ. Microbiol.* 55: 2545-2549.

Stephanou, E.G. 1992. α,ω-dicarboxylic acid salts and α,ω-dicarboxylic acids. Photooxidation products of unsaturated fatty acids, present in marine aerosols and marine sediments. *Naturwiss.* 79: 28-131.

Stephanou, E.G., and N. Stratigakis. 1993. Oxocarboxylic and α,ω-dicarboxylic acids: photooxidation products of biogenic unsaturated fatty acids present in urban aerosols. *Environ. Sci. Technol.* 27: 1403-1407.

Stieb, M., and B. Schink. 1985. Anaerobic oxidation of fatty acids by *Clostridium bryantii* sp. nov., a sporeforming, obligately syntrophic bacterium. *Arch. Microbiol.*140: 387-390.

Styrvold, O.B., and A.R. Strøm. 1984. Dimethylsulphoxide and trimethylamine oxide respiraton of Proteus vulgaris. *Arch. Microbiol.* 140: 74-78.

Subba-Rao, R.V., and M. Alexander. 1982. Effect of sorption on mineralization of low concentrations of aromatic compounds in lake water samples. *Appl. Environ. Microbiol.* 44: 659-668.

Subba-Rao, R.V., H.E. Rubin, and M. Alexander. 1982. Kinetics and extent of mineralization of organic chemicals at trace levels in freshwater and sewage. *Appl. Environ. Microbiol.* 43: 1139-1150.

Sun, Y., and J.J. Pignatello. 1993. Organic intermediates in the degradation of 2,4-dichlorophenoxyacetic acid by Fe^{3+}/H_2O_2 and $Fe^{3+}/H_2O_2/UV$. *J. Agric. Food Chem.* 41: 1139-1142.

Sutherland, J.B., A.L. Selby, J.P. Freeman, F.E. Evans, and C.E. Cerniglia. 1991. Metabolism of phenanthrene by *Phanerochaete chrysosporium*. *Appl. Environ. Microbiol.* 57: 3310-3316.

Suzuki, T. 1978. Enzymatic methylation of pentachlorophenol and its related compounds by cell-free extracts of *Mycobacterium* sp. isolated from soil. *J. Pesticide Sci.* 3: 441-443.

Suzuki, T., K. Tanaka, I. Matsubara, and S. Kinoshita. 1969. Trehalose lipid and alpha-branched-beta-hydroxy fatty acid formed by bacteria grown on n-alkanes. *Agr. Biol. Chem.* 33: 1619-1627.

Szewzyk, R., and N. Pfennig. 1987. Complete oxidation of catechol by the strictly anaerobic sulfate-reducing *Desulfobacterium catecholicum* sp. nov. *Arch. Microbiol.* 147: 163-168.

Taeger, K., H.-J. Knackmuss, and E. Schmidt. 1988. Biodegradability of mixtures of chloro- and methylsubstituted aromatics: simultaneous degradatioin of 3-chlorobenzoate and 3-methylbenzoate. *Appl. Microbiol. Biotechnol.* 28: 603-608.

Tam, A.C., R.M. Behki, and S.U. Khan. 1987. Isolation and characterization of an S-ethyl-*N,N*-dipropylthiocarbamate-degrading *Arthrobacter* strain and evidence for plasmid-associated S-ethyl-*N,N*-dipropylthiocarbamate degradation. *Appl. Environ. Microbiol.* 53: 1088-1093.

Tasaki, M., Y. Kamagata, K. Nakamura, and E. Mikami. 1991. Isolation and characterization of a thermophilic benzoate-degrading, sulfate-reducing bacterium, Desulfotomaculum thermobenzoicum sp. nov. *Arch. Microbiol.* 155: 348-352.

Taylor, B.F. 1983. Aerobic and anaerobic catabolism of vanillic acid and some other methoxy-aromatic compounds by *Pseudomonas* sp. strain PN-1. *Appl. Environ. Microbiol.* 46: 1286-1292.

Taylor, B.F., and D.C. Gilchrist. 1991. New routes for aerobic biodegradation of dimethylsulfoniopropionate. *Appl. Environ. Microbiol.* 57: 3581-3584.

Taylor, B.F., and M.J. Heeb. 1972. The anaerobic degradation of aromatic compounds by a denitrifying bacterium. Radioisotope and mutant studies. *Arch. Microbiol.* 83: 165-171.

Taylor, B.F., J.A. Amador, and H.S. Levinson. 1993. Degradation of *meta*-trifluoro-methylbenzoate by sequential microbial and photochemical treatments. *FEMS Microbiol. Lett.* 110: 213-216.

Taylor, S.L., L.S. Guthertz, M. Leatherwood, and E.R. Lieber. 1979. Histamine production by *Klebsiella pneumoniae* and an incident of scombroid fish poisoning. *Appl. Environ. Microbiol.* 37: 274-278.

Thayer, J.R., and M.L. Wheelis. 1982. Active transport of benzoate in *Pseudomonas putida*. *J. Gen. Microbiol.* 128: 1749-1753.

Thomas, A.W., J. Lewington, S. Hope, A.W. Topping, A.J. Weightman, and J.H. Slater. 1992. Environmentally directed mutations in the dehalogenase system of *Pseudomonas putida* strain PP3. *Arch. Microbiol.* 158: 176-182.

Topp, E., R.L. Crawford, and R.S. Hanson. 1988. Influence of readily metabolizable carbon on pentachlorophenol metabolism by a pentachlorophenol-degrading *Flavobacterium* sp. *Appl. Environ. Microbiol.* 54: 2452-2459.

Tschech, A., and G. Fuchs. 1989. Anaerobic degradation of phenol via carboxylation to 4-hydroxybenzoate: in vitro study of isotope exchange betyween $^{14}CO_2$ and 4-hydroxybenzoate. *Arch. Microbiol.* 152: 594-599.

Tuazon, E.C., H. MacLeod, R. Atkinson, and W.P.L. Carter. 1986. α dicarbonyl yields from the NO_x-air photooxidations of a series of aromatic hydrocarbons in air. *Environ. Sci. Technol.* 20: 383-387.

Valli, K., and M.H. Gold. 1991. Degradation of 2,4-dichlorophenol by the lignin-degrading fungus *Phanerochaete chrysosporium. J. Bacteriol.* 173: 345-352.

Valli, K., H. Wariishi, and M.H. Gold. 1992a. Degradation of 2,7-dichlorodibenzo-*p*-dioxin by the lignin- degrading basidiomycete *Phanerochaete chrysosporium. J. Bacteriol.* 174: 2131-2137.

Valli, K., B.J. Brock, D.K. Joshi, and M.H. Gold. 1992b. Degradation of 2,4-dinitrotoluene by the lignin-degrading fungus *Phanerochaete chrysosporium. Appl. Environ. Microbiol.* 58: 221-228.

Van der Kooij, D., A. Visser, and J.P. Oranje. 1980. Growth of *Aeromonas hydrophila* at low concentrations of substrates added to tap water. *Appl. Environ. Microbiol.* 39: 1198-1204.

Van der Kooij, D., and W.A.M. Hijnen. 1984. Substrate utilization by an oxalate-consuming *Spirillum* species in relation to its growth in ozonated water. *Appl. Environ. Microbiol.* 47: 551-559.

Van der Kooij, D., J.P. Oranje, and W.A.M. Hijnen. 1982. Growth of *Pseudomonad aeruginosa* in tap water in relation to utilization of substrates at concentrations of few micrograms per liter. *Appl. Environ. Microbiol.* 44:1086-1095.

van der Meer, J.R., T.N.P. Bosma, W.P. de Bruin, H. Harms, C. Holliger, H.H.M. Rijnaarts, M. E. Tros, G. Schraa, and A.J.B. Zehnder. 1992. Versatility of soil column experiments to study biodegradation of halogenated compounds under environmental conditions. *Biodegradation* 3: 265-284.

van der Meer, J.R., A.R.W. van Neerven, E.J. de Vries, W.M. de Vos, and A.J.B. Zehnder. 1991. Cloning and characterization of plasmid-encoding genes for the degradation of 1,2-dichloro-, 1,4-dichloro-, and 1,2,4-trichlorobenzene of *Pseudomonas* sp. strain P51. *J. Bacteriol.* 173: 6-15.

van Ginkel, C.G., H.G. J. Welten, and J.A.M. de Bont. 1987. Oxidation of gaseous and volatile hydrocarbons by selected alkene-utilizing bacteria. *Appl. Environ. Microbiol.* 53: 2903-2907.

Vandenbergh, P.A., and R.L. Cole. 1986. Plasmid involvement in linalool metabolism by *Pseudomonas fluorescens. Appl. Environ. Microbiol.* 52: 939-940.

Vandenbergh, P.A., and A.M. Wright. 1983. Plasmid involvement in acyclic isoprenoid metabolism by *Pseudomonas putida. Appl. Environ. Microbiol.* 45: 1953-19 55.

Vandenbergh, P.A., R.H. Olsen, and J.E. Colaruotolo. 1981. isolation and genetic character-ization of bacteria that degrade chloroaromatic compounds. *Appl. Environ. Microbiol.* 42: 737-739.

Vannelli, T., and A.B. Hooper. 1992. Oxidation of nitrapyrin to 6-chloropicolinic acid by the ammonia-oxidizing bacterium *Nitrosòmonas europaea. Appl. Environ. Microbiol.* 58: 2321-2325.

Vannelli, T., and A.B. Hooper. 1993. Reductive dehalogenation of the trichloromethyl group of nitrapyrin by the ammonia-oxidizing bacterium *Nitrosomonas europaea*. *Appl. Environ. Microbiol.* 59: 3597-3601.

Véron, M., and L. Le Minor. 1975. Nutrition et taxonomie des *Enterobacteriaceae* et bactéries voisines. III. Caractéres nutritionnels et différenciation des groupes taxonomiques. *Ann. Microbiol.* (Inst. Pasteur) 126B: 125-147.

Wackett, L.P., G.A. Brusseau, S.R. Householder, and R.S. Hanson. 1989. Survey of microbial oxygenases: trichloroethylene degradation by propane-oxidizing bacteria. *Appl. Environ. Microbiol.* 55: 2960-2964.

Wada, H., K. Sendoo, and Y. Takai. 1989. Rapid degradation of γ-HCH in the upland soil after multiple application. *Soil Sci. Plant Nutr.* 35: 71-77.

Walker, J.D., and R. S. Pore. 1978. Growth of Prototheca isolates on *n*-hexadecane and mixed-hydrocarbon substrate. *Appl. Environ. Microbiol.* 35: 694-697.

Walker, N. 1973. Metabolism of chlorophenols by *Rhodotorula glutinis*. *Soil Biol. Biochem.* 5: 525-530.

Ward, B.B. 1987. Kinetic studies on ammonia and methane oxidation by *Nitrosococcus oceanus*. *Arch. Microbiol.* 147: 126-133.

Wayne, L.G. et al. 1991. Fourth report of the cooperative, open-ended study of slowly growing mycobacteria by the international working group on mycobacterial taxonomy. *Int. J. Syst. Bacteriol.* 41: 463-472.

Wedemeyer, G. 1967. Dechlorination of 1,1,1-trichloro-2,2-*bis*[p-chlorophenyl]ethane by *Aerobacter aerogenes*. *Appl. Microbiol.* 15: 569-574.

Weiner, J.H., D.P. MacIsaac, R.E. Bishop, and P.T. Bilous.1988. Purification and properties of *Escherichia coli* dimethyl sulfoxide reductase, an iron-sulfur molybdoenzyme with broad substrate specificity. *J. Bacteriol.* 170: 1505-1510.

Whited, G.M., and D.T. Gibson.1991. Separation and partial characterization of the enzymes of the toluene-4-monooxygenase catabolic pathway in *Pseudomonas mendocina* KR1. *J. Bacteriol.* 173: 3017-3020.

Widdel, F., G.-W. Kohring, and F. Mayer. 1983. Studies on dissimilatory sulfate-reducing bacteria that decompose fatty acids. *Arch. Microbiol.* 134: 286-294.

Williams, P.A., and M.J. Worsey. 1976. Ubiquity of plasmids in coding for toluene and xylene metabolism in soil bacteria: evidence for the existence of new TOL plasmids. *J. Bacteriol.* 125: 818-828.

Williams, R.J., and W.C. Evans. 1975. The metabolism of benzoate by *Moraxella* species through anaerobic nitrate respiration. *Biochem. J.* 148: 1-10.

Wolfe, N.L., R.G. Zepp, and D.F. Paris. 1978a. Carbaryl, propham and chloropropham: a comparison of the rates of hydrolysis and photolysis with the rates of biolysis. *Water Res.* 12: 565-571.

Wolfe, N.L., R.G. Zepp, and D.F. Paris. 1978b. Use of structure-reactivity relationships to estimate hydrolytic persistence of carbamate pesticides. *Water Res.* 12: 561-563.

Wolt, J.D., J.D. Schwake, F.R. Batzer, S.M. Brown, L.H. McKendry, J.R. Miller, G.A. Roth, M.A. Stanga, D. Portwood, and D.L. Holbrook. 1992. Anaerobic aquatic degradation of flumetsulam [*N*-(2,6-difluorophenyl)-5-methyl[1,2,4]trizolo[1,5a]pyrimidine-2-sulfona-mide]. *J. Agric. Food Chem.* 40: 2302-2308.

Wong, A.S., and D.G. Crosby. 1981. Photodecomposition of pentachlorophenol in water. *J. Agric. Food Chem.* 29: 125-130.

Woodburn, K.B., F.R. Batzer, F.H. White, and M.R. Schultz. 1993. The aqueous photolysis of triclorpyr. *Environ. Toxicol. Chem.* 12: 43-55.

Wyndham, R.C. 1986. Evolved aniline catabolism in *Acinetobacter calcoaceticus* during continuous culture of river water. *Appl. Environ. Microbiol.* 51: 781-789.

Yamauchi, T., and T. Handa. 1987. Characterization of aza heterocyclic hydrocarbons in urban atmospheric particulate matter. *Environ. Sci. Technol.* 21: 1177-1181.

Yasahura, A., and M. Morita. 1988. Formation of chlorinated aromatic hydrocarbons by thermal decomposition of vinylidene chloride polymer. *Environ. Sci. Technol.* 22: 646-650.

Yoshinaga, D.H., and H.A. Frank. 1982. Histamine-producing bacteria in decomposing skipjack tuna (*Katsuwonus pelamis*). *Appl. Environ. Microbiol.* 44: 447-452.

Zaidi, B.R., Y. Murakami, and M. Alexander. 1988. Factors limiting success of inoculation to enhance biodegradation of low concentrations of organic chemicals. *Environ. Sci. Technol.* 22: 1419-1425.

Zepp, R.G., and P.F. Schlotzhauer. 1983. Influence of algae on photolysis rates of chemicals in water. *Environ. Sci. Technol.* 17: 462-468.

Zepp, R.G., G.L. Baugham, and P.A. Scholtzhauer. 1981a. Comparison of photochemical behaviour of various humic substances in water. I. Sunlight induced reactions of aquatic pollutants photosensitized by humic substances. *Chemosphere* 10: 109-117.

Zepp, R.G., G.L. Baugham, and P.A. Scholtzhauer. 1981b. Comparison of photochemical begavious of vatious humic substances in water: II. Photosensitized oxygenations. *Chemosphere* 10: 119-126.

Zhang, Y., and R.M. Miller. 1992. Enhanced octadecane dispersion and biodegradation by a *Pseudomonas* rhamnolipid surfactant (biosurfactant). *Appl. Environ. Microbiol.* 58: 3276-3282.

Zhao, H., D. Yang, C.R. Woese, and M.P. Bryant. 1990. Assignment of *Clostridium bryantii* to *Syntrophospora bryantii* gen. nov., comb. nov. on the basis of a 16S rRNA sequence analysis of its crotonate-grown pure culture. *Int. J. Syst. Bacteriol.* 40: 40-44.

Ziegler, K., K. Braun, A. Böckler, and G. Fuchs. 1987. Studies on the anaerobic degradation of benzoic acid and 2-aminobenzoic acid by a denitrifying *Pseudomonas* strain. *Arch. Microbiol.* 149: 62-69.

Zoro, J.A., J.M. Hunter, G. Eglinton, and C.C. Ware. 1974. Degradation of p,p'-DDT in reducing environments. *Nature* 247: 235-237.

Zylstra, G.J., L.P. Wackett, and D.T. Gibson. 1989. Trichloroethylene degradation by *Escherichia coli* containing the cloned *Pseudomonas puitida* F1 toluene dioxygenase genes. *Appl. Environ. Microbiol.* 55: 3162-3166.

CHAPTER 5

Persistence: Experimental Aspects

SYNOPSIS

Experimental procedures for investigating the microbial degradation and transformation of xenobiotics are outlined, together with a critical analysis of the limitations inherent in conventional tests for determining ready biodegradability. A brief description of procedures for elective enrichment is given including the basic components of growth media together with salient features of their composition and use. Attention is directed to problems associated with low water solubility, toxicity, or volatility of the substrates. An outline is given of the various types of experiments that have been designed to assess biodegradability and biotransformation including the application of both batch and continuous culture procedures. Examples are provided of the use of pure cultures and stable consortia, microcosms and larger scale field systems. A number of procedures for investigating metabolic pathways are given including the use of mutants, the application of stable isotopes, and *in vivo* studies using nuclear magnetic resonance. The difficulty in assigning organisms of metabolic interest to taxonomic groups is noted in addition to the generally limited existing knowledge of their genetic systems.

INTRODUCTION

A general overview of the processes that determine the fate and persistence of xenobiotics in the environment has been presented in Chapter 4, but it is important to anchor this to the experimental methods on which the conclusions from such investigations must ultimately be based. That is the aim of the present chapter.

Like organic chemistry on which it critically depends, the study of microbial metabolism is a relatively young discipline, not much more than one hundred years old. And the cardinal experimental procedures for isolation of microorganisms for studies on the metabolism of xenobiotics remain those of elective enrichment pioneered by Winogradsky, Beijerinck, Kluyver, van Niel, and their successors, backed up by the use of pure cultures using procedures developed by Koch in his classic investigations on anthrax. The major innovation has been the development of generally applicable procedures introduced by Hungate (1969) for the isolation of strictly anaerobic organisms. There have, however, been major developments in methods for the elucidation of metabolic pathways: (1) the availability of isotopically labeled compounds — in particular, ^{14}C; (2) the application of genetic procedures; and (3) the application of modern analytical procedures and physical methods for structure determination that have been outlined in greater detail in Chapter 2. In studies of microbial metabolism, the advantages resulting from the requirement for only extremely small quantities of material needed for gas-chromatographic quantification and gas chromatographic-mass spectrometric identification can hardly be overestimated.

In the following sections, an attempt will be made to provide a critical outline of experimental aspects of investigations directed to biodegradation and biotransformation with particular emphasis on outstanding issues to which sufficient attention has not always been paid and which have not, therefore, received ultimate resolution. Before proceeding further, it is desirable to define clearly some operational terms in addition to those that have been given in Chapter 4.

Ready biodegradability refers to the situation in which the test compound is totally degraded (under aerobic conditions to CO_2, H_2O, etc.) within the time span of a standardized test usually lasting 5 d, 7 d, or 28 d.

Inherent biodegradability is applied when the compound *may* be degraded, though not under the standard conditions generally used; their degradation may require, for example, pre-exposure to the xenobiotic.

Recalcitrant is a valuable concept (Alexander 1975) that has been applied to compounds which have not been demonstrably degraded under the conditions used for their examination.

Biotransformation is applied to situations in which, even though degradation is not achieved, minor structural modifications of the test compound have occurred; illustrative examples have already been given in Chapter 4.

Rigid boundaries between these terms should not, however, be drawn, since all of them are operational rather than absolute.

5.1 DETERMINATION OF READY BIODEGRADABILITY

Because of the central role that estimates of biodegradability play in environmental impact assessments, a great deal of effort has been devoted to developing standardized test procedures (Gerike and Fischer 1981). In spite of this, conventional tests for biodegradability under aerobic conditions retain some questionable, or even undesirable, features from an environmental point of view. Attention is, therefore, drawn to two valuable critiques of widely used procedures (Howard and Banerjee 1984; Battersby 1990). Some of the important issues in the design of such tests are, therefore, only briefly summarized here.

The inoculum — For assessment of biodegradation in freshwater systems that have been most extensively examined, the inoculum is generally taken from municipal sewage treatment plants and is, therefore, dominated by microorganisms that have been subjected to selection primarily for their ability to use readily degraded substrates. Whereas these organisms are clearly valuable in evaluating the persistence of compounds that might be poorly degraded in such treatment systems and, therefore, might be discharged unaltered into the environment, they are not necessarily equally suited to investigations on the degradability of possibly recalcitrant substances in natural ecosystems that are dominated by microorganisms adapted to a different environment.

Concentration of the substrate — In conventional assay systems, substrate concentrations are generally used at levels appropriate to measurements of the uptake of oxygen or the evolution of carbon dioxide, and these concentrations greatly exceed those likely to be encountered in receiving waters; attention has been drawn to this issue in Chapter 4 (Section 4.5.2). Application to poorly water-soluble (De Morsier et al. 1987) or to volatile substrates may be difficult or even impossible due to the need for the high substrate concentrations. In addition, high substrate concentrations may be toxic to the test organism and, thereby, provide false-negative results. Use of ^{14}C-labeled substrates and measurement of the evolution of $^{14}CO_2$ enables very much lower substrate concentrations to be used, and the wider application of this procedure is limited primarily by the availability of labeled substrates, though an increasingly wider range of industrially important compounds including agrochemicals is becoming commercially available.

The end-points — For aerobic degradation, uptake of oxygen or the evolution of carbon dioxide is most widely used. Use of the concentration of dissolved organic carbon may present technical problems when particulate matter is present, though analysis of dissolved inorganic carbon in a closed system has been advocated (Birch and Fletcher 1991) and may simultaneously overcome problems with poorly soluble or volatile compounds.

For anaerobic degradation, advantage has been taken of methane production (Birch et al. 1989; Battersby and Wilson 1989). Whereas this may be valuable in the context of municipal sewage treatment plants, it is more questionable whether this is generally a valid parameter in investigations concerned with anaerobic degradation in natural ecosystems in view of the extensive evidence for anaerobic degradation by non-methanogenic bacteria such as sulfate-reducing anaerobes; relevant examples are given in Chapter 6 (Sections 6.7.1 and 6.7.3).

Design of experiments — Probably most investigations have been carried out in conventional batch cultures, but attention should be drawn to an attractive and flexible procedure using a cyclone fermentor (Liu et al. 1981).

Metabolic limitations — In the most widely applied procedures, the test system is restricted in flexibility by the salinity and pH requirements of the test organisms, but probably the most serious limitation of all these test systems is that no account is taken of biotransformation reactions, nor is identification of their products routinely attempted. A good example in which this has, however, been carried out is presented by a study of the degradation of sodium dodecyltriethoxy sulfate under mixed-culture die-away conditions using acclimated cultures (Griffiths et al. 1986). The metabolites were identified, and the kinetics of their synthesis compared with the degradation pathways elucidated in investigations using pure cultures (Hales et al. 1982; Hales et al. 1986). This procedure could advantageously receive more general application.

Application to marine systems — In view of the substantial quantities of xenobiotics that enter the marine environment, surprizingly little effort has been directed to this problem. The degradation of several structurally diverse substrates including nitrilotriacetate, 3-methylphenol, and some chlorobenzenes was evaluated from the rates of incorporation of ^{14}C-labeled substrates into biomass and production of $^{14}CO_2$. These were used to evaluate the differences between freshwater, estuarine, and marine environments and revealed the difficulty of correlating rates with characteristics of the microbial community (Bartholomew and Pfaender 1983). Recent studies have used both dissolved organic carbon and oxygen uptake as parameters (Nyholm and Kristensen 1992) or analysis for specific compounds at low substrate concentration which are proposed as "simulation tests" (Nyholm et al. 1992). Both these tests used the indigenous organisms present in seawater and, thereby, provided a valuable degree of relevance even though they inevitably encountered the variability in the nutritional status — particularly for organic carbon — of seawater. It is, therefore, hazardous to extrapolate results from freshwater to marine systems since these factors, together with the low temperatures that characterize ocean water, introduce complexities that have not been systematically investigated.

5.2 ISOLATION AND ELECTIVE ENRICHMENT

It is only seldom that it has been possible to obtain bacteria with a desired metabolic capability directly from natural habitats. Almost always large numbers of other organisms are present so that some form of selection or enrichment is generally adopted before metabolic studies are attempted. Use of antibiotics or even more drastic procedures using alkali or hypochlorite have been used only infrequently except for the isolation of pathogenic bacteria such as *Mycobacterium tuberculosis*. Although valuable results have been obtained from experiments using metabolically stable mixed cultures, the problems of repeatability limit their application. Probably most metabolic studies on xenobiotics have, therefore, been carried out with pure cultures of organisms. With the possible exception of anaerobic bacteria and, in particular, methanogens, relatively few of these organisms have been isolated from samples of municipal sewage sludge. Most have been obtained after elective enrichment of natural samples of water, sediments, or soils. This methodology was developed by Beijerinck and Winogradsky and has been extensively exploited in the pioneering investigations carried out by the Delft school and their successors over many years. In the present context, one of its particularly attractive features is the inherent degree of environmental realism introduced by its application, and the flexibility, whereby virtually any environmental condition can be mimicked, must be considered as one of the most attractive features of this procedure. There is extensive evidence for the existence of enrichment in the natural environment, and a number of examples illustrating its operation have been given in Chapter 4 (Section 4.5.4), but possibly the last word should be left to one of the pioneers of its application:

"But once an elective culture method for a particular microbe is available, it may safely be concluded that this organism will also be found in nature under conditions corresponding in detail to those of the culture, and that it will carry out the same transformations." (van Niel 1955)

5.2.1 General Procedures

In its simplest form, the procedure consists of elective enrichment of the micro-organisms in an environmental sample by growth at the expense of a single compound serving as the sole source of carbon and energy; successive transfer into fresh medium after growth has occurred, followed by isolation of the appropriate organisms. Some of the experimental details are briefly described in the following paragraphs. The three successive stages used in isolating the desired organisms are outlined first, followed in Section 5.2.2 by a more extensive discussion of media.

1. An appropriate mineral medium containing the organic compound which is to be studied is inoculated with a sample of water, soil, or sediment; in studies of the environmental fate of a xenobiotic in a specific ecosystem, samples are generally taken from the area putatively contaminated with the given compound so that a degree of environmental relevance is automatically incorporated. Recent attention has, in addition, been directed to pristine environments, and the question of adaptation or pre-exposure has already been discussed in Chapter 4 (Section 4.5.4).

   If the test compound is to serve as a source of sulfur, nitrogen, or phosphorus, these elements must clearly be omitted from the medium, and an appropriate carbon source must be available either from the xenobiotic under investigation or supplied by another intentionally added substrate. In such experiments, glassware must be scrupulously cleaned to remove interfering traces of, for example, detergents which may contain residues of all of these nutrients and which could, therefore, compromise the outcome of the experiment. The following may be given as illustration of enrichments designed to obtain organisms using organic compounds as sources of N, S, or P — though not necessarily able to use the test substrates as carbon sources.

   - Nitrophenols as N-source using succinate as C-source (Bruhn et al. 1987).
   - 2-Chloro-4,6-diamino-1,3,5-triazine as N-source and lactate as C-source (Grossenbacher et al. 1984) or 2-chloro-4-aminoethyl-6-amino-1,3,5-triazine as N-source and glycerol as C-source (Cook and Hütter 1984).
   - Arylsulfonic acids as S-source and succinate or glycerol as C-sources (Zürrer et al. 1987).
   - 2-(1-Methylethyl)amino-4-hydroxy-6-methylthio-1,3,5-triazine as S-source and glucose as C-source (Cook and Hütter 1982).
   - Glyphosate (N-phosphonomethylglycine) as P-source and glucose as C-source (Talbot et al. 1984).

   These procedures may, of course, result in the dominance of organisms that carry out only biotransformation of the xenobiotic, although the biodegradation of many of these compounds has also been demonstrated using the same or other organisms; relevant illustrations are given in Chapter 6.

2. The cultures are then incubated under relevant conditions of temperature, pH, and oxygen concentration, and after growth has occurred, successive transfer to fresh medium is carried out. Interfering particulate matter will, thereby, be

removed by dilution, and a culture suitable for isolation may be obtained. Incubation is generally carried out in the dark, though for the isolation of phototrophic organisms, illumination at suitable wavelength and intensity must obviously be supplied. There are no rigid rules on how many transfers should be carried out, but especially for anaerobic organisms, sufficient time should elapse between transfers to allow growth of these often slow-growing organisms. As a result, the enrichment may take up to a year or even longer.

3. After metabolically stable cultures have been obtained, pure cultures of the relevant organisms may then be obtained by either of three basic procedures:

- By preparing serial dilution of the culture in a suitable buffer medium and spreading portions onto solid media (agar plates or roll tubes for anaerobic bacteria) containing the organic compound as source of carbon, sulfur, nitrogen, or phosphorus. The plates (or tubes) are then incubated under appropriate conditions; after growth has taken place, single colonies are then selected and pure cultures obtained by repeated re-streaking on the original defined medium. Use of complex media or substrate analogues may introduce serious ambiguities since overgrowth of unwanted, rapidly-growing organisms may occur. Considerable difficulty may be experienced when "spreading" organisms conceal the desired organism, and this may make the isolation of single colonies a tedious procedure.

- Serial dilution may be carried out in a defined liquid growth medium and the dilutions incubated under suitable conditions; successive transfers are then made from the highest dilution showing growth. This experimentally tedious procedure may indeed be obligatory for organisms such as *Thermomicrobium fosteri* — a name no longer accepted by the authors (Zarilla and Perry 1984) — which are unable to produce colonies on agar plates (Phillips and Perry 1976) and has been quite extensively used for anaerobic bacteria in which the liquid medium is replaced by a soft-agar medium.

- Mechanical methods may be used for the isolation of filamentous organisms. For example, washing on the surface of membrane filters or micromanipulation on the surface of agar medium (Skerman 1968) has occasionally been employed, and an ingenious procedure that uses an electron microscope grid for preliminary removal of other organisms has been described for use under anaerobic conditions (Widdel 1983).

Lack of repeatability in the results of metabolic studies using laboratory strains which have been maintained by repeated transfer for long periods under non-selective conditions may be encountered; these strains may have lost their original metabolic capabilities, and this may be particularly prevalent when the strains carry catabolic plasmids which may be lost under such conditions. For these reasons, strains should be maintained in the presence of suitable cryoprotectant such as glycerol at low temperatures ($-70°C$ or in liquid N_2) as soon as possible after isolation. Freeze-drying is also widely adopted.

In some cases, difficulty may be experienced in isolating pure cultures with the desired metabolic capability, and the mixed enrichment cultures or consortia must be used for further studies. For example, although an enrichment culture effectively degraded atrazine (2-chloro-4-ethylamine-6-isopropylamine-1,3,5-triazine), none of the 200 pure cultures isolated from this were able to use the substrate as N-source (Mandelbaum et al. 1993). A probably frequent situation that is illustrated in Chapter 4 (Section 4.4.1) is when two (or more) organisms cooperate in degradation of the substrate. Considerable effort may then be required to determine the appropriate combination of organisms. For example, an enrichment with 4-chloroacetophenone yielded 8 pure strains, although none of these could degrade the substrate in pure culture. All pairwise combinations were then analyzed, and this revealed that the degradation of the substrate was accomplished by strains of an *Arthrobacter* sp. and a *Micrococcus* sp.; these were then used to elucidate the metabolic pathway (Havel and Reineke 1993). Attention is drawn to the fact that, in this case, isolation of the pure strains from the enrichment culture was carried out using a complex medium under non-selective conditions.

These batch procedures for enrichment and successive transfer may be replaced by the use of continous culture, and this may be particularly attractive when the test compound is toxic, when it is poorly soluble in water, or where the investigations is directed to substrate concentrations so low that clearly visible growth is not to be expected. The problem of isolating the relevant organisms still remains.

5.2.2 Basal Media

The choice of appropriate basal media is of cardinal importance, and a number of important practical considerations should be taken into account.

Basal mineral media — A plethora of basal media for the growth of freshwater organisms has been formulated and these may differ significantly, particularly in the concentrations of phosphate, while for anaerobic bacteria the inclusion of bicarbonate and a suitable reductant is standard practice; numerous examples of suitable media have been collected in *The Prokaryotes* (Balows et al. 1992). Clearly, if the organic substrate is to serve as a source of nitrogen or sulfur or phosphorus, these elements must be omitted from the basal medium. Otherwise, these inorganic nutrient requirements will generally be supplied by the following: S (generally as sulfate except for organisms such as chlorobia which require reduced sulfur as S^{2-}); N (generally as ammonium or nitrate except for N_2-fixing organisms); P (as phosphate); Mg^{2+}, Ca^{2+}, and lower concentrations of Na^+ and K^+. For marine organisms the basal medium is constructed to resemble natural seawater in the concentrations of Na^+, K^+, Mg^{2+}, Ca^{2+}, Cl^- and SO_4^{2-}, and a number of different formulations have been used (Taylor et al. 1981; Neilson 1980) though the exact composition does not seem to be critical. A few brief comments may, however, be inserted on the nitrogen source and the importance of its concentration.

1. Caution should be exercised regarding the inclusion of nitrate as a sole or supplementary nitrogen source. This may result in the synthesis of metabolites containing nitro substituents which are introduced by unknown mechanisms (Sylvestre et al. 1982; Omori et al. 1992); further details are given in Chapter 6 (Section 6.3.1).
2. The intrusion of chemical reactions between amines and reduction products of nitrate has already been noted in Chapter 4 (Section 4.1) and is noted again in Chapter 6 (Section 6.11.3).

3. The nitrogen status of the growth medium determines the levels of lignin peroxidases and manganese-dependent peroxidases that are synthesized in *Phanerochaete chrysosporium*. The role of Mn concentration is noted below, while the metabolic consequences of nitrogen concentration on the metabolism of polycyclic aromatic hydrocarbons are briefly discussed in Chapter 6 (Section 6.2.2).

Trace elements (generally B, Zn, Cu, Fe, Mn, Co, Ni, Mo), most of which are components of many enzymes and coenzymes are generally provided at low concentrations. Anaerobic bacteria are more fastidious in their trace-metal requirements so that routine addition is made of selenium as selenite and also of tungsten as tungstate (Tschech and Pfennig 1984). Selenium is required for the synthesis of active xanthine dehydrogenase in purine-fermenting clostridia (Wagner and Andreesen 1979) and of formate dehydrogenase in a number of organisms including methanogens (Jones and Stadtman 1981) and clostridia (Yamamoto et al. 1983). Addition of W is needed for the growth of various methanogens (Zellner et al. 1987; Winter et al. 1984) and of an anaerobic cellulolytic bacterium (Taya et al. 1985), for the synthesis of the carboxylic acid reductase in acetogenic clostridia (White et al. 1989), and is incorporated into some proteins of purinolytic clostridia (Wagner and Andreesen 1987). For the sake of completeness, it is worth noting the role of vanadium in the synthesis of alternative nitrogenase systems by several aerobic bacteria particularly those within the genus *Azotobacter* (Fallik et al. 1991); low yields of hydrazine are produced and suggest a different affinity of the nitrogenase for the N_2 substrate (Dilworth and Eady 1991).

A large number of different formulations of trace elements have been published; the A4 formulation (Arnon 1938) supplemented with Co^{2+} and Mo(VI) has been widely used, or one of the SL series of formulations developed by Pfennig and his co-workers particularly for the cultivation of anaerobic bacteria. SL 9 is typical of the later series of formulations (Tschech and Pfennig 1984). Somewhat conflicting views exist on the possibly deleterious effects resulting from the incorporation of complexing agents, particularly ETDA, so that their concentration should probably be kept to a minimum.

Buffering of the medium is usually achieved with phosphate, though excessive concentrations should be avoided since they cause problems of precipitation during sterilization by autoclaving. A number of organic buffers have been used effectively in different applications. In studies employing organic phosphorus compounds as P-sources, phosphate has been replaced by, for example, HEPES (Cook et al. 1978), and MOPS has been incorporated into media for growth of *Enterobacteriaceae* (Neidhardt et al. 1974). For media requiring low pH, MES has been used in a medium with extremely low phosphate concentration (Angle et al. 1991), and TRIS has been incorporated into media for growth of marine bacteria (Taylor et al. 1981). In all of these cases, it should be established that metabolic complications do not arise as a result of the ability of the organisms to use the buffer as sources of nitrogen or sulfur. Metal chelating agents such as NTA or ETDA may be used but their concentrations should be kept to a minimum in view of their potential toxicity; the former must obviously be omitted in studies of utilization of organic nitrogen as N-source since both NTA and its metal complexes are apparently quite readily degraded (Firestone and Tiedje 1975).

Vitamins such as thiamin, biotin, and vitamin B_{12} may be added. Once again, the requirements of anaerobes are somewhat greater, and a more extensive range of vitamins that includes pantothenate, folate, and nicotinate is generally employed. In some cases, additions of low concentrations of peptones, yeast extract, casamino acids, or rumen fluid may be used, though in higher concentrations, metabolic ambiguities may be introduced since these compounds may serve as additional carbon and/or nitrogen sources.

Mineral basal media may be sterilized by autoclaving, but for almost all organic compounds which are used as sources of C, N, S, or P, it is probably better to prepare concentrated stock solutions and sterilize these by filtration, generally using 0.2 μm cellulose nitrate or cellulose acetate filters. The same applies to solutions of vitamins and to solutions of bicarbonate and sulfide which are components of many media used for anaerobic bacteria.

The role of metal concentration:

1. Iron — Fe is required as a trace element, but it may also play a more subtle role in determining the biodegradability of a substrate that forms complexes with Fe. Two examples may be used as illustration of probably different underlying reasons for this effect.

   • A strain of *Agrobacterium* sp. was able to degrade ferric EDTA though not the free compound (Lauff et al. 1990). This may be due either to the adverse effect of free EDTA on the cells or the inability of the cells to transport the free compound. The former is supported by the established sensitivity (Wilkinson 1968) of some gram-negative organisms to ETDA and the increased surface permeability in enteric organisms exposed to EDTA (Leive 1968). These results are consistent with the requirement for high concentrations of Ca^{2+} in the enrichment medium used for isolating a mixed culture capable of degrading EDTA (Nörtemann 1992).

   • A strain of *Pseudomonas fluorescens* biovar II is able to degrade citrate whose metabolism requires access to the hydroxyl group. This group is, however, implicated in the tridentate ligand with Fe(II) so that this complex is resistant to degradation in contrast to the bidentate ligand with Fe(III) that has a free hydroxyl group and is readily degraded (Francis and Dodge 1993).

   These results may be viewed in the wider context of interactions between potential ligands of multifunctional xenobiotics and metal cations in aquatic environments and the subtle effects of the oxidation level of cations such as Fe.

   The Fe status of a bacterial culture has an important influence on synthesis of the redox systems of the cell since many of the electron transport proteins contain Fe. This is not generally evaluated systematically, but the degradation of tetrachloromethane by a strain of *Pseudomonas* sp. under denitrifying conditions clearly illustrated the adverse effect of Fe on the biotransformation of the substrate (Lewis and Crawford 1993; Tatara et al. 1993); this possibility should, therefore, be taken into account in the application of such organisms to bioremediation programs.

2. Manganese — The role of manganese concentration has seldom been explicitly examined in the context of biodegradation, although it is essential for the growth

of the purple non-sulfur anaerobic phototrophs *Rhodospirillum rubrum* and *Rhodopseudomonas capsulata* during growth with N_2 but not with glutamate (Yoch 1979). It does, however, play an essential role in the metabolic capability of the white-rot fungus *Phanaerochaete chrysosporium*. This organism produces two groups of peroxidases during secondary metabolism — lignin peroxidases and manganese-dependent peroxidases. Both of them are synthesized when only low levels of Mn (II) are present in the growth medium, whereas high concentrations of Mn result in repression of the synthesis of the lignin peroxidases and an enhanced synthesis of manganese-dependent peroxidases (Bonnarme and Jeffries 1990; Brown et al. 1990). Experiments with a nitrogen-deregulated mutant have shown that N-regulation of both these groups of peroxidases is independent of Mn(II) regulation (Van der Woude 1993).

The redox potential of media — Cultivation of strictly anaerobic organisms clearly requires that the medium be oxygen-free, but this is often not sufficient; the redox potential of the medium must generally be lowered to be compatible with that required by the organisms. This may be accomplished by addition of reducing agents such as sulfide, dithionite, or titanium(III) citrate, though any of these may be toxic, so that only low concentrations should be employed. Attention has been drawn to the fact that titanium(III) citrate-reduced medium may be inhibitory to bacteria during initial isolation (Wachenheim and Hespell 1984). Further comments on procedures are given in Section 5.2.4.

5.2.3 Organic Substrates

Although organic substrates such as carboxylic acids are sufficiently thermally stable that they may be sterilized by autoclaving with the basal media, many others including, for example, carbohydrates, esters, or amides are better prepared as concentrated stock solutions, sterilized by filtration through 0.22 μm filters and added to the sterile basal medium. A major problem arises when the desired organic substrates are poorly soluble in water or highly volatile or toxic. For volatile or gaseous substrates, sealed systems such as desiccators or sealed ampoules may be used. In addition, it has been found convenient to supply toxic substrate such as hydrogen cyanide in the gas phase (Harris and Knowles 1983), and attention should be directed to the requirement of many organisms for CO_2 that has been noted in Chapter 4 (Section 4.2). Serious problems may arise for substrates that are too toxic to be added in the free state at concentrations sufficient for growth. The toxicity of long chain aliphatic compounds with low water solubility has been examined in yeasts (Gill and Ratledge 1972), and the principles that emerged have been applied to circumventing toxicity in liquid media by supplying substrates such as toluene in an inert hydrophobic carrier (Rabus et al. 1993) or the water-soluble hexadecyltrimethylammonium chloride adsorbed on silica (van Ginkel et al. 1992). Some additional methods that have been applied to the preparation of solid media are described below.

Preparation of solid media — Although solid media have been prepared from silica gels, these have not been widely used. Agar for preparing solid medium should be of the highest quality and as free as possible from alternative carbon sources. It is generally preferable to autoclave agar separately from the mineral base; both are prepared at double the final concentration and mixed after autoclaving (Stanier et al. 1966). The problem of preparing plates for testing the metabolic capacity of poorly water-soluble substances is a serious one for which no universal solutions are available. Some of the techniques which have been advocated include the following:

1. Liquid hydrocarbons have been adsorbed on silica powder and dispersed in the agar medium; the silica itself may be autoclaved (Baruah et al. 1967), though this may be avoided by sterilizing the silica by heating and carrying out the adsorption and removal of solvent under aseptic conditions.
2. Solutions of the substrate have been prepared, for example, in acetone or diethyl ether and added to or spread over the surface of agar plates either before or after inoculation (Sylvestre 1980; Shiaris and Cooney 1983).
3. For compounds with relatively high volatility such as benzene, toluene, naphthalene, or other sufficiently volatile hydrocarbons, the substrate may be contained in a tube placed in the liquid medium (Claus and Walker 1964) or on the lid of a glass Petri dish (Söhngen 1913); for benzene and toluene, this also obviates problems with toxicity since the organisms are exposed to only low concentrations of the substrate.
4. A solution of the substrate in ethanol may be mixed with the bacterial suspension in agarose and poured over agar plates of the base medium (Bogardt and Hemmingsen 1992). This is the general procedure used with top agar in the Ames test (Maron and Ames 1983), although in this case dimethyl sulfoxide is generally used as the water-miscible solvent.

Volatile substrates — Gaseous or highly volatile substrates present a problem that may be overcome by the use of enclosed systems such as desiccators or sealed ampoules, and it has been found convenient to supply toxic volatile substrates such as hydrogen cyanide in the gas phase (Harris and Knowles 1983). This has also been employed for 4-chloroacetophenone (Havel and Reineke 1993) since the 4-chlorophenol produced is toxic to one of the components of the consortium. In this case, low concentrations of the toxic intermediate could also be maintained by adding gelatin to the medium, and this procedure facilitated growth of one of the components; this example has already been discussed in the context of enrichment procedures in Section 5.2.1 and as an example of metabolic interaction of organisms in Chapter 4 (Section 4.4.1), while metabolic details are given in Chapter 6 (Section 6.2.1). It should be noted that, particularly in sealed systems, it is important to satisfy the obligate requirement of many organisms for CO_2 (Chapter 4, Section 4.2).

Growth at the expense of alternative substrates — It may be found that growth does not occur on plates prepared with the compound which showed satisfactory growth in liquid medium, and some probable reasons for this have been outlined in Chapter 4 (Section 4.4). One additional possibility worth examining is that of attempting to grow the organisms with a potential metabolite, though it should be kept in mind that organisms may be unable to utilize compounds which are clearly established metabolites of the desired compound; examples are the inability of fluorescent pseudomonads which degrade aromatic compounds via *cis,cis* muconate to use this as a substrate (Robert-Gero et al. 1969) or of alkane-degrading bacteria to grow with the corresponding carboxylic acids which are the first metabolites in alkane degradation (Zarilla and Perry 1984). One plausible reason could be the lack of an effective transport system for the metabolite; another could be the failure of the metabolite to induce the enzymes necessary for its production. For example, whereas salicylate normally induces the enzymes required for the degradation of naphthalene, this is apparently not the case for a naphthalene-degrading strain of *Rhodococcus* sp. (Grund et al. 1992).

Use of complex media for isolating organisms after elective enrichment is, on the other hand, a potentially hazardous procedure. Media that are routinely used for non-metabolic studies in clinical laboratories generally contain high concentrations of peptones, yeast extract, or carbohydrates, and these may provide alternative carbon (and nitrogen) sources; their use may, therefore, result in only low selection pressure for the emergence of the desired organisms; overgrowth by undesired microorganisms may then take place.

5.2.4 Procedures for Anaerobic Bacteria

Substantial and increasing attention has been directed to the growth and isolation of anaerobic bacteria; in addition to the nutritional requirements noted above, their general requirement for CO_2 should be taken into consideration. Broadly, three types of experimental methodologies have been used:

1. Anaerobe jars containing a catalyst for the reaction between oxygen and hydrogen that is either added to, or generated within, the system; these systems have limitations in the kinds of experiment which can be carried out since, at some stage, exposure to air cannot be avoided, and they are not suitable for work with highly oxygen-sensitive organisms such as methanogens.
2. The classical Hungate technique (Hungate 1969) has been successfully used over many years and incorporates a number of features designed to minimize exposure to oxygen. This procedure enables incubation to be carried out under a variety of gas atmospheres and is designed to produce a redox potential in media that is suitable for growth. Roll-tubes have been used instead of Petri dishes, and, thereby, a strictly anoxic environment may be maintained during manipulation. A modification using serum bottles has been introduced (Miller and Wolin 1974), and roll-tubes have also been successfully used for isolation of anaerobic phycomycetous fungi from rumen fluid (Joblin 1981).
3. Anaerobe chambers of varying design have achieved increasing popularity since they enable standard manipulations to be carried out under anoxic conditions; these systems maintain a gas atmosphere of N_2, H_2, and CO_2 (generally 90:5:5) and include a heated catalyst for the maintainance of anaerobic conditions. They may employ either a glovebox design or free access through wrist bands and enable quite sophisticated experiments to be carried out and cultures maintained over lengthy periods. They can, therefore, be unequivocally recommended although their maintenance costs should not be underestimated.

5.3 DESIGN OF EXPERIMENTS ON BIODEGRADATION AND BIOTRANSFORMATION

There are four essentially different kinds of experiments which may be carried out. Two are laboratory-based, and two are field investigations:

1. Laboratory experiments using pure cultures or stable consortia.
2. Laboratory experiments using communities in microcosms simulating natural systems.
3. Field experiments in model ecosystems — pools or streams.
4. Large-scale field experiments under natural conditions.

It should be clearly appreciated that the objectives of these various investigations are rather different. The first two of them aim at elucidating the basic facts of metabolism, the products formed, and the kinetics of their synthesis; studies using pure cultures may ultimately be directed to studying more sophisticated aspects of the regulation and genetics of biodegradation. On the other hand, the third and fourth procedures are designed to obtain data of more direct environmental relevance and may profitably — and even necessarily — draw upon the results obtained using the first two procedures. While the degree of environmental realism increases from 1 through 4, so also do the experimental difficulties and the interpretative ambiguities; all of the procedures have clear advantages for specific objectives, and, indeed, it is highly desirable that several of them be combined.

There are significant differences in the control experiments that are possible in each of these systems. Before the quantifier *bio-* can be applied, the possibility of abiotic alteration of the substrate during incubation must be eliminated. On this point, only the first design lends itself readily to rigorous control, and even then there may be experimental difficulties. For experiments using cell suspensions, the obvious controls are incubation of the substrate in the absence of cells or using autoclaved cultures. Care should be exercised in the interpretation of the results, however, since some reactions may apparently be catalyzed by cell components in purely chemical reactions; the question may then legitimately be raised whether or not these are biochemically mediated. Two examples may be used as illustration of apparently chemically mediated reactions:

1. Dechlorination reactions of organochlorine compounds which have been referred to in Chapter 4 (Section 4.1.2) and are discussed again in Chapter 6 (Sections 6.4.4 and 6.6).
2. Reduction of aromatic nitro compounds by sulfide and catalyzed by extracellular compounds excreted into the medium during growth of *Streptomyces griseoflavus* (Glaus et al. 1992).

In experiments where small volumes of sediment suspensions are employed, autoclaving may significantly alter the structure of the sediment as well as introducing possibly severe analytical difficulties; in such cases, there seem few alternatives to incubation in the presence of toxic agents such as NaN_3 which has been used at a concentration of 2g/L. There remains, of course, the possibility that azide-resistant strains could emerge during prolonged incubation, and the occurrence of reactions between the substrate and azide must also be taken into consideration.

Only controls using inhibitors of microbial growth are applicable to microcosm experiments, and it seems unlikely that even these can realistically be applied to outdoor systems which therefore combine and generally fail to discriminate between abiotic and biotic reactions.

5.3.1 Pure Cultures and Stable Consortia

Different kinds of experimental procedures have been used, and these should be evaluated against the background presented in Chapter 4 (Section 4.4) that is devoted to situations in which several organisms or several substrates are simultaneously present. There are no essential differences in the design of experiments using pure cultures and those employing metabolically stable consortia. It should be emphasized, however, that even in the latter case, the experiments should be carried out under aseptic conditions; otherwise, interpretation of the results may be compromised by adventitious organisms.

Cell growth at the expense of the xenobiotic — In the simplest case, growth of the organism that has been isolated may be studied using the test substance that fulfills the nutritional requirement of the organism as the sole source of carbon, sulfur, nitrogen, or phosphorus. For a compound used as sole source of carbon and energy, the end-points could be growth and conversion of the substance into CO_2 under aerobic conditions or growth under anaerobic conditions accompanied by, for example, production of methane or sulfide from sulfate. In some studies, however, only diminution in the concentration of the initial substrate has been demonstrated, and this alone clearly does not constitute evidence for biodegradation; biotransformation is equally possible and should be taken into consideration. Ideally, use may be made of radio-labeled substrates followed by identification of the labeled products. ^{14}C, ^{35}S, ^{36}Cl, and ^{31}P have been used, though the relevant labeled products may not always be available commercially, and the required synthetic expertise may not be available in all laboratories. Further comments on the use of these isotopes together with the application of the non-radioactive isotopes ^{13}C and ^{19}F are given in Section 5.5.4.

Since considerable weight has been given to the environmental significance of biotransformation and the synthesis of toxic metabolites (Chapter 6, Section 6.11.5), it is particularly desirable to direct effort to the identification of such metabolites. This presents a substantially greater challenge than that of quantifying the original substrate for several reasons:

1. The structure of the metabolite will generally be unknown and must be predicted from knowledge of putative degradation pathways and confirmed by any of the methods outlined in Chapter 2 (Section 2.4.1).
2. The metabolite will frequently be more polar than the initial substrate so that specific procedures for extraction and analysis must generally be developed, and the pure compound must be available for quantification.
3. The metabolite may be transient with unknown kinetics of its formation and further degradation.

In practice, there is only one really satisfactory solution: the kinetics of the transformation must be followed. The justification for this substantial increase in effort is the dividend resulting in the form of a description of the metabolic pathway including the synthesis of possibly inhibitory metabolites. On the basis of this, it may then be possible to formulate generalizations on the degradation of other structurally related xenobiotics.

Use of dense cell suspensions — Dense cell suspensions have traditionally been used for experiments on the respiration of microbial cells at the expense of organic substrates, and they are equally applicable to experiments on biodegradation and biotransformation. Cells are grown in a suitable medium generally containing the test compound (or an alternative growth substrate), collected by centrifugation, washed in a buffer solution to remove remaining concentrations of the growth substrate and its metabolites, and resuspended in fresh medium before further exposure to the xenobiotic. For aerobic organisms, there are generally few experimental difficulties with three important exceptions:

1. For organisms which grow poorly in liquid medium, it may be difficult to obtain sufficient quantities of cells; in such cases, cultures may be grown on the surface of agar plates and the cells removed by scraping.
2. Organisms that have fastidious nutritional requirements may require undefined growth additives such as peptones, yeast extract, rumen fluid, or serum. Subsequent exposure to the xenobiotic may then be used to induce synthesis of the relevant catabolic enzymes. For example, the chlorophenol-degrading bacterium *Rhodococcus chlorophenolicus* has been grown in media containing yeast extract or rhamnose, and exposure to pentachlorophenol used to induced the enzymes required for the degradation of a wide range of chlorophenols (Apajalahti and Salkinoja-Salonen 1986).
3. Organisms such as actinomycetes may not produce well-suspended growth, and shaking, for example, in baffled flasks or in flasks with coiled-wire inserts may be advantageous in partially overcoming this problem.

For anaerobic bacteria, the same principles apply except that additional attention must be directed to preparing the cell suspensions. Use of an anaerobic chamber in which cultures can be transferred to tightly capped centrifuge tubes is virtually obligatory, and addition of an anaerobic indicator should be used to ensure that subsequent entrance of oxygen does not take place inadvertently; Oak Ridge centrifuge tubes are particularly suitable for centrifugation. Some investigations have used metabolically stable enrichment cultures to study biotransformation, and these are preferable to the use of sediment slurries since they avoid the ambiguities due to the presence of xenobiotics in the original soil or sediment and make possible reproducible experiments under defined conditions.

Use of immobilized cells — Cells may be immobilized on a number of suitable matrices in a reactor and the medium containing the test substrate circulated continuously. Although this methodology has been motivated by interest in biotechnology technology, it is clearly applicable to laboratory experiments on biodegradation and biotransformation, and these could readily be carried out under sterile conditions. This procedure has been used, for example, to study the biodegradation of 4-nitrophenol (Heitkamp et al. 1990), pentachlorophenol (O'Reilly and Crawford

1989), and 6-methylquinoline (Rothenburger and Atlas 1993). A few additional comments may be inserted.

1. It is possible to carry out experiments with immobilized cells in essentially non-aqueous media (Rothenburger and Atlas 1993), and this might prove an attractive strategy for compounds with limited water solubility provided that solvents can be found that are compatible with the solubility of the substrate and the sensitivity of the cells to organic solvents.
2. Encapsulated cells have been successfully used for the commercial biosynthesis of a number of valuable compounds such as aminoacids, and this technique could readily be adapted to investigations on biodegradation. This methodology offers the advantage that the metabolic activity of the cells can be maintained over long periods of time so that a high degree of reproducibility in the experiments is guaranteed, and the stability of such systems may be particularly attractive in studies of recalcitrant compounds.
3. With appropriate experimental modifications, the various procedures could be adapted to study biodegradation under anaerobic conditions; for example, sparging with air could be replaced by the use of a gas mixture containing appropriate concentrations of CO_2 and H_2 in an inert gas such as N_2.

Application of continuous culture procedures — The use of these has been very briefly noted in Chapter 4; they have been particularly valuable in studies using low concentrations of xenobiotics and for the isolation of consortia that have been used in elucidating metabolic interactions between the various microbial components. In many cases, consortia containing several organisms are obtained even though only a few of their members are actively involved in the metabolism of the xenobiotic. It is possible that the low substrate concentrations that have been used in these experiments favor selection for organisms able to take advantage of the lysis products from such cells that do not play a direct role in the degradation of the xenobiotic. Three examples may suffice as illustration.

1. Enrichment in a two-stage chemostat with parathion (O,O-diethyl-O-(4-nitrophenyl)-phosphorothioate) as the sole source of carbon and sulfur resulted in a community which was stable for several years (Daughton and Hsieh 1977). It should be noted that, on account of the toxicity of parathion to the culture, only low substrate concentrations could be used, and this methodology is ideally adapted to such situations. Degradation was accomplished by two organisms, *Pseudomonas stutzeri* and *P. aeruginosa*, whereas the third organism in the stable community had no defined function: *P. stutzeri* functioned only in ester hydrolysis which is the first step in the degradation of parathion. This was the first demonstration of degradation of parathion by a metabolically defined microbial consortium, though degradation by a culture consisting of nine organisms had already been demonstrated (Munnecke and Hsieh 1976).
2. Chemostat enrichment was carried out with a mixture of linear alkylbenzene sulfonates as the sole sources of carbon and sulfur at a concentration of 10 mg/L and resulted in the development of a four-component consortium (Jiménez et al. 1991). Three of the organisms were apparently necessary to accomplish this

apparently straightforward degradation, though the isolation procedure that used a complex medium with glucose as carbon source is not entirely unequivocal. A similar situation arises with hexadecyltrimethylammonium chloride from which three strains that could grow with the substrate were obtained again after streaking on yeast-glucose medium (van Ginkel et al. 1992).

3. Chemostat enrichment with 2-chloropropionamide yielded a community of at least six organisms: one of these, a *Mycoplana* sp. carried out hydrolysis of the amide, while various other components used the resulting free acid for growth. An interesting observation was that after prolonged incubation at a dilution rate of $0.01h^{-1}$, a single strain of *Pseudomonas* sp. capable of growth solely on 2-chloropropionamide as carbon source could be isolated (Reanney et al. 1983).

There are some general conclusions which may plausibly be drawn from the results of all these experiments:

1. The reactions necessary for degradation were apparently relatively simple and would be expected to be accessible to single organisms.
2. The first stages for two of these degradations were straightforward hydrolytic reactions.
3. In all cases, organisms with undefined metabolic functions were present and probably fulfilled an important role in providing complex organic substrates in the form of cell lysis products or nutritional requirements (Chapter 4, Section 4.4.1).

Reaction sequences used for the degradation of xenobiotics in natural systems may, therefore, be more complex than might plausibly be predicted on the basis of studies with pure cultures using relatively high substrate concentrations.

Attention should be drawn to experiments in which solutions of the substrate in a suitable mineral medium are percolated through soil that is used as the source of inoculum. This is one of the classical procedures of soil microbiology and has been exploited to advantage in studies on the degradation of a range of chlorinated contaminants in groundwater (van der Meer et al. 1992). Apart from the fact that this mimics closely the natural situation and incorporates the features inherent in any enrichment methodology, this procedure offers a degree of flexibility that enables systematic exploration of the following:

1. The effect of varying redox conditions since by altering the gas phase experiments can readily be carried out under aerobic, microaerophilic, or anaerobic conditions.
2. The effect of substrate concentration and the important issue of the existence or otherwise of threshold concentrations below which degradation is not effectively accomplished: this has been discussed in Chapter 4 (Section 4.5.2).
3. The influence of sorption/desorption on biodegradation that has been discussed in a wider context in Chapter 4 (Section 4.5.3).

Apart from its application to the specific problem of groundwater contamination, this procedure is potentially valuable for simulating bioremediation of contaminated soils.

Simultaneous presence of two substrates — Analogous to the fact that pure cultures of microorganisms seldom occur in natural ecosystems, it is very rare for a single organic substrate to exist in appreciable concentrations. The relevant microorganisms under natural situations are, therefore, exposed simultaneously to several compounds, and this situation can be simulated in laboratory experiments. The terminology has undergone a variety of different designations that have already been discussed in Chapter 4 (Section 4.4.2) so that only brief mention is justified here; although the term "cometabolism" has been used extensively, it has been applied to conflicting metabolic situations, and the pragmatic term "concurrent metabolism" (Neilson et al. 1985) offers an attractive alternative when more than one substrate is present. Two examples will be used to illustrate the application of this procedure to experiments in which the pathways for the biotransformation of different xenobiotics have been established.

1. Experiments have used cells with a metabolic capability that may plausibly be predicted as relevant to that of the xenobiotic. For example, elective enrichment failed to yield organisms able to grow at the expense of dibenzo-1,4-dioxan, but its metabolism could be studied in a strain of *Pseudomonas* sp. capable of growth with naphthalene (Klecka and Gibson 1979). Cells were grown with salicylate (1 g/L) in the presence of dibenzo-1,4-dioxan (0.5 g/L), and two metabolites of the latter were isolated: the *cis*-1,2-dihydro-1,2-diol and 2-hydroxydibenzo-1,4-dioxan. The former is consistent with the established dioxygenation of naphthalene and the role of salicylate as coordinate inducer of the relevant enzymes for conversion of naphthalene into salicylate.
2. An environmentally relevant situation may be simulated by the growth of an organism with a single substrate at a relatively high concentration and simultaneous exposure to a structurally unrelated xenobiotic present at a significantly lower concentration. A series of investigations has used growth substrates at concentrations of 200mg/L and xenobiotic concentrations of 100µg/L; it may reasonably be assumed that growth with compounds at the latter concentration is negligible. For example, during growth of a stable anaerobic enrichment culture with 3,4,5-trimethoxybenzoate, 4,5,6-trichloroguaiacol was transformed into 3,4,5-trichlorocatechol which was further dechlorinated to 3,5-dichlorocatechol (Neilson et al. 1987).

Use of unenriched cultures: undefined natural consortia — Laboratory experiments using natural consortia under defined conditions have particular value from several points of view. They are of direct environmental relevance and their use minimizes the ambiguities in extrapolation from the results of studies with pure cultures. They provide valuable verification of the results of studies with pure cultures and make possible an evaluating the extent to which the results of such studies may be justifiably extended to the natural environment.

It should be appreciated, however, that in some cases the habitats from which the inoculum was taken may already have been exposed to xenobiotics so that "natural" enrichment may already have taken place; this has been discussed briefly in Chapter 4 (Section 4.5.4).

Extensive studies — some of which have already been cited in Chapter 4 (Section 4.5.2) — on the effect of substrate concentration and of the bioavailability of the substrate to the appropriate microorganisms have employed samples of natural lake water supplemented with suitable nutrients. There are few additional details that need to be added since the experimental methods are straightforward and present no particular difficulties. Considerable use has also been made of a comparable methodology to determine the fate of agrochemicals in the terrestrial environment.

Because of the difficulty of obtaining pure cultures of anaerobic bacteria, extensive use has been made of anaerobic sediment slurries in laboratory experiments. In some of these, although no enrichment was deliberately incorporated, experiments were carried out over long periods of time in the presence of contaminated sediments, and adaptation of the natural flora to the xenobiotic during exposure in the laboratory may, therefore, have taken place. The design of these experiments may also inevitably result in interpretative difficulties. For example, although the results of experiments on the dechlorination of pentachlorophenol (Bryant et al. 1991) enabled elucidation of the pathways, this study also revealed one of the limitations in the use of such procedures: detailed interpretation of the kinetics of pentachlorophenol degradation using dichlorophenol-adapted cultures was equivocal due to carryover of phenol from the sediment slurries. It would, therefore, be desirable to avoid this ambiguity by using cultures in which the sediment is no longer present. A similar procedure was used to examine the biodegradation of acenaphthene and naphthalene under denitrifying conditions in soil-water slurries (Mihelcic and Luthy 1988), though, in this case, only analyses for the concentrations of the initial substrates were carried out. In both of these examples, growth of the degradative organisms was supported at least partly by organic components of the soil and sediment, so that the physiological state of the cells cannot be precisely defined.

5.3.2 Microcosm Experiments

Microcosms are laboratory systems generally consisting of tanks such as fish aquaria containing natural sediment and water. In those which have been most extensively evaluated, continuous flow systems are used; continuous measurement of $^{14}CO_2$ evolved from ^{14}C-labeled substrates may be incorporated, and recovery of both volatile and non-volatile metabolites is possible so that a material balance may be constructed (Huckins et al. 1984). It should be pointed out that the term microcosm has also been used to cover much smaller scale experiments that have been carried out under anaerobic conditions (Edwards et al. 1992) and to systems for evaluating the effect of toxicants on biota (Chapter 7, Section 7.4.2). Three examples may be given to illustrate different facets of the application of microcosms to study biodegradation.

1. Biodegradation of t-butylphenyl diphenyl phosphate was examined using sediments either from an uncontaminated site or from one having a history of chronic exposure to agricultural chemicals (Heitkamp et al. 1986). Mineralization was very much more extensive in the latter case but was inhibited by substrate concentrations exceeding 0.1 mg/L. Low concentrations of diphenyl phosphate, 4-t-butylphenol, and phenol indicated the occurrence of esterase activity, while the recovery of triphenyl phosphate suggested dealkylation by an unestablished pathway. A comparable study of naphthalene biodegradation (Heitkamp et al. 1987) found more rapid degradation when sediments chronically exposed to

petroleum hydrocarbons were used, and isolation of *cis*-1,2-dihydronaphthalene-1,2-diol, 1- and 2-naphthol, salicylate, and catechol confirmed the pathway established for the degradation of naphthalene. The results of both investigations illustrate the potential for a more extensive application of the procedure and at the same time the significance of preexposure to the xenobiotic.

2. One of the key issues in bioremediation is the survival of the organisms deliberately introduced into the contaminated system. A microcosm prepared from a pristine ecosystem was inoculated with a strain of *Mycobacterium* sp. that had a wide capacity for degrading PAHs, and this organism was used to study the degradation of 2-methylnaphthalene, phenanthrene, pyrene, and benzo[a]pyrene (Heitkamp and Cerniglia 1989). The test strain survived in the system with or without exposure to PAHs, but the addition of organic nutrients was detrimental to its maintenance. Clearly, an almost unlimited range of parameters could be varied to enable a realistic evaluation of the effectiveness of bioremediation in natural circumstances.

3. Concern has been expressed on the potential hazards from discharge into the environment of organisms carrying catabolic genes on plasmids. Investigations to which reference has been made in Chapter 4 (Section 4.8) (Jain et al. 1987; Fulthorpe and Wyndham 1989; Sobecky et al. 1992) used a set of microcosms to determine the conditions needed to preserve the genotype and its stability. Once again, the advantage of the technique is the ease of incorporating important variables that may be difficult to analyze in natural systems.

Truly field experiments on microbial reactions are extremely difficult to carry out, but a series of microcosm experiments on the substrates that may support anaerobic sulfate reduction approached quite closely this ideal situation (Parkes et al. 1989); the investigation used inhibition of sulfate reduction by molybdate to study the increase in the levels of a wide range of organic substrates endogenous in the sediments used. These included both a range of alkanoic acids and amino acids and very considerably increased the range of organic substrates able to support sulfate reduction. Additional biological aspects of this investigation are discussed in Chapter 8 (Section 8.5.2).

5.3.3 Experiments in Models of Natural Systems

It is extremely difficult to carry out field investigations in natural ecosystems with the rigor necessary to unravel metabolic intricacies, although such experiments have been successfully carried out in investigations aimed at determining the fate and persistence of agrochemicals in the terrestrial environment and in the context of bioremediation. Comparable experiments in natural aquatic ecosystems are much more difficult to design (Madsen 1991), but one example may be given to illustrate what can be accomplished if sufficient is known about the degradative pathways of the xenobiotic. Analysis of chlorobenzoates in sediments that had been contaminated with PCBs was used to demonstrate that the lower PCB congeners that had initially been produced by anaerobic dechlorination were subsequently degraded under aerobic conditions; the chlorobenzoates were transient metabolites, and their concentrations were extremely low since bacteria that could successfully degrade them were present in the sediment samples (Flanagan and May 1993). In general, however, simplified systems have been developed; these attempt to simulate critical segments of natural ecosystems in a clearly defined way. Outdoor model systems have been used, and two examples may be used to illustrate the kinds of data that can be assembled and the range of conclusions — and their limitations — that may be drawn from such experiments. Not only purely microbiological determinants of persistence may be revealed, but in addition, important data on the distribution and fate of the xenobiotic may be acquired (discussed in Chapter 7, Section 7.4.3). Attention has already been drawn to the general issue of partition among various environmental compartments in Chapter 3 and to the significance of bioavailability in Chapter 4 (Section 4.5.3).

1. Studies in an artificial stream system were designed to provide confirmation in a field situation of the results from laboratory experiments that had demonstrated the biodegradability of pentachlorophenol. Pentachlorophenol was added continuously to the system during 88 d, and its degradation followed (Pignatello et al. 1983, 1985). The results confirmed that pentachlorophenol was indeed degraded by the natural populations of microorganisms and, in addition, drew attention to the significance of both sediments and surfaces in the partitioning of pentachlorophenol between the phases within the system.

2. 4,5,6-Trichloroguaiacol was added continuously during several months to mesocosm systems simulating the Baltic Sea littoral zone; samples of water, sediment, and biota including algae were removed periodically for analysis both of the original substance and of metabolites identified previously in extensive laboratory experiments (Neilson et al. 1989). A complex of metabolic transformations of 4,5,6-trichloroguaiacol was identified, including O-methylation to 3,4,5-trichloroveratrole, O-demethylation to 3,4,5-trichlorocatechol, and partial dechlorination to a dichlorocatechol, and these metabolites were distributed among the various matrices in the system. Of particular significance was the fact that a material balance unequivocally demonstrated the role of the sediment phase as a sink for both the original substrate and the metabolites, so that a number of interrelated factors determined the fate of the initial substrate.

5.4 EXPERIMENTAL PROBLEMS: WATER SOLUBILITY, VOLATILITY, SAMPLING

For freely water-soluble substrates which have low volatility, there are few difficulties in carrying out the appropriate experiments described above. There is, however, increasing interest in xenobiotics such as polycyclic aromatic hydrocarbons (PAHs) and highly chlorinated compounds including, for example, PCBs which have only low water solubility; in addition, attention has been focused on volatile chlorinated aliphatic compounds such as the chloroethenes, dichloromethane and carbon tetrachloride. All of these substrates present experimental difficulties of greater or lesser severity.

1. Whereas suspensions of poorly water-soluble substrates can be used for experiments on the identification of metabolites, these methods are not suitable for kinetic experiments that necessitate the quantification of substrate concentrations.

In such cases, the whole sample must be sacrificed at each sampling time, and care must be taken to ensure that substrate concentrations in each incubation vessel are as far as possible equal. In addition, the whole sample must be extracted for analysis since representative aliquots cannot be removed. Solutions of the substrates in suitable organic solvents such as acetone, diethyl ether, methanol, or ethyl acetate may be prepared, sterilized by filtration, and suitable volumes dispensed into each incubation vessel. The solvent is then removed in a stream of sterile N_2 and the cell suspension added. As an alternative, solutions in any of these solvents or in others that are much less volatile, such as dimethyl sulfoxide and dimethyl formamide, have been added directly to media after sterilization by filtration. Since, however, the remaining concentration of the solvents may be appreciable, care must be taken to ensure that these are neither toxic nor compromise the results of the metabolic experiments. When the test substrate is a solid, it may be preferable to prepare saturated solutions in the basal medium, remove undissolved substrate by filtration through glass-fiber filters, and sterilize the solution by filtration before dispensing and adding cell suspensions.

2. Investigation of substrates with appreciable volatility — and many organic compounds including solids have significant vapor pressures at ambient temperatures — presents a greater experimental challenge especially if experiments are to be conducted over any length of time. Incubation vessels such as tubes or bottles may be closed with rubber stoppers or with teflon-lined crimp caps fitted with rubber seals, and these are particularly convenient for sampling with syringes. Even with teflon seals, however, these cap inserts may be permeable to compounds with appreciable vapor pressure, and sorption of the test substrate may also occur to a significant extent; both of these factors result in controls that display undesirable diminution in the concentration of the test substrate. A good illustrative example is provided by the results of a study with endosulfans and related compounds (Guerin and Kennedy 1992), and these support the importance even in laboratory experiments of gas/liquid partitioning that has been discussed in Chapter 3 (Section 3.4). Completely sealed glass ampoules may be used for less volatile compounds — though clearly not for highly volatile compounds — and sub-sampling cannot be carried out. For aerobic organisms, one serious problem with all these closed systems is that of oxygen limitation; the volume of the vessels should, therefore, be very much greater than that of the liquid phase, and conclusions from the experiments should recognize that microaerophilic conditions will almost certainly prevail during prolonged incubation. Such limitations clearly do not prevail for anaerobic organisms.

5.5 PROCEDURES FOR ELUCIDATION OF METABOLIC PATHWAYS

The general principles follow those used in all biochemical investigations, although some procedures are particularly well suited to microbial systems. The structure of the various metabolites and the order in which they are formed will reveal the broad outlines of the metabolic pathways followed during the degradation of the substrate. Subsequent refinements may involve isolating and characterizing the relevant enzymes, followed by investigation of their regulation. At all of these stages, the availability of mutants blocked at specific steps of the metabolic sequence is of the greatest value. Good examples of this are provided by the investigations of the metabolism of aromatic compounds which are discussed in considerable detail in Chapter 6 (Section 6.2). The specific details of the investigation must also take into account problems including the stability of the relevant enzymes, the availability of suitable enzyme assays, and access to the relevant mutants; the complete elucidation of a metabolic pathway is, therefore, a time-consuming and intellectually demanding operation of which only a few salient features can be presented briefly here.

5.5.1 The Principle of Sequential Induction

On the basis of the intensive and extensive investigations into microbial metabolism that have been carried out painstakingly over many years, it may now be possible to propose a hypothetical degradative pathway on the basis of plausible biochemical reactions; this pathway may then be confirmed by determining the structure of the intermediate metabolites and by further metabolic experiments using, for example, suitable isotopes (Section 5.5.4). Metabolic pathways for a wide range of structurally diverse compounds are discussed in detail in Chapter 6, so that only a general outline will be given here by way of illustration. The delineation of the pathways and the role of hypothetic metabolites may be confirmed — or eliminated — by application of the principle of simultaneous or sequential induction (Stanier 1947) in any of its many forms. Good illustrations may be taken from two early studies on biodegradation:

1. Strains of bacteria able to grow with naphthalene were unable to utilize 1- or 2-naphthol, phthalic acid, or catechol. This clearly suggested that these compounds were not *directly* involved in the degradation (Tausson 1927), and this conclusion has been amply confirmed in detailed subsequent studies.
2. For strains of bacteria during growth with toluene (Claus and Walker 1964), utilization of aromatic substrates was inducible and toluene-grown cells were assayed for oxidation of a number of possible intermediates; from the results of these experiments, it could be concluded that the degradation in these strains did not proceed through benzaldehyde and benzoic acid but by oxygenation to 3-methyl catechol. Degradation via benzoate has, of course, been established as an alternative in other organisms.

In both examples, further details of the mechanism of the ring-cleavage reaction that is discussed in Chapter 6 (Section 6.2) had to await chemical characterization and structural determination of the sometimes unstable intermediates.

The principle of sequential induction can be further extended to include the assay of specific enzymes when an inducer is added to cells grown with a non-inducing carbon source. For example, addition of DL-mandelate to a culture of *Pseudomonas putida* grown with succinate resulted in immediate synthesis of L-mandelate dehydrogenase that is the first enzyme of the degradative pathway but a delay before synthesis of the muconate lactonizing enzyme that is involved several stages later after cleavage of the aromatic ring (Hegeman 1966a,b).

5.5.2 Application of Mutants

The use of mutants has been extensively used in elucidating biosynthetic pathways in a number of widely different biological systems. The principles are equally applicable to metabolic studies using mutants defective in specific genes required for the complete degradation of the substrate, and these are essentially the same as those applied in sequential induction.

Mutagenesis may be accomplished by a number of procedures which differ both in the degree of killing and in the type of mutation induced; ethyl methanesulfonate has been used traditionally, though this has been largely superseded by N-methyl-N'-nitro-N-nitrosoguanidine which induces primarily base substitutions or by compounds related to substituted 5-aminoacridines which induce addition or deletion of one or more bases (Miller 1977). Use has also been made of a procedure whereby organisms are grown with a substrate and a structurally related halogenated analogue (Wigmore and Ribbons 1981). Selection of mutants with the desired lesions may be accomplished by replica plating onto basal media containing the initial substrate or substrates that are suspected as metabolites; colonies which grow on the latter, but not on the former, are then purified by restreaking. An attractive procedure employed in investigations on tryptophan metabolism by pseudomonads involved the use of media with a high concentration of tryptophan (1 g/L) together with a low concentration of asparagine (0.1 g/L) which permitted production of only small colonies from mutants defective in tryptophan degradation (Palleroni and Stanier 1964).

Mention should be made of two entirely different procedures that offer a wide range of potential application. Transposon mutagenesis offers a means of obtaining insertion mutants in gene functions for which there are no means of direct selection. Although this appears not to have been extensively used in studies of xenobiotic degradation, it has been used to study the genetics of the gentisate pathway in *Pseudomonas alcaligenes* (Tham and Poh 1993). Vector insertion mutagenesis is another procedure, and this has been applied to the elucidation of the pathway of nicotinate degradation in *Azorhizobium caulinodans* (Kitts et al. 1992).

The degradation of a wide range of aromatic compounds converges on catechol whose degradation, therefore, occupies a key position. Although details of the pathways will be presented in Chapter 6, a very brief sketch of the comprehensive investigations from the Berkeley laboratory that were directed to the elucidation of the pathways and their regulation will be used here to illustrate the various steps in the investigations that include good illustrations of the use of mutants. A number of complementary procedures were used of which the following were cardinal:

1. The use of purified enzymes for the synthesis of the appropriate metabolites.
2. Analysis of induction patterns after growing cells with different substrates and using appropriate assays to determine the levels of the relevant enzymes.
3. Isolation of mutants to study enzyme regulation.

Three typical examples of the application of these procedures will be given as illustration.

1. *Degradation of catechol and 3,4-dihydroxybenzoate* — The key observation was that the ring-cleavage product of catechol or 3,4-dihydroxybenzoate was β-ketoadipate that is formed by a series of lactonizations and rearrangements. The various steps were elucidated using pure samples of the proposed intermediates and enzyme preparations to study induction patterns. Mutants were then used to elucidate the regulation of the pathways: *cis,cis*-muconate is the inducer for the catechol pathway, and β-ketoadipate for the 3,4-dihydroxybenzoate pathway (Ornston 1966).

2. *Degradation of L-mandelate* — Mandelate is degraded by successive stages to benzoate which is then metabolized by the pathways outlined above for 3,4-dihydroxybenzoate. Details of the degradation were elucidated by similar means except that the task was considerably simplified by the commercial availability of most of the intermediates (Hegeman 1966a,b).

3. *Degradation of L-tryptophan* — Mutants defective in the suite of enzymes required for degradation of the substrate were isolated: for initial cleavage of the ring (tryptophan pyrrolase), for hydrolysis of the formyl group of the cleavage product (formylkynurenine formamidase), and for conversion of the hydrolysis product, kynurenine into anthranilic acid (kynureninase). These were used to provide a detailed picture of the regulation of the degradative pathway and to demonstrate that of the three potential inducers for the degradation of L-tryptophan, it is kynurenine that coordinately induces synthesis of all three enzymes (Palleroni and Stanier 1964).

In studies examining regulation such as those discussed above, assays of enzymatic activity play a key role. There are no simple rules for the preparation of cell extracts, and some enzymes are notoriously unstable. In these cases, use of toluene-permeabilized cells may be employed. Whole cells may be disrupted in a number of ways, for example, by grinding with abrasives such as Al_2O_3 — following the original method of Buchner who used a mixture of quartz sand and diatomite — by sonication, or by application of pressure. Probably the last procedure is generally preferable, and it possesses the additional advantage that cell disruption can readily be carried out in an anaerobic atmosphere.

5.5.3 Use of Metabolic Inhibitors

As an alternative to the use of mutants, metabolic inhibitors may be used to interrupt metabolic pathways; even transient intermediates may then be accumulated and provide evidence for the details of the consecutive steps. A wide range of compounds has been used in investigations on electron transport pathways and bioenergetics and the inhibition of sulfate reduction by molybdate has been used (Parkes et al. 1989). A single example will be given to illustrate their use in delineating an unusual degradative pathway for the degradation of 3-methylphenol carried out by a methanogenic consortium (Londry and Fedorak 1993). Three compounds were used as selective inhibitors:

1. 6-Fluoro-3-methylphenol brought about the accumulation of 4-hydroxy-2-methylbenzoate.

2. 3-Fluorobenzoate resulted in the transient accumulation of 4-hydroxy-benzoate.
3. Addition of bromoethanesulfonate that inhibits the synthesis of methane caused the accumulation of benzoate.

In addition, the corresponding fluorinated metabolites were identified, and this enabled construction of the complete pathway for the degradation that included the unusual reductive loss of a methyl group from the aromatic ring; these reactions are discussed in a wider context in Chapter 6 (Section 6.7.3).

5.5.4 Use of Synthetic Isotopes

The use of isotopically substrates labeled in specific positions makes it possible to follow the fate of individual atoms in the molecule during the microbial degradation. Under optimal conditions, both the kinetics of the degradation and of the formation of metabolites may be followed — provided, of course, that samples of the labeled metabolites are available. Many of the classical studies on the microbial metabolism of carbohydrates, carboxylic acids, and aminoacids used radioactive ^{14}C-labeled substrates and specific chemical degradation of the metabolites to determine the position of the label. The method is indeed obligatory for distinguishing between degradative pathways when the same products are produced from the substrate by different pathways. A good example is provided by the β-methylaspartate and hydroxyglutarate fermentations of glutamate, both of which produce butyrate but which can clearly be distinguished by the use of [4-^{14}C]-glutamate (Buckel and Barker 1974). An attempt will be made to illustrate briefly the application of a wider range of isotopes without going into the details which are provided in the references that are cited.

Carbon (^{14}C and ^{13}C) — Traditional use has been made of the isotope ^{14}C which has the convenience of being radioactive, and details need hardly be given here. Illustrative examples include the elucidation of pathways for the anaerobic degradation of amino acids and purines that are discussed in greater detail in Chapter 6 (Sections 6.7.1 and 6.7.4). On the other hand, increasing application has been made of ^{13}C using high-resolution Fourier transform NMR in whole-cell suspensions; this is equally applicable to molecules containing the natural ^{19}F or the synthetic ^{31}P nuclei, and further details of its application are given in Section 5.5.5.

Sulfur (^{35}S) and chlorine (^{36}Cl) — Although quite extensive use of ^{35}S has been made in studies on the degradation of alkyl sulfonates (Hales et al. 1986), ^{36}Cl has achieved only limited application because of technical difficulties resulting from the low specific activities and the synthetic inaccessibility of appropriately labeled substrates. One of the few examples of its application to the degradation of xenobiotics is provided by a study of the anaerobic dechlorination of hexachlorocyclohexane isomers (Jagnow et al. 1977), the results of which will be discussed later in Chapter 6 (Section 6.4.1).

Hydrogen (^{2}H) and oxygen (^{18}O) — The radioactive isotope ^{3}H has been extensively used for studies on the uptake of xenobiotics into whole cells, although the intrusion of exchange reactions and the large isotope effect renders this isotope possibly less attractive for general metabolic studies. Both deuterium ^{2}H — labeled substrates and oxygen $^{18}O_2$ and $^{18}OH_2$ have been used in metabolic studies since essentially pure labeled compounds are increasingly readily available, and access to the necessary mass spectrometer facilities has increased over the years. Deuterium labelling has been invaluable in studying rearrangement reactions involving protons; for example, it has been used to reveal the operation of the NIH-shift during metabolism of [4-^{2}H]ethylbenzene by the monooxygenase system from *Methylococcus capsulatus* (Dalton et al. 1981), of [2,2′,3,3′,5,5′,6,6′-^{2}H]biphenyl by *Cunninghamella echinulata* (Smith et al. 1981), and of [1-^{2}H]naphthalene and [2-^{2}H]naphthalene by *Oscillatoria* sp. (Narro et al. 1992).

^{18}O has been used effectively in investigation of oxidations to determine the source of oxygen and the number of oxygen atoms incorporated during metabolism under both aerobic and anaerobic conditions. Three typical examples are given:

1. During the biodegradation of 2,4-dinitrotoluene by a strain of *Pseudomonas* sp., two atoms of oxygen are incorporated from $^{18}O_2$ during the formation of 4-methyl-5-nitrocatechol by dioxygenation with loss of nitrite (Spanggord et al. 1991).
2. During degradation of 2-chloroacetophenone by a strain of *Alcaligenes* sp., one atom of $^{18}O_2$ is incorporated into 2-chlorophenol formed from the 2-chlorophenyl acetate that is initially formed by a Baeyer-Villiger monooxygenation (Higson and Focht 1990).
3. Benzene and toluene are anaerobically hydroxylated to phenol and 4-hydroxytoluene, and experiments with $H_2{}^{18}O$ showed that the oxygen atoms come from water (Vogel and Grbic-Galic 1986).

In experiments involving the use of both of these isotopes, care should be taken to exclude chemical exchange reactions involving potentially labile C-H or C-O bonds since these reactions could seriously compromise the conclusions. A good illustration of the pitfalls in such investigations is shown by a study on the dechlorination of pentachlorophenol by the dehalogenase from a strain of *Arthrobacter* sp. The initial reaction in the degradation of pentachlorophenol is mediated by a pentachlorophenol dehalogenase that produces tetrachloro-1,4-dihydroxybenzene. Experiments using the enzyme showed that ^{18}O is incorporated into this metabolite only after incubation with $H_2{}^{18}O$ and not with $^{18}O_2$; in fact, the labelling occurs as a result of exchange between the initially formed unlabeled metabolite and $H_2{}^{18}O$. An unambiguous elucidation of the mechanism of the reaction is not, therefore, possible since even if ^{18}O had been incorporated during the reaction with $^{18}O_2$, exchange with the excess $H_2{}^{16}O$ in the medium would yield an unlabeled product (Schenk et al. 1990).

5.5.5 Application of NMR and EPR to Whole-Cell Suspensions

Application of NMR — The advantages of this technique which is non-destructive is that unambigious structural assignment can be made to metabolites during their synthesis in the spectrometer tubes; these may be equipped with gas inlets which enable studies to be carried out under virtually any metabolic conditions. In addition, there are no experimental restrictions in handling radioactive material, though for studies of carbon metabolism, ^{13}C-labeled substrates must be available; possibly, however, these are no longer significantly less accessible than their ^{14}C analogues. On the other hand, the relatively low sensitivity of the method has hitherto precluded identification of metabolites which may be formed transiently only in low concentration.

Carbon ^{13}**C** — Interpretative difficulties may arise from the inherent design of NMR experiments that may necessitate the use of high substrate concentrations due to the relatively low sensitivity of the procedure. A single illustrative example may be given of the occurrence of artifacts that may be encountered; during a study of the metabolism of mandelate by *Pseudomonas putida* (Halpin et al. 1981), benzyl alcohol was unexpectedly identified when experiments were carried out at high substrate concentration (50 mM). This was, however, subsequently shown to be due to the action of a non-specific alcohol dehydrogenase under the anaerobic conditions prevailing at the high substrate concentration used for the identification of the metabolites (Collins and Hegeman 1984).

^{13}C NMR using whole cells has been applied to the study of a number of relatively straightforward metabolic reactions involving small molecules, and these include the following that may be used as illustration of the potential of the procedure:

1. Glycolysis in *Saccharomyces cerevisiae* (den Hollander et al. 1986).
2. The reduction of dimethyl sulfoxide and trimethylamine-*N*-oxide by *Rhodobacter capsulatus* (*Rhodopseudomonas capsulata*) (King et al. 1987).
3. The metabolism of acetate and methanol in *Pseudomonas* sp. (Narbad et al. 1989).
4. Nicotinate and pyridine nucleotide metabolism in *Escherichia coli* and *Saccharomyces cerevisiae* (Unkefer and London 1984).
5. Combined use of ^{1}H and ^{13}C NMR was used in powerful combination in deducing the pathways of degradation of vinyl chloride (Castro et al 1992a; 1992b).

The procedure is particularly suited to the study of anaerobic transformations since there are no problems resulting from problems with oxygen limitation. A good example is provided by the application of ^{13}C NMR to the intricate relations of fumarate, succinate, propionate, and acetate in a syntrophic organism both in the presence (Houwen et al. 1991) and in the absence of methanogens (Stams et al. 1993; Plugge et al. 1993).

A single example will be given to illustrate both the strengths and the limitations of the technique. During the metabolism of 2-^{13}C-acetate in methylotrophic strains of *Pseudomonas* sp., it was shown that the substrate was converted into α, α-trehalose in isocitrate lyase-negative strains though not in one which synthesized this enzyme. In addition, an unknown compound was revealed by the in vitro experiments but was not present in perchloric acid extracts of the cells. Possibly more disturbing, however, was the fact that analysis of a strain during growth with ^{13}C-methanol did not reveal the presence of the intermediates known to be part of the serine pathway that functions in this organism (Narbad et al. 1989).

Fluorine ^{19}**F** — Application to fluorine-containing molecules is particularly attractive since naturally occurring fluorine is monoisotopic and since the range of chemical shifts in fluorine compounds is very much greater than the proton shifts for hydrogen-containing compounds. Although only a single example will be used as illustration, there is a vast potential for the application of ^{19}F NMR to metabolic studies of fluorine compounds. Substantial attention has been devoted to the metabolism of 5-fluorouracil and related compounds; for example, both anabolic reactions involving pyrimidine nucleotides and degradation to α-fluoro-β-alanine by the fungus *Nectria haematococca* have been successfully analyzed by ^{19}F NMR (Parisot et al. 1989, 1991), both in cell extracts and in whole mycelia. This methodology seems worthy of wider exploitation in the study of other groups of organofluorine compounds that are of industrial importance as agrochemicals and that have awakened increased environmental interest.

Phosphorus ^{31}P — Until recently, there has been only limited interest in the catabolism of organophosphorus compounds, although very considerable attention has been directed to anabolic reactions. Application of ^{31}P NMR has been used to examine nucleotide pools and transmembrane potential in bacteria after exposure to pentachlorophenol and to demonstrate the differences between *Escherichia coli* which does not degrade the substrate and a *Flavobacterium* sp. which is able to do so (Steiert et al. 1988). ^{31}P NMR has been used to examine the effect of pentachlorophenol on the energy metabolism of abalone (*Haliotis rufescens*) (Tjeerdema et al. 1991). One example may be given to illustrate the strengths and limitations of the technique. ^{31}P was used to examine the effect of ethanol on the metabolism of glucose by *Zymomonas mobilis*; whereas the sensitivity was sufficient to establish changes in nucleoside triphosphates during *in vivo* experiments, details of the changes in the various phosphorylated metabolites necessitated the use of perchloric acid cell extracts (Strohhäcker et al. 1993).

Application of EPR — Compared with the extensive application of NMR procedures, electron paramagnetic resonance (EPR) has been used only infrequently. One example may, however, be used to illustrate its application to elucidate the unexpected complexity of an apparently straightforward metabolic pathway; the metabolism of glycerol trinitrate by *Phanerochaete chrysosporium* does not proceed by simple hydrolysis to nitrate and glycerol but involves formation of nitric oxide bound to both non-heme and heme proteins (Servent et al. 1991). Indeed, interest in the biochemical significance of nitric oxide (Feldman et al. 1993) has led to the development of a spin-labeled EPR assay involving reaction with 1,2-bis(*iso*propylidene)cyclohexa-3,5-diene (Korth et al. 1992).

5.6 CLASSIFICATION AND IDENTIFICATION

It is unfortunately true that the level of classification of bacteria illustrating important pathways for the degradation of xenobiotics is sometimes rudimentary. As a consequence, the assignment of names may be based on the most slender evidence, and both they and the accompanying organisms may eventually find their way into national culture collections. One cannot help comparing the rigor applied to the taxonomy of major groups such as the aerobic pseudomonads and their relatives, the enterobacteriaceae or the mycobacteria, to the somewhat peripheral descriptions of many environmentally significant and interesting organisms.

In one sense, the problem is particularly acute for aerobic organisms, since there is such a wealth of information on previously accepted taxa so that an attempt at placing an unknown organism within acceptable circumscription of established taxa is indeed a formidable undertaking. By currently accepted standards, this would

necessitate not only nutritional and biochemical characterization but extensive studies of, for example, DNA/DNA hybridization, identification of fatty acids, cell wall
components, ubiquinones, or 16 S rRNA base sequencing. These are increasingly the
field of specialists, and laboratories well equipped for studies on metabolism may
lack both the facilities and the expertise for carrying out such investigations. It may
not be a scientifically acceptable defense, but the fact remains that the effort required
to assign an unknown organism — which might previously have been described, for
example, simply as *Pseudomonas* sp. — to the correct taxon may now simply not be
available. The creation of the genera *Comamonas* (Willems et al. 1991a), *Acidovorax*
(Willems et al. 1990), *Hydrogenophaga* (Willems et al. 1989), and *Variovorax*
(Willems et al. 1991b) for organisms that were originally assigned to the genus
Pseudomonas attests to the level of sophistication that is being applied and its success
in differentiating grossly similar genera.

An illustrative example of the magnitude of the problem may be taken from a
study which examined nineteen gram-negative aerobic organisms that degraded a
range of xenobiotics including 3-chlorobenzoate, a number of aromatic sulfonic
acids, 2,6-dinitrophenol, phosphonic acids, and 1,3,5-triazines (Busse et al. 1992).
All of them had DNA G+C contents in the range 61% to 68%, and most of them were
formerly unclassified. An extensive number of characters including the presence of
ubiquinones, patterns of soluble proteins, and DNA-DNA hybridization enabled
assignment of only eight out of the nineteen organisms to specific taxa — *Comamonas
testosteroni*, *C. acidovorans*, and species of the genus *Alcaligenes*. Comparison of
16S rRNA fragments suggested possible assignment of one of the remaining organisms to *Acidovorax facilis* and six others to "*Alcaligenes eutrophus*". Clearly, there
remain serious difficulties in successfully assigning such strains to established taxa,
and the problem is made more difficult by the current uncertainty surrounding the
nomenclature and taxonomic position of organisms previously assigned to the genera
Pseudomonas and *Xanthomonas*. An interesting summary of the problems surrounding assignment of a single organism, "*Pseudomonas maltophilia*," provides a good
illustration of the present confusion (Van Zyl and Syeyn 1992). Some of the salient
issues were as follows:

1. Problems arising from including "*Pseudomonas maltophilia*" in the genus
   Xanthomonas since the organism shows important differences from other members of the genus that includes many plant pathogens.
2. The divergence of opinion in interpreting results of DNA-DNA homology,
   analysis for ubiquinones, and oligonucleotide sequences of 16S rRNA.

The conclusion that the organism should be assigned to a new genus is plausibly
argued, and this has now been implemented under the name *Stenotrophomonas*
(Palleroni and Bradbury 1993). Although this is a taxonomically tidy solution, the
general adoption of this principle would inevitably lead to the proposal of a very large
number of new genera such as *Chelatobacter* and *Chelatococcus* for nitrilotriacetate-
degrading organisms (Aulin et al. 1993).

In summary, there seem to be at least three central difficulties in the taxonomic
assignment of organisms with established metabolic capabilities — and indeed to any
newly isolated organism:

1. The increasing level of sophistication nesessary for classification.
2. The apparent existence of relatively few relevant strains for comparison with new taxa; in many cases, only a few or even a single strain exists. Possible phenotypic differences within the taxon are not, therefore, available.
3. The proliferation of new names for existing taxa which make it increasingly difficult for the non-specialist to be informed of the correct synonymies.

Perhaps, one could hope that organisms with significant degradative capabilities are described in the literature with at least basic data of taxonomic relevance and that specialists be encouraged to carry out the studies — such as the one already described — which are needed for acceptable classification. It must, however, be accepted that with limited research funding, few of those engaged in studies of biodegradation possess either the means or the experience to take upon themselves this extra responsibility.

Possibly inevitably, a similar situation prevails for the genetics of these organisms: detailed maps exist, of course, for the two classic enteric organisms — *Escherichia coli* and *Salmonella typhimurium* — and for two species of *Pseudomonas*. Otherwise, systematic genetic studies encompassing a wide range of genetic markers hardly exist. With the possibility of cloning degradative genes for study in more suitable organisms, it seems possible that detailed genetic studies on metabolic regulation of xenobiotic degradation may be accomplished by alternative methods. Interest in the application of microorganisms to biotechnology has led to increased interest in the genetics of two other important groups — the methylotrophs and the rhodococci — though this does not currently reach the level of sophistication attained in the other groups of bacteria. Possibly the greatest impetus for such investigations will come from their application to various aspects of biotechnology.

REFERENCES

Alexander, M. 1975. Environmental and microbiological problems arising from recalcitrant molecules. *Microbial Ecol.* 2: 17-27.

Angle, J.S., S.P. McGrath, and R.L. Chaney. 1991. New culture medium containing ionic concentrations of nutrients similar to concentrations found in soil solution. *Appl. Environ. Microbiol.* 57: 3674-3676.

Apajalahti, J.H.A., and M.S. Salkinoja-Salonen. 1986. Degradation of polychlorinated phenols by *Rhodococcus chlorophenolicus*. *Appl. Microbiol. Biotechnol.* 25: 62-67.

Arnon, D.I. 1938. Microelements in culture-solution experiments with higher plants. *Amer. J. Bot.* 25: 322-325.

Aulin, G., H.-J. Busse, T. Egli, T. El-Banna, and E. Stackebrandt. 1993. Description of the gram-negative, obligately aerobic, nitrilotriacetate (NTA)-utilizing bacteria as *Chelatobacter heintzii*, gen. nov., sp. nov., and *Chelatococcus asaccharovorans*, gen. nov., sp. nov. *Syst. Appl. Microbiol.* 16: 104-112.

Balows, A., H.G. Trüper, M. Dworkin, W. Harder, and K.-H. Schleifer (Eds.) 1992. *The Prokaryotes*. Springer-Verlag, Heidelberg.

Bartholomew, G.W., and F.K. Pfaender. 1983. Influence of spatial and temporal variations on organic pollutant biodegradation rates in an estuarine environment. *Appl. Environ. Microbiol.* 45: 103-109.

Baruah, J.N., Y. Alroy, and R.I. Mateles. 1967. Incorporation of liquid hydrocarbons into agar media. *Appl. Microbiol.* 15: 961.

Battersby, N.S. 1990. A review of biodegradation kinetics in the aquatic environment. *Chemosphere* 21: 1243-1284.

Battersby, N.S., and V. Wilson. 1989. Survey of the anaerobic biodegradation potential of organic chemicals in digesting sludge. *Appl. Environ. Microbiol.* 55: 433-439.

Birch, R.R., and R.J. Fletcher. 1991. The application of dissolved inorganic carbon measurements to the study of aerobic biodegradability. *Chemosphere* 23: 507-524.

Birch, R.R., C. Biver, R. Campagna, W.E. Gledhill, and U. Pagga. 1989. Screening chemicals for anaerobic degradability. *Chemosphere* 19: 1527-1550.

Bogardt, A.H., and B.B. Hemmingsen. 1992. Enumeration of phenanthrene-degrading bacteria by an overlay technique and its use in evaluation of petroleum-contaminated sites. *Appl. Environ. Microbiol.* 58: 2579-2582.

Bonnarme, P., and T.W. Jeffries. 1990. Mn(II) regulation of lignin peroxidases and manganese-dependent peroxidases from lignin-degrading white-rot fungi. *Appl. Environ. Microbiol.* 56: 210-217.

Brown, J.A., J.K. Glenn, and M.H. Gold. 1990. Manganese regulates expression of manganese peroxidase by *Phanerochaete chrysosporium*. *J. Bacteriol.* 172: 3125-3130.

Bruhn, C., H. Lenks, and H.-J. Knackmuss. 1987. Nitrosubstituted aromatic compounds as nitrogen source for bacteria. *Appl. Environ. Microbiol.* 53: 208-210.

Bryant, F.O., D.D. Hale, and J.E. Rogers. 1991. Regiospecific dechlorination of pentachlorphenol by dichlorophenol-adapted microorganisms in freshwater anaerobic sediment slurries. *Appl. Environ. Microbiol.* 57: 2293-2301.

Buckel, W., and H.A. Barker. 1974. Two pathways of glutamate fermentation by anaerobic bacteria. *J. Bacteriol.* 117: 1248-1260.

Busse, H.-J., T. El-Banna, H. Oyaizu, and G. Auling. 1992. Identification of xenobiotic-degrading isolates from the beta subclass of the Proteobacteria by a polyphasic approach including 16S rRNA sequencing. *Int. J. Syst. Bacteriol.* 42: 19-26.

Castro, C.E., D.M. Riebeth, and N.O. Belser. 1992a. Biodehalogenation: the metabolism of vinyl chloride by *Methylosinus trichosporium* OB-3b. A sequential oxidative and reductive pathway through chloroethylene oxide. *Environ. Toxicol. Chem.* 11: 749-755.

Castro, C.E., R.S. Wade, D.M. Riebeth, E.W. Bartnicki, and N.O. Belser. 1992b. Biodehalogenation: rapid metabolism of vinyl chloride by a soil *Pseudmonas* sp. Direct hydrolysis of a vinyl C-Cl bond. *Environ. Toxicol. Chem.* 11: 757-764

Claus, D., and N. Walker. 1964. The decomposition of toluene by soil bacteria. *J. Gen. Microbiol.* 36: 107-122.

Collins, J., and G. Hegeman. 1984. Benzyl alcohol metabolism by *Pseudomonas putida*: a paradox resolved. *Arch. Microbiol.* 138: 153-160.

Cook, A.M., and R. Hütter. 1982. Ametryne and prometryne as sulfur sources for bacteria. *Appl. Environ. Microbiol.* 43: 781-786.

Cook, A.M., and R. Hütter. 1984. Deethylsimazine: bacterial dechlorination, deamination and complete degradation. *J. Agric. Food Chem.* 32: 581-585.

Cook, A.M., C.G. Daughton, and M. Alexander. 1978. Phosphonate utilization by bacteria. *J. Bacteriol.* 133: 85-90.

Dalton, H., B.T. Golding, B.W. Waters, R. Higgins, and J.A. Taylor. 1981. Oxidations of cyclopropane, methylcyclopropane, and arenes with the mono-oxygenase system from *Methylococcus capsulatus*. *J. Chem. Soc. Chem. Comm.* 482-483.

Daughton, C.G., and D.P.H. Hsieh. 1977. Parathion utilization by bacterial symbionts in a chemostat. *Appl. Environ. Microbiol.* 34: 175-184.

De Morsier, A., J. Blok, P. Gerike, L. Reynolds, H. Wellens, and W.J. Bontinck. 1987. Biodegradation tests for poorly-soluble compounds. *Chemosphere* 16: 269-277.

den Hollander, J.A., T.R. Brown, K. Ugurbil, and R.G. Shulman. 1986. Studies of anaerobic and aerobic glycolysis in *Saccharomyces cerevisiae*. *Biochemistry* 25: 203-211.

Dilworth, M.J., and R.R. Eady. 1991. Hydrazine is a product of dinitrogen reduction by the vanadium-nitrogenase from *Azotobacter chroococcum*. *Biochem. J.* 277: 465-468.

Doudoroff, M. 1940. The oxidative assimilation of sugars and related substances by *Pseudomonas saccharophila* with a contribution to the direct respiration of di- and poly-saccharides. *Enzymologia* 9: 59-72.

Edwards, E.A., L.E. Williams, M. Reinhard, and D. Grbic-Galic. 1992. Anaerobic degradation of toluene and xylene by aquifer microrganisms under sulfate-reducing conditions. *Appl. Environ. Microbiol.* 58: 794-800.

Fallik, E., Y.-K. Chan, and R.L. Robson. 1991. Detection of alternative nitrogenases in aerobic gram-negative nitrogen-fixing bacteria. *J. Bacteriol.* 173: 365-371.

Feldman, P.L., O.W. Griffith, and D.J. Stuehr. 1993. The surprising life of nitric oxide. *Chem. Eng. News* 71 [51]: 26-39.

Firestone, M.K., and J.M. Tiedje. 1975. Biodegradation of metal-nitrilotriacetate complexes by a *Pseudomonas* species: mechanism of reaction. *Appl. Microbiol.* 29: 758-764.

Flanagan, W.P., and R.J. May. 1993. Metabolite detection as evidence for naturally occurring aerobic PCB degradation in Hudson River sediments. *Environ. Sci. Technol.* 27: 2207-2212.

Francis, A.J., and C.J. Dodge. 1993. Influence of complex structure on the biodegradation of iron-citrate complexes. *Appl. Environ. Microbiol.* 59: 109-113.

Fulthorpe, R.R., and R.C. Wyndham. 1989. Survival and activity of a 3-chlorobenzoate-catabolic genotype in a natural system. *Appl. Environ. Microbiol.* 55: 1584-1590.

Gerike, P., and W.K. Fischer. 1981. A correlation study of biodegradability determinations with various chemicals in various tests. II. Additional results and conclusions. *Ecotoxicol. Environ. Saf.* 5: 45-55.

Gill, C.O., and C. Ratledge. 1972. Toxicity of *n*-alkanes, *n*-alkenes, *n*-alkan-1-ols and *n*-alkyl-1-bromides towards yeasts. *J. Gen. Microbiol.* 72: 165-172.

Glaus, M.A., C.G. Heijman, R.P. Schwarzenbach, and J. Zeyer. 1992. Reduction of nitroaromatic compounds mediated by *Streptomyces* sp. exudes. *Appl. Environ. Microbiol.* 58: 1945-1951.

Griffiths, E.T., S.G. Hales, N.J. Russell, G.K. Watson, and G.F. White. 1986. Metabolite production during biodegradation of the surfactant sodium dodecyltriethoxy sulphate under mixed-culture die-away conditions. *J. Gen. Microbiol.* 132: 963-972.

Grossenbacher, H., C. Horn, A.M. Cook, and R. Hütter. 1984. 2-chloro-4-amino-1,3,5-triazine-6(5H)-one: a new intermediate in the biodegradation of chlorinated s-triazines. *Appl. Environ. Microbiol.* 48: 451-453.

Grund, E., B. Denecke, and R. Eichenlaub. 1992. Naphthalene degradation via salicylate and gentisate by *Rhodococcus* sp. strain B4. *Appl. Environ Microbiol.* 58: 1874-1877.

Guerin, T.F., and I.R. Kennedy. 1992. Distribution and dissipation of endosulfan and related cyclodienes in sterile aqueous systems: implication for studies on biodegradation. *J. Agric. Food Chem.* 40: 2315-2323.

Hales, S.G., G.K. Watson, K.S. Dodgson, and G.F. White. 1986. A comparative study of the biodegradation of the surfactant sodium dodecyltriethoxy sulphate by four detergent-degrading bacteria. *J. Gen. Microbiol.* 132: 953-961.

Hales, S.G., K.S. Dodgson, G.F. White, N. Jones, and G.K. Watson. 1982. Initial stages in the biodegradation of the surfactant sodium dodecyltriethoxy sulfate by Pseudomonas sp. strain DES1. *Appl. Environ. Microbiol.* 44: 790-800.

Halpin, R.A., G.D. Hegeman, and G.L. Kenyon. 1981. Carbon-13 nuclear magnetic resonance studies of mandelate metabolism in whole bacterial cells and in isolated, in vivo cross-linked enzyme complexes. *Biochemistry* 20: 1525-1533.

Harris, R., and C.J. Knowles. 1983. Isolation and growth of a *Pseudomonas* species that utilizes cyanide as a source of nitrogen. *J. Gen. Microbiol.* 129: 1005-1011.

Havel, J., and W. Reineke. 1993. Microbial degradation of chlorinated acetophenones. *Appl. Environ. Microbiol.* 59: 2706-2712.

Hegeman, G.D. 1966a. Synthesis of the enzymes of the mandelate pathway by *Pseudomonas putida*. I. Synthesis of enzymes of the wild type. *J. Bacteriol.* 91: 1140-1154.

Hegeman, G.D. 1966b. Synthesis of the enzymes of the mandelate pathway by *Pseudomonas putida*. II. Isolation and properties of blocked mutants. *J. Bacteriol.* 91: 1155-1160.

Heitkamp, M.A., and C.E. Cerniglia. 1989. Polycyclic aromatic hydrocarbon degradation by a *Mycobacterium* sp. in microcosms containing sediment and water from a pristine ecosystem. *Appl. Environ. Microbiol.* 55: 1968-1973.

Heitkamp, M.A., V. Camel, T.J. Reuter, and W.J. Adams. 1990. Biodegradation of p-nitrophenol in an aqueous waste stream by immobilized bacteria. *Appl. Environ. Microbiol.* 56: 2967-2973.

Heitkamp, M.A., J.P. Freeman, and C.E. Cerniglia. 1986. Biodegradation of tert-butylphenyl diphenyl phosphate. *Appl. Environ. Microbiol.* 51: 316-322.

Heitkamp, M.A., J.P. Freeman, and C.E. Cerniglia. 1987. Naphthalene biodegradation in environmental microcosms: estimates of degradation rates and characterization of metabolites. *Appl. Environ. Microbiol.* 53: 129-136.

Higson, F.K., and D.D. Focht. 1990. Bacterial degradation of ring-chlorinated acetophenones. *Appl. Environ. Microbiol.* 56: 3678-3685.

Houwen, F.P., C. Dijkema, A.J.M. Stams, and A.J.B. Zehnder. 1991. Propionate metabolism in anaerobic bacteria; determination of carboxylation reactions with ^{13}C-NMR spectroscopy. *Biochim. Biophys. Acta* 1056: 126-132.

Howard, P.H., and S. Banerjee. 1984. Interpreting results from biodegradability tests of chemicals in water and soil. *Environ. Toxicol. Chem.* 3: 551-562.

Huckins, J.N., J.D. Petty, and M.A. Heitkamp. 1984. Modular containers for microcosm and process model studies on the fate and effects of aquatic contaminants. *Chemosphere* 13: 1329-1341.

Hungate, R. E. 1969. A roll tube method for cultivation of strict anaerobes. In *Methods in Microbiology* (Eds. Norris and D.W. Ribbons), Vol. 3B p 117-132. Academic Press, New York.

Jagnow, G., K. Haider, and P.-C. Ellwardt. 1977. Anaerobic dechlorination and degradation of hexachlorocyclohexane by anaerobic and facultatively anaerobic bacteria. *Arch. Microbiol.* 115: 285-292.

Jain, R.K., G.S. Sayler, J.T. Wilson, L. Houston, and D. Pacia. 1987. Maintenance and stability of introduced genotypes in groundwater aquifer material. *Appl. Environ. Microbiol.* 53: 996-1002.

Jiménez, L., A. Breen, N. Thomas, T.W. Federle, and G.S. Sayler. 1991. Mineralization of linear alkylbenzene sulfonate by a four-member aerobic bacterial consortium. *Appl. Environ. Microbiol.* 57: 1566-1569.

Joblin, K.N. 1981. Isolation, enumeration, and maintenance of rumen anaerobic fungi in roll tubes. *Appl. Environ. Microbiol.* 42: 1119-1122.

Jones, J.B., and T.C. Stadtman. 1981. Selenium-dependent and selenium-independent formate dehydrogenase of *Methanococcus vannielii*. Separation of the two forms and characterization of the purified selenium-independent form. *J. Biol. Chem.* 256: 656-663.

King, G.F., D.J. Richardson, J.B. Jackson, and S.J. Ferguson. 1987. Dimethyl sulfoxide and trimethylamine-*N*-oxide as bacterial electron acceptors: use of nuclear magnetic resonance to assay and characterise the reductase system in *Rhodobacter capsulatus*. *Arch. Microbiol.* 149: 47-51.

Kitts, C.L., J.P. Lapointe, V.T. Lam, and R.A. Ludwig. 1992. Elucidation of the complete *Azorhizobium nicotinate* catabolism pathway. *J. Bacteriol.* 174: 7791-1197.

Klecka, G.M., and D.T. Gibson. 1979. Metabolism of dibenzo[1,4]dioxan by a *Pseudomonas* species. *Biochem. J.* 180: 639-645.

Korth, H.-G., K.U. Ingold, R. Sustmann, H. de Groot, and H. Sies. 1992. Tetramethyl-*ortho*-chinodimethan (NOCT-1), das erste Mitglied einer Familie maβgeschneiderter cheletroper Spinfänger für Stickstoffmonoxid. *Angew. Chem.* 104: 915-917.

Lauff, J.J., D.B. Steele, L.A. Coogan, and J.M. Breitfeller. 1990. Degradation of the ferric chelate of EDTA by a pure culture of an *Agrobacterium.* sp. *Appl. Environ. Microbiol.* 56: 3346-3353.

Leive, L. 1968. Studies on the permeability change produced in coliform bacteria by ethylenediaminetetraacetate. *J. Biol. Chem.* 243: 2373-2380.

Lewis, T.A., and R.L. Crawford. 1993. Physiological factors affecting carbon tetrachloride dehalogenation by the denitrifying bacterium *Pseudomonas* sp. strain KC. *Appl. Environ. Microbiol.* 59: 1635-1641.

Liu, D., W.M.J. Strachan, K. Thomson, and K. Kwasniewska. 1981. Determination of the biodegradability of organic compounds. *Environ. Sci. Technol.* 15: 788-793.

Londry, K.L., and P.M. Fedorak. 1993. Use of fluorinated compounds to detect aromatic metabolites from *m*-cresol in a methanogenic consortium: evidence for a demethylation reaction. *Appl. Environ. Microbiol.* 59: 2229-2238.

Madsen, E.L. 1991. Determining in situ biodegradation. Facts and challenges. *Environ. Sci. Technol.* 25: 1663-1673.

Mandelbaum, R.T., L.P. Wackett, and D.L. Allan. 1993. Mineralization of the *s*-triazine ring of atrazine by stable bacterial mixed cultures. *Appl. Environ. Microbiol.* 59: 1695-1701.

Maron, D.M., and B.N. Ames. 1983. Revised methods for the Salmonella mutagenicity test. *Mutat. Res.* 113: 173-215.

Mihelcic, J.R., and R.G. Luthy. 1988. Microbial degradation of acenaphthene and naphthalene under denitrification conditions in soil-water systems. *Appl. Environ. Microbiol.* 54: 1188-1198.

Miller, J.H. 1977. *Experiments in molecular genetics*. Cold Spring Harbor Laboratories, New York, U.S.A.

Miller, T.L., and M.J. Wolin. 1974. A serum bottle modification of the Hungate technique for cultivating obligate anaerobes. *Appl. Microbiol.* 27: 985-987.

Munnecke, D.M., and D.P.H. Hsieh. 1976. Pathways of microbial metabolism of parathion. *Appl. Environ. Microbiol.* 31: 63-69.

Narbad, A., M.J. Hewlins, and A. G. Callely. 1989. ^{13}C-NMR studies of acetate and methanol metabolism by methylotrophic *Pseudomonas* strains. *J. Gen. Microbiol.* 135: 1469-1477.

Narro, M.L., C.E. Cerniglia, C. Van Baalen, and D.T. Gibson. 1992. Evidence for an NIH shift in oxidation of naphthalene by the marine cyanobacterium *Oscillatoria* sp. strain JCM. *Appl. Environ. Microbiol.* 58: 1360-1363.

Neidhardt, F.C., P.L. Bloch, and D.F. Smith. 1974. Culture medium for enterobacteria. *J. Bacteriol.* 119: 736-747.

Neilson, A.H. 1980. Isolation and characterization of bacteria from the Swedish west coast. *J. Appl. Bacteriol.* 49: 215-223.

Neilson, A.H., A.-S. Allard, C. Lindgren, and M. Remberger. 1987. Transformations of chloroguaiacols, chloroveratroles and chlorocatechols by stable consortia of anaerobic bacteria. *Appl. Environ. Microbiol.* 53: 2511-2519.

Neilson, A.H., A.-S. Allard, and M. Remberger. 1985. Biodegradation and transformation of recalcitrant compounds pp. 29-86. In *Handbook of environmental chemistry* (Ed. O. Hutzinger), Vol. 2 Part C. Springer-Verlag, Berlin.

Neilson, A.H., H. Blanck, L. Förlin, L. Landner, P. Pärt, A. Rosemarin and M. Söderström. 1989. Advanced hazard assessment of 4,5,6-trichloroguaiacol in the Swedish Environment. In *Chemicals in the Aquatic Environment*, pp. 329-374. Ed. L. Landner, Springer-Verlag, Berlin.

Nörtemann, B. 1992. Total degradation of EDTA by mixed cultures and a bacterial isolate. *Appl. Environ. Microbiol.* 58: 671-676.

Nyholm, N., A. Damborg, and P. Lindgaard-Jörgensen. 1992. A comparative study of test methods for assessment of the biodegradability of chemicals in seawater — screening tests and simulation tests. *Ecotoxicol. Environ. Saf.* 23: 173-190.

Nyholm, N., and P. Kristensen. 1992. Screening methods for assessment of biodegradability of chemicals in seawater — results from a ring test. *Ecotoxicol. Environ. Saf.* 23: 161-172.

Omori, T., L. Monna, Y. Saiki, and T. Kodama. 1992. Desulfurization of dibenzothiophene by *Corynebacterium* sp. strain SY1. *Appl. Environ. Microbiol.* 58: 911-915.

O'Reilly, K.T., and R.L. Crawford. 1989. Degradation of pentachlorophenol by polyurethane-immobilized *Flavobacterium* cells. *Appl. Environ. Microbiol.* 55: 2113-2118.

Ornston, L.N. 1966. The conversion of catechol and protocatechuate to beta-ketoadipate by *Pseudomonas putida.* IV. Regulation. *J. Biol. Chem.* 241: 3800-3810.

Palleroni, N.J., and J.F. Bradbury. 1993. *Stenotrophomonas,* a new bacterial genus for *Xanthomonas maltophilia* (Hugh 1980) Swings et al. 1983. *Int. J. Syst. Bacteriol.* 43: 606-609.

Palleroni, N.J., and R.Y. Stanier. 1964. Regulatory mechanisms governing synthesis of the enzymes for tryptophan oxidation by *Pseudomonas fluorescens. J. Gen. Microbiol.* 35: 319-334.

Parisot, D., M.C. Malet-Martino, P. Crasnier, and R. Martino. 1989. [19]F nuclear magnetic resonance analysis of 5-fluorouracil metabolism in wild-type and 5-fluorouracil-resistant *Nectria haematococca. Appl. Environ. Microbiol.* 55: 2474-2479.

Parisot, D., M.C. Malet-Martino, R. Martino, and P. Crasnier. 1991. [19]F nuclear magnetic resonance analysis of 5-fluorouracil metabolism in four differently pigmented strains of *Nectria haematococca. Appl. Environ. Microbiol.* 57: 3605-3612.

Parkes, R.J., G.R. Gibson, I. Mueller-Harvey, W.J. Buckingham, and R.A. Herbert. 1989. Determination of the substrates for sulphate-reducing bacteria within marine and estuarine sediments with different rates of sulphate reduction. *J. Gen. Microbiol.* 135: 175-187.

Phillips, W.E., and J.J. Perry. 1976. *Thermomicrobium fosteri* sp. nov., a hydrocarbon-utilizing obligate thermophile. *Int. J. Syst. Bacteriol.* 26: 220-225.

Pignatello, J.J., L.K. Johnson, M.M. Martinson, R.E. Carlson, and R.L. Crawford. 1985. Response of the microflora in outdoor experimental streams to pentachlorophenol: compartmental contributions. *Appl. Environ. Microbiol.* 50: 127-

Pignatello, J.J., M.M. Martinson, J.G. Steiert, R.E. Carlson, and R.L. Crawford. 1983. Biodegradation and photolysis of pentachlorophenol in artificial freshwater streams. *Appl. Environ. Microbiol.* 46: 1024-1031.

Plugge, C.M., C. Dijkema, and A.J.M. Stams. 1993. Acetyl-CoA cleavage pathway in a syntrophic propionate oxidizing bacterium growing on fumarate in the absence of methanogens. *FEMS Microbiol. Lett.* 110: 71-76.

Rabus, R., R. Nordhaus, W. Ludwig, and F. Widdel. 1993. Complete oxidation of toluene under strictly anoxic conditions by a new sulfate-reducing bacterium. *Appl. Environ. Microbiol.* 59: 1444-1451.

Reanney, D.C., P.C. Gowland, and J.H. Slater. 1983. Genetic interactions among microbial communities. *Symp. Soc. Gen. Microbiol.* 34: 379-421.

Robert-Gero, M., M. Poiret, and R.Y. Stanier. 1969. The function of the beta-ketoadipate pathway in *Pseudomonas acidovorans. J. Gen. Microbiol.* 57: 207-214.

Rothenburger, S., and R.M. Atlas. 1993. Hydroxylation and biodegradation of 6-methylquinoline by pseudomonads in aqueous and nonaqueous immobilized-cell bioreactors. *Appl. Environ. Microbiol.* 59: 2139-2144.

Schenk, T., R. Müller, and F. Lingens. 1990. Mechanism of enzymatic dehalogenation of pentachlorophenol by *Arthrobacter* sp. strain ATCC 33790. *J. Bacteriol.* 172: 7272-7274.

Servent, D., C. Ducrorq, Y. Henry, A. Guissani, and M. Lenfant. 1991. Nitroglycerin metabolism by *Phanerochaete chrysosporium*: evidence for nitric oxide and nitrite formation. *Biochim. Biophys. Acta* 1074: 320-325.

Shiaris, M.P., and J.J. Cooney. 1983. Replica plating method for estimating phenanthrene-utilizing and phenanthrene-cometabolizing microorganisms. *Appl. Environ. Microbiol.* 45: 706-710.

Skerman, V.B.D. 1968. A new type of micromanipulator and microforge. *J. Gen. Microbiol.* 54: 287-297.

Smith, R.V., P.J. Davis, A.M. Clark, and S.K. Prasatik. 1981. Mechanism of hydroxylation of biphenyl by *Cunninghamella echinulata*. *Biochem. J.* 196: 369-371.

Sobecky, P.A., M.A. Schell, M.A. Moran, and R.E. Hodson. 1992. Adaptation of model genetically engineered microorganisms to lake water: growth rate enhancements and plasmid loss. *Appl. Environ. Microbiol.* 58: 3630-3637.

Spanggord, R.J., J.C. Spain, S.F. Nshino, and K.E. Mortelmans. 1991. Biodegradation of 2,4-dinitrotoluene by a *Pseudomonas* sp. *Appl. Environ. Microbiol.* 57: 3200-3205.

Stams, A.J.M., J.B. van Dijk, C. Dijkema, and C.M. Plugge. 1993. Growth of syntrophic propionate-oxidizing bacteria with fumarate in the absence of methanogenic bacteria. *Appl. Environ. Microbiol.* 59: 1114-1119.

Stanier, R.Y. 1947. Simultaneous adaptation: a new technique for the study of metabolic pathways. *J. Bacteriol.* 54: 339-348.

Stanier, R.Y., N.J. Palleroni, and M. Doudoroff. 1966. The aerobic pseudomonads: a taxonomic study. *J. Gen. Microbiol.* 43: 159-271.

Steiert, J.G., W.J. Thoma, K. Ugurbil, and R.L. Crawford. 1988. [31]P nuclear magnetic resonance studies of effects of some chlorophenols on *Escherichia coli* and a pentachlorophenol-degrading bacterium. *J. Bacteriol.* 170: 4954-4957.

Strohhäcker, J., A.A. de Graaf, S.M. Schoberth, R.M. Wittig, and H. Sahm. 1993. [31]P nuclear magnetic resonance studies of ethanol inhibition in *Zymomonas mobilis*. *Arch. Microbiol.* 159: 484-490.

Sylvestre, M. 1980. Isolation method for bacterial isolates capable of growth on *p*-chlorobiphenyl. *Appl. Environ. Microbiol.* 39: 1223-1224.

Sylvestre, M., R. Massé, F. Messier, J. Faiteux, J.G. Bisaillon and R. Beaudet. 1982. Bacterial nitration of 4-chlorobiphenyl. *Appl. Environ. Microbiol.* 44: 871-877.

Söhngen, N.L. 1913. Benzin, Petroleum, Paraffinöl und Paraffin als Kohlenstoff- und Energiequelle für Mikroben. *Centralbl. Bakteriol. Parisitenkd. Infektionskr. Zweite Abt.* 37: 595-609.

Talbot, H.W., L.M. Johnson, and D.M. Munnecke. 1984. Glyphosate utilization by Pseudomonas sp. and *Alkaligenes* sp. isolated from environmental sources. *Curr. Microbiol.* 10: 255-260.

Tatara, G.M., M.J. Dybas, and C.S. Criddle. 1993. Effect of medium and trace metals on kinetics of carbon tetrachloride transformation by *Pseudomonas* sp. strain KC. *Appl. Environ. Microbiol.* 59: 2126-2131.

Taya, M., H. Hinoki, and T. Kobayashi. 1985. Tungsten requirement of an extremely thermophilic cellulolytic anaerobe (strain NA 10). *Agric. Biol. Chem.* 49: 2513-2515.

Tausson, W.O. 1927. Naphthalin als Kohlenstoffquelle für Bakterien. *Planta* 4: 214-256.

Taylor, B.F., R.W. Curry, and E.F. Corcoran. 1981. Potential for biodegradation of phthalic acid esters in marine regions. *Appl. Environ. Microbiol.* 42: 590-595.

Tham, J.M.L., and C.L. Poh. 1993. Insertional mutagenesis, cloning and expession of gentisate pathway genes from *Pseudomonas alcaligenes* NCIB 9867. *J. Appl. Bacteriol.* 75: 159-163.

Tjeerdema, R.S., T. W.-M. Fan, R.M. Higashi, and D.G. Crosby. 1991. Effects of pentachlorophenol on energy metabolism in the abalone (*Haliotis rufescens*) as measured by in vivo ^{31}P NMR spectroscopy. *J. Biochem. Toxicol.* 6: 45-56.

Tschech, A., and N. Pfennig. 1984. Growth yield increase linked to caffeate reduction in *Acetobacterium woodii*. *Arch. Microbiol.* 137: 163-167.

Unkefer, C.J., and R.E. London. 1984. In vivo studies of pyridine nucleotide metabolism in *Escherichia coli* and *Saccharomyces cerevisiae* by carbon-13 NMR spectroscospy. *J. Biol. Chem.* 2311-2320.

van der Meer, J.R., T.N.P. Bosma, W.P. de Bruin, H. Harms, C. Holliger, H.H.M. Rijnaarts, M. E. Tros, G. Schraa, and A.J.B. Zehnder. 1992. Versatility of soil column experiments to study biodegradation of halogenated compounds under environmental conditions. *Biodegradation* 3: 265-284.

Van der Woude, M.W., K. Boominathan, and C.A. Reddy. 1993. Nitrogen regulation of lignin peroxidase and manganese-dependent peroxidase production is independent of carbon and manganese regulation in *Phanerochaete chrysosporium*. *Ach. Microbiol.* 160: 1-4.

van Ginkel, C.G., J.B. van Dijl, and A.G.M. Kroon. 1992. Metabolism of hexadecyl-trimethylammonium chloride in *Pseudomonas* strain B1. *Appl. Environ. Microbiol.* 58: 3083-3087.

van Niel, C.B. 1955. Natural selection in the microbial world. *J. Gen. Microbiol.* 13: 201-217.

Van Zyl, E., and P.L. Syeyn. 1992. Reinterpretation of the taxonomic position of *Xanthomonas maltophilia* and taxonomic criteria in this genus. Request for an opinion. *Int. J. Syst. Bacteriol.* 42: 193-198.

Vogel, T.M., and D. Grbic-Galic. 1986. Incorporation of water into toluene and benzene during anaerobic fermentative transformation. *Appl. Environ. Microbiol.* 52: 200-202.

Wachenheim, D.E., and R.E. Hespell. 1984. Inhibitory effects of titanium(III) citrate on enumeration of bacteria from rumen contents. *Appl. Environ. Microbiol.* 48: 444-445.

Wagner, R., and J.R. Andreesen. 1979. Selenium requirement for active xanthine dehydrgenase from *Clostridium acidiurici* and *Clostridium cylindrosporum*. *Arch. Microbiol.* 121: 255-260.

Wagner, R., and J.R. Andressen. 1987. Accumulation and incorporation of [185]W-tungsten into proteins of *Clostridium acidiurici* and *Clostridium cylindrosporum*. *Arch. Microbiol.* 147: 295-299.

White, H., G. Strobl, R. Feicht, and H. Simon. 1989. Carboxylic acid reductase: a new tungsten enzyme catalyses the reduction of non-activated carboxylic acids to aldehydes. *Eur. J. Biochem.* 184: 89-96.

Widdel, F. 1983. Methods for enrichment and pure culture isolation of filamentous gliding sulfate-reducing bacteria. *Arch. Microbiol.* 134: 282-285.

Wigmore, G.J., and D.W. Ribbons. 1981. Selective enrichment of *Pseudomonas* spp. defective in catabolism after exposure to halogenated substrates. *J. Bacteriol.* 146: 920-927.

Wilkinson, S.G. 1968. Studies on the cell walls of pseudomonas species resistant to ethylenediaminetetra-acetic acid. *J. Gen. Microbiol.* 54: 195-213.

Willems, A., J. Busse, M. Goor, B. Pot, E. Falsen, E. Jantzen, B. Hoste, M. Gillis, K. Kersters, G. Auling, and J. de Ley. 1989. *Hydrogenophaga*, a new genus of hydrogen-oxidizing bacteria that includes *Hydrogenophaga flava* comb. nov. (formerly *Pseudomonas flava*), *Hydrogenophaga palleronii* (formerly *Pseudomonas palleronii*), *Hydrogenophaga pseudoflava* (formerly *Pseudomonas pseudoflava* and *"Pseudomonas carboxydoflava"*), and *Hydrogenophaga taeniospiralis* (formerly *Pseudomonas taeniospiralis*). *Int. J. Syst. Bacteriol.* 39: 319-333.

Willems, A., E. Falsen, B. Pot, E. Jantzen, B. Hoste, P. Vandamme, M. Gillis, K. Kersters, and J. de Ley. 1990. *Acidovorax*, a new genus for *Pseudomonas facilis*, *Pseudomonas delafieldii*, EF group 13, EF group 16, and several clinical isolates, with the species *Acidovorax facilis* comb. nov., *Acidovorax delafieldfii* comb. nov., and *Acidovorax temperans* sp. nov. *Int. J. Syst. Bacteriol.* 40: 384-398.

Willems, A., B. Pot, E. Falsen, P. Vandamme, M. Gillis, K. Kersters, and J. de Ley. 1991a. Polyphasic taxonomic study of the amended genus *Comamonas*: relationship to *Aquaspirillum aquaticum*, E. Falsen group 10, and other clinical isolates. *Int. J. Syst. Bacteriol.* 41: 427-444.

Willems, A., J. de Ley, M. Gillis, and K. Kersters.1991b. *Comomonadaceae*, a new family encompassing the acidovorans rRNA complex, including *Variovorax paradoxus* gen. nov., comb. nov., for *Alcaligenes paradoxus* (Davis 1969). *Int. J. Syst. Bacteriol.* 41: 445-450.

Winter, J., C. Lerp, H.-P. Zabel, F. X. Wildenauer, H. König, and F. Schindler. 1984. *Methanobacterium wolfei*, sp. nov., a new tungsten-requiring, thermophilic, autotrophic methanogen. *System. Appl. Microbiol.* 5: 457-466.

Yamamoto, I., T. Saiki, S.M. Liu, and L.G. Ljungdahl. 1983. Purification and properties of NADP-dependent formate dehydrogenase from *Clostridium thermoaceticum*, a tungsten-selenium-iron protein. *J. Biol. Chem.* 258: 1826-1832.

Yoch, D.C. 1979. Manganese, an essential trace element for N_2 fixation by *Rhodospirilllum rubrum* and *Rhodopseudomonas capsulata*: role in nitrogenase regulation. *J. Bacteriol.* 140: 987-995.

Zarilla, K.A., and J.J. Perry. 1984. Thermoleophilum album gen. nov. and sp. nov., a bacterium obligate for thermophily and *n*-alkane substrates. *Arch. Microbiol.* 137: 286-290.

Zellner, G., C. Alten, E. Stackebrandt, E.C. de Macario, and J. Winter. 1987. Isolation and characterization of *Methanocorpusculum parvum*, gen. nov., spec. nov., a new tungsten-requiring, coccoid methanogen. *Arch. Microbiol.* 147: 13-20.

Zürrer, D., A.M. Cook, and T. Leisinger. 1987. Microbial desulfonation of substituted naphthalenesulfonic acids and benzenesulfonic acids. *Appl. Environ. Microbiol.* 53: 1459-1463.

CHAPTER 6

Pathways of Biodegradation and Biotransformation

SYNOPSIS

An attempt is made to describe the various pathways used by microorganisms to degrade or transform a wide range of xenobiotics. Most of the major structural groups are considered including aliphatic, alicyclic, aromatic, and heterocyclic compounds, including those with oxygen, sulfur, nitrogen, phosphorus, or halogen substituents. Although organochlorine compounds have received most attention, an attempt has been made also to include organobromine compounds; the degradation of organofluoro compounds is discussed separately, since these compounds differ significantly even in their chemical properties from those of the other halogens. Reactions carried out by both aerobic and anaerobic bacteria — and to a lesser extent, by yeasts and fungi — are considered, although no attempt is made to provide details of the relevant enzyme systems. Attention has been directed to the various pathways that may be used by different organisms for the degradation of a given xenobiotic. Investigations using aerobic bacteria have almost invariably been exemplified from the results of experiments using pure cultures, whereas for anaerobic bacteria, this has been supplemented by results using mixed cultures or stable consortia. Some examples are given of a few biotransformation reactions that may be relevant to biotechnology and of the environmental significance of the biotransformations of xenobiotics which result in metabolites more toxic than their precursors. Finally, an attempt has been made to classify the reactions involved in the degradation of xenobiotics on the basis of well-established chemical transformations, and specific reference has been made to the appropriate sections in which these reactions are discussed in more detail.

INTRODUCTION

It is desirable to explain both the motivation and the objectives of this chapter since some of the material has already been presented from a different perspective in preceding chapters: Chapter 4 attempted to provide a general background with a microbiological emphasis, while Chapter 5 filled this out with an outline of procedures for carrying out the appropriate experiments. This chapter attempts a survey of the pathways by which a range of structurally diverse xenobiotics are degraded or transformed by microorganisms; the emphasis is on reactions mediated by bacteria which are the most effective agents in carrying out biodegradation in most natural aquatic ecosystems.

It is appropriate to begin by underscoring the two rather different — and possibly conflicting — approaches to addressing problems of biodegradation and biotransformation and to which attention has already been briefly drawn. These concern the *level* at which assessments of biodegradability are carried out.

On the one hand, conventional tests for assessing ready biodegradability do not provide an adequate base for determining what occurs after release of the compound into natural ecosystems, even though they may be adequate for assessing biodegradability in the municipal treatment systems from which the inocula were taken. Indeed, the effort directed to developing standardized test systems may even have been counterproductive to environmental relevance for reasons that have been outlined in more detail in Chapter 5 (Section 5.1).

On the other hand, the comprehensive investigations which have been pursued on the physiology, biochemistry, genetics, and regulation of biodegradation cannot realistically be incorporated even into an advanced hazard assessment except in a very few instances.

An additional problem arises from the immense structural range of organic compounds that are used industrially or have been incorporated into commercial products. The skill of the organic chemist is seemingly unlimited and with the inevitable need for new compounds which attempt to avoid the undesirable consequences of their traditional counterparts, the number of compounds — as well as their structural diversity — seems unlikely to diminish.

There is an enormous literature on the microbial degradation and transformation of organic compounds, and it would be attractive to take advantage of this to construct generalizations on the pathways used for the degradation of broad structural classes of xenobiotics. That is the objective of this chapter. This approach has been illustrated by Alexander (1981), and this chapter attempts to provide details that were not possible within the space of that seminal review. In addition, this procedure would have a predictive capability that is not restricted to compounds which have already been investigated. Although structure-activity relationships are useful for classifying existing data, they have an inevitably restricted potential for application to completely novel chemical structures. Support for this mechanistic approach is provided by its success in assessing the biodegradability of 50 structurally diverse xenobiotics (Boethling et al. 1989).

This chapter does not attempt to encompass the enzymology of the reactions involved in the degradation of xenobiotics, so that the word "pathways" is more appropriate than "mechanisms": however desirable, discussions of enzymology lie beyond both the scope of the present work and the competence of the author. A few parenthetical comments on the enzymology of the reactions have, however, been made if they elucidate the scope and the generality of the reactions under consideration. Some important details of the reaction pathways involved in degradation have been deliberately omitted in the figures used to illustrate the various sequences; for example, (a) even when the degradation of carboxylic acids takes place through initial formation of the coenzyme-A esters, sequences have depicted the free carboxylic acids and (b) in some cases although the structures of intermediates have not been rigorously determined, these have been included to illustrate more clearly the structural relationships between the initial substrate and the various metabolites.

The presentation is made on the basis of the chemical structure of xenobiotics and is dominated by examples of reactions carried out by aerobic and anaerobic bacteria and — to a lesser extent — aerobic fungi and yeasts; some examples of biotransformation reactions carried out by other microorganisms are given in Chapter 4 (Section 4.3) and by higher organisms in Chapter 7 (Section 7.5). Although anaerobic fungi are known and are certainly important in rumen metabolism (Mountfort 1987), their existence in other habitats does not appear to have been established, and their potential for degrading xenobiotics does not seem to have been explored. Since the emphasis is on degradative *pathways*, less attention will be devoted to the taxonomic delineation of the various organisms except where these belong to less common taxa. Reference has already been made in Chapter 5 (Section 5.6) to the serious problems that have been encountered in attempting to classify bacteria of established degradative importance. In addition, no attempt has been made to provide the currently acceptable taxonomic assignment of the organisms that are involved, and the designations used by the authors have been retained with only a few exceptions. Except for the simplest reaction sequences, structural representations of the various pathways are given in the form of flow-diagrams rather than by using conventional chemical nomenclature; it is hoped that the reactions are, thereby, more clearly perceived in geometrical terms particularly to those who are not organic chemists and who are understandably repelled by the seemingly barbarous complexity and apparent incomprehensibility of systematic organic chemical nomenclature.

In the following sections, an account will be presented of the pathways by which xenobiotics are degraded by microorganisms. At the same time, it is essential to bear in mind certain fundamental aspects of the microbiology and biochemistry of the cells carrying out these reactions and, in particular, the role of metabolites that are required for biosynthesis on which continued growth and replication of the organisms ultimately depend.

1. If an organic compound is to support growth and replication of an organism, it must also provide the necessary metabolic energy and serve as the source of carbon (and, in some cases, also nitrogen or sulfur or phosphorus) for the synthesis of cell material. Details of these metabolic reactions are not given here, and a good account may be found, for example, in Mandelstam, McQuillan and Dawes (1982). These reactions then determine the extent to which the constituent atoms of xenobiotics are incorporated into the global carbon, nitrogen, sulfur, or phosphorus cycles; these are not discussed here, and reference may be made to the valuable account of carbon cycles into which the products from the degradation of xenobiotics are incorporated (Hagedorn et al. 1988). Whereas the functional operation of these reactions is a prerequisite for biodegradation, biotransformation may be accomplished by non-growing cells or in cells growing at the expense of more readily degradable substrates; this has been discussed in Chapter 4 (Section 4.4.2).

2. Just as there is no single pathway universally used for the catabolism of simple substrates such as glucose, there are no unique pathways for the degradation of a given xenobiotic. The following examples may be used to illustrate the

considerable differences in the pathways used for the degradation of xenobiotics by bacteria and by fungi or even by different taxa of bacteria.

- The degradation of DDT by *Phanerochaete chrysosporium* (Bumpus and Aust 1987) and by *Aerobacter aerogenes* (Wedemeyer 1967).
- The degradation of 2,4-dichlorophenol by *Ph. chrysosporium* (Valli and Gold 1991) and by a strain of *Acinetobacter* sp. (Beadle and Smith 1982).
- The degradation of quinoline by pseudomonads and by *Rhodococcus* sp. (Schwarz et al. 1989).
- The degradation of tryptophan by *Pseudomonas fluorescens* that takes place via the β-ketoadipate pathway and by *Ps. acidovorans* that utilizes the quinoline pathway (Stanier 1968).

3. There is no absolute distinction between the degradative pathways used by aerobic and by anaerobic bacteria. Simple reductions are carried out by organisms with a strictly aerobic metabolism; these include, for example, reductive dechlorination of phenolic compounds by *Rhodococcus chlorophenolicus* (Apajalahti and Salkinoja-Salonen 1987a), reduction of hydroxylated pyrimidines by *Pseudomonas stutzeri* (Xu and West 1992), and degradation of anthranilate by a strain of *Pseudomonas* sp. that is able to use this as a source of both carbon and nitrogen and degrades the substrate by initial reactions involving the reduction of the aromatic ring (Altenschmidt and Fuchs 1992b). This should not, however, be interpreted to imply that the underlying cellular metabolism of aerobic and anaerobic microorganisms is necessarily comparable.

4. Although a synopsis of the reactions used by microorganisms for the degradation and transformation of organic compounds is given in Section 6.12, it may be valuable to provide some general comments at this stage. The basic reactions known in organic chemistry provide a suitable background for rationalizing most biochemical reactions — addition, elimination, substitution, oxidation, reduction, and rearrangement — and all of these can be mediated by microorganisms although, for example, degradation involving addition reactions is rather unusual. The degradation of aliphatic (and alicyclic) and aromatic (including heterocyclic) compounds has been treated separately in this chapter, since both their chemistry and their microbial degradation pathways differ significantly. The following categorical summary may illustrate the broad types of reactions that are most commonly encountered and may serve as a prelude to the more detailed discussions of individual groups of compounds that follow.

- Most organic xenobiotics are relatively highly reduced compounds so that their degradation to CO_2 and H_2O inevitably involves introduction of oxygen into the molecule either by dioxygenation or by hydroxylation. Whereas dioxygenation is clearly restricted to aerobic conditions, hydroxylation can be accomplished both aerobically and anaerobically. It should, of course, be appreciated that oxidation can occur under anaerobic conditions provided that a redox balance is preserved within the system. Methanogenesis is the terminal — though complex — step in the reduction of the precursors (CO_2 or acetate) that are produced by the degradation of more complex substrates.
- Under aerobic conditions, dehydrogenations may be involved, and these may also function under anaerobic conditions.

- Depending on the oxidation level, compounds containing N, P, O, or S atoms may be either oxidized or reduced.
- The degradation of compounds carrying substituents such as halogen may occur by elimination or by displacement reactions, and these may be either reductive, oxidative, or hydrolytic.
- Rearrangements are particularly important among anaerobic bacteria where they involve coenzyme-B_{12}, and the rearrangement of the substituents on aromatic rings (the NIH shift) is well established, particularly among fungi.

A cardinal issue for the successful biodegradation of xenobiotics is the bioenergetics of these reactions, although this aspect is not discussed here. Whereas the synthesis of ATP under aerobic conditions is at least formally straightforward, a much greater range of *mechanisms* operates under anaerobic conditions where ATP may be generated, for example, from intermediate acyl phosphates (generally acetyl phosphate), carbamyl phosphate, or 10-formyl tetrahydrofolate.

As has already been emphasized, citations to the literature are eclectic rather than complete. Comprehensive reviews of many of the groups of compounds have been provided in the books and in the review articles that are given at the beginning of the reference list in Chapter 4 (Section 4.10.1), and these should be consulted for further details.

6.1 AEROBIC DEGRADATION OF NON-AROMATIC HYDROCARBONS

6.1.1 Alkanes

There is an enormous literature on the microbial degradation of alkanes; this has been motivated by aims as diverse as the utilization of microorganisms for the production of single-cell protein or their application to combating oil spills. Both the number and the taxonomic range of microorganisms are equally impressive, and they include many different taxa of bacteria, yeasts and fungi. Extensive reviews that cover most aspects have been presented (Ratledge 1978; Britton 1984).

The simplest alkane is, of course, methane, but the pathways for its degradation and assimilation do not reflect this structural simplicity. In outline the pathway of degradation is straightforward and involves successive oxidation to methanol, formaldehyde, and formate, but the cells must also be capable of synthesizing cell material from the substrate so that some fraction of the C_1 metabolites must also be assimilated. Several distinct pathways have been described, and these are merely summarized here since a comprehensive and elegant presentation of the details has been given (Anthony 1982).

1. The ribulose *bis*phosphate pathway for the assimilation of CO_2 which is identical to the Benson-Calvin cycle used by photosynthetic organisms.
2. The ribulose monophosphate cycle for the incorporation of formaldehyde.
3. The serine pathway for the assimilation of formaldehyde.

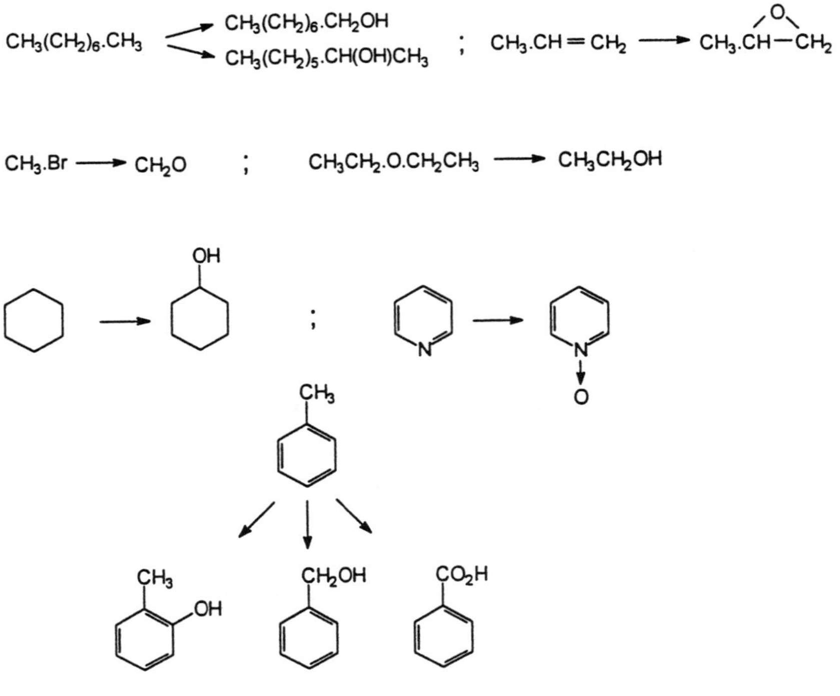

$CH_3(CH_2)_6.CH_3 \longrightarrow$ $CH_3(CH_2)_6.CH_2OH$
$CH_3(CH_2)_5.CH(OH)CH_3$; $CH_3.CH = CH_2 \longrightarrow CH_3.CH-CH_2$ (epoxide)

$CH_3.Br \longrightarrow CH_2O$; $CH_3CH_2.O.CH_2CH_3 \longrightarrow CH_3CH_2OH$

Figure 6.1 Examples of the reactions catalyzed by methane monooxygenase.

$R.CH_2.CH_2.CH_2.CH_3 \longrightarrow R.CH_2.CH_2.CH_2.CH_2OH \longrightarrow R.CH_2.CH_2.CH_2.CHO$

$\longrightarrow R.CH_2.CH_2.CH_2.CO_2H \longrightarrow R.CH_2.CO_2H + CH_3.CO_2H \longrightarrow \longrightarrow$

Figure 6.2 Outline of the metabolism of alkanes.

One additional aspect is the wide spectrum of substrates which can be metabolized by the methane monooxygenase system, and some illustrative examples are given in Figure 6.1. Attention has already been drawn in Chapter 4 (Section 4.3.1) to the similarity of this enzyme to that involved in the oxidation of ammonia, while the broad substrate specificity of cyclohexane oxygenase is noted again in Section 6.1.2.

The initial reactions involved in the metabolism of higher alkanes ($> C_1$) are formally similar to those used for the metabolism of methane. Enzymatically, however, the position may be more complex than would appear since, for example, a number of distinct alcohol and fatty acid dehydrogenases have been isolated from an *Acinetobacter* sp. during the metabolism of hexadecane (Singer and Finnerty 1985a,b). Further degradation of the resulting carboxylic acid involves a β-oxidation with successive loss of acetate residues (Figure 6.2). A structurally wide range of hydrocarbons may be degraded by microorganisms including linear alkanes with both even numbers of carbon atoms up to at least C_{30}, some odd numbered alkanes including the plant wax $C_{29} H_{60}$ (Hankin and Kolattukudy 1968), and branched alkanes such as pristane (2,6,10,12-tetramethylpentadecane) (McKenna and Kallio 1971; Pirnik et al. 1974). A number of details merit brief comment.

1. In some cases, reaction between the initially formed alkanol and its oxidation product, the alkanoic acid, may produce esters which are resistant to further degradation (Kolattukudy and Hankin 1968).
2. For complete degradation and assimilation of the products into anabolic pathways, the cells must clearly be capable of synthesizing the appropriate enzymes. When β-oxidation results in the production of acetate, cells must be capable of synthesizing the enzymes of the glyoxylate cycle. When odd-membered alkanes are oxidized, propionate is also produced, and its further degradation may follow a number of alternative pathways (Figure 6.3) (Wegener et al. 1968).
3. Oxidation of compounds such as pristane proceeds by both β-oxidation and ω-oxidation (McKenna & Kallio 1971; Pirnik et al. 1974) (Figure 6.4).
4. The existence of chain branching may present an obstacle to degradation, although this can be circumvented by a carboxylation pathway (Figure 6.5) (Fall et al. 1979) that is formally comparable to that illustrated above for the degradation of propionate.
5. A number of substituted 2,2-bisphenylpropanes are degraded by oxidation and cleavage at the quaternary carbon atom (Figure 6.6) (Lobos et al. 1992), although this is probably facilitated by the presence of the phenyl rings and is, therefore, not typical for compounds containing quaternary-substituted carbon groups.
6. An unusual pathway has been proposed for the degradation of *n*-alkanes to the carboxylic acids by a *Pseudomonas* sp. under anaerobic conditions; this involves initial dehydrogenation and hydroxylation followed by successive oxidations (Figure 6.7) (Parekh et al. 1977).

The degradation of alkanoic acids by β-oxidation has been noted parenthetically above, but alternative pathways may occur. For example, the metabolism of hexanoic acid by strains of *Pseudomonas* sp. may take place by ω-oxidation with subsequent formation of succinate and 2-tetrahydrofuranylacetate as a terminal metabolite (Kunz and Weimer 1983). In a strain of *Corynebacterium* sp., the specificities of the relevant catabolic enzymes are consistent with the production of dodecanedioic acid by ω-oxidation of dodecane but not of hexadecanedioic acid from hexadecane (Broadway et al. 1993).

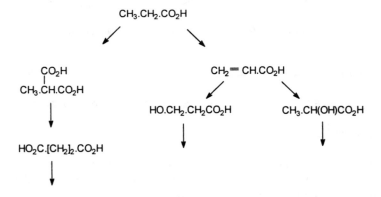

Figure 6.3 Pathways for the biodegradation of propionate.

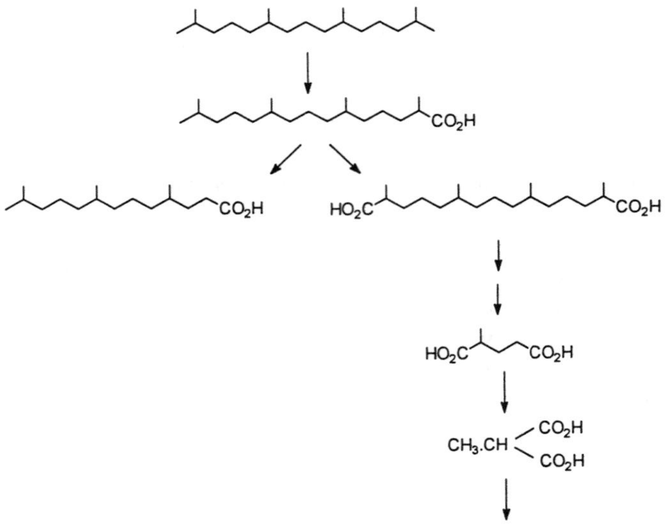

Figure 6.4 Pathways for the biodegradation of pristane.

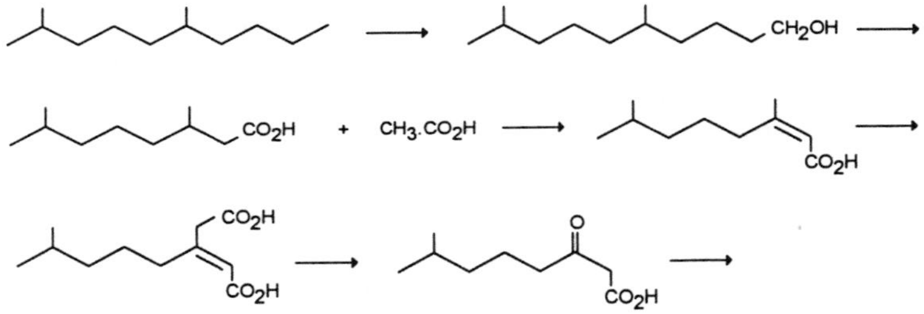

Figure 6.5 Carboxylation pathway for the biodegradation of branched chain alkanes.

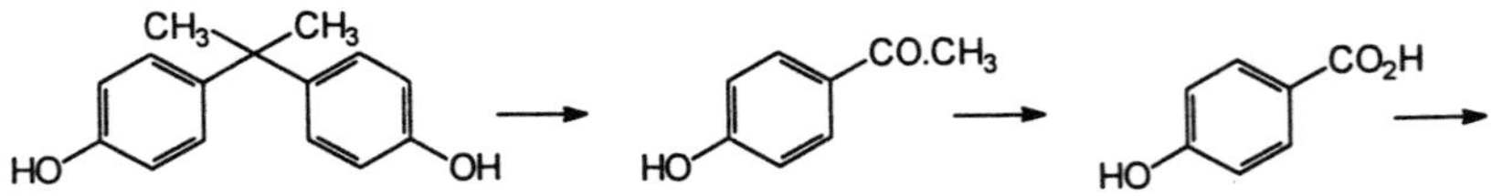

Figure 6.6 Biodegradation of 2,2-bisphenylpropane.

$$R-CH_2-CH_3 \longrightarrow R.CH=CH_2 \longrightarrow R.CH_2.CH_2OH \longrightarrow R.CH_2.CO_2H$$

Figure 6.7 Biodegradation of an alkane under anaerobic conditions.

6.1.2 Cycloalkanes

Reviews of the degradation of alicyclic compounds have been given by one of the pioneers (Trudgill 1978, 1984), and these should be consulted for further details; only a bare outline will, therefore, be given here.

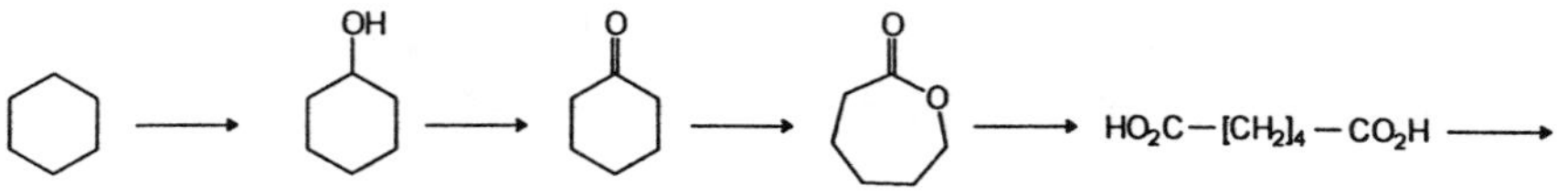

Figure 6.8   Biodegradation of cyclohexane.

Even though the first two steps in the oxidation of cycloalkanes are formally similar to those used for degradation of linear alkanes, it was some time before pure strains of microorganisms were isolated that could grow with cycloalkanes or their simple derivatives. The degradation of cyclohexane has been examined in detail (Stirling et al. 1977; Trower et al. 1985), and there are two critical steps in its degradation: (1) hydroxylation of the ring and (2) subsequent cleavage of the alicyclic ring that involves insertion of oxygen in a reaction formally similar to the Baeyer-Villiger persulfate oxidation. The pathway is illustrated for cyclohexane (Figure 6.8) (Stirling et al. 1977), and a comparable one operates also for cyclopentanol (Griffin and Trudgill 1972), while the enantiomeric specificity of this oxygen-insertion reaction has been examined in a strain of camphor-degrading *Pseudomonas putida* (Jones et al. 1993). Attention has already been drawn in Chapter 4 (Section 4.3.1) to the wide metabolic versatility of cyclohexane oxygenase (Branchaud and Walsh 1985) which is reminiscent of that of methane monooxygenase. Cyclohexylacetate is degraded to cyclohexanone by elimination of the side-chain after hydroxylation at the ring junction (Ougham and Trudgill 1982), but the cyclopropane ring in *cis*-11,12-methyleneoctadecanoate that is a lipid constituent of *Escherichia coli* is degraded by *Tetrahymena pyriformis* using the alternative ring opening pathway (Figure 6.9) (Tipton and Al-Shathir 1974).

An alternative and unusual pathway for the degradation of cyclohexanecarboxylate has also been found in which the ring is dehydrogenated to 4-hydroxybenzoate before ring cleavage (Figure 6.10) (Blakley 1974; Taylor and Trudgill 1978). The degradation of polyhydroxylated cyclohexanes such as quinate and shikimate also involves aromatic intermediates (Ingledew et al. 1971), though in these examples a mechanism for the formation of the aromatic ring by elimination reactions is more readily rationalized (Figure 6.11).

Comparable oxidations are also used for the degradation of alicyclic compounds containing several rings such as terpenes and sterols. For example, some of them are part of the sequence of reactions involved in the degradation of the bicyclic monoterpene camphor (Ougham et al. 1983). The degradation of sterols and related compounds has been extensively studied and for bile acids, involves a complex sequence of reactions that illustrate additional metabolic possibilities. For compounds oxygenated at C_3, initial reactions lead to the formation of the 1,4-diene-3-one, but the critical reaction that results in cleavage of the B-ring is hydroxylation at C_9 with formation of the 9,10-*seco* compound under the driving force of aromatization of the A-ring (Figure 6.12) (Leppik 1989).

At the same time, attention should be directed to numerous transformation reactions — generally hydroxylations, oxidations of alcohols to ketones or dehydrogenations — of both terpenes and sterols that have been accomplished by microorganisms especially fungi; this interest has been motivated by the great interest of the pharmaceutical industry in the products (Smith et al. 1988), and some of these reactions are discussed briefly in Section 6.11.2.

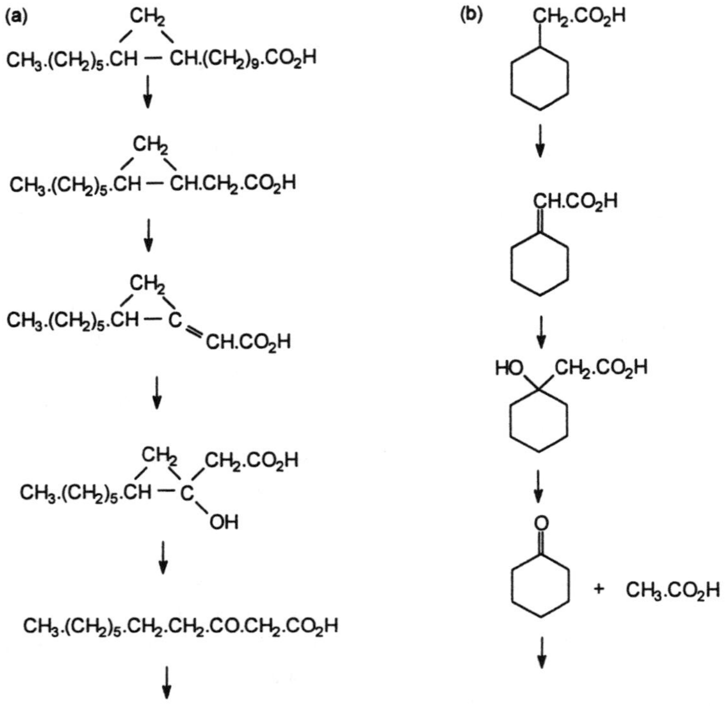

Figure 6.9 Biodegradation of **(a)** cyclopropylacetate and **(b)** cyclohexylacetate.

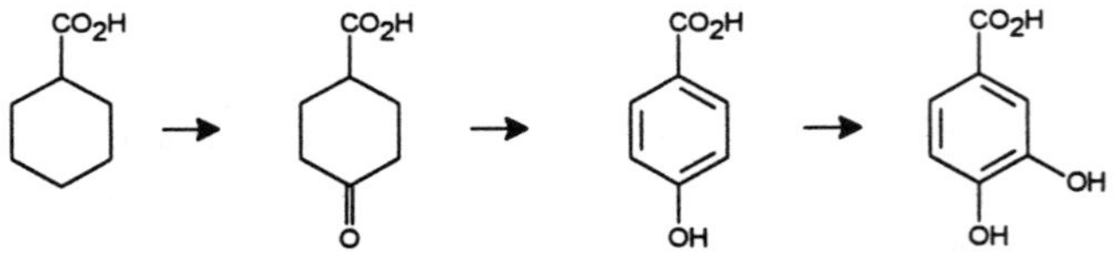

Figure 6.10 Alternative pathway for the biodegradation of cyclohexane carboxylates.

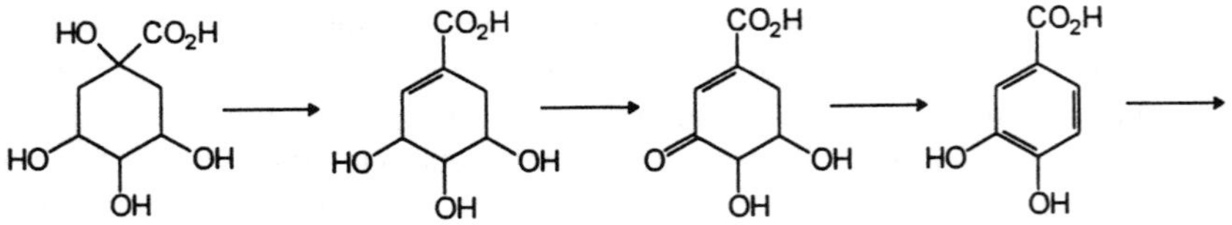

Figure 6.11 Biodegradation of quinate.

6.1.3 Alkenes

There are two different kinds of investigations which have been carried out: (1) on growth of microorganisms at the expense of alkenes and (2) on biotransformations resulting in the synthesis of epoxides. Studies have demonstrated growth, for example,

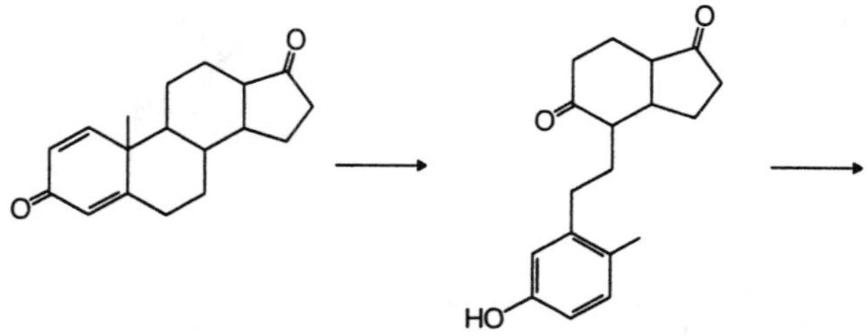

Figure 6.12   Biodegradation of C_3-oxygenated bile acids.

at the expense of propene and butene (van Ginkel and de Bont 1986), and an interesting observation is the pathway for the degradation of intermediate *n*-alkenes produced by an aerobic organism under anaerobic conditions (Parekh et al. 1977) that has already been noted in Section 6.1.1. Although the generality of this pathway remains unknown, it is clearly possible that such degradations might be accomplished even by aerobic bacteria under anoxic conditions and, for example, the degradation of hexadecane may be accomplished at quite low oxygen concentrations (Michaelsen et al. 1992). Attention should also be drawn to the possibility that intermediate metabolites may be incorporated into biosynthetic pathways; for example, hexadecene is oxidized by the fungus *Mortierella alpina* by ω-oxidation (Shimizu et al. 1991), but the lipids contain carboxylic acids containing with both 18 and 20 carbon atoms including the unusual polyunsaturated acid whose structure is shown in Figure 6.13.

The degradation of the initially formed epoxides is quite complex, and two distinct pathways have been observed:

1. Hydrolysis to the diol followed by dehydration to the aldehyde and oxidation to the carboxylic acid; this is the pathway used by a propene-utilizing species of *Nocardia* (de Bont et al. 1982).
2. Alternatively, the aldehyde may be formed directly from the epoxide, and this is clearly the case for the metabolism of styrene by a strain of *Xanthobacter* sp. (Hartmans et al. 1989).

Epoxides may be formed from alkenes during degradation by *Pseudomonas oleovorans*, but oct-1,2-epoxide is not further transformed, and degradation of oct-1-ene takes place by ω-oxidation (May and Abbot 1973; Abbott and Hou 1973). Considerable attention has, however, been directed to the epoxidation of alkenes on account of industrial interest in these compounds as intermediates. The wide metabolic capability of methane monooxygenase has been noted above and has been applied to the epoxidation of C_2, C_3, and C_4 alkenes (Patel et al. 1982). A large number of propane-utilizing bacteria are also effective in carrying out the epoxidation

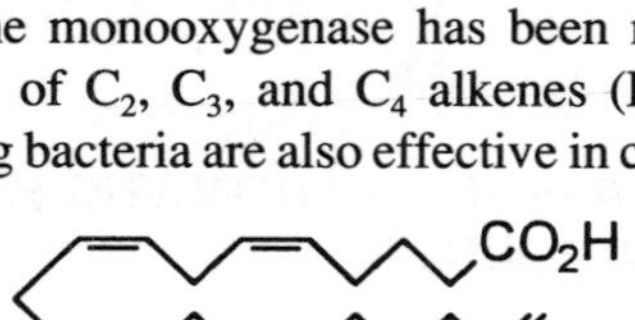

Figure 6.13   Polyunsaturated carboxylic acid synthesized during the biotransformation of hexadecene.

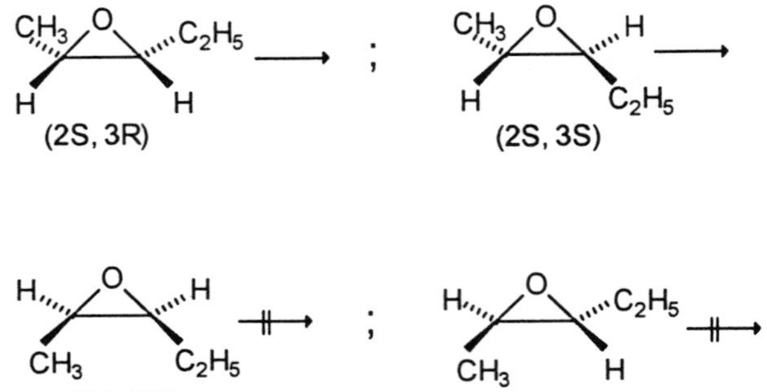

Figure 6.14 Biodegradability of enantiomeric of epoxides of *cis*- and *trans*-pent-2-enes.

$$HC \equiv C.CH_2.CH_2OH \longrightarrow HC \equiv C.CH_2.CHO \longrightarrow HC \equiv CH_2.CO_2H$$

$$\longrightarrow CH_3.CO.CH_2.CO_2H \longrightarrow CH_3.CO_2H$$

Figure 6.15 Aerobic biodegradation of but-3-ynol.

of alkenes (Hou et al. 1983). Especially valuable is the possibility of using microorganisms for resolving racemic mixtures of epoxides; this has been realized for *cis*- and *trans*-2,3-epoxypentanes using a *Xanthobacter* sp. which is able to degrade only one of the pairs of enantiomers leaving the other intact (Figure 6.14) (Weijers et al. 1988).

6.1.4 Alkynes

The degradation of alkynes has been the subject of sporadic but effective interest during many years so that the pathway has been clearly delineated. It is quite distinct from those used for alkanes and alkenes and is a reflection of the enhanced nucleophilic character of the alkyne C-C triple bond. The primary step is, therefore, hydration of the triple bond followed by ketonization of the initially formed enol. This reaction operates during the degradation of acetylene itself (de Bont and Peck 1980), acetylene carboxylic acids (Yamada and Jakoby 1959), and more complex alkynes (Figure 6.15) (Van den Tweel and de Bont 1985). It is also appropriate to note that the degradation of acetylene by anaerobic bacteria proceeds by the same pathway (Schink 1985b).

6.2 AEROBIC DEGRADATION OF AROMATIC HYDROCARBONS AND RELATED COMPOUNDS

The degradation of aromatic hydrocarbons has attracted interest over many years, for at least four reasons:

1. They are significant components of creosote and tar (Sundström et al. 1986) that have traditionally been used for wood preservation.
2. They are components of unrefined oil, and there has been serious concern over the hazard associated with their discharge into the environment after accidents at sea.
3. Many of the polycyclic representatives have been shown to be human carcinogens.
4. There has been increased concern over their presence in the atmosphere as a result of combustion processes and consequent air pollution.

Although growth at the expense of aromatic hydrocarbons has been known for many years (Söhngen 1913; Tausson 1927; Gray and Thornton 1928), it was many years later before details of the ring-cleavage reactions began to emerge. Two converging lines of investigations have examined in detail (1) the degradation of the monocyclic aromatic hydrocarbons benzene and toluene and (2) the degradation of oxygen-substituted compounds such as benzoate, hydroxybenzoates, and phenols. As a result of this activity, the pathways of degradation and their regulation are now known in considerable detail, and ever increasing attention has been directed to the degradation of polycyclic aromatic hydrocarbons. Since many of these metabolic sequences recur in the degradation of a wide range of aromatic compounds, a brief sketch of the principal reactions may conveniently be presented here. Extensive reviews that include almost all aspects have been given (Hopper 1978; Cripps and Watkinson 1978; Ribbons and Eaton 1982; Gibson and Subramanian 1984). Developments in regulatory aspects have been presented (Rothmel et al. 1991; van der Meer et al. 1992; Parales and Harwood 1993).

It is important at the outset to appreciate two important facts.

1. For complete degradation of an aromatic hydrocarbon to occur, it is necessary that the products of ring oxidation and cleavage can be further degraded to molecules that enter anabolic and energy-producing reactions.
2. Essentially different mechanisms operate in bacteria and fungi, and these differences have important consequences. In bacteria, the initial reaction is carried out by dioxygenation and results in the synthesis of a *cis* 1,2-dihydro-1,2-diol which is then dehydrogenated to a catechol before ring cleavage by the further action of dioxygenases. In fungi, on the other hand, the first reaction is monooxygenation to an epoxide followed by hydrolysis to a *trans*-1,2-dihydro-1,2-diol and dehydration to a phenol; ring cleavage of polycyclic aromatic hydrocarbons does not generally occur in fungi, so that these reactions are essentially biotransformations. These reactions are schematically illustrated in Figure 6.16. It should be noted, however, that both fungi and yeasts are able to degrade simpler oxygenated aromatic compounds such as vanillate (Ander et al. 1983) (Figure 6.17) and 3,4-dihydroxybenzoate (Cain et al. 1968). It may also be noted that the degradation of 3,4-dihydroxybenzoate by the yeast *Trichosporon cutaneum* proceeds initially by a pathway different from that used by bacteria and involves hydroxylative decarboxylation to 1,2,4-trihydroxybenzoate prior to ring cleavage (Anderson and Dagley 1980).

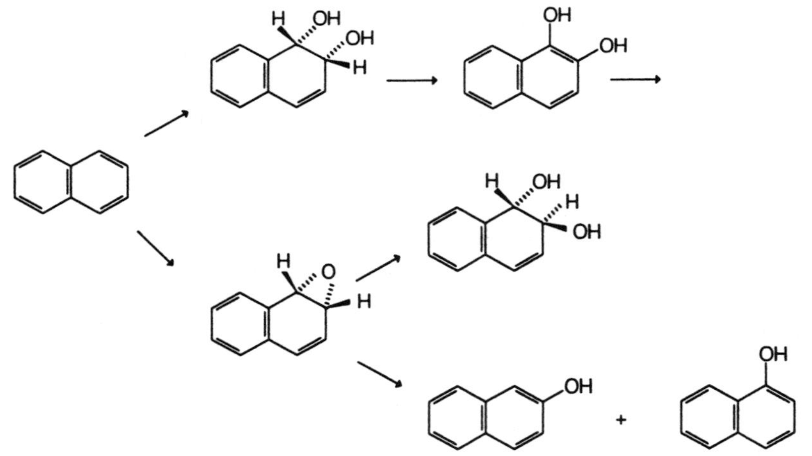

Figure 6.16 Alternate pathways for the oxidative metabolism of naphthalene by microorganisms.

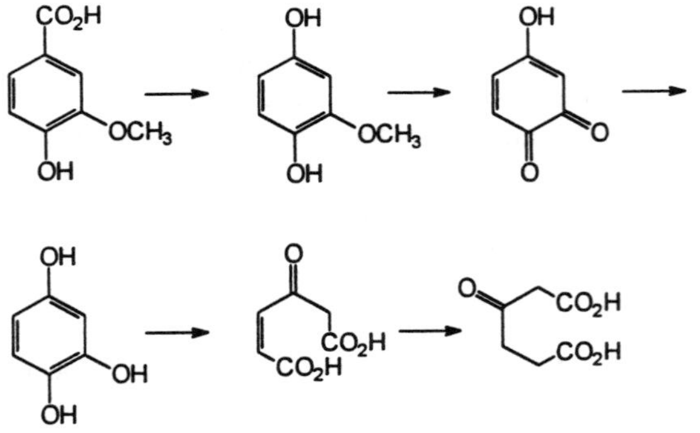

Figure 6.17 Biodegradation of vanillic acid by fungi.

6.2.1 Bacterial Degradation of Monocyclic Aromatic Compounds

Benzene and alkyl benzenes — Details of the metabolism of benzene and alkylated benzenes have been established as a result of the classic studies of David Gibson and his collaborators (Gibson et al. 1968, 1970). The key intermediate is catechol (Figure 6.18), although the details of its further degradation depend on the organism; for example, whereas in *Pseudomonas putida*, this is mediated by an extradiol 2,3-dioxygenase (Figure 6.19a), an intradiol 1,2-dioxygenase is involved (Figure 6.19b) in a species of *Moraxella* (Högn and Jaenicke 1972). The significance of which of these pathways is followed in substituted benzenes is discussed again in Section 6.5.1, since it has particular significance in situations when two different substrates are simultaneously present.

For monoalkylated benzenes, there are two additional factors: (1) the genes may be either chromosomal or carried on plasmids, and (2) oxidation may be initiated either on the aromatic ring or at the alkyl substituent. For example, toluene may be degraded by several different pathways.

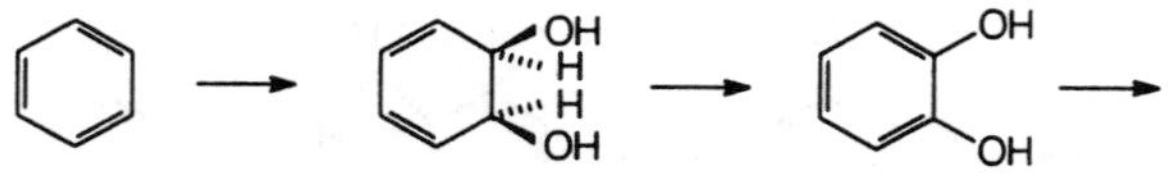

Figure 6.18 Biodegradation of benzene by *Pseudomonas putida*.

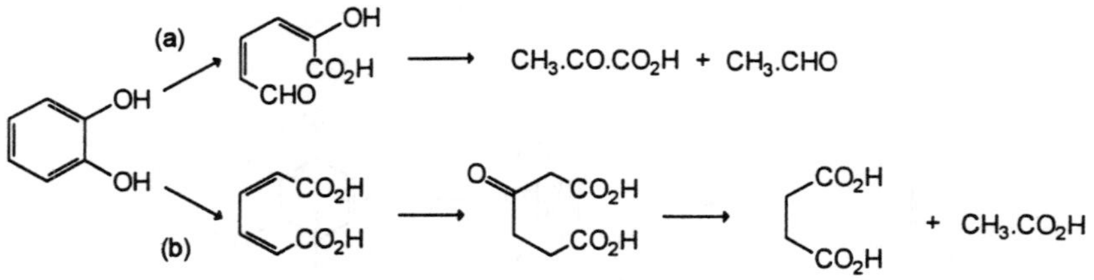

Figure 6.19 Metabolism of catechol **(a)** by catechol 2,3-dioxygenase in *Pseudomonas putida* and **(b)** by catechol 1,2-dioxygenase in *Moraxella* sp.

1. When the catabolic genes are carried on the TOL plasmid, degradation takes place by sequential side-chain oxidation of the methyl group to a carboxylate (Abril et al. 1989; Whited et al. 1986), followed by dioxygenation of the resulting benzoate to catechol that is cleaved by 2,3-dioxygenation (Figure 6.20a). The enzymes encoded by this plasmid have a relaxed specificity which is consistent with the ability of organisms carrying the plasmid to degrade other alkyl benzenes such as xylenes and 1,2,4-trimethyl benzenes.

2. In the second sequence, degradation is mediated by a 2,3-dioxygenase reaction with the methyl group intact (Figure 6.20b), and this pathway is followed in the metabolism of alkylated benzenes such as ethylbenzene and *iso*propylbenzene (Eaton and Timmis 1986).

3. Other possibilities involving ring monooxygenation of toluene have been found in *Pseudomonas cepacia* G4 that produces 2-methylphenol (Shields et al. 1989) and in *Pseudomonas mendocina* KR that produces 4-methylphenol (Whited and Gibson 1991).

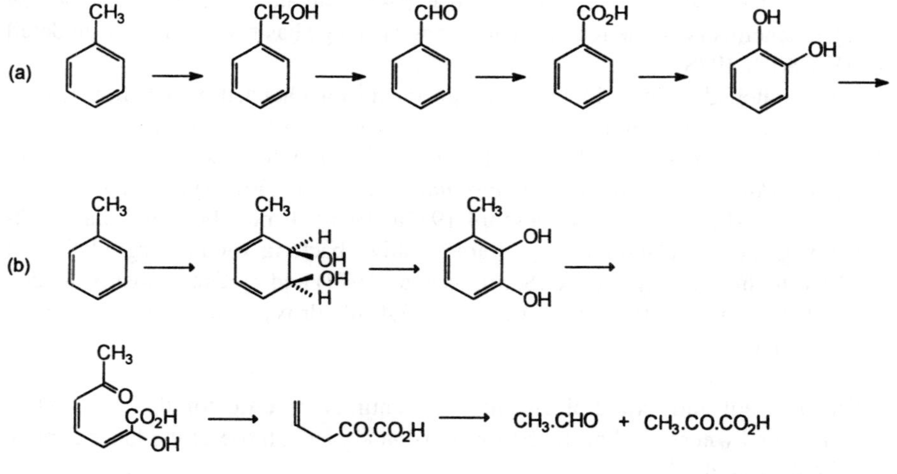

Figure 6.20 Biodegradation of toluene **(a)** by side-chain oxidation and **(b)** with the methyl group intact.

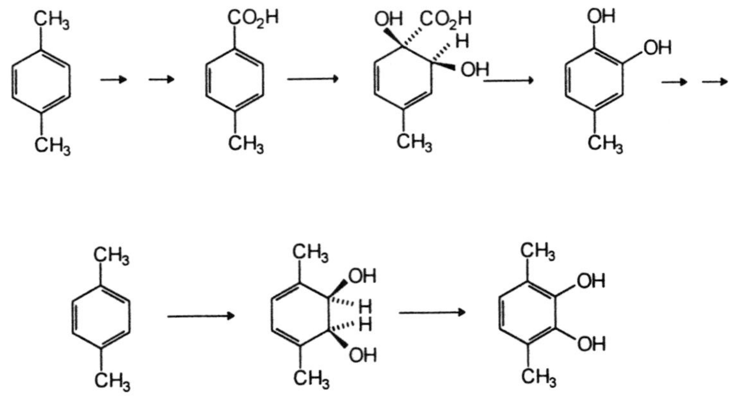

Figure 6.21   Biodegradation of 1,4-dimethylbenzene.

The degradation of dialkylbenzenes such as the dimethylbenzenes (xylenes) depends critically on the position of the methyl groups (Baggi et al. 1987); two distinct pathways have been found for the 1,4-isomer (Davey and Gibson 1974; Gibson et al. 1974), and these are illustrated in Figure 6.21.

Phenols and benzoates — Considerable effort has been devoted to the bacterial metabolism of oxygenated compounds including phenol, catechol, benzoate, and hydroxybenzoates which are much more readily degraded than the parent hydrocarbons, and some of the details have been tacitly assumed in the foregoing discussion. The ring cleavage reactions are generally mediated by dioxygenases after formation of 1,2-dihydroxy compounds, but there are important differences in the pathways used by different groups of organisms.

1. The pathways and their regulation during the degradation of catechol and 3,4-dihydroxybenzoate in *Pseudomonas putida* have been elucidated in extensive studies (Ornston 1966). In this organism, metabolism proceeds by a 1,2-dioxygenase ring cleavage to produce β-ketoadipate (Figure 6.22), and the stereochemistry of the reactions after ring cleavage has been examined in detail (Kozarich 1988).
2. Pseudomonads of the acidovorans group, on the other hand, use a 4,5-dioxygenase system to produce pyruvate and formate (Wheelis et al. 1967) (Figure 6.23a).
3. The third alternative for the ring cleavage of 3,4-dihydroxybenzoate is exemplified in *Bacillus macerans* and *B. circulans* that use a 2,3-dioxygenase to accomplish this (Figure 6.23b) (Crawford 1975a, 1976). It may be noted that a 2,3-dioxygenase is elaborated by Gram-negative bacteria for the degradation of 3,4-dihydroxyphenylacetate (Sparnins et al. 1974) and by Gram-positive bacteria for the degradation of L-tyrosine via 3,4-dihydroxyphenylacetate (Sparnins and Chapman 1976).

The intradiol and extradiol enzymes are entirely specific for their respective substrates, and whereas all of the first group contain Fe^{3+}, those of the latter contain Fe^{2+} (Wolgel et al. 1993).

In some cases, hydroxylation to 1,4-dihydroxy compounds activates the ring to oxidative cleavage. This alternative pathway is followed during the degradation of

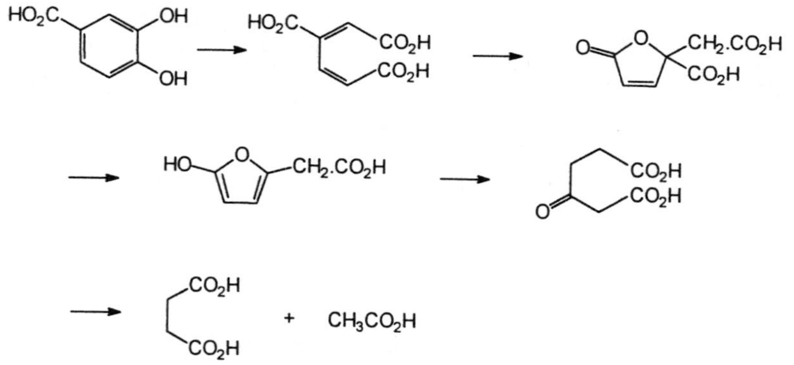

Figure 6.22   The β-ketoadipate pathway.

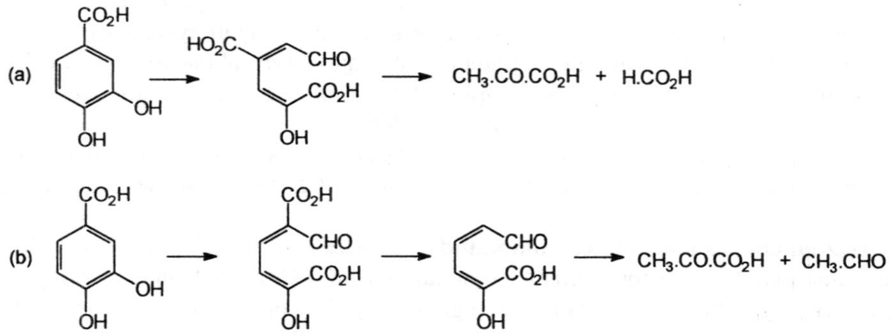

Figure 6.23   Biodegradation of 3,4-dihydroxybenzoate mediated **(a)** by a 4,5-dioxygenase system in *Pseudomonas acidovorans* and **(b)** by a 2,3-dioxygenase system in *Bacillus macerans*.

3-methylphenol, 3-hydroxybenzoate, and salicylate by a number of bacteria including species of *Pseudomonas* and *Bacillus* and involves gentisate (2,5-dihydroxybenzoate) as an intermediate; ring cleavage then produces pyruvate and fumarate or maleate (Crawford 1975b; Poh and Bayly 1980) (Figure 6.24). This pathway may plausibly be involved in the degradation even of benzoate itself by a denitrifying strain of *Pseudomonas* sp. in which the initial reaction is the formation of 3-hydroxybenzoate (Altenschmidt et al. 1993). The gentisate pathway is used for the degradation of salicylate produced from naphthalene by a *Rhodococcus* sp. (Grund et al. 1992), rather than by the more usual sequence involving hydroxylative decarboxylation of salicylate to catechol. Gentisate is also formed in an unusual rearrangement reaction from 4-hydroxybenzoate by a strain of *Bacillus* sp. (Crawford 1976) that is formally analogous to the formation of 2,5-dihydroxyphenylacetate from 4-hydroxyphenylacetate by *Pseudomonas acidovorans* (Hareland et al. 1975). It may be noted that the formal hydroxylation of 4-methylphenol to 4-hydroxybenzyl alcohol before conversion to 3,4-dihydroxybenzoate and ring cleavage is accomplished by initial dehydrogenation to a quinone methide followed by hydration (Hopper 1988).

   Although benzoate is generally metabolized by oxidative decarboxylation to catechol followed by ring cleavage, non-oxidative decarboxylation may also occur:

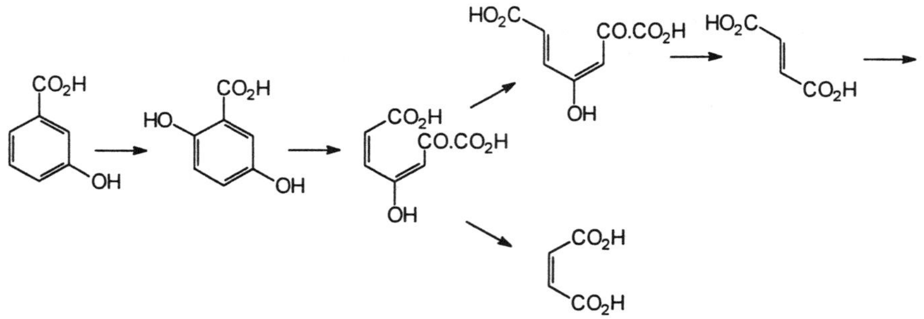

Figure 6.24   The gentisate pathway.

1. Strains of *Bacillus megaterium* transform vanillate to guaiacol by decarboxylation (Crawford and Olson 1978).
2. The degradation of *o*-phthalate takes place by initial dioxygenation to 4,5-dihydroxyphthalate followed by decarboxylation to 3,4-dihydroxybenzoate (Pujar and Ribbons 1985), and the degradation of 5-hydroxy*iso*phthalate takes place similarly via 4,5-dihydroxy*iso*phthalate and decarboxylation to 3,4-dihydroxybenzoate (Elmorsi and Hopper 1979).
3. A number of decarboxylations of aromatic carboxylic acids by facultatively anaerobic Enterobacteriaceae have been noted in Chapter 4 (Section 4.3.1).

Acetophenones and related compounds — There are two quite different pathways that may be used for aromatic compounds with a C_2 side-chain containing a carbonyl group adjacent to the benzene ring; this includes not only acetophenones but also reduced compounds that may be oxidized to acetophenones.

1. The mandelate pathway in *Pseudomonas putida* proceeds by successive oxidation to benzoyl formate and benzoate that is further metabolized via catechol and the β-ketoadipate pathway (Figure 6.25a) (Hegeman 1966). A comparable pathway is used by a strain of *Alcaligenes* sp. that degrades 4-hydroxyacetophenone via 4-hydroxybenzoyl methanol to 4-hydroxybenzoate; this is further metabolized to β-ketoadipate via 3,4-dihydroxybenzoate (Figure 6.25b) (Hopper et al. 1985).
2. An entirely different sequence is followed during the metabolism of acetophenone by gram-positive strains of *Arthrobacter* sp. and *Nocardia* sp. (Cripps et al. 1978) and of 4-hydroxyacetophenone and 4-ethylphenol by *Pseudomonas putida* strain JD1 (Darby et al. 1987). Acetophenone is converted by a Baeyer-Villiger oxidation to phenyl acetate that is hydrolyzed to phenol and then hydroxylated to catechol before ring cleavage (Figure 6.25c); similarly, 4-hydroxyacetophenone is oxidized to 4-hydroxyphenyl acetate that is hydrolyzed to 1,4-dihydroxybenzene before ring cleavage to β-ketoadipate. A mixed culture of an *Arthrobacter* sp. and a *Micrococcus* sp. is able to degrade 4-chloroacetophenone by an analogous sequence via 4-chlorophenyl acetate, 4-chlorophenol and 4-chlorocatechol (Havel and Reineke 1993). Formally similar Baeyer-Villiger monooxygenation of cycloalkanones has been noted previously in Section 6.1.2.

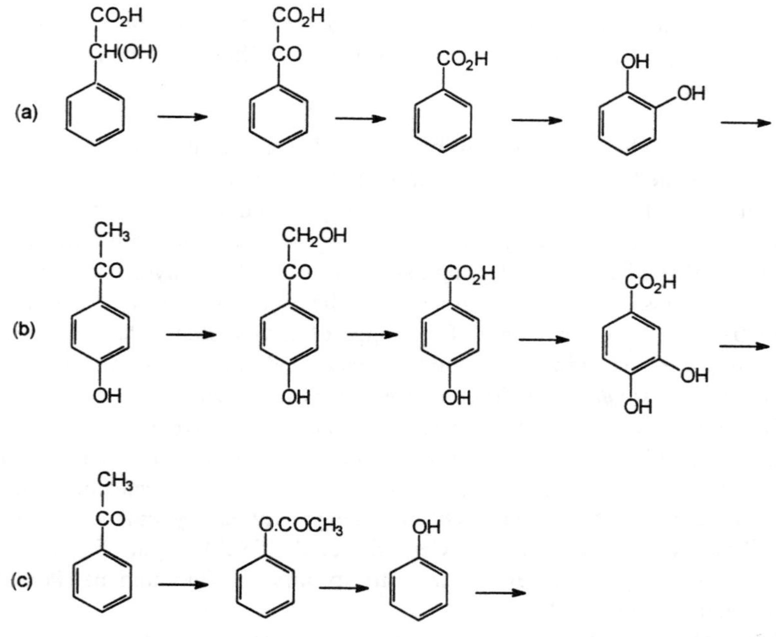

Figure 6.25 Degradation of **(a)** mandelate and **(b)** acetophenone by side-chain oxidation pathways and **(c)** acetophenone by Baeyer-Villiger-type monooxygenation.

Although it can be concluded that reactions formally similar to those that have been considered in this section are involved in the aerobic degradation of a wide range of aromatic compounds, it may be convenient to summarize here some of the major exceptions to ring dioxygenation:

1. Some heterocyclic aromatic compounds — particularly those containing nitrogen — may be degraded by reactions involving hydroxylation rather than dioxygenation before rupture of the rings (Section 6.3.1)
2. For halogenated phenols, a number of alternatives to direct hydroxylation (Beadle and Smith 1982) followed by ring cleavage are available (Section 6.5.1).

6.2.2 Metabolism of Polycyclic Aromatic Hydrocarbons by Fungi and Yeasts

Fungal metabolism of polycyclic aromatic hydrocarbons has been studied particularly extensively with *Cuninghamella elegans*. The analogy between the metabolic pathways used by fungi and by higher organisms (Smith and Rosazza 1983) coupled to concern with polycyclic aromatic hydrocarbons as human carcinogens has undoubtedly stimulated this interest. Although there is no reason to doubt that the reactions carried out by *C. elegans* are representative of those carried out by other fungi, there is one important exception: the white-rot fungus *Phanerochaete chrysosporium* that can partially mineralize a number of polycyclic aromatic hydrocarbons including phenanthrene and pyrene (Bumpus 1989; Hammel et al. 1992).

There are a number of aspects of the metabolism of polycyclic aromatic hydro-carbons by fungi which are worth noting since they differ from the reactions medi-ated by bacteria.

1. The phenol which is formed from the initially produced *trans* dihydrodiol may be conjugated to form sulfate esters or glucuronides (Cerniglia et al. 1982a; Golbeck et al. 1983), and the less common glucosides have also been identified; 1-phenanthreneglucopyranoside is produced from phenanthrene by *Cunninghamella elegans* (Cerniglia et al. 1989) and 3-(8-hydroxyfluoranthene)-glucopyranoside from fluoranthene by the same organism (Pothuluri et al. 1990). As a further example of the range of carbohydrates that may be conju-gated, the xylosylation of 4-methylguaiacol and vanillin by the basisiomycete *Coriolus versicolor* (Kondo et al. 1993) may be given.

2. The biotransformation of a considerable number of polycyclic aromatic hydro-carbons has been examined using *C. elegans*; reactions are generally confined to oxidation of the rings with formation of phenols or catechols, and ring cleavage does not take place. Different rings may be oxygenated, for example, in 7-methylbenz[a]anthracene (Cerniglia et al. 1982b) (Figure 6.26), or oxida-tion may take place in several rings, for example, in fluoranthene (Pothuluri et al. 1990) (Figure 6.27).

3. Although the biotransformation of polycyclic aromatic hydrocarbons by fungi bears a rather close resemblance to that carried out by mammalian systems (Smith and Rosazza 1983), there is one very significant difference — and that is the stereochemistry of the products; *trans*-1,2-dihydroxy-1,2-dihydroanthracene and *trans*-1,2-dihydroxy-1,2-dihydrophenanthrene are formed from the hydro-carbons by *C. elegans*, but these dihydrodiols have the *S,S*-configuration in contrast to the *R,R*-configuration of the metabolites from rat liver microsomes (Cerniglia and Yang 1984). It has become clear, however, that the situation among a wider range of fungi is much less straightforward. For example, the *trans*-9,10-dihydrodiol produced by *Phanerochaete chrysosporium* was pre-dominantly the 9*S*,10*S*-enantiomer, whereas those produced by *Cunninghamella elegans* and by *Syncephalastrum racemosum* were dominated by the 9*R*,10*R*-enantiomers (Sutherland et al. 1993). Comparable differences were also ob-served for the *trans*-1,2-dihydrodiols and *trans*-3,4-dihydrodiols so that gener-alizations on the stereoselectivity of these reactions are currently unwarranted.

4. Lignin peroxidase from *Phanerochaete chrysosporium* may produce quinones from polycyclic aromatic hydrocarbons; for example, benzo[a]pyrene is metabo-lized to the 1,6-, 3,6-, and 6,12-quinones (Figure 6.28) (Haemmerli et al. 1986), and it is interesting to note that the same quinones are among the metabolites produced by fish from the same substrate (Little et al. 1984). Different products are formed from phenanthrene by the activity of lignin peroxidases and by the monooxygenase systems of *P. chrysosporium* that are synthesized under differ-ent nitrogen regimes; the peroxidases produce 2,2'-diphenic acid via phenan-threne-9,10-quinone (Hammel et al. 1992), whereas the monooxygenase pro-duces phenanthrene 3,4-oxide and phenanthrene 9,10-oxide that are further transformed into phenanthrene *trans*-dihydrodiols and phenanthrols (Sutherland et al. 1991).

5. In reactions involving monooxygenase systems, rearrangement of substituents may take place (Figure 6.29a); this is an example of the "NIH shift" which plays an important role in the metabolism of xenobiotics by mammalian systems (Daly et al. 1972) and has been observed in a few fungal (Faulkner and Woodcock 1965; Smith et al. 1981; Cerniglia et al. 1983) and bacterial (Dalton et al. 1981; Cerniglia et al. 1984) systems including the marine cyanobacterium *Oscillatoria* sp. (Narro et al. 1992) (Figure 6.29b,c).

The transformation of a few polycyclic hydrocarbons has also been investigated in yeasts. The metabolism of naphthalene, biphenyl, and benzo[a]pyrene has been examined in a strain of *Debaryomyces hansenii* and in number of strains of *Candida sp*. The results using *C. lipolytica* showed that the transformations were similar to those carried out by fungi; the primary reaction was formation of the epoxides that were then rearranged to phenols (Cerniglia and Crow 1981).

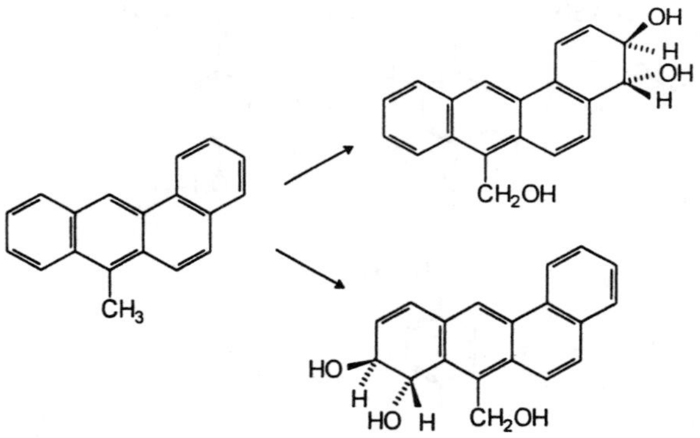

Figure 6.26 Biotransformation of 7-methylbenz[a]anthracene by *Cunninghamella elegans*.

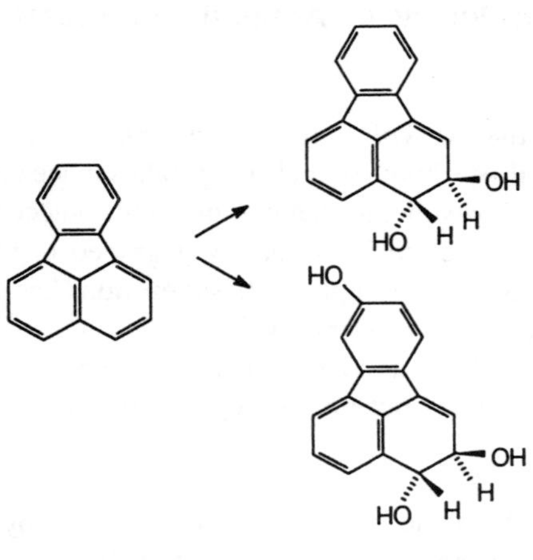

Figure 6.27 Alternative pathways for the biotransformation of fluoranthene by *Cunninghamella elegans*.

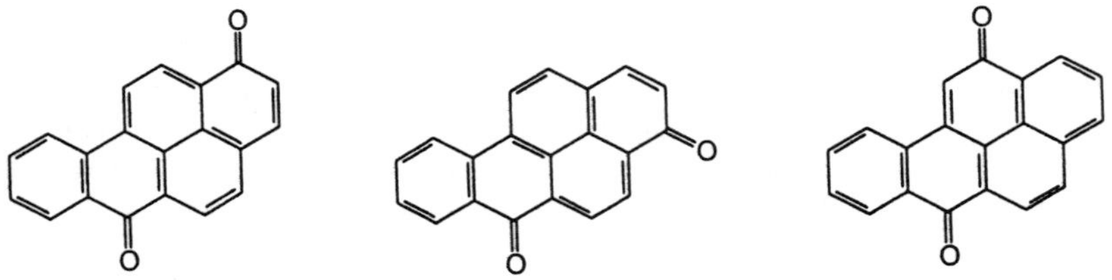

Figure 6.28 Quinones produced during the metabolism of benzo[a]pyrene by *Phanerochaete chrysosporium.*

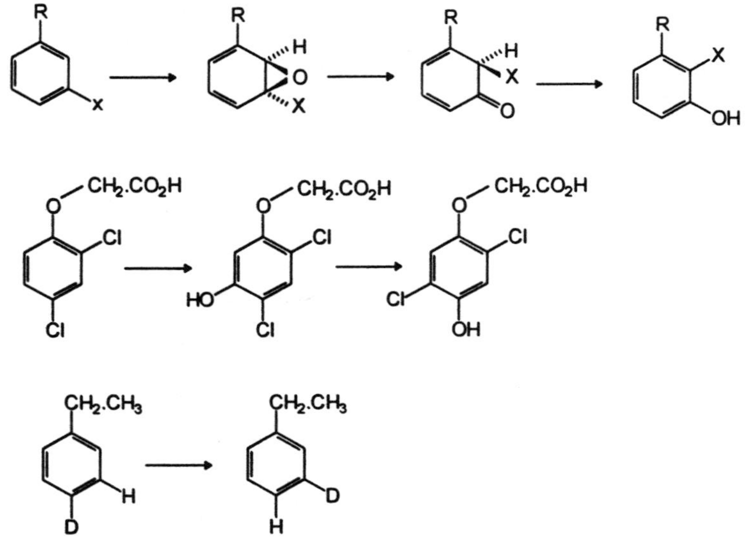

Figure 6.29   Examples of the NIH shift.

6.2.3 Metabolism of Polycyclic Aromatic Hydrocarbons by Bacteria

In contrast to the situation with fungi, bacteria may grow at the sole expense of polycyclic aromatic hydrocarbons; ring cleavage takes place after dioxygenation and dehydrogenation. For polycyclic aromatic compounds, successive ring degradation may occur, so that the structure is ultimately degraded to molecules which enter central anabolic pathways. An example of these reactions has already been given for the simplest representative — benzene itself.

There are a number of general conclusions that can be drawn from the extensive studies that have been carried out on the degradation of a wide range of aromatic hydrocarbons.

1. Degradation of naphthalene is readily carried out by many bacteria, and both the details of the initial steps (Jeffrey et al. 1975) and their enzymology have been elucidated. The enzymes for the two key steps — naphthalene dioxygenase

(Patel and Barnsley 1980; Ensley and Gibson 1983) and *cis* naphthalene dihydrodiol dehydrogenase (Patel and Gibson 1974) have been purified, while further details of the subsequent steps have been added (Eaton and Chapman 1992). The overall pathway is shown in Figure 6.30. Reference has been made above to the alternative gentisate pathway for the degradation of the intermediate salicylate.

2. Degradation becomes increasingly difficult for more highly condensed molecules and depends also on the annellation of the rings. Phenanthrene is more readily degraded than anthracene, and two different pathways are followed after cleavage of the peripheral ring; in one, the naphthalene pathway via salicylate is used (Evans et al. 1965), whereas in the other 1,2-phthalate and 3,4-dihydroxybenzoate are involved (Kiyohara and Nagao 1978; Kiyohara et al. 1976; Barnsley 1983) (Figure 6.31).

3. The situation with higher polycyclic aromatic hydrocarbons is more complex, and details of all the pathways have not been conclusively established; it seems, however, that degradation of all rings is not readily accomplished. For example, a strain of *Beijerinckia* sp., although unable to grow at the expense of benz[a]anthracene, readily formed 1-hydroxy-2-carboxyanthracene which was at least partly further degraded (Figure 6.32) (Mahaffey et al. 1988). For compounds with *peri*-fused structures such as fluoranthrene and pyrene, degradation has also been established. Degradation of pyrene by a *Mycobacterium* sp. has been established, and a plausible pathway has been proposed (Heitkamp et al. 1988) (Figure 6.33). One interesting and unusual observation for a bacterial system was the formation of both *cis* and *trans*-dihydrodiols; whereas the former is to be expected as the product of a dioxygenase system, the latter must be produced by a monooxygenase which might be comparable to that involved in the epoxidation of alkenes to epoxides also by a *Mycobacterium* sp. (Hartmans et al. 1991). The strain of *Mycobacterium* sp. that degraded pyrene also degraded fluoranthene with the initial formation of 1-acenaphthenone and of fluoren-9-one that was then further metabolized (Kelley et al. 1993).

4. The effect of substituents including halogen, sulfonate, and nitro will be discussed in more detail in Section 6.5.1, and in Sections 6.8.1 and 6.8.2, but a few general comments may usefully be inserted here. The presence of these electron-attracting substituents generally increases resistance to bacterial degradation though at least for naphthalenes the substituted compounds can be degraded. There are also significant differences depending on whether or not mechanisms exist for the early elimination of the substituents. Depending on its position, sulfonate may be eliminated at an early stage, and the resulting intermediate is then readily degraded. On the other hand, chlorine substituents are generally retained, and in compounds containing several aromatic rings, the ring with fewest chlorine substituents is degraded first; for example, 1,4-dichloro-naphthalene is degraded to 3,6-dichlorosalicylate (Figure 6.34) (Durham and Stewart 1987) and 2,4,4'-trichlorobiphenyl is degraded to 2,4-dichlorobenzoate (Furukawa et al. 1979). The degradation of polychlorinated biphenyls is discussed further in Section 6.5.1

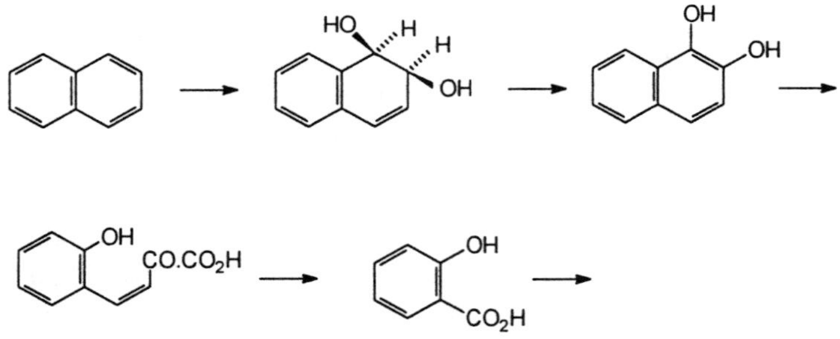

Figure 6.30 Biodegradation of naphthalene by bacteria.

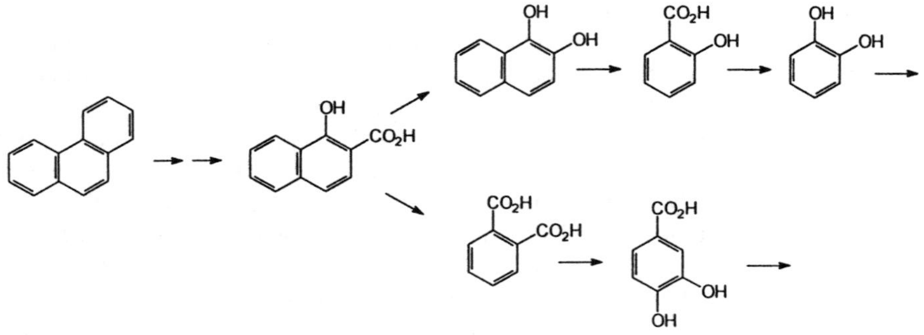

Figure 6.31 Alternative pathways for the biodegradation of phenanthrene.

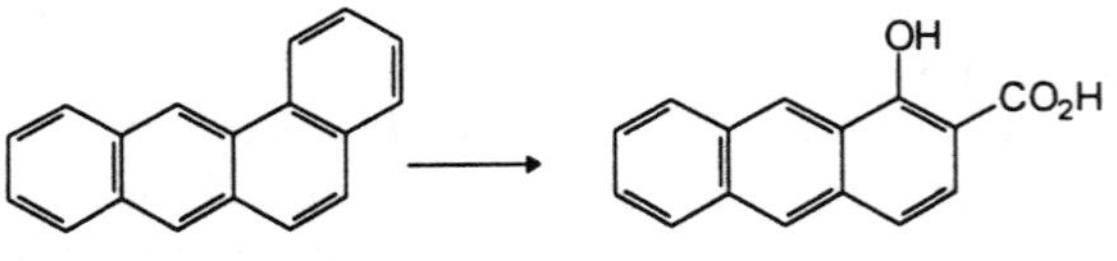

Figure 6.32 Biotransformation of benz[a]anthracene by *Beijerinckia* sp.

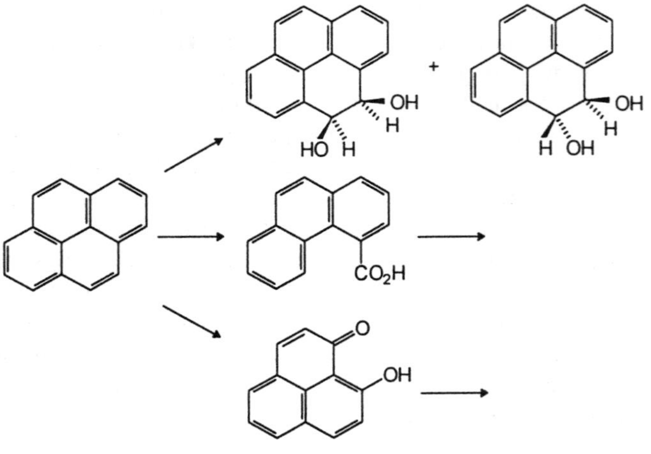

Figure 6.33 Biodegradation of pyrene.

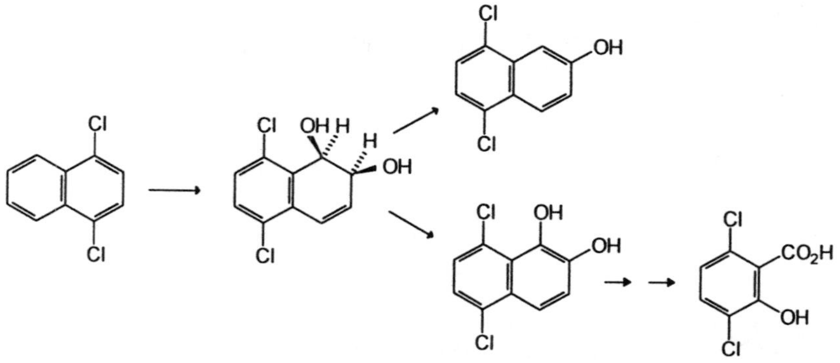

Figure 6.34   Biodegradation of 1,4-dichloronaphthalene.

6.3 AEROBIC DEGRADATION OF HETEROCYCLIC AROMATIC COMPOUNDS

Quite extensive investigations have been directed to the biodegradation of heterocyclic aromatic compounds, since a number of these are constituents of crude oil and creosote (Sundström et al. 1986). Some are used as agrochemicals, and many of them are important chemical intermediates. On the other hand, the polyhalogenated dibenzo-1,4-dioxins and dibenzofurans that have not been deliberately produced may be said to have achieved notoriety.

6.3.1 Reactions Mediated by Bacteria

The chemical reactivity of simple heterocyclic aromatic compounds varies widely; in electrophilic substitution reactions, thiophene is similar to benzene, and pyridine is less reactive than benzene, while furan and pyrrole are susceptible to polymerization reactions. Conversely, pyridine is more readily susceptible than benzene to attack by nucleophilic reagents. These differences are, to a considerable extent, reflected in the susceptibility of these compounds and their benzo analogues to microbial degradation. In contrast to the almost universal dioxygenation reaction used for the bacterial degradation of aromatic hydrocarbons, two broad mechanisms operate for heterocyclic aromatic compounds:

1. Hydroxylation of the ring by reactions formally comparable to those carried out by salicylate hydroxylase (White-Stevens et al. 1972), 4-hydroxybenzoate hydroxylase (Howell et al. 1972), or 2,4-dichlorophenol hydroxylase (Beadle and Smith 1982).
2. Dioxygenation presumably by enzymes formally comparable to those used for the degradation of aromatic hydrocarbons.

These reactions are not mutually exclusive, however, and both may operate in sequence.

Hydroxylation reactions — The degradation of furan-2-carboxylate (Trudgill 1969), thiophene-2-carboxylate (Cripps 1973), and pyrrole-2-carboxylate (Hormann

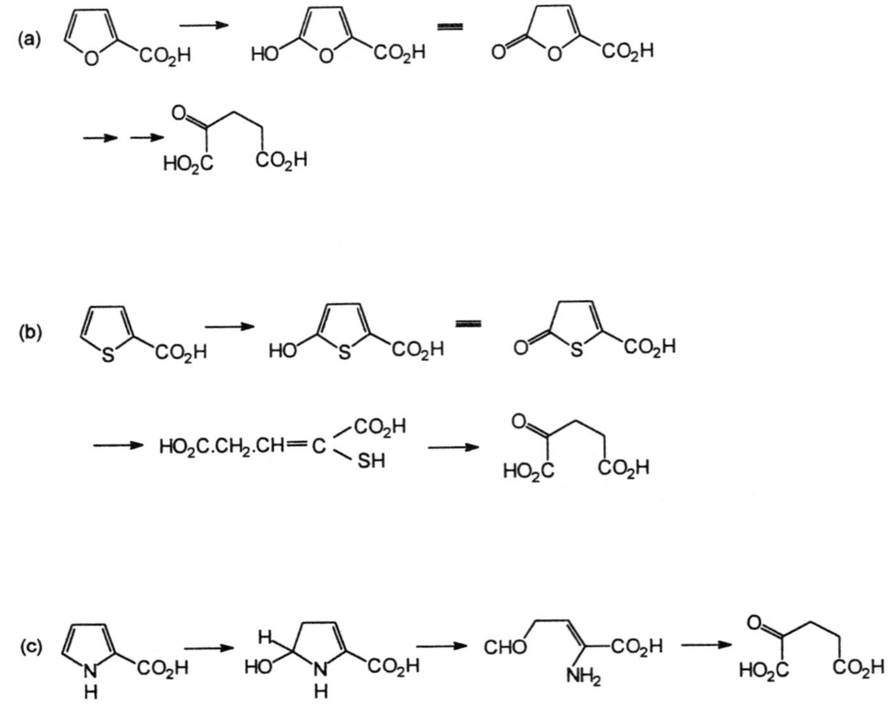

Figure 6.35 Biodegradation of **(a)** furan-2-carboxylate, **(b)** thiophene-2-carboxylate, and **(c)** pyrrole-2-carboxylate.

and Andreesen 1991) proceeds by initial ring hydroxylation; after ring cleavage, 2-ketoglutarate is produced from all of these compounds (Figure 6.35), and this then enters central anabolic and catabolic pathways.

The situation with pyridines is less uniform, and this is consistent with the significant chemical differences between the five- and six-membered N-heterocyclic systems. For pyridines, both hydroxylation, and the dioxygenation typical of truly aromatic compounds have been observed, and these are often accompanied by reduction of one of the double bonds in the pyridine ring. Some examples are given to illustrate the various metabolic possibilities.

1. The degradation of pyridine-4-carboxylate takes place by successive hydroxylations before reduction and ring cleavage (Kretzer and Andreesen 1991) (Figure 6.36).
2. Hydroxylation occurs as the initial step during the degradation of 4-hydroxypyridine and is followed by dioxygenation before ring cleavage (Figure 6.37) (Watson et al. 1974).
3. Degradation of pyridine itself by a species of *Nocardia* takes place by hydroxylation after initial reduction of the ring (Watson and Cain 1975).
4. Degradation of quinoline by pseudomonads and by a strain of *Rhodococcus* sp. has been investigated (Schwarz et al. 1989); the initial reaction was hydroxylation of the heterocyclic ring for all the organisms examined, although subsequent reactions were different for the gram-negative and gram-positive organisms (Figure 6.38a,b).

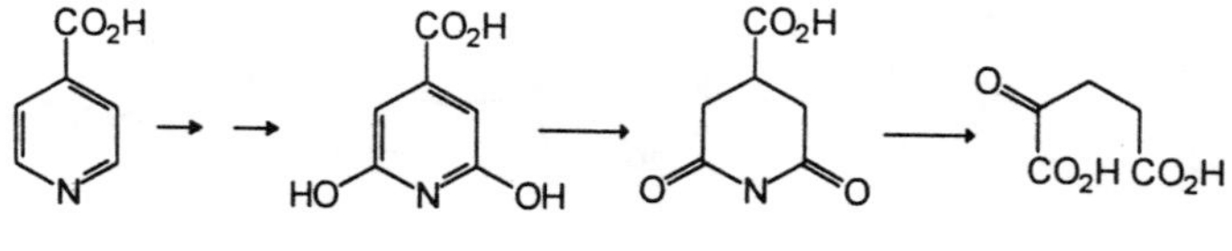

Figure 6.36 Biodegradation of pyridine-4-carboxylate.

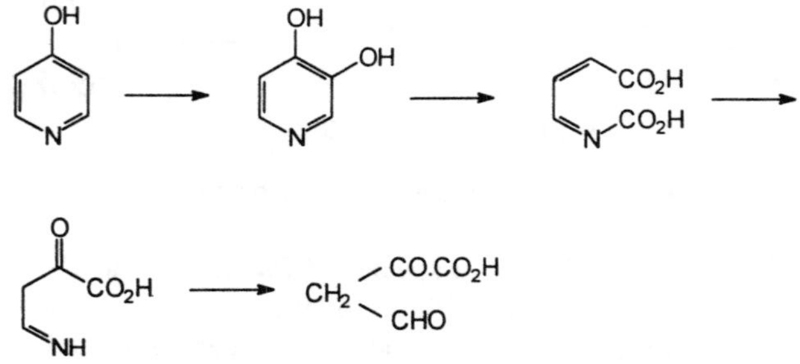

Figure 6.37 Biodegradation of 4-hydroxypyridine.

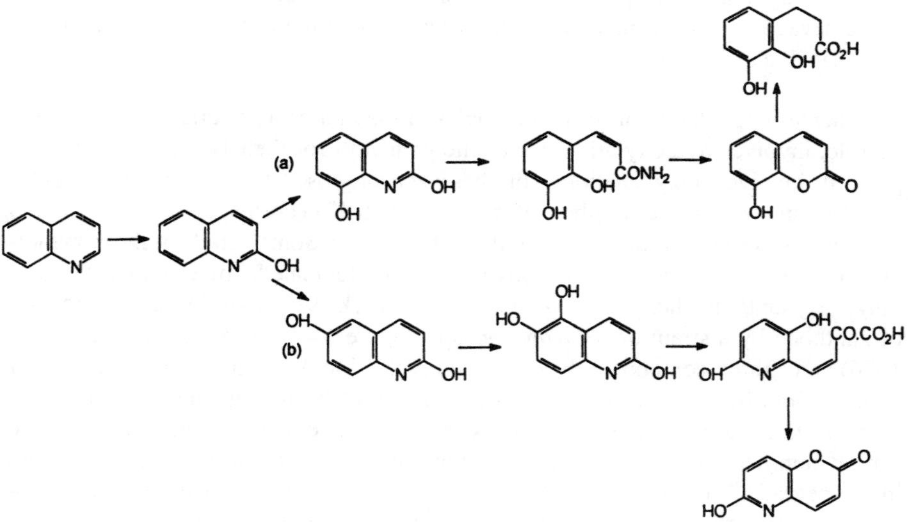

Figure 6.38 Alternative pathways for the biodegradation of quinoline.

It may, therefore, be concluded that hydroxylation of the pyridine ring is a widely distributed metabolic reaction, and this has been confirmed using $H_2{}^{18}O$ during the hydroxylation of both nicotinate and 6-hydroxynicotinate (Hirschberg and Ensign 1971).

Dioxygenation reactions — As noted above, the degradation of pyridines may involve either hydroxylation or dioxygenation reactions — and frequently both of them. Dioxygenation may be illustrated by three examples:

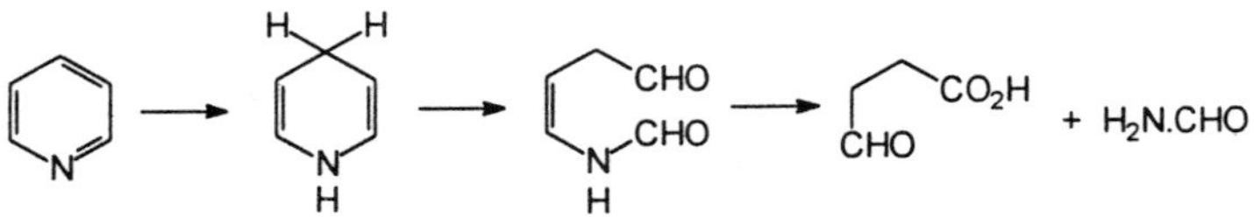

Figure 6.39 Biodegradation of pyridine.

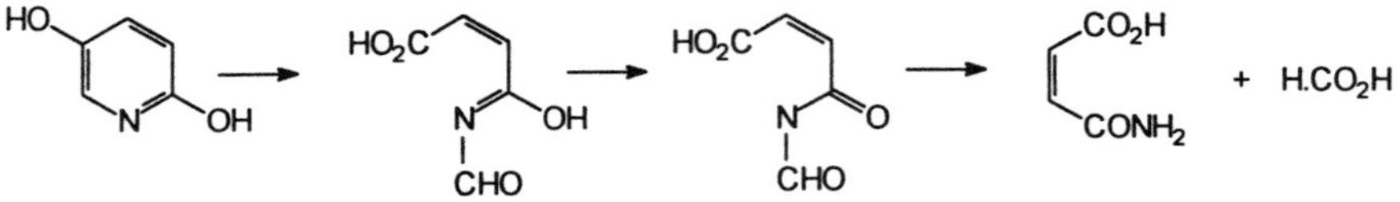

Figure 6.40 Biodegradation of 2,5-dihydroxypyridine.

1. The degradation of pyridine by a species of *Bacillus* (Watson and Cain 1975) produces formate and succinate (Figure 6.39).
2. The degradation of 2,5-dihydroxypyridine has been examined using $^{18}O_2$ and $H_2^{18}O$ (Gauthier and Rittenberg 1971) and shown to involve the incorporation of both oxygen atoms of oxygen, one each into formate and maleamate (Figure 6.40).
3. The degradation of quinol-4-one involves initial hydroxylation at C_3, but ring cleavage is then mediated by a dioxygenase with the formation of anthranilate (Block and Lingens 1992).

Whereas degradation of the carboxylates of the monocyclic furan, thiophene and pyrrole, involves hydroxylation, metabolic pathways for their benzo analogues apparently involve dioxygenation. This has been proposed for indole (Figure 6.41) (Fujioka and Wada 1968), dibenzofuran (Figure 6.42) (Fortnagel et al. 1990), and dibenzo-1,4-dioxin (Figure 6.43) (Wittich et al. 1992). Some details of the enzymology of these dioxygenation reactions have been elucidated; for example, the 3,4-dihydroxyxanthone dioxygenase (Chen and Tomasek 1991) used for the degradation of xanthone by a strain of *Arthrobacter* sp. (Figure 6.44) (Tomasek and Crawford 1986) and a three-component dioxygenase involved in the angular dioxygenation of dibenzofuran by a strain of *Sphingomonas* sp. have been characterized (Bünz and Cook 1993). The pathways for these oxygen heterocyclic compounds require little further comment except to note the occurrence of the angular dioxygenation reaction in the degradation of dibenzofuran. This is supported by the isolation of the 1,10-dihydrodiol of fluorene-9-one during degradation of fluorene by a dibenzofuran-degrading strain of *Brevibacterium* sp.(Engesser et al. 1989) and the overall pathway by the identification of a trihydroxybiphenyl from dibenzofuran before further degradation to salicylate (Strubel et al. 1991). The key enzyme for ring fission of 2,2′,3-trihydrozybiphenyl has been biochemically and genetically analyzed in a strain of *Sphingomonas* sp. (Happe et al. 1993). The oxidation of benzo-1,4-dioxin by cultures of a strain of *Sphingomonas* sp. grown with diphenyl ether is noted in Section 6.9.1.

A greater diversity of pathways for the degradation of dibenzothiophene has emerged, and the following alternatives are worth specific comment:

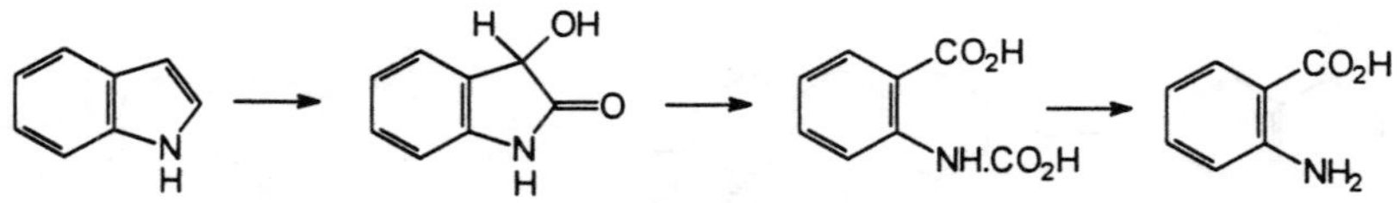

Figure 6.41 Biodegradation of indole.

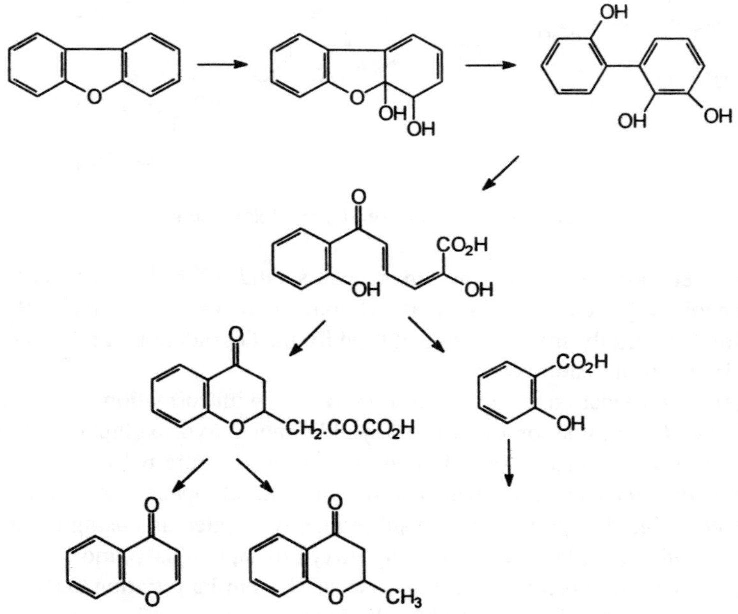

Figure 6.42 Biodegradation of dibenzofuran.

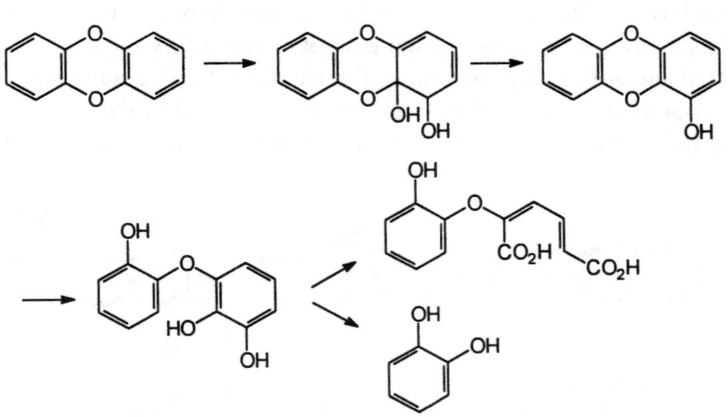

Figure 6.43 Biodegradation of dibenzo-1,4-dioxin.

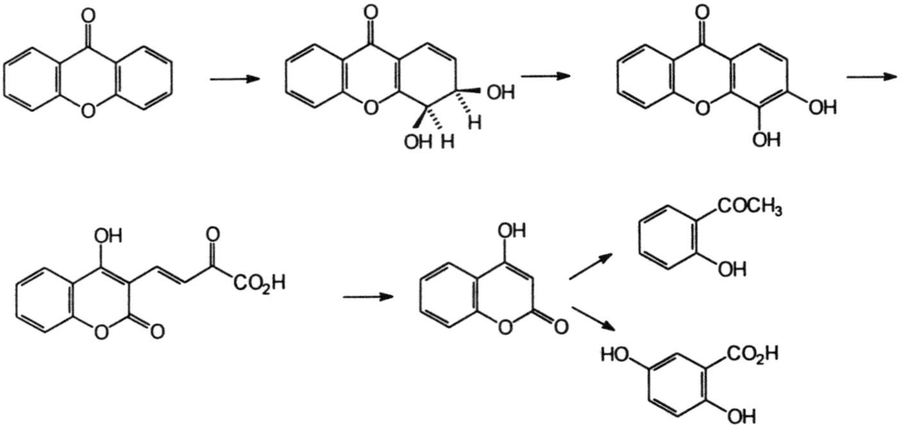

Figure 6.44 Biodegradation of xanthone.

1. Dioxygenation of one of the rings occurs, and after dehydrogenation to a catechol, ring cleavage takes place (Figure 6.45) (Kodama et al. 1973); this pathway is strictly analogous to that used for the degradation of naphthalene and the isoelectronic anthracene.
2. Successive oxidation at the sulfur atom occurs with formation of the sulfoxide followed by elimination of sulfite to yield either 2-hydroxybiphenyl (Omori et al. 1992) or benzoate (van Afferden et al. 1990) (Figure 6.46).
3. The *Corynebacterium* sp. which utilizes dibenzothiophene as a sulfur source produces 2-hydroxybiphenyl and subsequently nitrates this using the nitrate in the growth medium to form two hydroxynitrobiphenyls (Omori et al. 1992) (Figure 6.47a). This reaction is reminiscent of a similar nitration that takes place during the metabolism of 4-chlorobiphenyl (Figure 6.47b) (Sylvestre et al. 1982).

Reductive reactions — In many of the degradations of pyridine that have been noted successive hydroxylation and reduction reactions are involved. A purely reductive pathway is used even by some aerobic organisms for the degradation of hydroxylated and aminated pyrimidines. For example, the degradation of uracil and related compounds that are used as N-sources for the growth of *Pseudomonas stutzeri* takes place by a reductive pathway with the formation of the 4,5-dihydro compounds

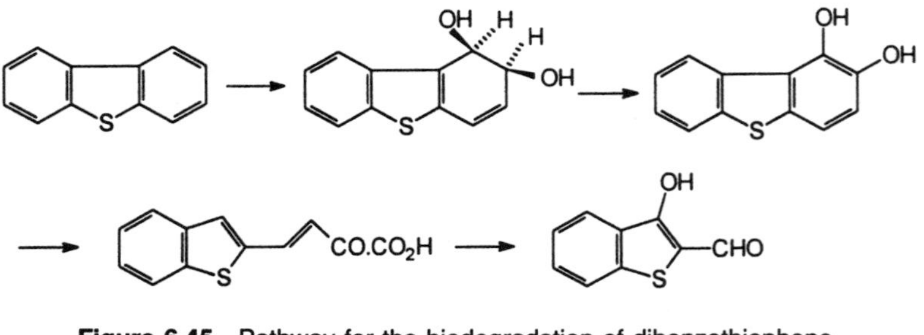

Figure 6.45 Pathway for the biodegradation of dibenzothiophene.

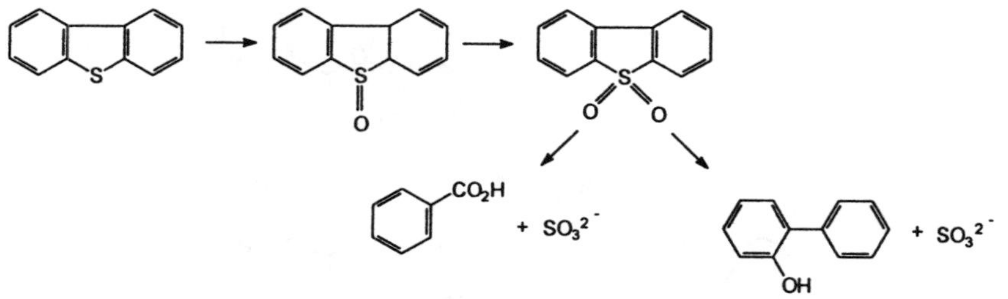

Figure 6.46   Pathway for the biodegradation of dibenzothiophene.

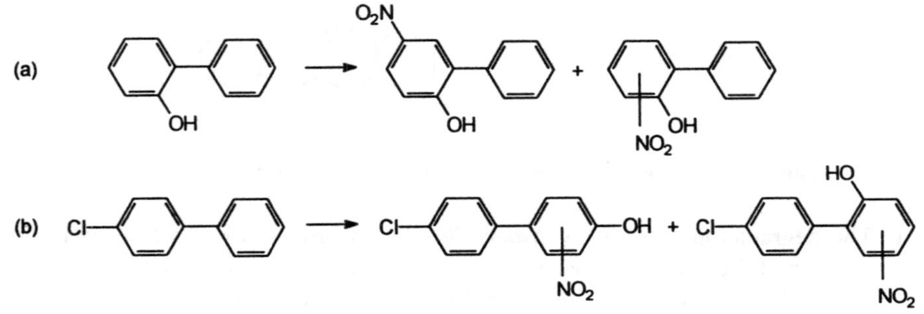

Figure 6.47   Formation of nitrohydroxybiphenyls during metabolism **(a)** of 2-hydroxybiphenyl produced from dibenzothiophene and **(b)** of 4-chlorobiphenyl.

before hydrolytic ring cleavage to form N-carbamoyl-β-alanine (Xu and West 1992). This pathway (Figure 6.89) is also used during pyrimidine degradation by anaerobic clostridia and is discussed again in Section 6.7.3

Hydrolytic reactions resulting in ring cleavage — Derivatives of 1,3,5-triazine are important herbicides so that attention has been directed to their persistence particularly in the terrestrial environment. Some examples have been given in Chapter 5 (Section 5.2) of the utilization of amino and thiol substituents of 1,3,5-triazines as sources of nitrogen and sulfur; subsequent degradation of these hydroxylated metabolites is mediated by hydrolytic reactions which result in the cleavage of the ring with the formation of CO_2 and NH_4^+ (Jutzi et al. 1982).

6.3.2 Reactions Mediated by Fungi

By analogy with their ability for biotransformation of polycylic aromatic hydrocarbons, fungi are capable of carrying out a number of transformations of heterocyclic aromatic compounds though — as for the hydrocarbons — these do not generally result in the total degradation of the substrate. Three reactions of flavanones may be used to represent the broad classes of biotransformation reaction which have been observed: dehydrogenation, hydroxylation, and ring scission (Figure 6.48) (Ibrahim and Abul-Hajj 1990).

A number of other rather unusual reactions have, however, been observed during the degradation of oxygen heterocyclic compounds, and these are worth illustrating:

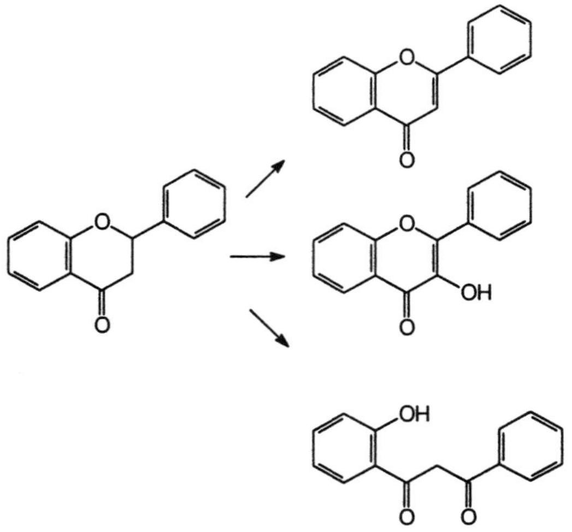

Figure 6.48 Alternative pathways for the biodegradation of flavanones.

1. The degradation of rutin by *Aspergillus flavus* proceeds by hydrolysis to the aglycone followed by release of the unusual metabolite carbon monoxide (Figure 6.49) (Simpson et al. 1963; Krishnamurty and Simpson 1970).
2. Degradation of 5-hydroxy*iso*flavones by strains of *Fusarium* sp. involves initial reduction of the heterocyclic ring before further oxidative fission (Figure 6.50) (Willeke and Barz 1982).
3. Reductive fission of the dihydrofuran ring of 6a-hydroxyinermin is an early step in the total degradation of pisatin by strains of *Fusarium oxysporium* (Figure 6.51) (Fuchs et al. 1980).

In view of current interest in the persistence of recalcitrant organochlorine compounds, it is appropriate to note the degradation of 2,7-dichlorobenzo-1,4-dioxin by *Phanerochaete chrysosporium* (Valli et al. 1992a). Degradation involves dechlorination and ring scission as the initial reactions (Figure 6.52); the products are then degraded by pathways similar to those used for the degradation of 2,4-dichlorophenol by the same organism (Valli and Gold 1991).

6.4 DEGRADATION OF HALOGENATED ALKANES AND ALKENES

6.4.1 Elimination Reactions

By analogy with chemical reactions, the biodegradation of halogenated alkanes may take place by elimination, by nucleophilic displacement, or by reduction; in contrast to chemical reactions, however, the first two do not occur simultaneously, and indeed elimination is relatively uncommon. It is, however, one of the initial reactions in the degradation of DDT by the facultatively anaerobic bacterium *Aerobacter aerogenes* (Figure 6.53) (Wedemeyer 1967), and the recovery of the elimination product (DDE) from environmental samples long after restriction on the

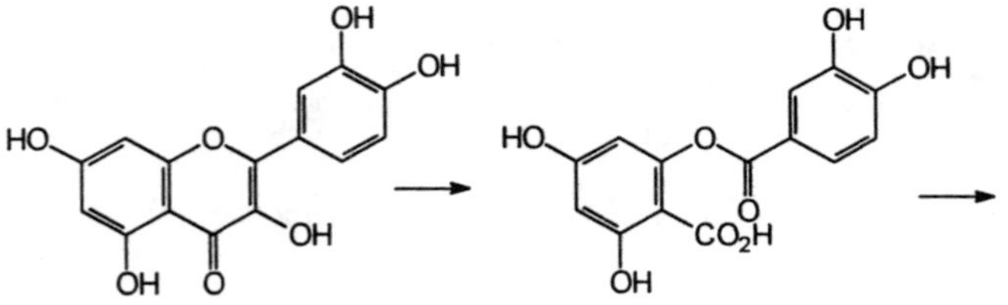

Figure 6.49   Biodegradation of rutin by *Aspergillus flavus*.

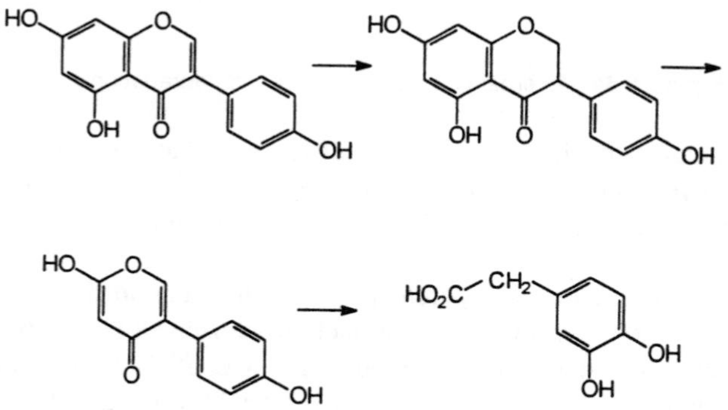

Figure 6.50   Biodegradation of a 5-hydroxy*iso*flavone by *Fusarium* sp.

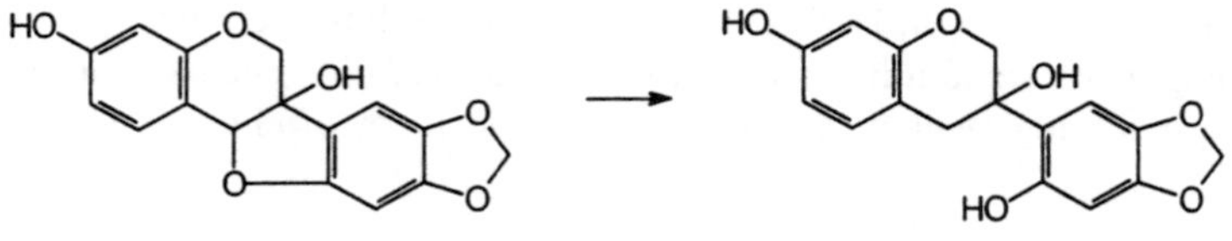

Figure 6.51   Biodegradation of 6a-hydroxyinermin by *Fusarium oxysporium*.

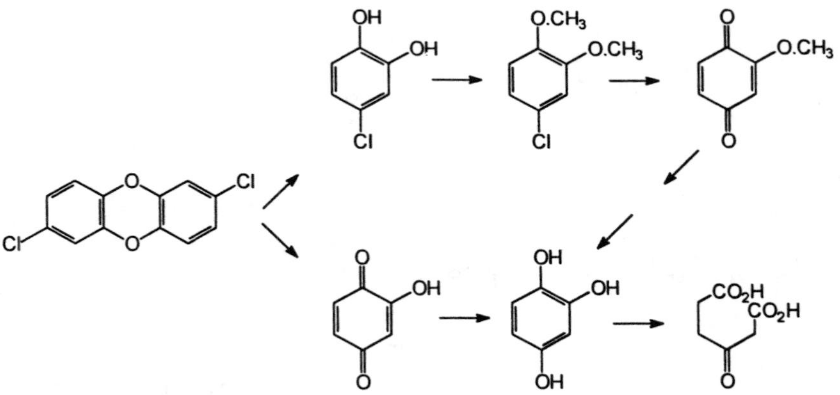

Figure 6.52   Biodegradation of 2,7-dichlorodibenzo-1,4-dioxin.

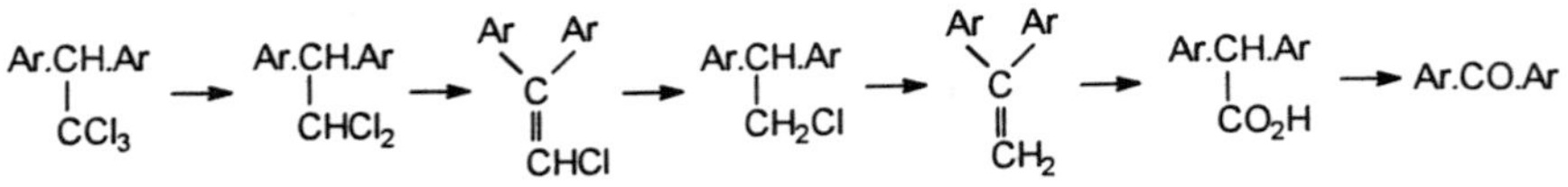

Ar.CH.Ar → Ar.CH.Ar → Ar.CH.Ar → Ar.CH.Ar → Ar.CH.Ar → Ar.CO.Ar
| | | |
CCl₃   CHCl₂   CH₂Cl   CO₂H

Figure 6.53 Biodegradation of DDT by *Aerobacter aerogenes*. Ar = 4-chlorophenyl.

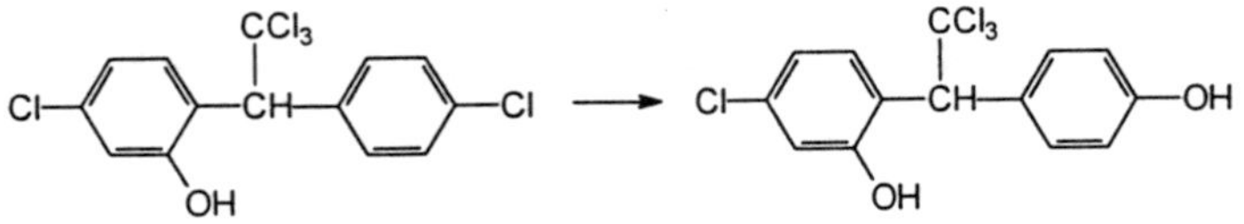

Figure 6.54 Alternative pathway for the biodegradation of DDT.

use of DDT suggests the high degree of persistence of that metabolite. It should also be noted, however, that degradation of DDT may also occur by hydroxylation of the ring and displacement of the aromatic ring chlorine atom by hydroxyl (Figure 6.54) (Massé et al. 1989).

An elimination reaction is apparently one of the steps in the degradation of γ-hexachlorocyclohexane from which pentachlorobenzene (Tu 1976) or γ-2,3,4,6-tetrachlorocyclohex-1-ene may be formed (Jagnow et al. 1977) (Figure 6.55a). The formation of both 2,5-dichlorophenol and 2,4,5-trichlorophenol during the aerobic degradation of γ-hexachlorocyclohexane by *Pseudomonas paucimobilis* presumably occurs by comparable elimination reactions (Sendoo and Wada 1989), and further details of the transformation that produces also 1,2,4-trichlorobenzene have been provided (Nagasawa et al. 1993a). The degradation is, however, more complex since these compounds are terminal metabolites, and degradation occurs by a sequence of reactions involving sequential hydrolytic displacement of chloride (Figure 6.55b)

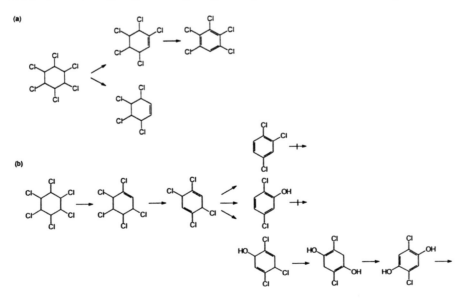

Figure 6.55 Pathway for the **(a)** biotransformation of γ-hexachlorocyclohexane and **(b)** biodegradation of γ-hexachlorocyclohexane.

$$Br.CH_2.CH_2.Br \longrightarrow CH_2 = CH_2 \quad ; \quad Br.CH = CH.Br \longrightarrow CH \equiv CH$$

Figure 6.56 Metabolism of 1,2-dibromoethane.

$$Cl.CH_2.CH_2.Cl \longrightarrow Cl.CH_2.CH_2OH \longrightarrow Cl.CH_2CHO \longrightarrow Cl.CH_2.CO_2H \longrightarrow HOCH_2.CO_2H$$

Figure 6.57 Biodegradation of 1,2-dichloroethane.

(Nagasawa et al. 1993b). The hydrolase has been cloned, and in a constructed strain of *E. coli* that overexpresses the dehalogenase gene, a range of chloroalkanes was dechlorinated (Nagata et al. 1993). Since the elimination reactions are not themselves dependent on the presence of oxygen, they may occur under anaerobic conditions. For example, pure cultures of strictly anaerobic methanogenic bacteria produce ethene from 1,2-dibromoethane and ethyne from 1,2-dibromoethane (Figure 6.56) (Belay and Daniels 1987).

6.4.2 Hydrolytic Reactions

Displacement of halogen by hydroxyl is a widely distributed reaction in the degradation of haloalkanes and haloalkanoates. An apparently simple pathway involving two displacement steps is illustrated in Figure 6.57, but it should be emphasized that the enzymology of hydrolytic dehalogenation is much more complex than might appear; for example, two different dehalogenases are involved in the dechlorination of 1,2-dichloroethane and of chloroacetate (van den Wijngaard et al. 1992). Indeed, a number of distinct dehalogenases exist, and they may differ significantly in their substrate specificity (Scholtz et al. 1988; Sallis et al. 1990) in respect to chain length and the influence of halogen atoms at the ω-position. The degradation of 2,2-dichloropropionate involves dehalogenation to pyruvate, but even here two different dehalogenases are synthesized (Allison et. al. 1983). A strain of *Pseudomonas* sp. is able to degrade a wide range of 1-bromoalkanes having 6 to 10 carbon atoms (Shochat et al. 1993), though the range of unsubstituted or chloroalkanes that is degraded is quite restricted.

It should also be pointed out that although all these reactions are formally simple nucleophilic displacements of the chlorine atoms by hydroxyl groups, a different mechanism clearly operates in the degradation of dichloromethane by *Hyphomicrobium* sp. since the enzyme is glutathione-dependent (Stucki et al. 1981) and the reaction presumably involves at least two steps.

6.4.3 Monooxygenase Systems

Degradation of halogenated alkenes by direct displacement of halogen is not expected on purely chemical grounds, although this reaction apparently occurs during the degradation of vinyl chloride by a strain of *Pseudomonas* sp. that carries out direct hydrolysis to acetaldehyde followed by mineralization to CO_2 (Castro et al. 1992b). Although the degradation of 3-chlorocrotonate could take place by a comparable reaction, it seems more plausible that degradation involves initial addition of the elements of water to the α,β-unsaturated ketone followed by elimination of chloride (Figure 6.58) (Kohler-Staub and Kohler 1989).

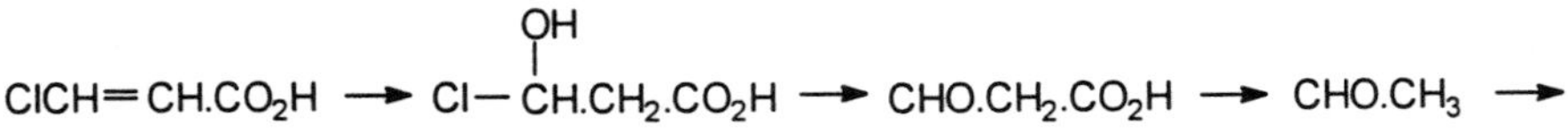

$$CICH=CH.CO_2H \longrightarrow CI-\overset{\overset{\displaystyle OH}{|}}{CH}.CH_2.CO_2H \longrightarrow CHO.CH_2.CO_2H \longrightarrow CHO.CH_3 \longrightarrow$$

Figure 6.58 Biodegradation of 3-chlorocrotonic acid.

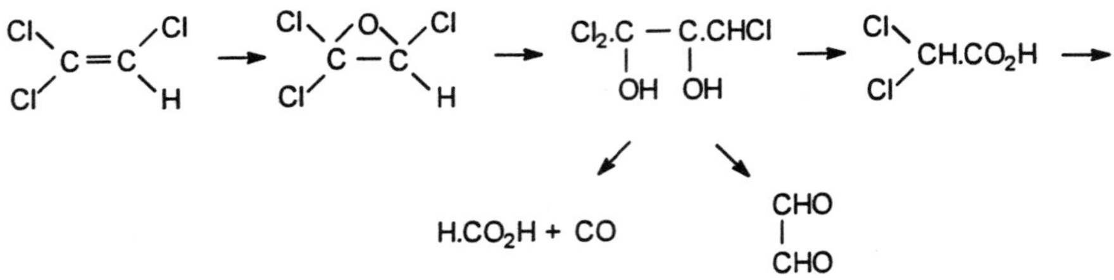

Figure 6.59 Biodegradation of trichloroethene.

Unactivated halogenated alkenes may be degraded by a completely different pathway that involves a monooxygenase system. Attention has already been drawn to the remarkable spectrum of structures which are amenable to attack by the monooxygenase system of methanotrophic and methylotrophic bacteria; the degradation of trichloroethene provides a good example of the operation of this system (Figure 6.59) (Little et al. 1988). This degradation has also been used in Chapter 4 (Section 4.7) to illustrate the problem presented by the synthesis of the toxic metabolite carbon monoxide. In addition, 2,2,2-trichloroacetaldehyde is produced during oxidation of trichloroethene by several methanotrophs and undergoes a dismutation to form trichloroethanol and trichloroacetate (Newman and Wackett 1991); at least formally, the transformation of trichloroethene to 2,2,2-trichloroacetaldehyde is analogous to an NIH shift (Section 6.2.2). As might be expected, a number of haloalkanes including dichloromethane, chloroform, 1,1-dichloroethane, and 1,2-dichloroethane may also be degraded by the soluble methane monooxygenase system of *Methylosinus trichosporium* (Oldenhius et al. 1989), though *cis* and *trans* 1,2-dichloroethene formed stable epoxides, and tetrachloroethene was resistant to attack. The aerobic degradation of vinyl chloride by *Mycobacterium aurum* also proceeds by initial formation of an epoxide mediated by an alkene monooxygenase (Hartmans and de Bont 1992), and this reaction has also been demonstrated to occur with *Methylosinus trichosporum* even though subsequent conversion to glycollate involves purely chemical reactions (Castro et al. 1992a).

An interesting observation has been made with camphor-grown cells of a strain of *Pseudomonas putida* which putatively contain cytochrome P-450 monooxygenase; trichloroethane was degraded by a dominant aerobic pathway to chloroacetate and glyoxylate and *simultaneously* by a minor reductive reaction which must also involve an elimination reaction with the formation of vinyl chloride (Figure 6.60) (Castro and Belser 1990).

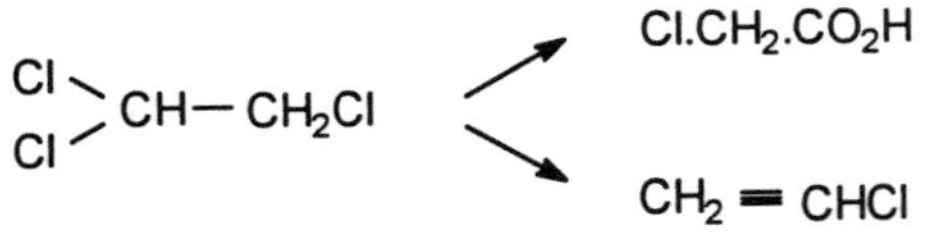

Figure 6.60 Biodegradation of trichloroethane by *Pseudomonas putida*.

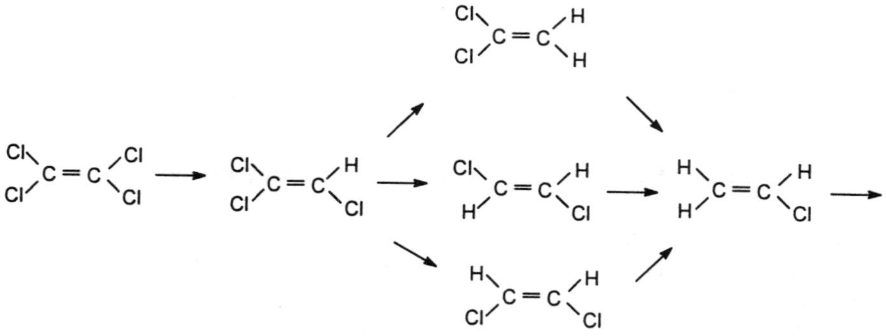

$$Cl_2C=CH_2$$

$$Cl_2C=CCl_2 \longrightarrow Cl_2C=CHCl \longrightarrow ClHC=CHCl \longrightarrow H_2C=CHCl \longrightarrow$$

$$ClHC=CHCl$$

Figure 6.61 Anaerobic dechlorination of tetrachloroethene.

6.4.4 Reductive Dehalogenation Reactions of Halogenated Aliphatic and Alicyclic Compounds

Considerable effort has been devoted to the anaerobic transformation of polychlorinated C_1 alkanes and C_2 alkenes in view of their extensive use as industrial solvents and their identification as widely distributed groundwater contaminants. Early experiments which showed that tetrachloroethene was transformed into vinyl chloride (Vogel and McCarty 1985) (Figure 6.61) aroused concern, though it has now been shown that complete dechlorination can occur under some conditions. A number of investigations have examined the persistence of tetrachloroethene under anaerobic conditions with the result that widely differing pathways for biotransformation have emerged. Under methanogenic conditions, a strain of *Methanosarcina* sp. transformed this into trichloroethene (Fathepure and Boyd 1988), though in the presence of suitable electron donors such as methanol, complete reduction of tetrachloroethene to ethene may be achieved in spite of the fact that the dechlorination of vinyl chloride appeared to be the rate-limiting step (Freedman and Gossett 1989). Complete dechlorination of high concentrations of tetrachloroethene in the absence of methanogenesis and with methanol as electron donor has been achieved (DiStefano et al. 1991), though these conditions will clearly seldom occur in natural environments. Reductive transformation to the fully reduced ethane has also been observed in a fixed-bed reactor (De Bruin et al. 1992). Some of the unresolved issues are illustrated by the different effects of electron donors on the partial dechlorination of tetrachloroethene in anaerobic microcosms (Gibson and Sewell 1992).

An organism that is able to use methyl chloride as energy source and converts this into acetate has been isolated (Traunecker et al. 1991), although it appears that different enzyme systems are involved in the dehalogenation of this substrate and in the *O*-demethylation of methoxylated aromatic compounds such as vanillate (Meßmer et al. 1993); further comment on these *O*-demethylation reactions by acetogenic bacteria is given in Section 6.7.3. Cultures of a number of anaerobic bacteria are able to dechlorinate tetrachloromethane, and *Acetobacterium woodii* formed dichloromethane as the final chlorinated metabolite by successive dechlorination, although CO_2 was also produced by an unknown mechanism (Egli et al. 1988). By formally similar dechlorinations, a strain of *Clostridium* sp. transformed 1,1,1-trichloroethane to 1,1-dichloroethane and tetrachloromethane successively to trichloromethane and dichloromethane (Gälli and McCarty 1989).

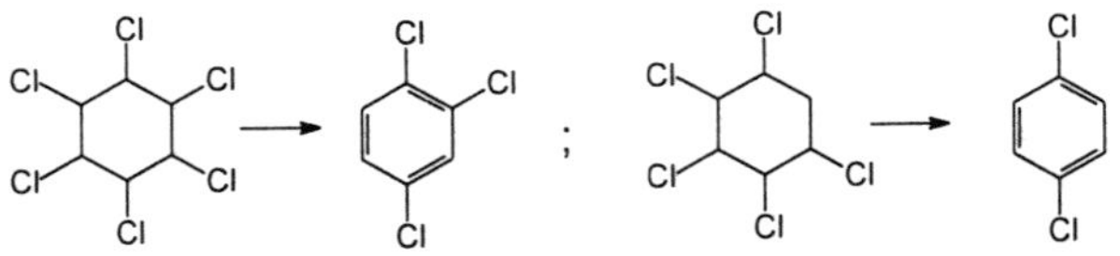

Figure 6.62 Biotransformation of γ-hexachlorocyclohexene.

Reductive dechlorination in combination with the elimination of chloride has been demonstrated in a strain of *Clostridium rectum* (Ohisa et al. 1982); γ-hexachloro-cyclohexene formed 1,2,4-trichlorobenzene, and γ-1,3,4,5,6-pentachlorocyclohexene formed 1,4-dichlorobenzene (Figure 6.62). It was suggested that this reductive dechlorination is coupled to the synthesis of ATP, and this possibility has been clearly demonstrated during the dehalogenation of 3-chlorobenzoate coupled to the oxidation of formate in *Desulfomonile tiedjei* (Mohn and Tiedje 1991). Combined reduction and elimination has also been demonstrated in methanogenic cultures that transform 1,2-dibromoethane to ethene and 1,2-dibromoethene to ethyne (Belay and Daniels 1987).

From an environmental point of view, it therefore appears that the complete degradation of chlorinated alkenes and alkanes will often require the operation of both anaerobic and aerobic steps; for example, partial or complete dehalogenation may occur under anaerobic conditions, and aerobic degradation of the partially dechlorinated metabolites such as dichloromethane (La Roche and Leisinger 1991) and vinyl chloride (Castro et al. 1992a, 1992b; Hartmans and de Bont 1992) may then subsequently take place.

A mechanistic basis for such dechlorinations has been formulated using the results of experiments with 1,2-dichloroethane which is transformed to ethene and chloroethane by cell suspensions of methanogenic bacteria. Further investigations showed that cobalamin, or factor F_{430}, or boiled cell extracts from *Methanosarcina barkeri* catalyzed this transformation in the presence of Ti(III) citrate as electron donor (Holliger et al. 1992a). In addition, crude cell extracts from *Methanobacterium thermoautotrophicum* carried out the same transformation using H_2 as electron donor (Holliger et al. 1992b). These observations have been incorporated into a model scheme for the dechlorination reaction, and it is reasonable to assume that these reactions and the purely abiotic reactions outlined in Chapter 4 (Section 4.1.2) have a common mechanistic basis.

6.5 AEROBIC DEGRADATION OF HALOGENATED AROMATIC COMPOUNDS

6.5.1 Bacterial Systems

On account of their potentially adverse environmental consequences including their toxicity and their potential for concentration in biota, enormous activity has been devoted to studies on the biodegradation of halogenated — and particularly chlorinated — aromatic compounds including hydrocarbons, phenols, anilines, and carboxylic acids. Many of these are constituents of valuable commercial products or are important intermediates in the synthesis of agrochemicals; in addition, a range of structurally diverse chlorinated aromatic compounds is produced during production of bleached pulp using conventional technologies with molecular chlorine (refs in Neilson et al. 1991). Attention is drawn to reviews by one of the major contributors to this field (Reineke 1984, 1988).

Halogenated aromatic hydrocarbons — Ultimately, ring cleavage of halogenated aromatic hydrocarbons must occur if degradation — rather than merely biotransformation — is to be accomplished. The pathways outlined in Section 6.2.3 for unsubstituted hydrocarbons are generally followed, but some additional issues may be illustrated by the following examples.

1. The overall pathway via a *cis* dihydrodiol followed by dehydrogenation to the catechol and ring cleavage is generally followed and has already been illustrated with 1,4-dichloronaphthalene. A strain of *Pseudomonas* sp. able to degrade a range of polychlorinated biphenyl congeners had both 2,3-dioxygenase and 3,4-dioxygenase activity, and four of the open reading frames were homologous to components of toluene dioxygenase (Erickson and Mondelo 1992). At least in gram-positive organisms that degrade PCBs by the 2,3-dioxygenase system, there are, however, several non-homologous enzymes with a narrow substrate specificity (Asturias and Timmis 1993). Whereas, however, this dioxygenation pathway is used for the degradation of 1,2,4-trichlorobenzene (Figure 6.63), a chlorine atom is eliminated during production of the chlorocatechol during degradation of 1,2,4,5-tetrachlorobenzene (Figure 6.64) (Sander et al. 1991).

2. Attention has already been briefly drawn to the fact that where alternatives exist, the ring with fewer halogen substituents is degraded first. This has already been illustrated for the degradation of 1,4-dichloronaphthalene and is much more extensively supported by the results from experiments that examined a large number of PCB congeners in which the less highly substituted ring is almost invariably degraded to a carboxylic acid apparently by 2:3-cleavage of the intermediate catechol; for example, in the 2,4,5-, 2,5,2'-, and 2,3,4,5- congeners (Figure 6.65). On the other hand, care should be exercised in attempting to promulgate rules of widespread applicability to PCBs since both steric and electronic factors are important determinants of biodegradability; congeners with substituents at the 2- or 6-positions of either or both rings are apparently less susceptible to degradation than other isomers (Furukawa et al. 1979). Further complexities are illustrated by studies on the degradation of 2,4'-dichloro- and 4,4'-dichlorobiphenyl in which both dehalogenation and dioxygenation occur (Ahmad et al. 1991). All of these results clearly show that it is not merely the number of substituents but also their orientation that are determinative factors.

3. The ultimate products are generally chlorinated muconic acids from monocyclic aromatic compounds and chlorobenzoic acids from PCBs. The unusual metabolite 2,4,5-trichloroacetophenone has, however, been isolated (Bedard et al. 1987) from the degradation of 2,4,5,2',4',5'-hexachlorobiphenyl (Figure 6.66) by *Alcaligenes eutrophus* H850 which has an usually wide spectrum of degradative activity for PCB congeners. The essential reaction in its formation seems analogous to that in which acetophenone is produced from cinnamate by a pseudomonad

(Hilton and Cain 1990); this presumably involves addition of the elements of water to the α,β-unsaturated C=C bond, followed by oxidation and decarboxylation. The metabolism of chloroacetophenones (Havel and Reineke 1993) has been discussed in Section 6.2.1.

Halogenated aromatic compounds carrying additional substituents — The metabolism of a wide range of halogenated aromatic compounds bearing additional

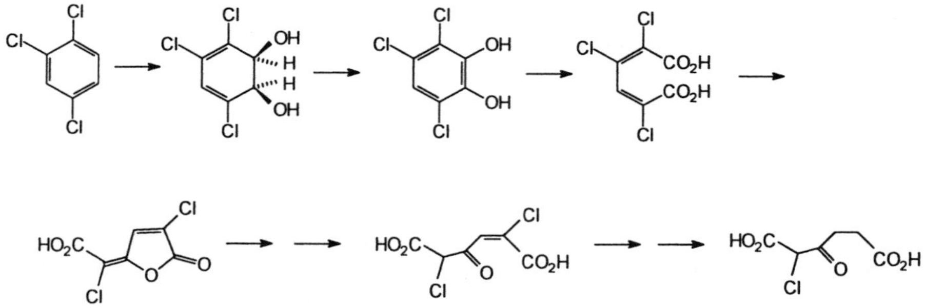

Figure 6.63 Biodegradation of 1,2,4-trichlorobenzene.

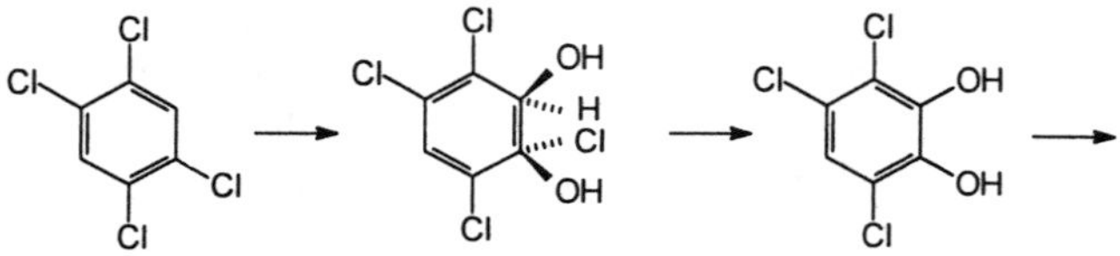

Figure 6.64 Biodegradation of 1,2,4,5-tetrachlorobenzene.

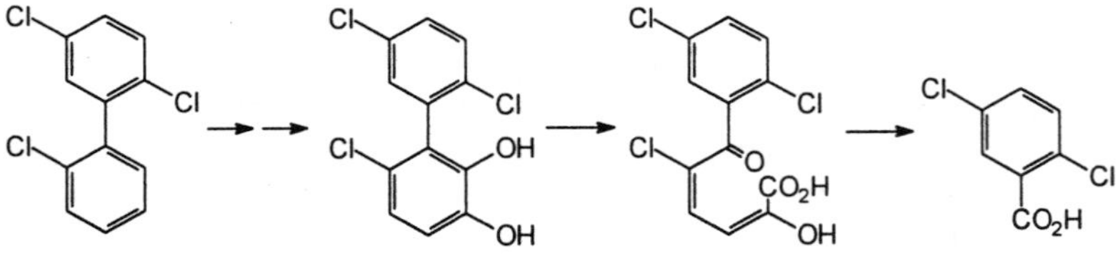

Figure 6.65 Biodegradation of 2,2′,5-trichlorobiphenyl.

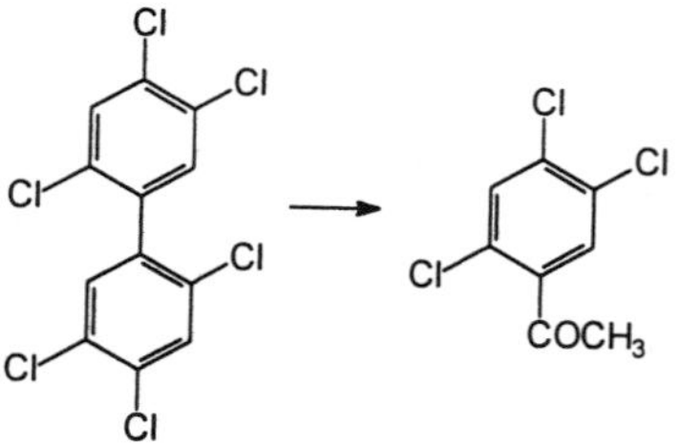

Figure 6.66 Biodegradation of 2,4,5,2′,4′,5′-hexachlorobiphenyl.

substituents such as hydroxyl or carboxyl groups proceeds by two distinct pathways differing on whether or not halogen is removed from the ring before disrupture. These possibilities may be defined more specifically:

1. Direct oxidation of the ring may be mediated by dioxygenation or hydroxylation followed by dehydrogenation and disruption of the ring; halogen is subsequently eliminated during further degradation of the acyclic product.
2. Direct elimination of halide from the ring may occur by the hydrolytic or reductive displacement of halogen; the products then undergo oxidation followed by disruption of the aromatic ring.

Halogen elimination after ring cleavage — There are a number of important details which merit fuller discussion.

1. Mechanisms of ring oxidation — In those cases where the loss of halogen takes place after ring cleavage, halogenated phenols with three or less halogen substituents are converted into halogenated catechols by hydroxylase systems (Beadle and Smith 1982), whereas all other aromatic compounds such as halogenated carboxylic acids, halogenated amines, and most halogenated hydrocarbons are converted by dioxygenation initially into *cis*-dihydrodiols before dehydrogenation to halogenated catechols which therefore occupy a central position in the metabolism of all these compounds (Figure 6.67). Hydroxylation and dioxygenation are not, however, mutually exclusive since the toluene dioxygenase from *Pseudomonas putida* F1 hydroxylates both phenol and 2,5-dichlorophenol with the introduction of only one atom of oxygen (Spain et al. 1989). In addition, dioxygenation may also result in the elimination of halogen. This has been illustrated for 1,2,4,5-tetrachlorobenzene (Figure 6.64) 6.44 (Sander et al. 1991) and apparently also occurs during the degradation of both 2,3-dichloro- and 2,5-dichlorobenzoate (Hickey and Focht 1990) so that a chlorine atom is lost before production of the chlorocatechols (Figure 6.68). Subsequent degradation follows established pathways in which chloride is eliminated from muconic acids.
2. Mechanisms for the ring-cleavage of chlorocatechols — Whereas the degradation of unsubstituted catechol may proceed as illustrated previously by formation of muconate, this represents only one of the possibilities: the reaction in which the bond between the two oxygen-bearing atoms is broken with the formation of a dicarboxylic acid has been termed the *ortho, endo*, intradiol, or 1,2-cleavage — of which the last two seem most descriptive and appropriate since the enzyme carrying out this reaction is designated as a catechol-1,2-dioxygenase. Alternatively, the bond between one of the oxygen-bearing atoms and the adjacent unsubstituted atom may be broken with the formation of a monocarboxylic acid and an aldehyde, and by analogy has been designated the *meta, exo*, extradiol, or 2,3-cleavage. The two pathways are compared in Figure 6.69; which of them is followed depends both on the organism and on the substrate that is being metabolized. In addition, there are differences in the details of the 1,2-pathway between prokaryotic and eukaryotic cells, and these refinements have been discussed in detail (Cain et al. 1968; Cain 1988).

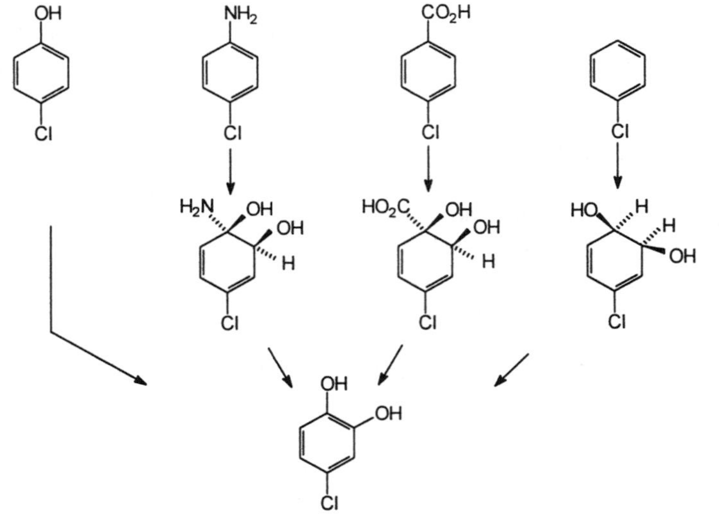

Figure 6.67 Convergence of the pathways for the biodegradation of 4-chlorophenol, 4-chloroaniline, 4-chlorobenzoate, and chlorobenzene.

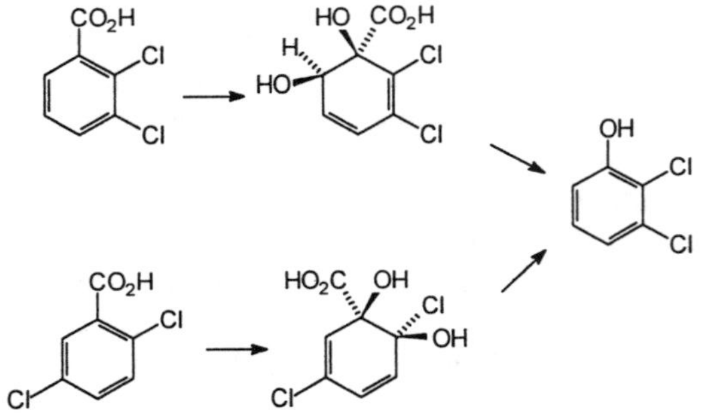

Figure 6.68 Biodegradation of dichlorobenzoates.

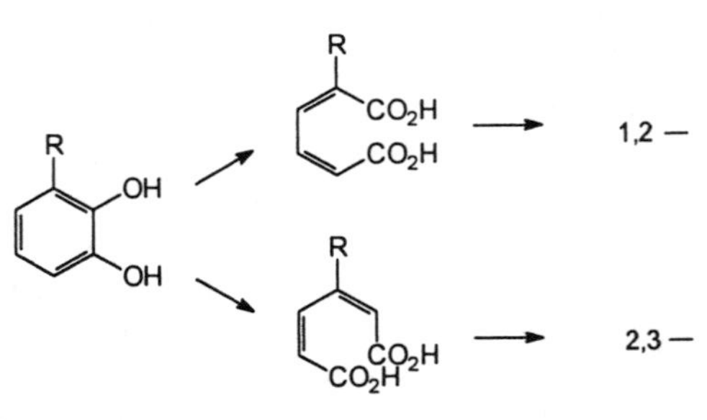

Figure 6.69 Ring-cleavage pathways for the biodegradation of substituted catechols.

Which of the pathways is followed may be critical in determining the degradability of halogenated compounds, since one or other of them may result in the synthesis of toxic metabolites. In the example given in Figure 6.70, the 1,2-pathway is used instead of the 2,3-pathway that would result in the synthesis of the toxic acyl chloride

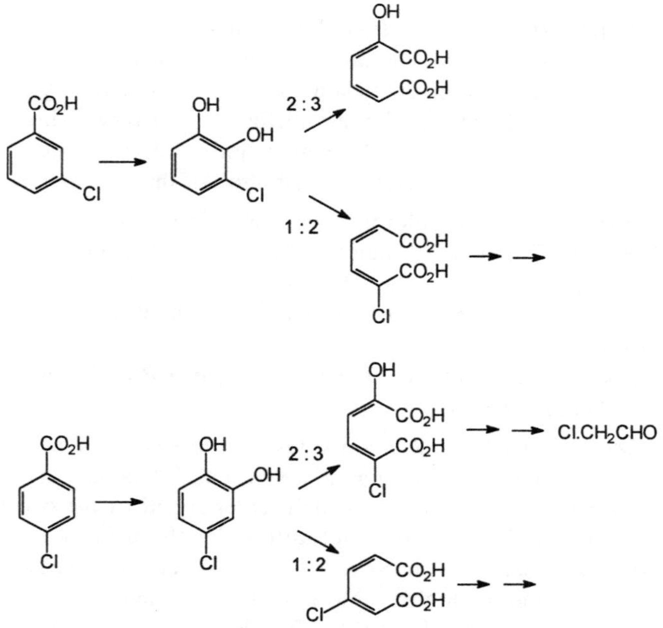

Figure 6.70 Ring-cleavage pathways for the biodegradation of chlorocatechols formed from chlorobenzoates.

from 3-chlorobenzoate or chloroacetaldehyde from 4-chlorobenzoate. There is an additional problem which has important implications for biotechnology: the situation in which two substrates such as a chlorinated and an alkylated aromatic compound are simultaneously present. The 2,3-pathway is generally used for the degradation of alkylbenzenes (Figure 6.71), but this may be incompatible with the degradation of chlorinated aromatic compounds since the 3-chlorocatechol produced inhibits the activity of the catechol-2,3-oxygenase (Klecka and Gibson 1981; Bartels et al. 1984). There do exist, however, mutant strains which have successfully reconciled this incompatibility (Taeger et al. 1988; Pettigrew et al. 1991).

Loss of halogen before ring cleavage — An attempt is made to summarize and illustrate the various pathways that may be used for the elimination of halogen from intact aromatic rings.

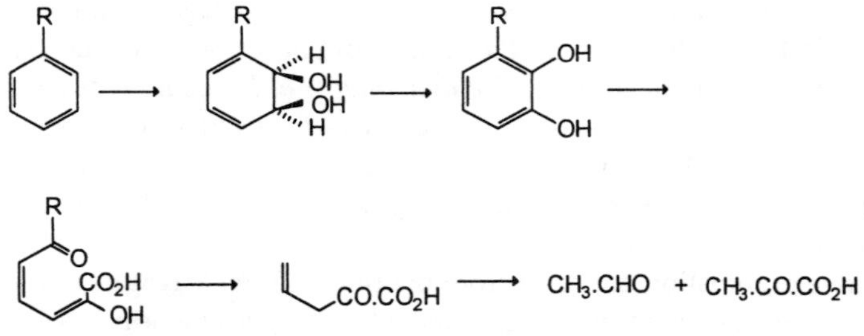

Figure 6.71 Biodegradation of alkylbenzenes.

1. Hydrolytic elimination of halide from the ring before cleavage — In contrast to the situation for monochloro- and dichlorophenols, phenols with three or more chlorine substituents undergo degradation by pathways involving displacement of the chlorine atoms by hydroxyl groups before ring cleavage. This reaction has been shown to be the initial step in the degradation of a number of chlorophenols, and a few examples may be given as illustration of this alternative:

   • The degradation of pentachlorophenol in *Rhodococcus chlorophenolicus* (Apajalahti and Salkinoja-Salonen 1987b) and in a strain of *Flavobacterium* sp. (Steiert and Crawford 1986) (Figure 6.72).
   • The degradation of 2,4,6-trichlorophenol by a strain of *Azotobacter* sp. (Li et al. 1991).
   • The degradation of 2,4,5-trichlorophenoxyacetate by *Pseudomonas cepacia* (Haugland et al. 1990).

   It has been shown that the hydroxylase (hydrolase) is a monooxygenase requiring O_2 in both *Flavobacterium* sp. (Xun et al. 1992a) and in *Rhodococcus chlorophenolicus* (Uotila et al. 1992); in the latter organism, the system converting the initial hydrolysis product tetrachlorohydroquinone to 1,2,4-trihydroxybenzene does not, however, require O_2. The occurrence of this hydroxylation reaction in highly chlorinated compounds is consistent with mechanistic arguments that this reaction would be favored by the low electron density on the aromatic rings created by strongly electron-withdrawing substituents. This is supported by two sets of observations:

   • The pentachlorophenol hydroxylase from *Flavobacterium* sp. (Xun et al. 1992b) is able to hydroxylate a range of other substituted phenols with elimination of, for example, chloride, bromide, iodide, cyanide, and nitrite from the 4-position.
   • Loss of fluoride from pentafluorophenol and bromide from pentabromophenol is catalyzed by the hydroxylase from *Rhodococcus chlorophenolicus* (Uotila et al. 1992).
   • The metabolism of 1,3,5-triazine herbicides containing chlorine substituents also involves hydrolytic displacement of chloride (Cook 1987). The inducible enzyme has been purified from *Rhodococcus corallinus* (Mulby et al. 1994) and is also capable of carrying out deamination of amino-substituted 1,3,5-triazines.

   These investigations have dealt exclusively with chlorine compounds, and the conclusions may not necessarily be extrapolated to bromine compounds. For example, whereas a strain of *Pseudomonas aeruginosa* degrades 2-bromobenzoate by a hydrolytic pathway with the initial formation of salicylate (Higson and Focht 1990), a mutant strain of *Pseudomonas putida* accumulates 4-bromo-*cis*-2,3-dihydroxy-cyclohexa-4,6-diene-1-carboxylate from 4-bromobenzoate (Taylor et al. 1987). It would, therefore, seem premature to draw any general conclusion from these results which may depend both on the fact that bromine is involved and on its position on the ring.

2. Reductive elimination of halide from the ring before cleavage — In addition to hydrolytic replacement of halogen, reductive displacement has been shown to occur during the degradation of a few aromatic compounds; for example, pathways for the degradation of pentachlorophenol (Apajalahti and Salkinoja-Salonen

1987b) (Figure 6.72) and of 2,4-dichlorobenzoate (van den Tweel et al. 1987) (Figure 6.73) involve both hydrolytic and reductive reactions. A comparable reductive dechlorination is used by a pentachlorophenol-degrading *Flavobacterium* sp., and the enzyme has been purified; the successive dechlorinations to form 2-chloro-1,4-dihydroxybenzene are probably mediated by a glutathione *S*-transferase system (Xun et al. 1992c). It should be emphasized, however, that there is no compelling evidence to suggest so far that the dechlorination reaction bears any relation to the formally comparable dechlorination reactions that take place under strictly anaerobic conditions and that are discussed further in Section 6.6.

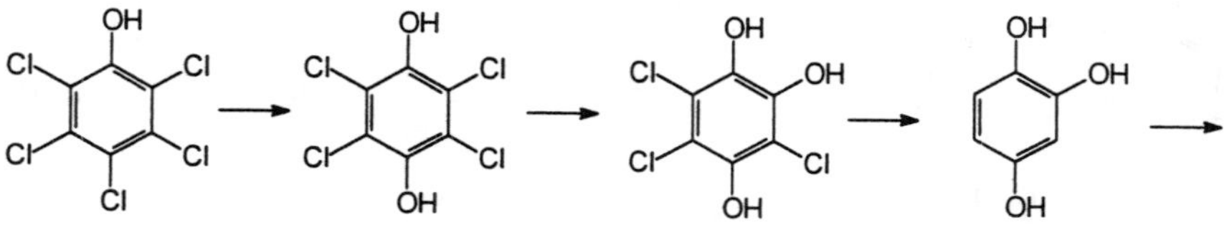

Figure 6.72 Biodegradation of pentachlorophenol.

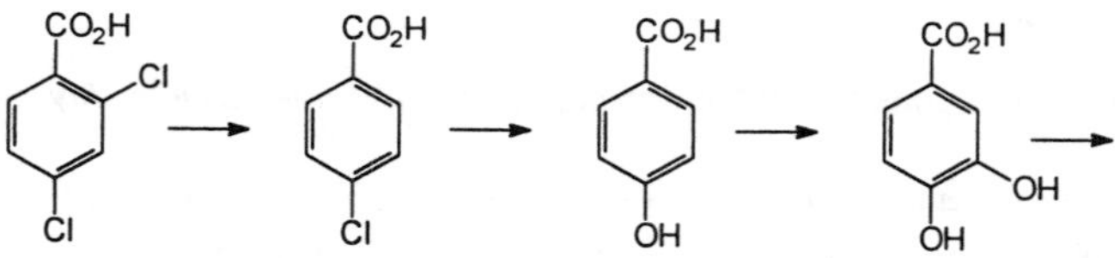

Figure 6.73 Reductive steps in the biodegradation of 2,4-dichlorobenzoate.

6.5.2 Fungal Systems

Although these have been less exhaustively investigated than their bacterial counterparts, the results of these investigations have revealed some interesting and significant features.

1. The NIH shift (Daly et al. 1972) with translocation of chlorine has been demonstrated during the biotransformation of 2,4-dichlorophenoxyacetate by *Aspergillus niger* (Figure 6.74) (Faulkner and Woodcock 1965). The NIH shift is not restricted to fungi since it has also been demonstrated with protons — though not apparently with other substituents — in prokaryotic systems (Section 6.2.2).
2. Degradation of 2,4-dichlorophenol has been demonstrated with the white rot basidiomycete *Phanerochaete chrysosporium* under conditions of nitrogen limitation and apparently involves both lignin peroxidase and manganese-dependent peroxidase activities (Valli and Gold 1991). The reaction proceeds by a pathway involving a series of oxidation and reductions (Figure 6.75) which is entirely different from the sequence that is employed by bacteria, and an essentially similar sequence is involved in the degradation of 2,4,5-trichlorophenol that is accomplished more rapidly, possibly due to less interference from polymerization reactions (Joshi and Gold 1993). It should be noted that, in addition, 2-chloro-1,4-dimethoxybenzene is produced from 2,4-dichlorophenol and

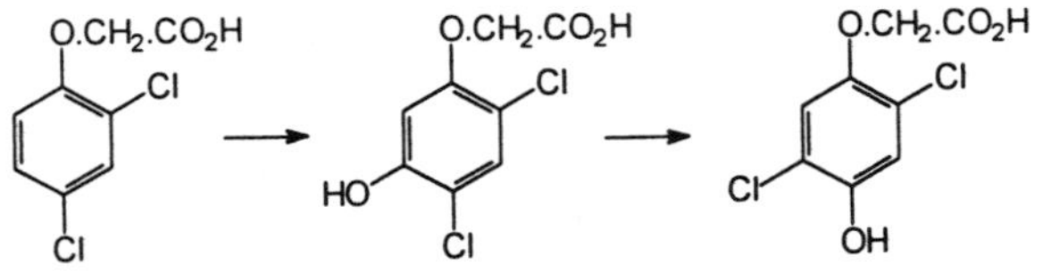

Figure 6.74 NIH shift during the metabolism of 2,4-dichlorophenoxyacetate by *Aspergillus niger*.

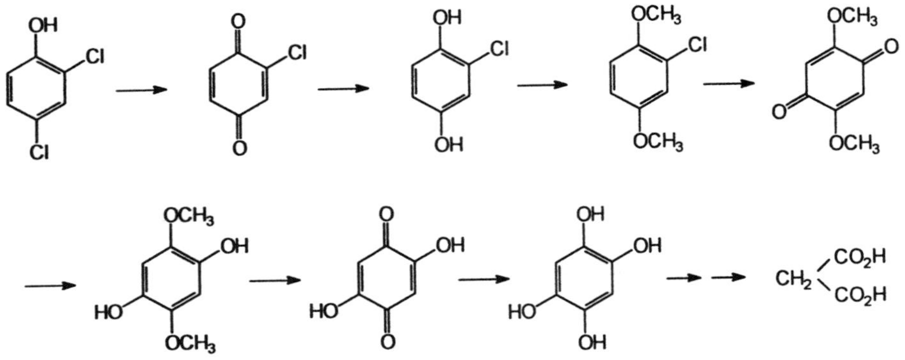

Figure 6.75 Biodegradation of 2,4-dichlorophenol by *Phanerochaete chrysosporium*.

2,5-dichloro-1,4-dimethoxybenzene from 2,4,5-trichlorophenol. 2,4,5-trichlorophenol also gives rise to small amounts of a dimeric product that has been tentatively identified as 2,2′-dihydroxy-3,5,6,3′,5′,6′-hexachlorobiphenyl. A formally comparable sequence of reactions is used for the degradation of 2,4-dinitrotoluene (Valli et al. 1992b) that is discussed in Section 6.8.2.

6.6 ANAEROBIC METABOLISM OF HALOGENATED AROMATIC COMPOUNDS

There has been substantial interest in the fate of chlorinated aromatic compounds under anaerobic conditions since some of these such as PCBs, polychlorinated phenols, polychlorinated catechols, and polychlorinated anilines have been recovered from anaerobic sediment samples. The persistence of these compounds is, therefore, determined by the activity of anaerobic dehalogenating bacteria, and extensive effort has, therefore, been devoted to isolating the relevant organisms.

So far, however, only two organisms have been obtained in pure culture:

1. *Desulfomonile tiedjei* is a sulfate-reducing bacterium capable of reducing 3-chlorobenzoate to benzoate (DeWeerd et al. 1990) during which ATP is synthesized by coupling proton translocation to dechlorination (Dolfing 1990; Mohn and Tiedje 1991). Cells induced by growth with 3-chlorobenzoate were able to partially dechlorinate polychlorinated phenols specifically at the 3-position, whereas the monochlorophenols were apparently resistant to dechlorination (Mohn and Kennedy 1992).

2. A spore-forming organism that does not reduce sulfate and generally dechlorinates chlorophenols preferentially at the 2-position (Madsen and Licht 1992) has also been isolated.

It is important to appreciate a significant difference between most experiments whose results have been discussed hitherto and those that are summarized in this section. Compared with the large number of pure strains of aerobic bacteria that have been used for studies in biodegradation, relatively few strains of anaerobic bacteria that are capable of degrading halogenated aromatic compounds have been isolated in pure culture; metabolic studies have, therefore, of necessity used mixed cultures. The design of these investigations differ considerably, and it is important to distinguish between the various experimental conditions since these may critically affect the interpretation of the results:

1. Unenriched slurries containing both bacteria and significant amounts of sediment or sludge.
2. Enrichment cultures obtained after only a small number of successive transfers.
3. Stable enrichment cultures which have been obtained after successive transfer over extended periods of time and in which organic matter from the sludge or sediment has been removed by dilution.

Although the use of metabolically stable mixed cultures enables an acceptable degree of repeatability to be attained, the use of suspensions or slurries containing sediment or metabolites from previous additions of enrichment substrates introduces undesirable ambiguity in the interpretaton of analytical results. Inevitably, however, most examples of dechlorination depend on the results of studies with mixed cultures. There are several important issues which have emerged from these quite extensive investigations, and these merit brief discussion.

1. Although complete dechlorination of polyhalogenated compounds under anaerobic conditions has been observed, the most common situation is that in which only partial dehalogenation occurs; all of these reactions are, therefore, strictly biotransformations. Illustrative examples include pentachlorophenol (Figure 6.76à) (Mikesell and Boyd 1986), hexachlorobenzene (Figure 6.76b) (Fathepure et al. 1988), 3,4,5-trichlorocatechol (Figure 6.76c) (Allard et al. 1991), 2,3,5,6-tetrachlorobiphenyl (Figure 6.76d) (Van Dort and Bedard 1991), chloroanilines (Figure 6.76e) (Kuhn et al. 1990), and chlorobenzoates (Figure 6.76f) (Gerritse et al. 1992). Collectively, these results seem to imply that the less highly substituted congeners are more resistant to dechlorination. In general, this is the reverse of the situation pertaining under aerobic conditions, and this suggests that the complete degradation of these polyhalogenated compounds will probably involve both kinds of reactions; a similar situation has been suggested for halogenated alkanes and alkenes (Section 6.4.4). For polyhalogenated compounds that have been partitioned into the sediment phase, partial anaerobic dechlorination to less highly chlorinated congeners is, therefore, likely to occur; if these were subsequently released into the water mass, aerobic degradation could presumably occur provided that the compounds existed in a state accessible to the relevant microorganisms.

2. The physiology of a mixed culture is inevitably incompletely defined, and organisms differing as basically as methanogenic and sulfidogenic bacteria may be simultaneously present. Considerable attention has, therefore, been devoted to the role of sulfate in dechlorination reactions, and two essentially different situations have emerged.

- Degradation of 4-chlorophenol and 2,4-dichlorophenol has been demonstrated both in methanogenic cultures in the presence of sulfate (Kohring et al. 1989) and in non-methanogenic cultures reducing sulfate (Häggblom and Young 1990).
- On the other hand, the rate of dechlorination of a series of chloroanilines by methanogenic groundwater mixed cultures was diminished by the presence of sulfate (Kuhn et al. 1990), and the dechlorination of pentachlorophenol was inhibited by the presence of sulfate (Madsen and Amand 1991).
- The effect of adding sulfate to enrichment cultures has been examined and has been found to be counter selective to the development of cultures with dechlorination capability (Genther et al. 1989; Allard et al. 1992). Presumably, in these cases, sulfate effectively competes for reducing equivalents provided by the organic substrate. On the other hand, addition of sulfate to pure cultures of *Desulfomonile tiedjei* (DeWeerd et al. 1991) or to mixed cultures from several enrichments which possessed dechlorination capability (Allard et al. 1992), did not inhibit dechlorination.

Although considerable effort has been directed to the dechlorination of aromatic compounds, only a few studies have examined debromination. A few examples will be given to illustrate the probably widespread distribution of this plausible reaction:

1. Aquifer slurries under methanogenic conditions debrominated the agrochemical bromacil (5-bromo-3-*sec*-butyl-6-methyluracil) (Adrian and Suflita 1990).
2. Marine sediment slurries debrominated the naturally occurring 2,4-dibromophenol and also 2,4,6-tribromophenol to phenol that was then further degraded (King 1988).
3. Freshwater sediment slurries debrominated 2-bromobenzoate to benzoate that was then degraded by methanogenic organisms (Horowitz et al. 1983).
4. Metabolically stable enrichment cultures reduced the aldehyde group and debrominated 5-bromovanillin to 4-methylcatechol and 3-bromo-4-hydroxybenzaldehyde to 4-hydroxybenzoate (Neilson et al. 1988a).
5. Debromination of polybrominated biphenyls has been observed in cultures obtained from sediments contaminated with the corresponding biphenyls (Morris et al. 1992).

6.7 REACTIONS CARRIED OUT BY ANAEROBIC BACTERIA OTHER THAN DEHALOGENATION

The comments made at the beginning of Section 6.6 apply equally to some of the reactions described in this section and, in particular, to the degradation of aromatic compounds that is discussed in Section 6.7.3. In addition, attention should be drawn to the widely differing physiologies of the organisms that include strictly fermentative organisms, phototrophic organisms, and organisms that use nitrate or sulfate or carbonate as electron acceptors. The sole unifying feature of the reactions is that they do not involve incorporation of molecular oxygen.

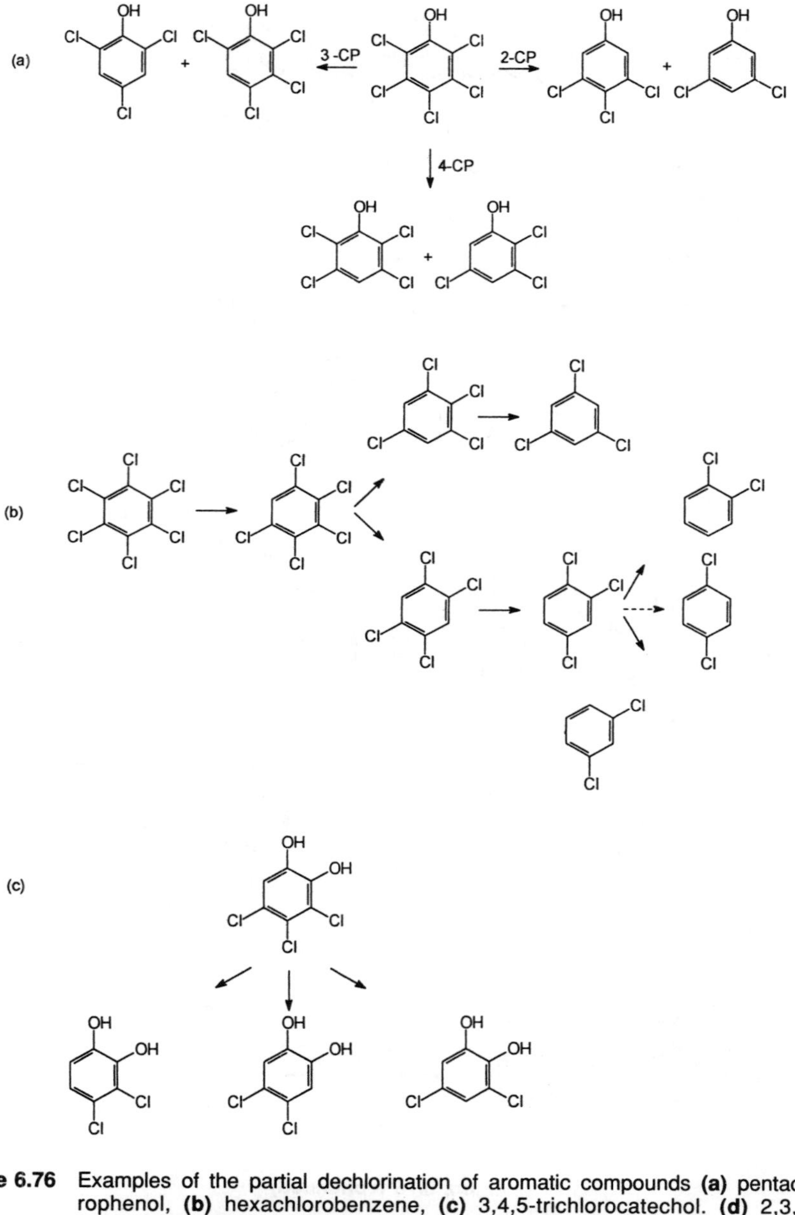

Figure 6.76 Examples of the partial dechlorination of aromatic compounds **(a)** pentachlorophenol, **(b)** hexachlorobenzene, **(c)** 3,4,5-trichlorocatechol. **(d)** 2,3,5,6-tetrachlorobiphenyl, **(e)** chloroanilines, **(f)** 2,4,6-trichlorobenzoate.

6.7.1 Aliphatic Compounds

The degradation of aliphatic carboxylic acids is of great ecological importance since compounds such as acetate, propionate, or butyrate may be the terminal fermentation products of organisms degrading more complex compounds including carbohydrates, proteins, and lipids (Zeikus 1980; Mackie et al. 1991). Degradation

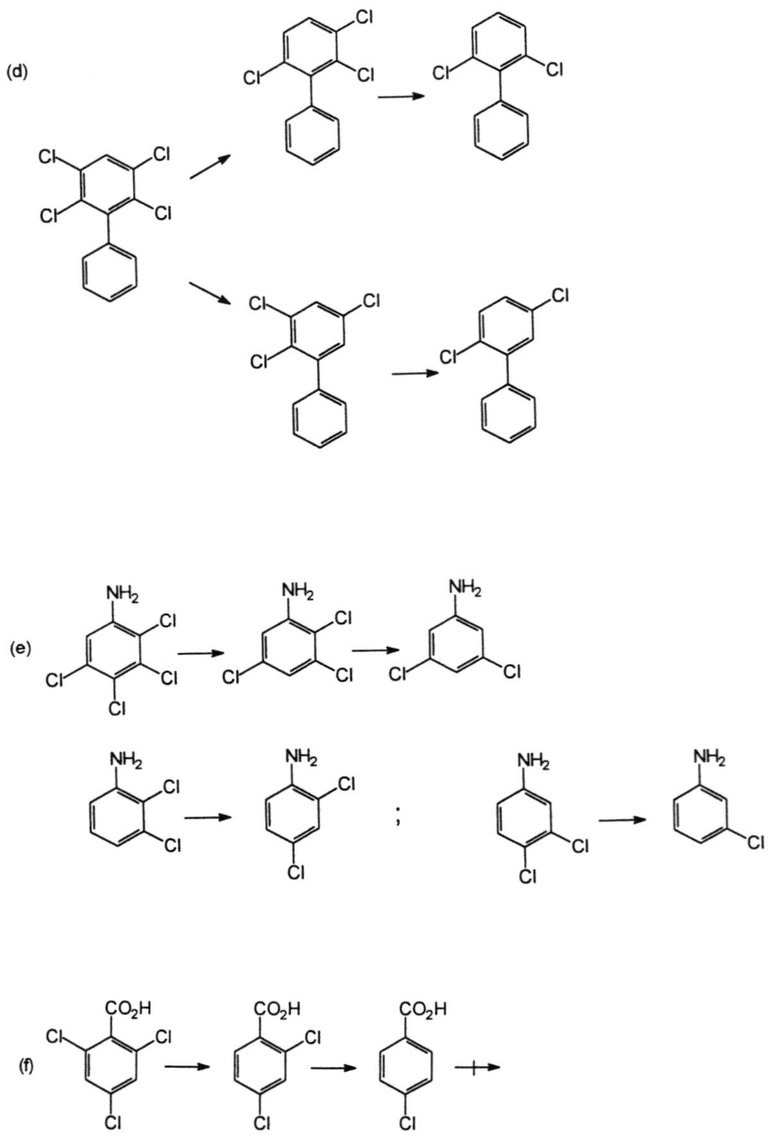

Figure 6.76 (continued)

of aliphatic carboxylic acids by sulfate-reducing bacteria was traditionally restricted to lactate and its near relative, pyruvate, but recent developments have radically altered the situation and increased the spectrum of compounds which can be oxidized to CO_2 at the expense of sulfate reduction. In the following paragraphs, an attempt will be made to present a brief summary of the anaerobic degradation of the main groups of aliphatic compounds. Greatest attention has been directed to compounds with functional groups and, in particular, alkanoic acids and amino acids.

Alkanes — Complete oxidation of hexadecane by a sulfate-reducing bacterium has been reported (Aeckersberg et al. 1991) and presents a particularly interesting addition to the range of highly reduced compounds which can be oxidized under anaerobic conditions.

Alkenes — Degradation of hex-1-ene has been observed in a methanogenic consortium (Schink 1985a) that converted the substrate to methane, and a plausible pathway involving hydration and oxidation was suggested.

Alkynes — Acetylene supports the growth of *Pelobacter acetylenicus* (Schink 1985b) and undergoes initial hydration to acetaldehyde followed by dismutation into acetate and ethanol.

Alkanols — Oxidation of methanol — though not incorporation into cellular material — has been observed in a sulfate-reducing bacterium (Braun and Stolp 1985), while ethanol may be converted into methane by *Methanogenium organophilum* (Widdel 1986; Frimmer and Widdel 1989), and oxidation of primary alkanols has been demonstrated in *Acetobacterium carbinolicum* (Eichler and Schink 1984). Secondary alcohols such as propan-2-ol and butan-2-ol may be used as hydrogen donors for methanogenesis with concomitant oxidation to the corresponding ketones (Widdel et al. 1988). An NAD-dependent alcohol dehydrogenase has been purified from *Desulfovibrio gigas* that can oxidize ethanol, and it has been shown that the enzyme does not bear any relation to classical alcohol dehydrogenases (Hensgens et al. 1993). The metabolism of 1,2-diols has attracted considerable attention and, in particular, that of glycerol in view of its ubiquity as a component of lipids. Widely different pathways have been found of which two are given as illustration:

1. *Anaerovibrio glycerini* ferments glycerol to propionate (Schauder and Schink 1989) and *Desulfovibrio carbinolicus* to 3-hydroxypropionate (Nanninga and Gottschal 1987).
2. *Desulfovibrio alcoholovorans* converts glycerol to acetate and 1,2-propandiol to acetate and propionate (Qatibi et al. 1991).

Alkylamines — Few investigations have been devoted to the degradation of alkylamines, though there has been considerable interest in the metabolism of trimethylamine, choline, and glycine betaine as a source of methane in marine sediments (King 1984). Two aspects of the metabolism of N-alkyl compounds may be used as illustration.

1. Methane formation from trimethylamine by *Methanosarcina barkeri* has been demonstrated (Hippe et al. 1979; Patterson and Hespell 1979), and the metabolically versatile organisms *Methanococcoides methylutens* (Sowers and Ferry 1983) and *Methanolobus tindarius* (Konig and Stetter 1982) that use methylamines and methanol for methane formation have been described.
2. The metabolism of betaine that is an important osmoregulatory solute in many organisms has been studied, and different metabolic pathways have been revealed.
   - Betaine is demethylated to dimethylglycine by *Eubacterium limosum* (Müller et al. 1981) and by strains of *Desulfobacterium* sp. (Heijthuijsen and Hansen 1989a).

- Betaine is fermented by a strain of *Desulfuromonas acetoxidans* with the production of trimethylamine and acetate (Heijthuijsen and Hansen 1989b), and the same products are formed by *Clostridium sporogenes* in a Stickland reaction with alanine, valine, leucine, or *iso*leucine (Naumann et al. 1983). Further comments on the Stickland reaction are given later.

Alkanoic acids — For a number of reasons, a great deal of effort has been directed to the degradation of alkanoic acids; acetate, propionate, and butyrate are fermentation products of carbohydrates and are metabolites of the aerobic degradation of alkanes and related compounds, while long-chain acids are produced by the hydrolysis of lipids. Studies on the degradation of alkanoic acids have been carried out using both pure cultures and syntrophic associations that have been discussed in Chapter 4 (Section 4.4.1).

For the degradation of acetate, two different reactions may take place: oxidation to CO_2 and dismutation to methane and CO_2. To some extent, as will emerge, segments of both pathways are at least formally similar, although the mechanisms for anaerobic degradation of these apparently simple compounds are quite subtle. Degradation of the short-chain carboxylic acids acetate and butyrate can be accomplished by *Desulfotomaculum acetoxidans* (Widdel and Pfennig 1981), and of acetate by *Desulfuromonas acetoxidans* (Pfennig and Biebl 1976) and species of *Desulfobacter* (Widdel 1987), while propionate is degraded by species of *Desulfobulbus* (Widdel and Pfennig 1982; Samain et al. 1984). A sulfur-reducing organism with a much wider degradative capability than *Desulfuromonas acetoxidans* has been isolated (Finster and Bak 1993), and this organism is capable of accomplishing the complete oxidation of, for example, propionate, valerate, and succinate.

The oxidation of acetate under anaerobic conditions can occur by two completely different pathways, both of which have been investigated in detail, and the enzymology has been delineated (Thauer et al. 1989).

1. Oxidation may take place by a modified tricarboxylic acid cycle in which the production of CO_2 is coupled to the synthesis of NADPH and reduced ferrodoxin, and the dehydrogenation of succinate to fumarate is coupled to the synthesis of reduced menaquinone. This pathway is used, for example, by *Desulfuromonas acetoxidans* and in modified form by *Desulfobacter postgatei*.
2. On the other hand, dissimilation of acetate may take place by reversal of the pathway used by organisms such as *Clostridium thermoaceticum* for the synthesis of acetate from CO_2. In the degradation of acetate, the pathway involves a dismutation in which the methyl group is successively oxidized via methyl tetrahydrofolate to CO_2 while the carbonyl group is oxidized via bound carbon monoxide. Such THF-mediated reactions are of great importance in the anaerobic degradation of purines that will be discussed in Section 6.7.4.

Acetate may also be converted into methane by a few methanogens belonging to the genus *Methanosarcina*. The methyl group is initially converted into methyltetrahydromethanopterin (corresponding to methyltetrahydrofolate in the acetate oxidations discussed above) before reduction to methane via methyl-coenzyme M; the carbonyl group of acetate is oxidized via bound CO to CO_2.

Both the synthesis of propionate and its metabolism may take place under anaerobic conditions, and in *Desulfobulbus propionicum*, the degradation plausibly takes place by reversal of the steps used for its synthesis from acetate (Stams et al. 1984): carboxylation of propionate to methylmalonate followed by coenzyme B_{12}-mediated rearrangement to succinate which then enters the tricarboxylic acid cycle. Growth of syntrophic propionate-oxidizing bacteria in the absence of methanogens has been accomplished using fumarate as the sole substrate (Plugge et al. 1993). Fumarate plays a central role in the metabolism of this organism since it is produced from propionate via methylmalonate and succinate, and fumarate itself is metabolized by the acetyl-CoA cleavage pathway via malate, oxalacetate, and pyruvate. This has already been noted in Chapter 4 (Section 4.4.1).

The degradation of long-chain carboxylic acids is important in the anaerobic metabolism of lipids, and an extensive compilation of the organisms that can accomplish this has been given (Mackie et al. 1991). This capability has been demonstrated in syntrophic bacteria in the presence of hydrogen-utilizing bacteria; for example, β-oxidation of C_4 to C_8, and C_5 to C_7 carboxylic acids was carried out by the *Syntrophomonas wolfei* association (McInerney et al. 1981) and of C_4 to C_{10}, and C_5 to C_{11} by the *Clostridium bryantii* syntroph (Stieb and Schink 1985). Acetate and propionate were the respective terminal products from the even- and odd-numbered acids. Pure cultures of many sulfate-reducing bacteria are also able to carry out comparable reactions (Mackie et al. 1991).

In summary, the anaerobic degradation of alkanoic acids may truly be described as ubiquitous and is carried out by organisms with widely different taxonomic affinity, both in pure culture and in syntrophic associations.

Amino acids — The degradation pathways of amino acids have been examined in detail particularly in clostridia (Barker 1981), and these investigations have revealed a number of important reactions not encountered in other degradations.

1. Coenzyme B_{12}-mediated rearrangements have been elucidated as an important reaction in the degradation of glutamate and ornithine and subsequently led to a detailed investigation into the role of this rearrangement in other reactions (Barker 1972). An outline of the β-methylaspartate pathway for the degradation of glutamate by *Clostridium tetanomorphum* is given in Figure 6.77.
2. The transformation of aromatic amino acids has been examined, and it has been shown (Elsden et al. 1976; D'Ari and Barker 1985) that in *Cl. difficile* these compounds are possible biogenic sources of 4-methyl phenol, and that in an organism designated "*Cl. aerofoetidum*" toluene may be produced (Pons et al. 1984); the postulated reactions are schematically shown in Figure 6.78.
3. An unusual dismutation reaction involving pairs of amino acids has been studied extensively in *Cl. sporogenes* (Stickland 1934, 1935a,b). The reaction can be carried out by many other clostridia and has been summarized (Barker 1961). The products from proline and alanine (Stickland 1935a) illustrate the reaction and are shown in Figure 6.79, and the mechanism by which glycine is reduced to acetate by glycine reductase from *Cl. sticklandii* has been elucidated (Arkowitz and Abeles 1989). The involvement of betaine in the reaction has already been noted. Formally similar dismutations have been suggested for the biotransformation of γ-hexachlorocyclohexane (Ohisa et al.1980) and of 5-chlorovanillin (Neilson et al. 1988a).

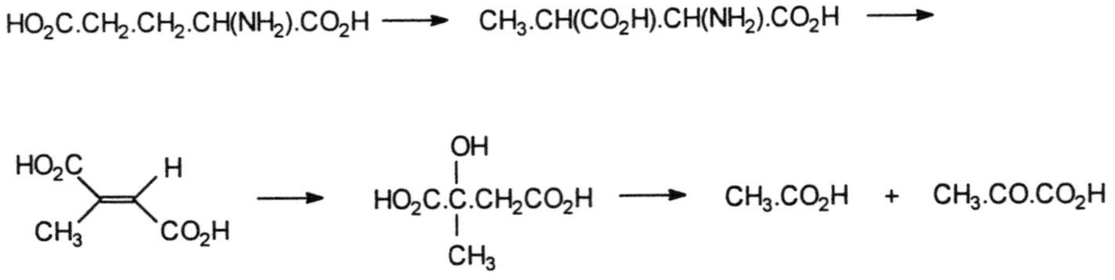
HO2C.CH2.CH2.CH(NH2).CO2H
CH3.CH(CO2H).CH(NH2).CO2H
HO2C
H
CH3
CO2H
OH
HO2C.C.CH2CO2H
CH3
CH3.CO2H
+
CH3.CO.CO2H

Figure 6.77 Biodegradation of glutamate by *Clostridium tetanomorphum.*

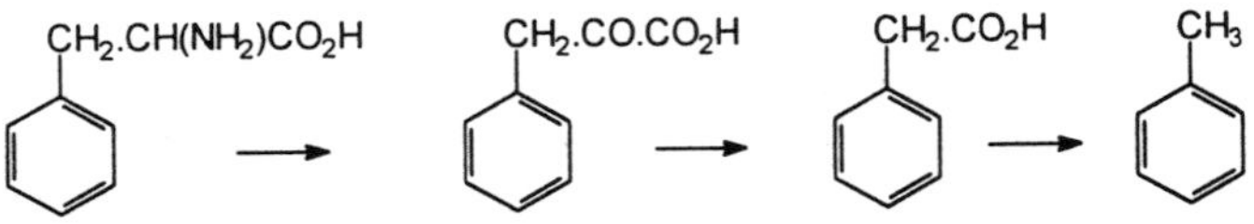
CH2.CH(NH2)CO2H
CH2.CO.CO2H
CH2.CO2H
CH3

Figure 6.78 Metabolism of phenylalanine.

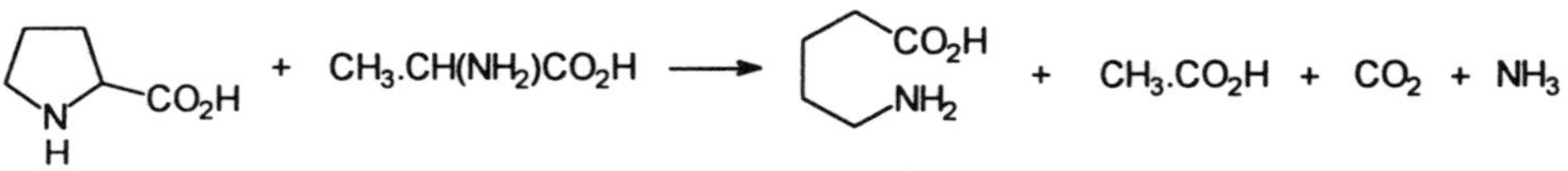
N
H
CO2H
+
CH3.CH(NH2)CO2H
CO2H
NH2
+
CH3.CO2H
+
CO2
+
NH3

Figure 6.79 Exemplification of the Stickland reaction between proline and alanine.

In summary, therefore, pathways exist for the metabolism of a wide range of aliphatic compounds, and a net of metabolic interactions operate for their complete degradation to CO_2 and cell material.

6.7.2 Biotransformation of Alicyclic Compounds Containing Several Rings

The transformation of sterols and bile acids has been examined quite thoroughly in the context of their intestinal metabolism, and a number of reactions that are otherwise quite unusual have been observed, most frequently in organisms belonging to the genus *Eubacterium.* These reactions may be very briefly summarized:

1. Reduction of the $\Delta^{4,5}$ bond with production of 5β-reduced compounds (Mott et al. 1980).
2. Reductive dehydroxylation of 7α-hydroxy bile acids (Masuda et al. 1984) and 16α-hydroxy and 21-hydroxy corticosterols (Bokkenheuser et al. 1980).

6.7.3 Aromatic Compounds

Considerable effort has been devoted to the anaerobic degradation of aromatic compounds, in particular to their conversion to methane. In spite of this, however, pure cultures of fermentative organisms have not always been achieved so that many investigations have used mixed cultures with the result that the proposed pathways are inevitably speculative. This section will summarize first reactions carried out by pure cultures — several of them phototrophic or denitrifying organisms — and then those accomplished by various types of consortia.

It seems clear that benzoate occupies a central position in the anaerobic degradation of both phenols and alkaryl hydrocarbons and that carboxylation, hydroxylation, and reductive dehydroxylation are important — and less expected — reactions.

Pure cultures — An attempt is made to provide a spectrum of the reactions carried out by pure cultures of a wide range of bacteria under anaerobic conditions.

1. Aromatic methoxy groups may be used to support the growth of methoxybenzoates (Bache and Pfennig 1981); acetogenic organisms such as *Acetobacterium woodii* use the methyl group for growth, and sulfate-reducing organisms can carry out the demethylation reaction even in the absence of sulfate (Tasaki et al. 1992). The methyl group may alternatively be converted into butyrate by *Clostridium pfennigii* (Krumholz and Bryant 1985) or into methanethiol and dimethyl sulfide by inorganic sulfide in the growth medium (Bak et al. 1992). The details of this complex demethylation reaction have been examined in a strain of a sulfide-methylating homoacetogenic bacterium that degraded gallate by the same pathway as that employed by *Pelobacter acidigallici* (Kreft and Schink 1993). The metabolism of 4-hydroxy-3,5-dimethoxybenzoate by an organism that has an obligate dependence on H_2 (Liu and Suflita 1993) indicates additional complexities of these apparently straightforward O-demethylation reactions.

2. 3,4,5-trihydroxybenzoate is degraded by *Pelobacter acidigallici* to acetate and CO_2 (Schink and Pfennig 1982) and by *Eubacterium oxidoreducens* in the presence of exogenous H_2 or formate to acetate, butyrate, and CO_2 (Krumholz et al. 1987). The degradation has been studied in detail and takes place by the unusual pathway (Figure 6.80) (Brune and Schink 1992; Krumholz and Bryant 1988). The formation of phloroglucinol in *P. acidigallici* involves a series of intramolecular hydroxyl transfer reactions (Brune and Schink 1990) (Figure 6.81), and the reaction in *E. oxidoreducens* also involves 1,2,3,5-tetrahydroxybenzene although details of the pathway may be different (Haddock and Ferry 1993). Subsequent reduction and ring cleavage follow a pathway comparable to that involved in the degradation of resorcinol by species of clostridia (Tschech and Schink 1985), although an alternative pathway involving direct hydrolysis has been observed in denitrifying bacteria (Gorny et al. 1992).

3. Vanillate is transformed by *Clostridium thermoaceticum* to catechol and phenol, and the CO_2 produced relieves the requirement for supplemental CO_2 (Hsu et al. 1990); the metabolically produced CO_2 is then able to enter a sequence of reactions which results in acetate synthesis.

4. Aniline and catechol are degraded to CO_2 by *Desulfobacterium anilini* (Schnell et al. 1989), and catechol is slowly degraded by *D. catecholicum* (Szewzyk and Pfennig 1987) and by a strain of *Desulfotomaculum* sp. (Kuever et al. 1993). Aniline is carboxylated to 4-aminobenzoate which is then reductively deaminated to benzoate (Schnell and Schink 1991), and it seems possible that by analogy with phenol itself (Section 6.7.3), catechol is carboxylated to 3,4-dihydroxybenzoate which is consistent with the CO_2-dependence of the catechol-degrading *Desulfotomaculum* sp. (Kuever et al. 1992): dehydroxylation to benzoate, which is then further degraded, may then take place. By analogy with the initial steps for the degradation of aniline, 2-aminobenzoate is degraded

under denitrifying conditions by a *Pseudomonas* sp. to benzoate which is then reduced to cyclohexene-1-carboxylate (Lochmeyer et al. 1992).

5. Aromatic aldehydes may be involved in both reductive and oxidative reactions under anaerobic conditions, and in some cases the carboxylic acid is further decarboxylated.

   • Oxidation of substituted benzaldehydes to benzoates at the expense of sulfate reduction has been demonstrated in strains of *Desulfovibrio* sp., although the carboxylic acids produced were apparently stable to further degradation (Zellner et al. 1990). Vanillin was, however, used as a substrate for growth by a strain of *Desulfotomaculum* sp. and was metabolized via vanillate and catechol (Kuever et al. 1992). Different aldehyde oxidoreductases have been isolated from *Clostridium thermoaceticum and Cl. formicoaceticum* (White et al. 1993); whereas both cinnamate and cinnamaldehyde were good substrates for the Mo-containing enzyme from the latter, benzoate was an extremely poor substrate. The W-containing enzyme from *Cl. formicoaceticum,* however, displays high activity towards a greater range of substituted benzoates (White et al. 1991).

   • On the other hand, cell extracts of *Cl. formicoaceticum* reduce benzoate to benzyl alcohol at the expense of carbon monoxide (Fraisse and Simon 1988), and metabolically stable enrichment cultures of anaerobic bacteria carried out a reaction in which aromatic hydroxyaldehydes were concomitantly transformed into the corresponding carboxylic acid and the benzyl alcohol (Neilson et al. 1988a); in some cases, the carboxylic acid was decarboxylated to produce phenols or catechols (Figure 6.82). A comparable decarboxylation accompanied by dechlorination in freshwater sediments has been demonstrated with 3-chloro-4-hydroxybenzoate that produces phenol before carboxylation to benzoate and further degradation (Zhang and Wiegel 1992).

6. Benzoate metabolism has been studied in the anaerobic phototroph *Rhodopseudomonas palustris*; a benzoate-coenzyme A ligase is induced (Geissler et al. 1988), and this seems to be an essential reaction for the activation of benzoic and other carboxylic acids. Two cyclohexadiene carboxylates have been identified (Figure 6.83) (Gibson and Gibson 1992), but although these seem plausible intermediates, their direct significance is uncertain in the absence of unequivocal evidence.

7. The metabolism of cinnamate and ω-phenylalkane carboxylates has been studied in *Rhps. palustris* (Elder et al. 1992) and for growth with the higher homologues, additional CO_2 was necessary. The key degradative reaction was β-oxidation; for compounds with chain lengths of 3-, 5-, and 7-carbon atoms, benzoate was formed and further metabolized, but for the even-numbered compounds with 4-,6-, and 8-carbon atoms, phenylacetate was a terminal metabolite.

8. Strains of denitrifying bacteria have been shown to degrade toluene in the absence of oxygen using N_2O as electron acceptor (Schocher et al. 1991), and the data are consistent with a pathway involving successive oxidation of the ring methyl group with the formation of benzoate. The details of this pathway involving benzyl alcohol and benzaldehyde have been clearly demonstrated with a strain of *Pseudomonas* sp. under denitrifying conditions (Altenschmidt and

Fuchs 1992a), but attention should be drawn to alternative pathways that include, for example, (a) hydroxylation to 4-hydroxytoluene followed by oxidation of the methyl group and dehydroxylation (Rudolphi et al. 1991) or (b) carboxylation to phenyl acetate and then oxidation via phenylglyoxylate to benzoate which is an established pathway for phenyl acetate itself (Dangel et al. 1991). Terminal metabolites identified as benzyl succinate and benzyl fumarate have been identified during metabolism of toluene by a denitrifying organism, and it has been suggested that by analogy, degradation of toluene proceeds by condensation with acetate to form phenylpropionate and benzoate before ring cleavage (Evans et al. 1992); the same terminal metabolites are also produced during anaerobic degradation of toluene by sulfate-reducing enrichment cultures (Beller et al. 1992). The subsequent pathway for the degradation of benzoate remains, however, incompletely unresolved though various alicyclic compounds are apparently involved. These plausible intermediates had been proposed earlier (Evans and Fuchs 1988), but clear evidence in support of the initial steps has required considerable additional effort (Koch and Fuchs 1992). An entirely different pathway presumably operates in *Desulfobacula toluolica* since degradation via ring or side-chain hydroxylation or carboxylation seems to be unlikely (Rabus et al. 1993).

9. Phenol is transformed into benzoyl coenzyme-A by a denitrifying strain of *Pseudomonas* sp., and it has been suggested that phenyl phosphate may be the actual substrate for carboxylation (Lack and Fuchs 1992). The central role of 4-hydroxybenzoyl-CoA in the metabolism of phenol, 4-methylphenol, 4-hydroxyphenylacetate, and related aromatic compounds by a denitrifying strain of *Pseudomonas* sp. is supported by the isolation and purification of the dehydroxylating enzyme (Brackmann and Fuchs 1993), and some plausible mechanistic arguments have been put forward.

10. Phenylacetate and 4-hydroxyphenylacetate are oxidized sequentially under anaerobic conditions by a denitrifying strain of *Pseudomonas* sp. to the phenylglyoxylates and benzoate (Mohamed et al. 1993).

11. The degradation of hydroquinone takes place by carboxylation followed by dehydroxylation to benzoate that was further degraded by reduction of the aromatic ring (Gorny and Schink 1994).

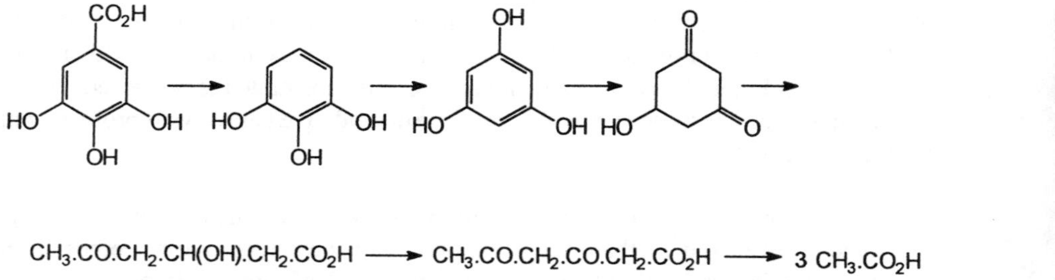

$$CH_3.CO.CH_2.CH(OH).CH_2.CO_2H \longrightarrow CH_3.CO.CH_2.CO.CH_2.CO_2H \longrightarrow 3\ CH_3.CO_2H$$

Figure 6.80 Pathway for the biodegradation of 3,4,5-trihydroxybenzoate.

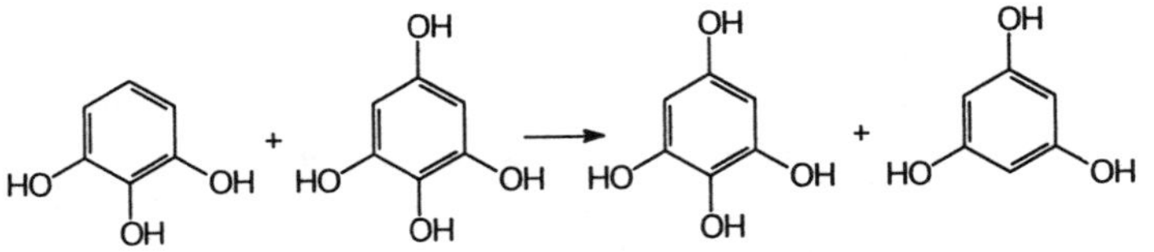

Figure 6.81   Intramolecular hydroxyl group transfers in the biodegradation of 1,2,3-trihydroxybenzoate.

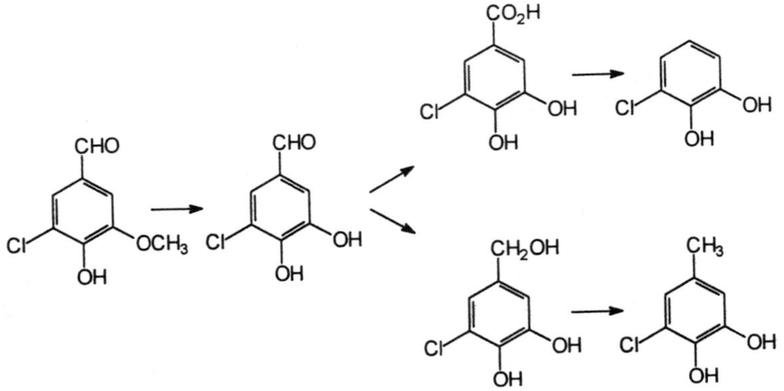

Figure 6.82   Biotransformation of 5-chlorovanillin.

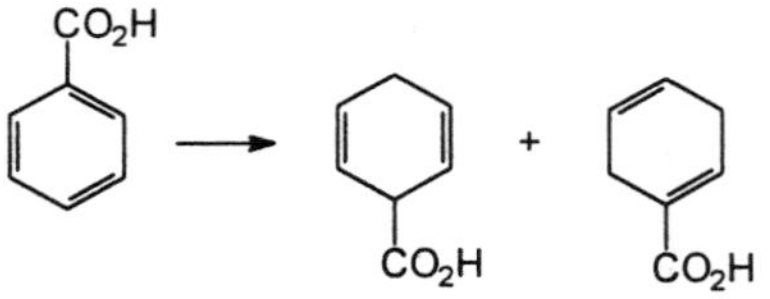

Figure 6.83   Anaerobic biodegradation of benzoate.

Mixed Cultures — There is a substantial literature dealing with the anaerobic degradation of compounds, such as benzoate and phenol, and more recently of benzene, toluene, and the xylenes by mixed cultures. There can be no doubt that such compounds can be degraded to methane in some cases or to aliphatic carboxylic acids and CO_2 in others. It is virtually impossible, however, to present a detailed account of the reactions involved since no pure cultures have been investigated, and both methanogenic and non-methanogenic cultures have been examined. Instead, a summary of the individual reactions which have been demonstrated is presented, but an attempt to synthesize these to suggest complete degradative pathways is probably not currently justified.

1. Benzene and toluene are anaerobically hydroxylated to phenol and 4-hydroxytoluene by mixed methanogenic cultures, and the oxygen atom comes from water (Vogel and Grbic-Galic 1986). On the other hand, sulfidogenic mixed cultures degraded toluene and xylene to CO_2 (Edwards et al. 1992). The mechanisms may well resemble or be identical to those used by denitrifying bacteria that were discussed above.

2. Carboxylation of phenols has been demonstrated under a number of conditions, and this reaction is presumably analogous to that demonstrated in pure cultures of *Desulfobacterium anilini* during metabolism of aniline and in denitrifying bacteria during metabolism of phenol. The following examples are used to illustrate some important details of these reactions.

- Phenol is carboxylated by a defined obligate syntrophic consortium to benzoate which is then degraded to acetate, methane, and CO_2 (Knoll and Winter 1989).
- 2-methylphenol is carboxylated by a methanogenic consortium to 4-hydroxy-3-methylbenzoate which was dehydroxylated to 3-methylbenzoate although this was stable to further transformation (Bisaillon et al. 1991).
- 3-methylphenol is carboxylated to 2-methyl-4-hydroxybenzoate by a methanogenic enrichment culture before degradation to acetate (Roberts et al. 1990); ^{14}C-labeled bicarbonate produced carboxyl-labeled acetate, while ^{14}C-methyl labeled 3-methylphenol yielded methyl-labeled acetate. 2-methylbenzoate formed by dehydroxylation of 2-methyl-4-hydroxybenzoate was not, however, further metabolized (Figure 6.84a). A similar reaction occurs with a sulfate-reducing mixed culture (Ramanand and Suflita 1991).
- An unusual reaction occurs during the degradation of 3-methylphenol by a methanogenic consortium (Londry and Fedorak 1993); carboxylation to 2-methyl-4-hydroxybenzoate took place as in the preceding example, but further metabolism involved loss of the methyl group with the formation of methane before dehydroxylation to benzoate (Figure 6.84b).

It may be concluded from all these observations that whereas benzoate produced by the carboxylation of phenols can be degraded, dehydroxylation with the formation of substituted benzoates will frequently produce stable terminal metabolites.

3. Benzoate has been identified as a transient metabolite during methanogenic degradation of ferulate (Figure 6.85) (Grbic-Galic and Young 1985), toluene (Grbic-Galic and Vogel 1987), and styrene (Figure 6.86) (Grbic-Galic et al.1990), and there seems little doubt that benzoate can be degraded anaerobically though details of the mechanism have not finally been established in methanogenic organisms.

6.7.4 Heterocyclic Aromatic Compounds

Nitrogen-containing heterocyclic compounds — Whereas the metabolism of a structurally diverse range of heterocyclic aromatic compounds has been examined under aerobic conditions, a much more limited range has been investigated under anaerobic conditions, and these investigations have apparently centered on the N-heterocyclic compounds. Demonstration of the possibility of anaerobic degradation of purines belongs to the golden age of microbiology and appropriately was discovered in Beijerinck's laboratory in Delft. Liebert (1909) obtained a pure culture of an organism that was able to grow anaerobically with 2,6,8-trihydroxypurine (uric acid) and which he named *Bacillus acidi urici* (Leibert 1909), although the organism is now known as *Clostridium acidurici*. Subsequently, several other purinolytic clostridia have been isolated, and the details of their metabolism of purines has been extensively delineated by Barker and his collaborators.

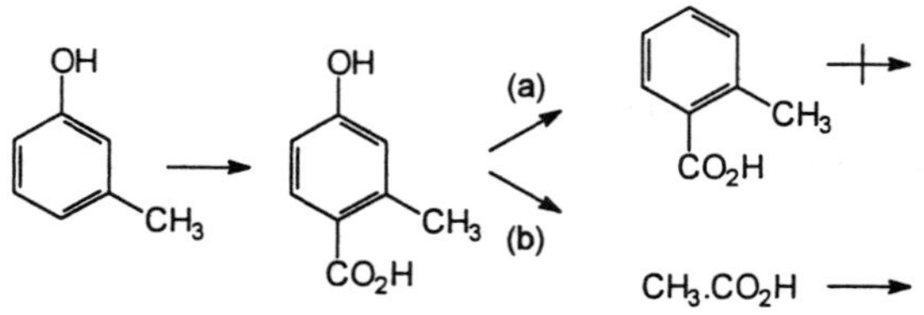

Figure 6.84 Anaerobic biodegradation of 3-methylphenol.

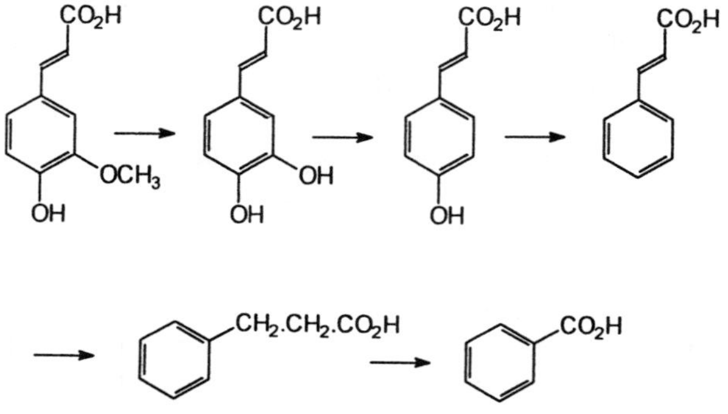

Figure 6.85 Anaerobic biodegradation of ferulate.

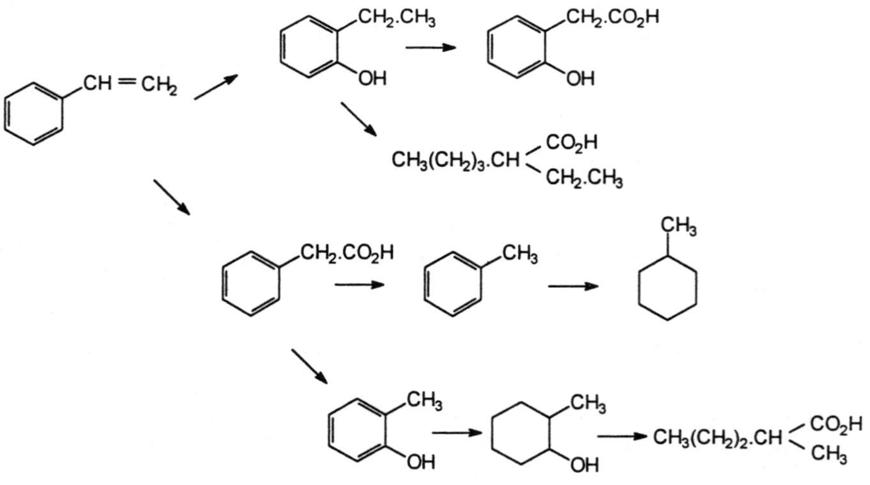

Figure 6.86 Anaerobic biodegradation of styrene.

Detailed investigations using pure cultures have elucidated the degradation pathways, and their enzymology is known in considerable detail. In the following paragraphs, the pathways used for the degradation of pyridines, pyrimidines, and purines by anaerobic bacteria will be briefly summarized. Complete mineralization may not always be achieved, although this is clearly the case for nicotinate that is degraded by *Desulfococcus niacinii* and for the degradation of indoles by *Desulfobacterium indolicum* (Bak and Widdel 1986).

Broadly, four different types of reactions have been found.

1. Hydroxylation of the heterocyclic ring — The metabolism of nicotinate has been extensively studied in clostridia, and the details of the pathway (Figure 6.87) have been delineated in a series of studies (Kung and Tsai 1971). The degradation is initiated by hydroxylation of the ring, and the level of nicotinic acid hydroxylase is substantially increased by the addition of selenite to the medium (Imhoff and Andreesen 1979). The most remarkable feature of the pathway is the mechanism whereby 2-methylene-glutarate is converted into methylitaconate by a coenzyme B_{12}-mediated reaction (Kung and Stadtman 1971). This is representative of a group of related reactions which have been discussed in a classic review (Barker 1972). Whereas in *Clostridium barkeri*, the end-products are carboxylic acids, CO_2, and ammonium, the anaerobic sulfate-reducing *Desulfococcus niacinii* degrades nicotinate completely to CO_2 (Imhoff-Stuckle and Pfennig 1983) though the details of the pathway remain unresolved.

2. Ring-cleavage reactions: the degradation of substituted purines — Very substantial effort has been devoted to elucidating the details of the anaerobic degradation of purines containing hydroxyl and amino groups; the most studied group of organisms are the clostridia including *Clostridium acidurici, Cl. Cylindrospermum*, and *Cl. purinilyticum* (Schiefer-Ullrich et al. 1984), but attention should also be drawn to the non-spore-forming *Eubacterium angustum* (Beuscher and Andreesen 1984) and *Peptostreptococcus barnesae* (Schieffer-Ullrich and Andreesen 1985). Many of the basic investigations on the mechanisms of purine degradation by clostridia were carried out by Barker and his colleagues (Barker 1961), and more recent developments have been presented by Dürre and Andreesen (1983). One of the significant findings was that of selenium dependency due to its requirement for the synthesis of several critical enzymes — xanthine dehydrogenase, formate dehydrogenase, and glycine reductase. The pathways for the degradation of purines containing amino and/or hydroxyl groups converge on the synthesis of xanthine (2,6-dihydroxypurine) and are followed by its degradation to formiminoglycine (Figure 6.88). This compound is then used for the synthesis of glycine and 5-formimino-tetrahydrofolate whose further metabolism to formate results in the synthesis of ATP. The energy requirements of the cell are supplemented by the contribution of ATP produced during the reduction of glycine to acetate in an unusual reaction catalyzed by glycine reductase (Arkowitz and Abeles 1989). An essentially similar pathway is used by *Methanococcus vannielii* that uses a number of purines as nitrogen sources (DeMoll and Auffenberg 1993).

3. Reduction of the heterocyclic ring — Reduction of the pyrimidine ring has been shown to be the first step in the degradation of orotic acid by *Clostridium (Zymobacterium) oroticum* (Lieberman and Kornberg 1955) and of uracil by *Cl. uracilicum* (Campbell 1960); the pathways are illustrated in Figure 6.89 and are otherwise unexceptional. Comparable pathways used for the degradation of substituted pyrimidines by aerobic bacteria have already been noted in Section 6.3.1.

4. Cleavage of the iminazole ring of purines — Whereas the degradation of substituted purines generally takes place by initial cleavage of the pyrimidine ring with the formation of formiminoglycine as a key metabolite, an alternative

pathway exists for *Clostridium purinilyticum*. This organism degrades a wide range of purines and was shown to be obligately dependent on selenium for growth (Dürre et al. 1981); under conditions of selenium starvation, this organism degrades uric acid by cleavage of the iminazole ring to produce 5,6-diaminouracil which is then degraded to formate, acetate, glycine, and CO_2 (Dürre and Andreesen 1982).

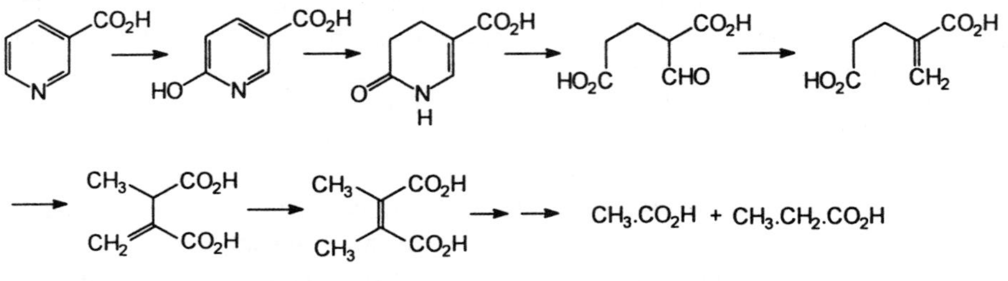

Figure 6.87   Anaerobic biodegradation of nicotinate.

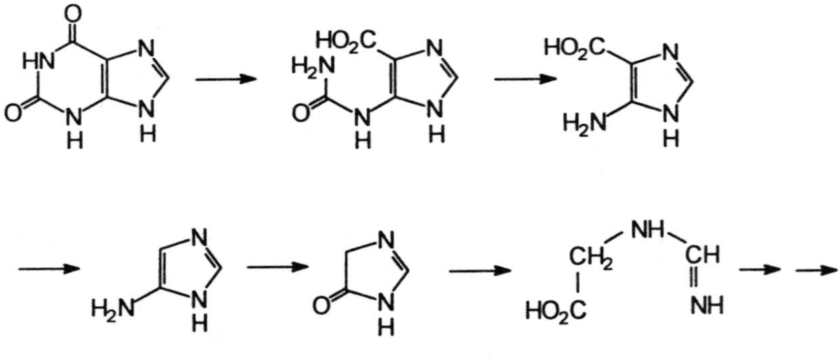

Figure 6.88   Anaerobic biodegradation of xanthine.

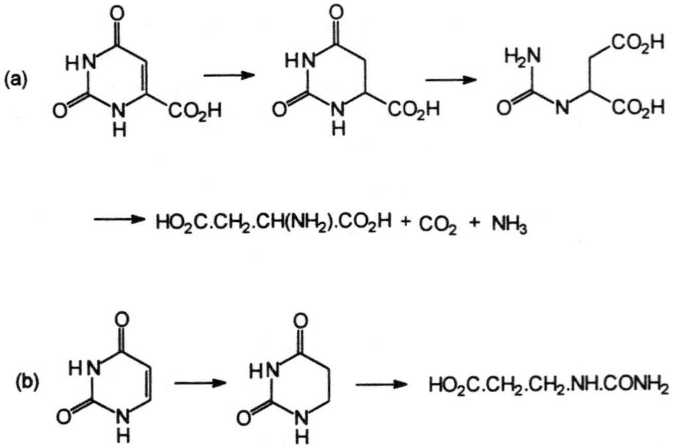

Figure 6.89   Anaerobic biodegradation of **(a)** orotic acid and **(b)** uracil.

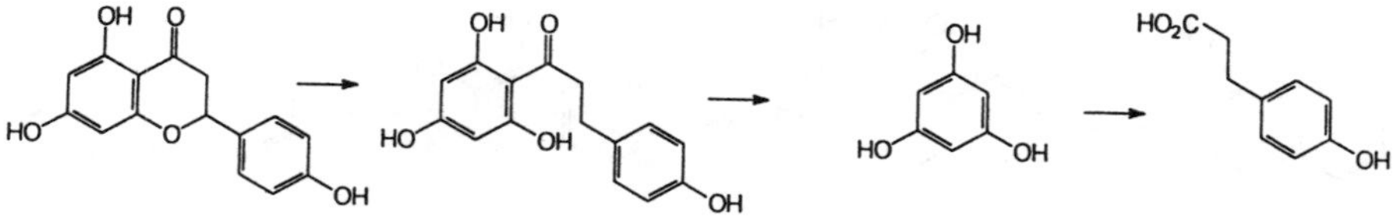

Figure 6.90 Anaerobic biodegradation of flavanoids.

Oxygen-containing heterocyclic compounds — Probably all higher plants synthesize flavanoids of widely varying structure, and these compounds make a significant contribution to the food intake of both herbivores and man; their metabolism has, therefore, been studied under anaerobic conditions which prevail in their digestive tracts. Many of these compounds exist naturally as glycosides, and these are readily hydrolyzed to the aglycones. The primary degradative reaction is generally reductive cleavage of the heterocyclic ring of the aglycones with the formation of aromatic carboxylic acids and phenolic compounds which are then further metabolized. This reduction has been observed in strains of a *Clostridium* sp. (Winter et al. 1989) and a strain of *Butyrivibrio* sp. (Cheng et al. 1971; Krishnamurty et al. 1970), and an illustrative example is given in Figure 6.90. It should be noted that such reactions are not carried out exclusively by anaerobic bacteria since they have been observed also during the aerobic metabolism of the pentahydroxyflavone quercitin by strains of *Rhizobium loti* and *Bradyrhizobium* sp. (Rao et al. 1991).

6.8 AEROBIC DEGRADATION OF AROMATIC COMPOUNDS CONTAINING NITRO OR SULFONATE GROUPS

Aromatic sulfonates are important structural elements of many industrially important dyestuffs, while aromatic nitro compounds are industrially important both as intermediates for the synthesis of a wide range of compounds, including dyestuffs and pharmaceutical products, and as explosives. Increased attention has therefore been directed to their biodegradation, particularly in biological treatment systems.

6.8.1 Aromatic Sulfonates

Whereas chlorine is only exceptionally removed from aromatic rings during dioxygenation, this is more generally the case for carboxyl and sulfonate groups. The pathway for the degradation of aromatic sulfonates has been elucidated in a detailed study (Cain and Farr 1968) and is illustrated for 4-toluenesulfonate in Figure 6.91. The basic reaction is a dioxygenation with subsequent elimination of sulfite as illustrated for naphthalene-1-sulfonate (Figure 6.92), and the dioxygenase has been purified from a naphthalene-2-sulfonate-degrading pseudomonad (Kuhm et al. 1991). Two variants of this pathway have emerged.

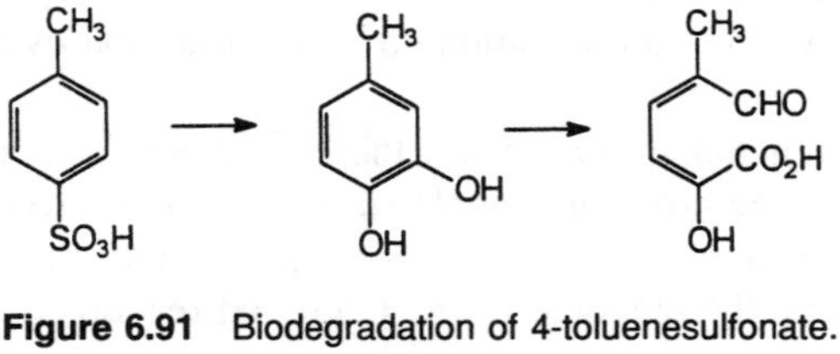

Figure 6.91 Biodegradation of 4-toluenesulfonate.

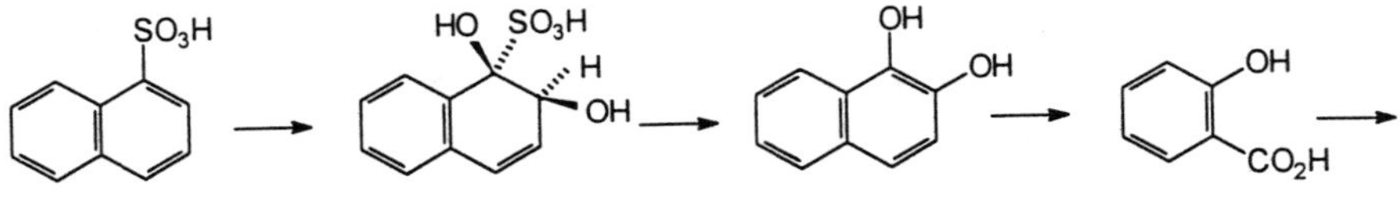

Figure 6.92 Biodegradation of naphthalene-1-sulfonate.

1. Oxidation of the methyl group in 4-toluenesulfonate to a carboxyl group may precede elimination of sulfite (Figure 6.93) (Locher et al. 1991).
2. Considerable attention has been directed to the degradation of naphthalene-sulfonates, and one additional reaction has emerged for the elimination of sulfite. The pathway for the degradation of naphthalene 1,6- and 2,6-disulfonates involves the expected elimination of the 1- or 2-sulfonate groups with the formation of 5-sulfosalicylate; this is then, however, converted into 2,5-dihydroxy-benzoate by what formally may be represented as a hydroxylation with elimination of sulfite (Figure 6.94) (Wittich et al. 1988).

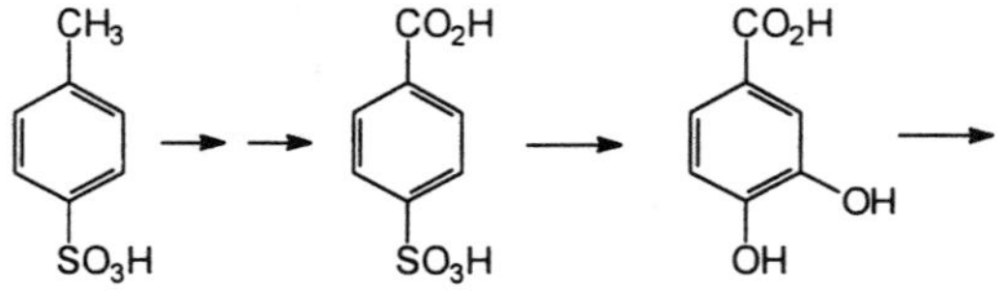

Figure 6.93 Alternative pathway for the biodegradation of 4-toluenesulfonate.

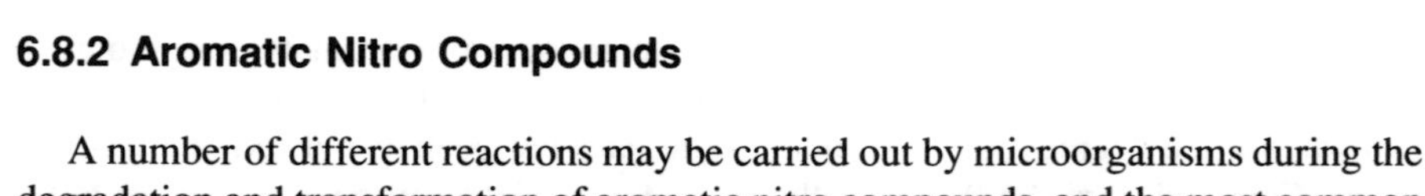

Figure 6.94 Biodegradation of naphthalene-1,6-disulfonate.

6.8.2 Aromatic Nitro Compounds

A number of different reactions may be carried out by microorganisms during the degradation and transformation of aromatic nitro compounds, and the most common of these are as follows:

1. Elimination of nitrite followed by ring cleavage.
2. Reduction to the amine followed in some cases by acetylation.
3. Partial reduction to phenylhydroxylamines followed by rearrangement.
4. Reactions involving dimerization and intramolecular cyclization.

It will become apparent that in some cases, these reactions are merely biotransformations and that they are therefore tangential to degradation. It is worth pointing out that — as is the case for halogen substituents — the position of the nitro groups may be critical in determining the outcome of the microbial reactions. An attempt will be made to illustrate the various possibilities, and examples will be drawn from the metabolism of aromatic nitro compounds by both bacteria and fungi.

Elimination of nitrite — This is a rather widely distributed reaction among bacteria and may be mediated by both dioxygenation and monooxygenation (or hydroxylation) reactions.

1. The term monooxygenation has been applied pragmatically to reactions carried out by bacteria in which a single oxygen atom is incorporated with loss of nitrite, even though the reaction may, in fact, be catalyzed by a hydroxylase. The bacterial degradation of both 2- and 4-nitrophenol involve formation of dihydroxybenzenes, possibly via quinones (Figure 6.95) (Zeyer and Kocher 1988; Spain and Gibson 1991), though the pathway for the degradation of 3-nitrophenol — as might be expected on mechanistic grounds — is apparently different (Zeyer and Kearney 1984).

2. Dioxygenation is apparently involved as the first step in the degradation of dinitro compounds and is formally analogous to the elimination of sulfite from aromatic sulfonates. Nitrocatechols are initially formed from 1,3-dinitrobenzene, and nitrite is then eliminated in a second step (Figure 6.96) (Dickel and Knackmuss 1991; Spanggord et al. 1991).

3. An unusual reaction involving reductive elimination of nitrite has been observed in cultures of a *Pseudomonas* sp. that can use 2,4,6-trinitrotoluene as N-source (Duque et al. 1993). The substrate is transformed by successive loss of nitro groups with the formation of toluene; although this product cannot be metabolized by this strain, it can be degraded by a transconjugant containing the TOL plasmid from *Pseudomonas putida*.

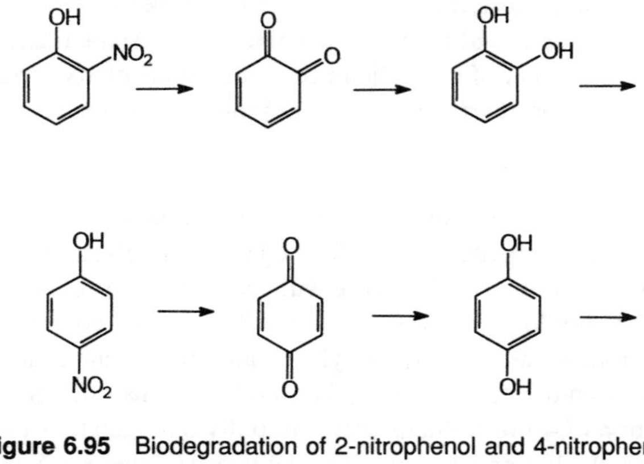

Figure 6.95 Biodegradation of 2-nitrophenol and 4-nitrophenol.

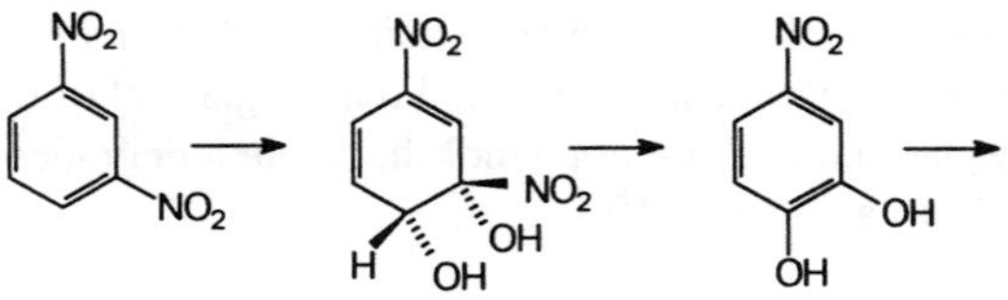

Figure 6.96 Biodegradation of 1,3-dinitrobenzene.

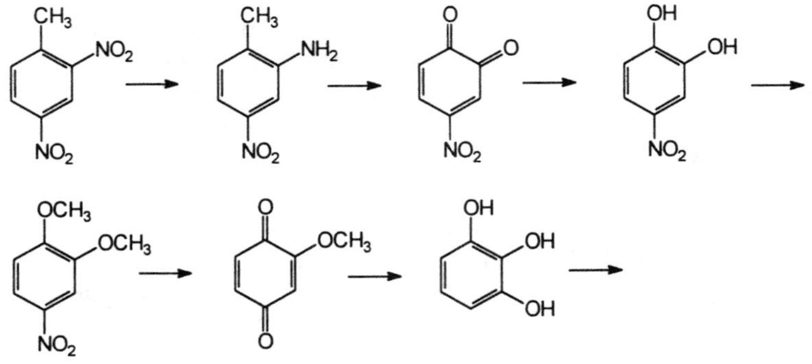

Figure 6.97   Biodegradation of 2,4-dinitrotoluene by *Phanerochaete chrysosporium*.

Degradation of 2,4-dinitrotoluene by *Phanerochaete chrysosporium* has been investigated (Valli et al. 1992b) and involves operation of both the mangenese-dependent peroxidase and lignin peroxidase systems; after the initial reduction of one nitro group, further degradation proceeds by a pathway (Figure 6.97) reminiscent of that used by the same organism for the degradation of 2,4-dichlorophenol (Valli and Gold 1991).

Reduction of nitro groups — Reduction of nitro groups may occur under either aerobic or anaerobic conditions, and the complete sequence of reduction products is produced from 2,6-dinitrotoluene by *Salmonella typhimurium* strain TA 98 (Sayama et al. 1992): 2-nitroso-6-nitrotoluene, 2-hydroxylamino-6-nitrotoluene, and 2-amino-6-nitrotoluene. Complete reduction of the nitro group to an amine may not necessarily occur, however, and partial reduction of the nitro group has been demonstrated. During the metabolism of 4-chloronitrobenzene by a strain of the yeast *Rhodosporodinium* sp., the hydroxylamine that is formed may undergo rearrangement (Corbett and Corbett 1981). Hydroxylamines may also be intermediates during the degradation of a number of aromatic nittro compounds by bacteria: in all of these cases, the hydroxylamine rearranges to an aminophenol before ring cleavage with loss of NH_4^+.

1. During the degradation of 4-nitrobenzoate by *Comamonas acidovorans,* 4-nitroso and 4-hydroxylaminobenzoate were formed successively, and the latter was then metabolized to 3,4-dihydroxybenzoate with the elimination of NH_4^+ (Groenewegen et al. 1992). It should be noted that these cells were not adapted to growth with either 4-aminobenzoate or 4-hydroxybenzoate that are alternative plausible intermediates. A comparable pathway is used by strains of *Pseudomonas* sp. for the degradation of 4-nitrotoluene that is initially oxidized to 4-nitrobenzoate via 4-nitrobenzyl alcohol and 4-nitrobenzaldehyde (Haigler and Spain 1993; Rhys-Williams et al. 1993).

2. The biodegradation of nitrobenzene by a strain of *Pseudomonas pseudoalcaligenes* takes place by an enzymatically mediated rearrangement of the initially formed phenylhydroxylamine to 2-aminophenol that is then degraded by extradiol ring cleavage (Nishino and Spain 1993).

It is appropriate to summarize some of the metabolic complexities for this apparently straightforward reaction.

1. The amines may not be further metabolized by the organism carrying out the reduction so that their formation is really a biotransformation; this is noted again later. The amines may not even lie on the degradative pathway; this was clearly the case for the aerobic metabolism of nitrobenzoates (Cartwright and Cain 1959), for some dinitro compounds (Schackmann and Müller 1991), and for 4-nitrobenzoate (Groenewegen et al. 1992).

2. The established carcinogenicity of many aromatic amines has led to concern over the health risk associated with exposure to nitrated aromatic compounds; this has motivated studies on their reductive metabolism by intestinal and other anaerobic bacteria; a range of pure cultures has been examined including *Veilonella alkalescens* (McCormick et al. 1976), strains of clostridia, and a *Eubacterium* sp. (Rafii et al. 1991) and a range of strains of methanogenic and sulfidogenic bacteria and clostridia (Gorontzy et al. 1993).

3. In some aerobic bacteria, the amines are acetylated presumably to less toxic metabolites (Van Alfen and Kosuge 1974; Beunink and Rehm 1990).

4. As in the case noted above with 2,6-dinitrotoluene, reduction to the amine presumably proceeds in stages, and the intermediate hydroxylamines may undergo chemical rearrangement; the metabolism of 4-chloronitrobenzene by a strain of the yeast *Rhodosporodinium* sp. illustrates this possibility as well as the complete reduction to the amine followed by acetylation (Figure 6.98) (Corbett and Corbett 1981).

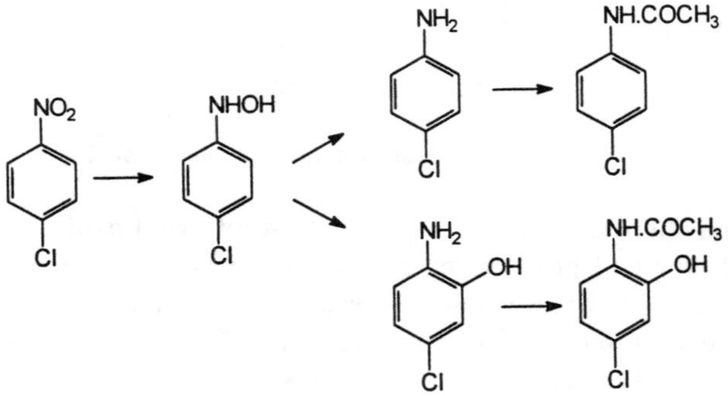

Figure 6.98 Metabolism of 4-chloronitrobenzene by *Rhodosporodinium* sp.

Biotransformation reactions with retention of the nitro substituents — A number of biotransformation reactions in which reduction of the nitro group occurs have already been noted, but two examples are given in which the nitro group is retained.

1. The biotransformation of nitrotoluenes has been examined in toluene-grown cells of pseudomonads which contain a toluene dioxygenase. The products depended on the relative position of the substituents and clearly illustrated the various metabolic reactions that may be mediated by this dioxygenase; 2-nitro- and 3-nitrotoluene were converted into the corresponding benzyl alcohols, whereas 4-nitrotoluene produced 3-methyl-6-nitrocatechol and 2-methyl-5-nitrophenol (Figure 6.99) (Robertson et al. 1992).

2. The degradation of phenols containing several nitro groups has revealed an unusual reaction. Although this occurs with the elimination of nitrite — not apparently by any of the pathways illustrated above — terminal metabolites are formed by reduction of the aromatic ring. 4,6-dinitrohexanoate is produced from 2,4-dinitrophenol (Lenke et al. 1992) and 2,4,6-trinitrocyclohexanone from 2,4,6-trinitrophenol (Lenke and Knackmuss 1992). The reductive pathways may plausibly be attributed to the presence of the strongly electron-withdrawing nitro groups which facilitate what are formally hydride reductions. The formation of 2,4-dinitrophenol from 2,4,6-trinitrophenol by *Rhodococcus erythropolis* is consistent with a pathway involving loss of nitrite from a Meisenheimer-like hydride complex (Lemke and Knackmuss 1992), and an analogous complex has been identified from a strain of *Mycobacterium* sp. during metabolism of 2,4,6-trinitrotoluene (Vorbeck et al. 1994).

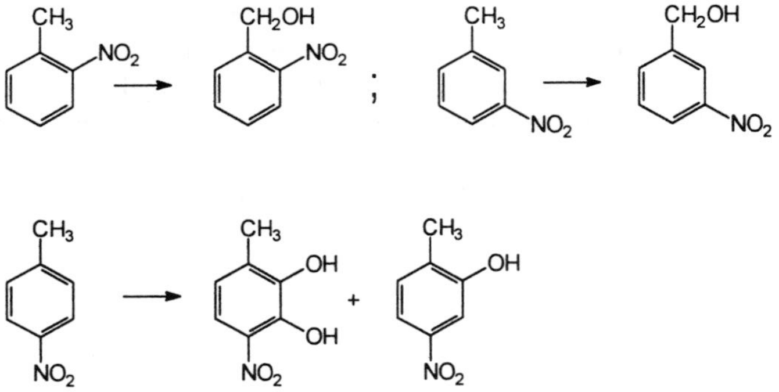

Figure 6.99 Biotransformation of nitrotoluenes.

Interactions between reduction products at different oxidation levels — The formation of coupling products by interactions between metabolites at different levels of oxidation may reasonably be presumed to account for the identification of azoxy compounds as biotransformation products of 2,4-dinitrotoluene by the fungus *Mucrosporium* sp. (Figure 6.100) (McCormick et al. 1978) and by a *Pseudomonas* sp. that uses 2,4,6-trinitrotoluene as a N-source (Duque et al. 1993).

Intramolecular reactions — Cyclization to form benziminazoles may take place in compounds in which the nitro group is vicinal to an amino or substituted amine group. This has been demonstrated as one of the pathways during the fungal metabolism of dinitramine (Figure 6.101) (Laanio et al. 1973) and in a mixed bacterial culture which metabolized 2-nitroaniline to 2-methylbenziminazole (Hallas and Alexander 1983).

6.8.3 Aromatic Azo Compounds

Aromatic azo compounds are components of many commercially important dyes and pigments, so that attention has been directed to their degradation and transformation. Possibly the most significant discovery on the metabolism of aromatic azo compounds that had implications which heralded the age of modern chemotherapy concerned the bactericidal effect of the azo dye Prontosil. It was shown that the effect

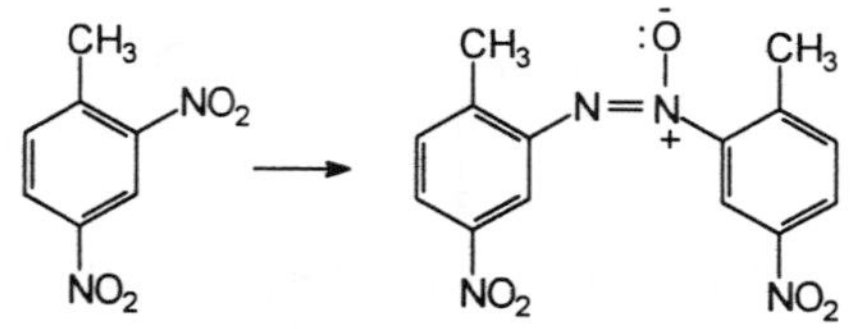

Figure 6.100 Metabolism of 2,4-dinitrotoluene by *Mucrosporium* sp.

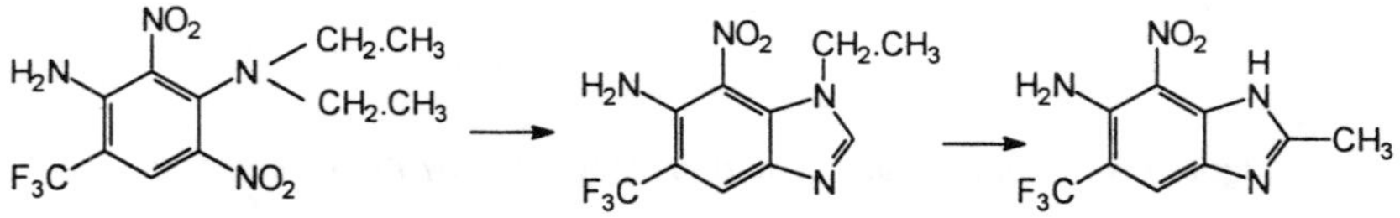

Figure 6.101 Biodegradation of dinitramine.

of Prontosil *in vivo* was in fact due to the action of its transformation product sulfanilamide, which is an antagonist of 4-aminobenzoate that is required for the synthesis of the vitamin folic acid. Indeed, this reduction is the typical reaction involved in the first stage of the biodegradation of aromatic azo compounds. The reaction is readily accomplished under anaerobic conditions (Haug et al. 1991), and the azoreductase and nitroreductase from *Clostridium perfringens* apparently involve the same protein (Rafii and Cerniglia 1993). Bacterial azoreductases have, however, been purified also from aerobic organisms including strains of *Pseudomonas* sp. adapted to grow at the expense of azo dyes (Zimmermann et al. 1984). The amines resulting from all these reductive transformations then enter well-established metabolic pathways for the degradation of anilines (McClure and Venables 1986; Fuchs et al. 1991).

6.9 ALIPHATIC COMPOUNDS CONTAINING OXYGEN, NITROGEN, SULFUR, AND PHOSPHORUS

6.9.1 Ethers and Amines

Aerobic pathways — Degradation of the simplest representatives under aerobic conditions has been investigated in the course of comprehensive studies on the metabolic potential of methylotrophic bacteria. These reactions have been admirably summarized in a book (Anthony 1982) to which reference may be made for details. Only two simple examples will therefore be used for illustration:

1. Diethyl ether is metabolized by monooxidation at C_1 to ethanol and acetaldehyde.
2. By successive formation of formaldehyde, tetramethylammonium salts are oxidized either directly to formaldehyde or indirectly via N-methylglutamate; the C_1 metabolites then enter the conventional pathways for their further degradation to CO_2.

In contrast, *Aminobacter aminovorans* (*Pseudomonas aminovorans*), although it is able to utilize methylamine and trimethylamine, is unable to use methane, methanol, or

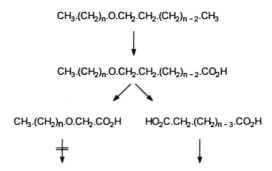

Figure 6.102 Biodegradation of di-*n*-heptyl ether by *Acinetobacter* sp.

dimethylamine (Urakami et al. 1992). The potential complexities of these apparently straightforward monooxygenation reactions is revealed by the fact that although the degradation of nitrilotriacetate takes place by successive loss of glyoxylate (Cripps and Noble 1973; Firestone and Tiedje 1978), the monooxygenase system consists of two components, both of which are necessary for hydroxylation (Uetz et al. 1992).

Degradation of symmetrical long-chain dialkyl ethers may be used to illustrate an entirely different metabolic pathway. The di-*n*-heptyl-, di-*n*-octyl-, di-*n*-nonyl-, and di-*n*-decyl ethers are degraded by a strain of *Acinetobacter* sp. to two different groups of metabolites (Figure 6.102):

1. To *n*-heptan-, *n*-octan-, *n*-nonan-, and *n*-decanol-1-acetic acids which are not metabolized further.
2. To glutaric, adipic, pimelic, and suberic acids which serve as sources of carbon and energy; these compounds are formed by terminal oxidation followed by an unusual oxidation at the carbon atom β to the ether bond (Modrzakowski and Finnerty 1980).

The possible persistence in the environment of alkylethoxy sulfates has led to extensive investigations of their degradability. Although the role of alkyl sulfatase activity is noted in Section 6.11.1, ether-cleavage has been shown to be the major pathway for the degradation of dodecyltriethoxy sulfate (Hales et al. 1986).

Three groups of xenobiotics containing aryl ether groups have been extensively used and have given rise to environmental concern.

1. The considerable interest in the persistence of chlorinated phenoxyalkanoates — and of phenoxyacetates, in particular, that have been used as agrochemicals — has stimulated studies on the degradation of aromatic ethers. It should, however, be pointed out that considerably greater attention has been devoted to elucidating the subsequent steps that culminate in the fission of the aromatic ring that have been discussed in Section 6.5.1. The first step in the degradation of phenoxyalkanoates is dealkylation to the corresponding phenol with the formation of glyoxylate (Evans et al. 1971; Pieper et al. 1988). Although this reaction involves monooxygenation of the aromatic ring (Gamar and Gaunt 1971), the enzyme is, in fact, an α-ketoglutarate-dependent dioxygenase (Fukumori and Hausinger 1993).

2. Concern has also been expressed over the extensive use of alkylphenol polyethoxylates that are used as detergents, since some of the metabolites are apparently persistent compounds; so although the alkyl phenols may be formed as a result of complete oxidation of the polyethoxylate side chains, partially degraded metabolites are, in some cases, apparently resistant to further degradation (Ball et al. 1989).

3. Halogenated derivatives of diphenyl ether have been used as pesticides so that attention has been directed to this class of compounds that includes, in principle, also structures such as dibenzofuran and dibenzo-1,4-dioxin. The degradation of diphenyl ether itself by *Pseudomonas cepacia* has been examined (Pfeifer et al. 1989, 1993) and yields 2-pyrone-6-carboxylate as a stable end-product; it may be formed from the initially produced 2,3-dihydroxydiphenyl ether in a reaction formally analogous (Figure 6.103) to that in which 3-*O*-methylgallate is converted to 2-pyrone-4,6-dicarboxylate by 3,4-dihydroxybenzoate 4,5-dioxygenase in pseudomonads (Kersten et al. 1982). The degradation of diphenyl ether by a strain of *Sphingomonas* sp. proceeded with degradation of both rings (Schmidt et al. 1992), and cells grown with diphenyl ether were able to oxidize dibenzo-1,4-dioxin to 2-(2-hydroxyphenoxy)-*cis,cis*-muconate. After adaptation to growth with 4,4'-difluorodiphenyl ether, the organism grew with the corresponding chloro, but not the bromo, compound (Schmidt et al. 1993). In general, however, some degree of recalcitrance seems to be associated with substituted diaryl ethers.

Anaerobic pathways — The anaerobic degradation of polyethylene glycol (PEG) has been investigated in a variety of organisms including *Pelobacter venetianus* (Schink and Stieb 1983), an *Acetobacterium* sp. (Schramm and Schink 1991), and *Desulfovibrio desulfuricans* and a *Bacteroides* sp. (Dwyer and Tiedje 1986). The initial product is acetaldehyde, which is formed in two stages by the action of a diol dehydratase and a polyethylene glycol acetaldehyde lyase (Figure 6.104) (Frings et al.1992) which is apparently found in all PEG-degrading anaerobic bacteria.

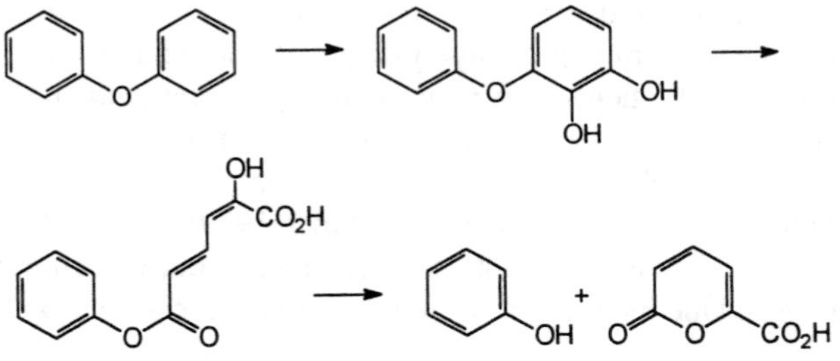

Figure 6.103 Biodegradation of diphenyl ether.

$$HO.(CH_2.CH_2.O)_n.CH_2.CH_2.OH \longrightarrow HO.(CH_2.CH_2.O)_n.CH(OH).CH_3$$

$$\longrightarrow HO.(CH_2.CH_2.O)_{n-1}CH_2.CH_2OH + CH_3.CHO$$

Figure 6.104 Anaerobic biodegradation of polyethylene glycol.

6.9.2 Aliphatic Nitro Compounds

In contrast to aromatic nitro compounds, little attention appears to have been directed to their aliphatic counterparts. Two examples may be used to illustrate aerobic and anaerobic reactions.

1. 2-nitropropane is oxidized by *Hansenula mrakii*, and the enzyme has been purified; this is a dioxygenase that converts two molecules of the substrate to two molecules of acetone with the elimination of nitrite (Kido and Soda 1976).
2. The plant toxins 3-nitropropionic acid and 3-nitropropanol are converted into the corresponding amines by mixed cultures of ruminant microorganisms, and the β-alanine resulting from the former is then further degraded (Anderson et al. 1993).

6.9.3 Sulfides, Disulfides, and Related Compounds

Interest in the possible persistence of aliphatic sulfides has arisen from the fact that they are produced in marine anaerobic sediments and from an appreciation of the possible implication of dimethylsulfide in climate alteration (Charlson et al. 1987) that has been noted briefly in Chapter 4 (Section 4.2).

Sulfides and related compounds may be degraded by a number of quite different pathways which may be illustrated by the following examples:

1. Anaerobic reduction of dimethyl sulfide with production of methane (Oremland et al. 1989), and of dibenzyl disulfide to toluenethiol and finally toluene (Miller 1992).
2. Elimination reactions have been implicated in the degradation of a variety of simple sulfur compounds:
   - The formation of pyruvate from cysteine is mediated by a desulfhydrase (Figure 6.105a) (Kredich et al. 1973), and pyruvate is also formed by a similar reaction from aminoethylcysteine (Rossol and Pühler 1992).
   - The formation of ethene from methionine by *Escherichia coli* takes place by an elimination reaction (Figure 6.105b) (Ince and Knowles 1986), though considerable complexities in the control and regulation of this reaction have emerged (Mansouri and Bunch 1989). It should be noted that the synthesis of ethene in plants proceeds by an entirely different reaction via *S*-adenosyl methionine and 1-aminocyclopropane-1-carboxylate (Kende 1989).
   - Elimination is one of the pathways used for the degradation of dimethylsulfoniopropionate and has been demonstrated in a strain of *Clostridium* sp. (Figure 6.105c) (Wagner and Stadtman 1962).
3. Dimethyldisulfide is degraded by autotrophic sulfur bacteria with the formation of sulfate and CO_2 which then enters the Benson-Calvin cycle (Smith and Kelly 1988). On the other hand, dimethyl sulfide and dimethyl sulfoxide are degraded by a strain of *Hyphomicrobium* sp. by pathways involving the formation from both carbon atoms of formaldehyde which subsequently enters the serine pathway

(Figure 6.106) (Suylen et al. 1986). The key enzyme is methyl mercaptan oxidase which converts methyl sulfide into formaldehyde, sulfide, and peroxide (Suylen et al. 1987). A strain of *Thiobacillus* sp. metabolizes dimethyl sulfide by an alternative pathway involving transfer of the methyl group, probably to tetrahydrofolate by a cobalamin carrier (Visscher and Taylor 1993). Oxygen is not involved in the removal of the methyl groups so that the reaction may proceed anaerobically.

4. Dimethylsulfoniopropionate is *S*-demethylated to 3-methylthiopropionate by a strain of *Desulfobacterium* sp. (Van der Maarel 1993) that also degrades betaine to dimethylglycine by a formally analogous reaction, and both 3-methylthiopropionate and methyl sulfide are produced by anoxic marine sediments (Kiene and Taylor 1988).

6.9.4 Phosphonates

Considerable interest has been directed to the biodegradation of phosphonates which contain the rather unusual C-P bond, since a number of them are naturally

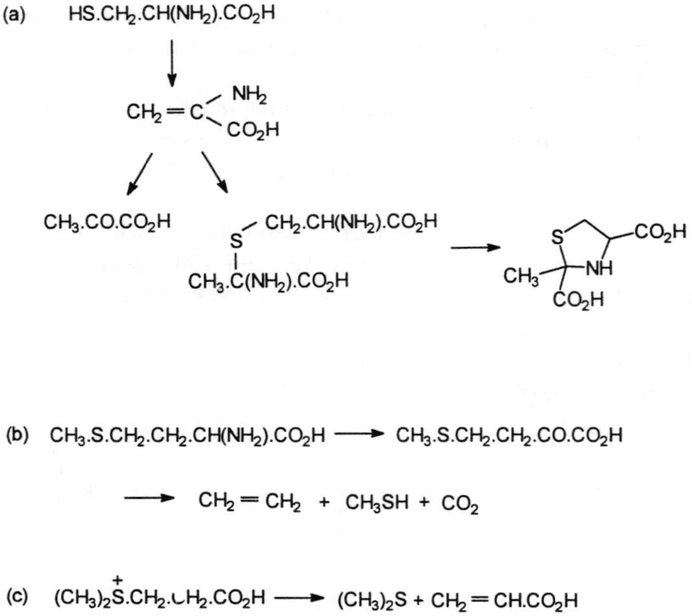

Figure 6.105 Examples of elimination reactions during metabolism of aliphatic sulfur compounds (a) cysteine (b) methionine (c) dimethylsulfoniopropionate.

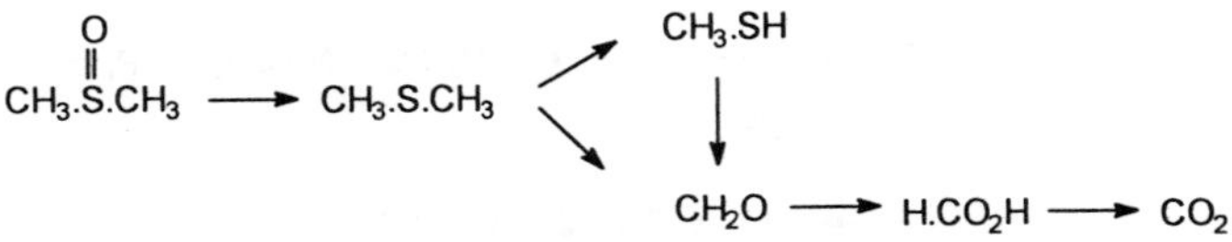

Figure 6.106 Biodegradation of dimethyl sulfoxide by *Hyphomicrobium* sp.

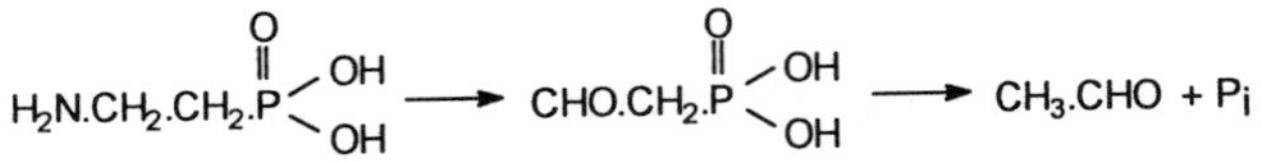

$$H_2N.CH_2.CH_2.P\begin{matrix}O\\\|\end{matrix}\begin{matrix}OH\\OH\end{matrix} \longrightarrow CHO.CH_2.P\begin{matrix}O\\\|\end{matrix}\begin{matrix}OH\\OH\end{matrix} \longrightarrow CH_3.CHO + P_i$$

Figure 6.107 Biodegradation of 2-aminomethylphosphonate.

$$\begin{matrix}CH_3\\CH_3\end{matrix}P\begin{matrix}O\\\end{matrix}OH \longrightarrow CH_3-P\begin{matrix}O\\\|\end{matrix}\begin{matrix}OH\\OH\end{matrix} + CH_4 \longrightarrow P_i + CH_4$$

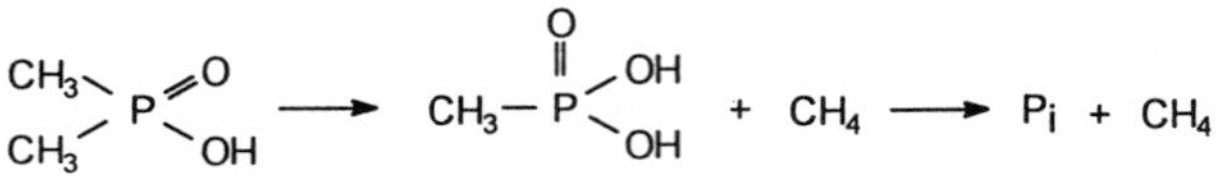

Figure 6.108 Reductive biodegradation of alkyl phosphonates and phosphites.

occurring compounds, and one of them — glyphosate — has been extensively used as a herbicide.

Many bacteria are able to use phosphonates as a source of phosphorus (Cook et al. 1978a; Schowanek and Verstraete 1990), and this necessarily involves cleavage of the C-P bond with the formation of inorganic phosphate. In *Bacillus cereus*, 2-aminomethylphosphonate is initially oxidized to 2-phosphonoacetaldehyde (La Nauze and Rosenberg 1968) before cleavage of the C-P bond (Figure 6.107) (La Nauze et al. 1970). Alkyl phosphonates and phosphites are degraded by a number of bacteria by a pathway in which the alkyl groups are reduced to alkanes (Figure 6.108) (Wackett et al. 1987).

In view of its importance as a herbicide, the degradation of glyphosate has been investigated in a number of organisms, and two pathways have been elucidated, differing in the stage at which the C-P bond is cleaved:

1. Loss of a C_2 fragment — formally glyoxylate — with the formation of aminomethylphosphonate (Pipke and Amrhein 1988) which may be further degraded by cleavage of the C-P bond to methylamine and phosphate (Figure 6.109a) (Jacob et al. 1988).
2. Initial cleavage of the C-P bond with the formation of sarcosine which is then metabolized to glycine (Figure 6.109b) (Pipke et al. 1987; Liu et al. 1991).

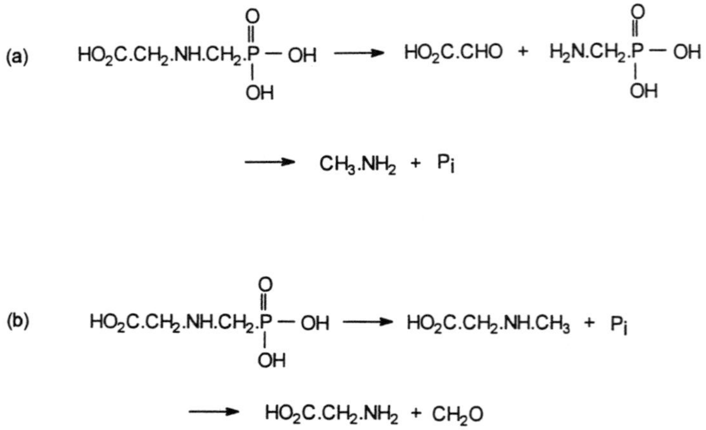

(a) $$HO_2C.CH_2.NH.CH_2.P\begin{matrix}O\\\|\end{matrix}OH \longrightarrow HO_2C.CHO + H_2N.CH_2.P\begin{matrix}O\\\|\end{matrix}OH$$
$$\qquad\qquad\qquad\quad OH\qquad\qquad\qquad\qquad\qquad\qquad\qquad OH$$

$$\longrightarrow CH_3.NH_2 + P_i$$

(b) $$HO_2C.CH_2.NH.CH_2.P\begin{matrix}O\\\|\end{matrix}OH \longrightarrow HO_2C.CH_2.NH.CH_3 + P_i$$
$$\qquad\qquad\qquad\quad OH$$

$$\longrightarrow HO_2C.CH_2.NH_2 + CH_2O$$

Figure 6.109 Alternative pathways for the biodegradation of glyphosate.

The gene cluster required for the utilization of phosphonates is induced in *Escherichia coli* by phosphate limitation, and genetic evidence suggests a connection between the metabolism of phosphonates and phosphites. On the basis of this, the interesting suggestion has been made that there may exist a phosphorus redox cycle and that phosphorus is involved not only at the +5 oxidation level but at lower oxidation levels (Metcalf and Wanner 1991).

6.10 ORGANOFLUORINE COMPOUNDS

These deserve a section to themselves, since fluorine differs so much from the other halogens, even from that of its nearest relative, chlorine. There are two outstanding features of fluorine: (1) its high electron-attracting power or electronegativity and (2) its small atomic radius so that the length of the C-F bond is 1.36Å compared with 1.755Å for the C-Cl bond; it is, therefore, closer to that of a normal saturated C-H bond with a length of 1.09Å. Organofluorine compounds have achieved enormous industrial significance in view of their generally great chemical stability (e.g., polytetrafluoroethylene) or their volatility (e.g., fluoroalkanes used as refrigerants), while some of them are valuable pharmaceutical products. Indeed, there has been increasing interest in perfluorinated aliphatic compounds as surfactants (Dams 1993) and in perfluoropolyethers (Guarini et al. 1993). Attention has been drawn in Chapter 2 (Section 2.5) to some unusual fluorinated by-products from industrial activity. In addition, some organofluorine compounds are toxic, including simple naturally occurring compounds such as fluoroacetate, whereas others such as the phosphorofluoridates have been deliberately synthesized as nerve poisons with potential military application. Systematic investigation of the biodegradability of organofluorine compounds appears not to have been made, but an attempt will be made to summarize the principles which have hitherto emerged.

6.10.1 Aliphatic Fluoro Compounds

Very few details are available on the biodegradation of aliphatic fluoro compounds. Although a pseudomonad has been shown to degrade fluoroacetate (Meyer et al. 1990) and the enzymology has been examined (Goldman 1965), lack of definitive information on other compounds and the limited extent of the effort directed to studying the degradation of aliphatic fluoro compounds suggests caution in concluding that these are recalcitrant molecules — even though chemical evidence would suggest their substantial resistance to biodegradation.

6.10.2 Aromatic Fluoro Compounds

Substantial activity has been directed to the degradation particularly of fluorobenzoates under aerobic conditions. It should be noted that the ability to use these as sole source of carbon and energy is more restricted than for the corresponding chlorobenzoates, although they may be degraded under conditions of concurrent metabolism in the presence of suitable growth substrates. Pathways for the degradation of the chlorinated analogues may not be viable since, for example, 3-fluorocatechol

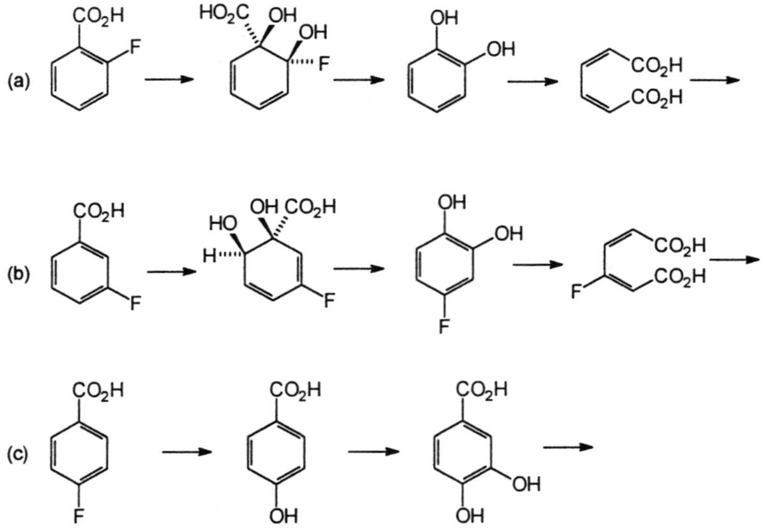

Figure 6.110 Pathways for the biodegradation of **(a)** 2-fluorobenzoate **(b)** 3-fluorobenzoate **(c)** 4-fluorobenzoate.

is not generally a good substrate for the catechol-cleaving enzymes. There are, therefore, alternative strategies that are used for the degradation of fluorobenzoates.

1. Elimination of fluoride from 2-fluorobenzoate may take place concomitantly with dioxygenation and decarboxylation (Figure 6.110a) (Engesser et al. 1980).
2. Degradation of 3-fluoro- and 4-fluorobenzote may proceed by dioxygenation with retention of fluorine which is subsequently eliminated from 3-fluoromuconate (Figure 6.110b) (Engesser et al. 1990a).
3. A strain of *Aureobacterium* sp. degrades 4-fluorobenzoate by hydrolytic elimination of fluoride and initial formation of 4-hydroxybenzoate (Figure 6.110c) (Oltmanns et al. 1989).
4. 4-hydroxybenzoate hydroxylase is able to catalyze loss of fluoride from 3,5-difluoro-4-hydroxybenzoate and 2,3,5,6-tetrafluoro-4-hydroxybenzoate with the formation of catechols (Husain et al. 1980), and in an apparently analogous reaction pentachlorophenol hydroxylase converts pentafluorophenol into 2,3,5,6-tetrafluoro-1,4-dihydroxybenzene (Uotila et al. 1992).

A range of 3-fluoro-substituted benzenes has been examined using a toluene-degrading strain of *Pseudomonas* sp. (Renganathan 1989), and two reactions were observed: one in which the fluorine was eliminated during dioxygenation and the other in which it was retained (Figure 6.111). The degradation of 5-fluorouracil by the fungus *Nectria haematococca* proceeds by the established pathway for the non-fluorinated compound with the formation of the stable α-fluoro-β-alanine (Parisot et al. 1991).

The degradation of aromatic fluoro compounds under anaerobic conditions appears to have attracted less attention, but a single example will be used as illustration;

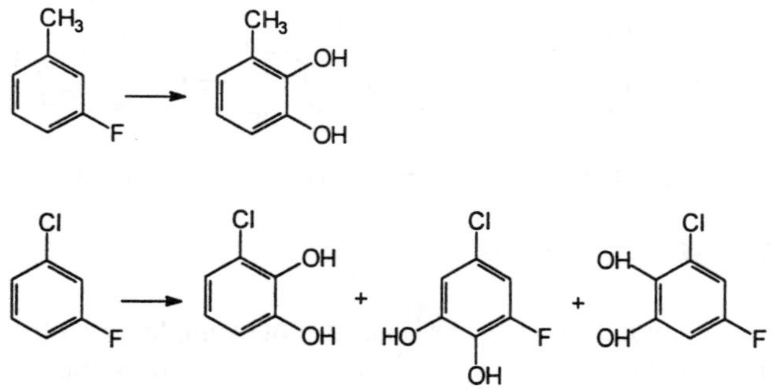

Figure 6.111 Biodegradation of 3-fluorotoluene and 3-chlorofluorobenzene.

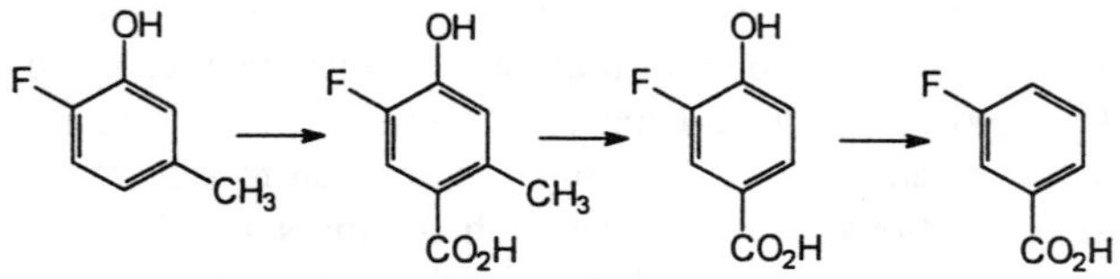

Figure 6.112 Biotransformation of 6-fluoro-3-methylphenol.

the pathway for the non-fluorinated analogue 3-methylphenol has already been noted in Section 6.7.3, and the principles used for delineating the pathway in Chapter 5 (Section 5.5.3). Under methanogenic conditions, 6-fluoro-3-methylphenol was degraded to 3-fluorobenzoate via 5-fluoro-4-hydroxy-2-methylbenzoate and 3-fluoro-4-hydroxybenzoate (Figure 6.112) (Londry and Fedorak 1993). Comparable carboxylation and dehydroxylation reactions have also been demonstrated with 2- and 3-fluorophenol that form 2- and 3-fluorobenzoate (Genther et al. 1990). It should be noted that in neither case was the fluorine atom eliminated even under these highly reducing conditions, so that the persistence of aromatic fluoro compounds under anaerobic conditions may, at least provisionally, be accepted.

6.10.3 Trifluoromethyl Compounds

The extraordinary chemical stability of the trifluoromethyl group is illustrated by the original synthesis of trifluoracetic acid by chromic acid oxidation of 3-trifluoromethylaniline (Swarts 1922); this recalcitrance is also revealed by the results of studies on the biodegradation of compounds containing the CF_3 group. Trifluoromethylbenzoates cannot support growth of aerobic bacteria, even though they are apparently effective substrates for the enzymes used for the metabolism of the methylbenzoates. Ring dioxygenation and dehydrogenation take place with the formation of trifluoromethyl catechols which may be further degraded with ring fission to form terminal metabolites in which the trifluoromethyl group remains intact (Figure 6.113) (Engesser et al. 1988 a,b); this has been confirmed in a more extensive study (Engesser et al. 1990b). Attention has already been directed in Chapter 4 (Section 4.1.1) to the degradability of 3-trifluoromethyl-4-nitrophenol photochemically (Carey and Cox 1981) and of 3- and 4-trifluoromethylbenzoates by sequential microbial and photochemical treatment (Taylor et al. 1993).

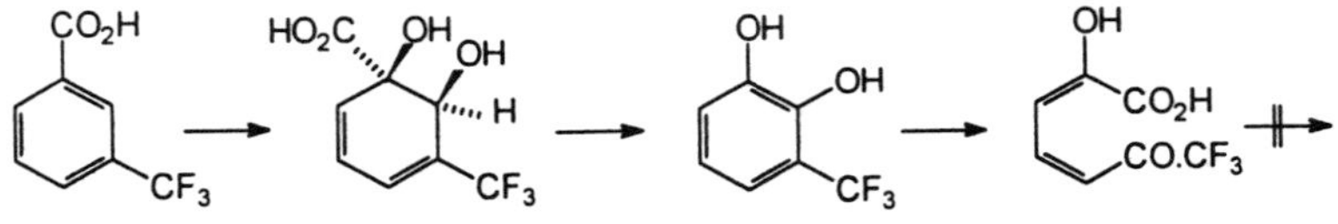

Figure 6.113   Metabolism of 3-trifluoromethylbenzoate.

6.11 BIOTRANSFORMATIONS

In the course of previous discussions, a number of examples of biotransformations have been illustrated, but it is convenient to summarize briefly some others that have not been discussed and that involve structural groups that are represented in diverse xenobiotics. Biotransformations have attracted interest for at least two reasons: first, from their biotechnological importance in the production of valuable organic compounds including pharmaceuticals, and second, because of their environmental significance. In addition, some biotransformation reactions are a response elicited from an organism by exposure to a toxicant and, therefore, represent a detoxification mechanism for the organism. Four sections are devoted to selected illustrations of biotransformation reactions, whereas the fifth attempts to illustrate the possible adverse environmental significance of some of the reactions.

6.11.1 Hydrolysis of Esters, Amides, and Nitriles

Hydrolytic activity towards a range of esters and related compounds is widely distributed among microorganisms, and reference has already been made to some of them. A somewhat brief summary may, therefore, suffice.

Carboxylic acid esters — Two examples may be given merely as illustration of the diversity of potential substrates: the hydrolysis of dibutyl phthalate (Eaton and Ribbons 1982) and of cocaine to benzoate and ecgonine methyl ester (Britt et al. 1992).

Sulfate esters — Alkyl sulfates and alkylethoxy sulfates have been extensively used as detergents so that concern has been expressed on their biodegradability; a review (Cain 1981) covers the degradation of a wide range of surfactants including both of these groups. It should also be appreciated that the hydrolysis of a range of naturally occurring sulfate esters may make a contribution to the sulfate present in aerobic soils (Fitzgerald 1976) quite apart from that contributed from the anthropogenic input of SO_x.

Long-chain unbranched aliphatic sulfate esters are generally degraded by initial hydrolysis to sulfate and the alkanol which is then degraded by conventional pathways; the alkylsulfatases show a diversity of specificity (Dodson and White 1983) generally for compounds with chain lengths >5, although organisms have been isolated that degrade short-chain (C_1 to C_4) primary alkyl sulfates (White et al. 1987). For alkylethoxy sulfates, a greater range of possibilities exists including ether-cleavage reactions that have been noted in Section 6.9.1, and direct removal of sulfate may be of less significance (Hales et al. 1986). It should be noted, however, that an unusual reaction may occur simultaneously: chain elongation of the carboxylic acid. For example, during degradation of dodecyl sulfate, lipids containing 14, 16, and 18 carbon atoms were synthesized (Thomas and White 1989). For chiral compounds such as octan-2-yl sulfate, hydrolysis proceeds with inversion of configuration by

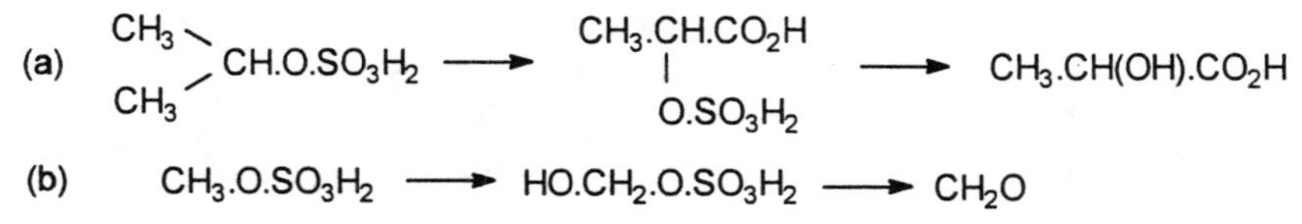

(a)
$$CH_3 \diagdown CH.O.SO_3H_2 \longrightarrow \begin{matrix} CH_3.CH.CO_2H \\ | \\ O.SO_3H_2 \end{matrix} \longrightarrow CH_3.CH(OH).CO_2H$$

(b) $CH_3.O.SO_3H_2 \longrightarrow HO.CH_2.O.SO_3H_2 \longrightarrow CH_2O$

Figure 6.114 Biodegradation of **(a)** propan-2-yl sulfate and **(b)** methyl sulfate.

cleavage of the alkyl-oxygen bond (Bartholomew et al. 1977). As an alternative pathway for degradation, oxidation may precede elimination of sulfate; examples are found in the degradation of propan-2-yl sulfate (Crescenzi et al. 1985) and of monomethyl sulfate (Figure 6.114) (Davies et al. 1990).

The hydrolysis of aryl sulfates has traditionally been a useful taxonomic character in the genus *Mycobacterium* (Wayne 1961). A test for the hydrolysis of phenolphthalein sulfate has been routinely incorporated into mycobacterial taxonomy, and a positive result in the 3-day aryl sulfatase test has been particularly valuable for distinguishing members of the rapidly growing *M. fortuitum* group which are potentially pathogenic to man; slow-growing *M. tuberculosis* and *M. bovis* are negative even after 10d. Regulation of the hydrolysis of phenolic sulfates is less straightforward than might appear from the nature of the reaction. The regulation of the synthesis of tyrosine sulfate sulfohydrolase has been examined in a strain of *Comamonas terrigena* (Fitzgerald et al. 1979), and both inducible and constitutive forms of the enzyme exist and are apparently distinct from the aryl sulfate sulfohydrolase which has found application in taxonomic classification and has been noted above.

Phosphate esters and related compounds — Concern over the persistence and the biodegradability of organophosphate and organophosphorothioates that are used as agrochemicals has stimulated studies into their degradation. Investigations have also been directed to the use of their degradation products as a source of phosphate for the growth of bacteria; a wide range of phosphates, phosphorothioates, and phosphonates has therefore been examined as suitable P-sources (Cook et al. 1978b). The first step in the degradation of all these phosphate and phosphorothioate esters is hydrolysis, and substantial effort has been directed to all of these groups; a summary of investigations in this important area has been given (Munnecke et al. 1982), and it is therefore sufficient merely to provide a typical illustration (Figure 6.115a). In addition, it is worth noting that the initial metabolites after hydrolysis such as the nitrophenols may be both toxic and apparently sometimes resistant to further degradation; this has been discussed in Chapter 4 (Section 4.4.1) and in Chapter 5 (Section 5.3.1).

Nitrate esters — Compared with the fairly numerous investigations on the microbial hydrolysis of carboxylic acid and sulfate and phosphate esters, data on the hydrolysis of nitrate esters is much more fragmentary. This has been clearly revealed in a review (White and Snape 1993) that summarizes existing knowledge on the microbial degradation of nitrate esters, including glycerol trinitrate and its close relatives, and the pharmaceutical products pentaerythritol tetranitrate and isosorbide 2,5-dinitrate. As an alternative to purely hydrolytic reactions, both reductive reactions that have been noted in Chapter 5 (Section 5.5.5) and reactions involving glutathione transferases seem to be important in eukaryotic microorganisms.

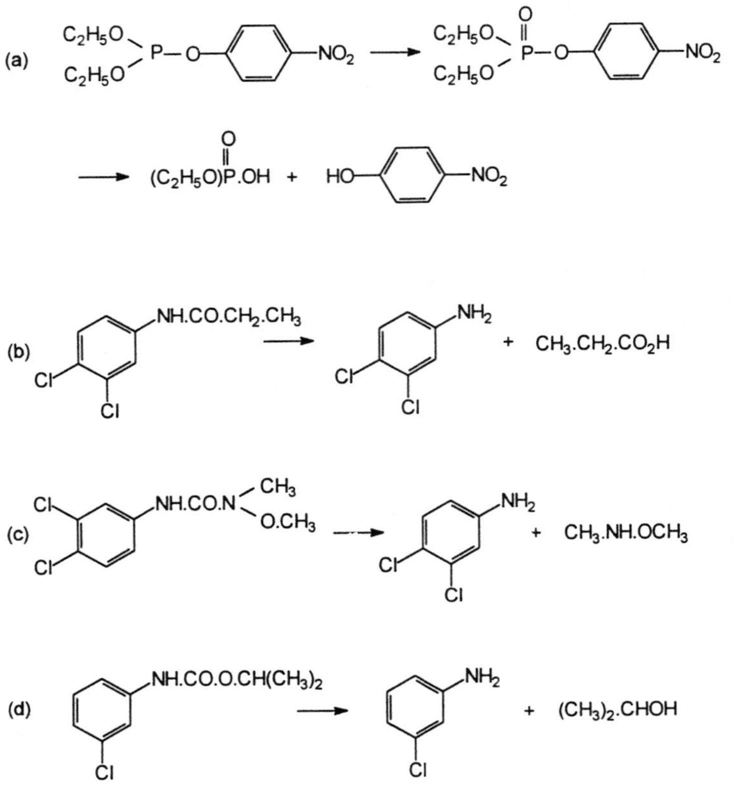

Figure 6.115 Hydrolytic pathways for the metabolism of **(a)** phosphate esters **(b)** acetanilides, **(c)** phenylureas, and **(d)** carbamates.

Amides and related compounds — A number of important agrochemicals are aromatic amides, carbamates, and ureas, and the the first step in their biodegradation is mediated by the appropriate amidases, carbamylases, and ureases (Figure 6.115b,c,d). It should, however, be pointed out that, for example, the chloroanilines that are formed from many of them as intial products may be substantially more resistant to further degradation. Application of tests for amidase activity, particularly using pyrazinamidase, has also been widely used in the classification of mycobacteria (Wayne et al. 1991). Sequential hydrolysis of nitriles to amides and carboxylic acids is well established both in aliphatic (Miller and Gray 1982; Nawaz et al. 1992) and in aromatic compounds (Harper 1977; McBride et al. 1986), though it should be noted that the herbicide bromoxynil may also be degraded by the elimination of cyanide from the ring with the initial formation of 2,6-dibromohydroquinone (Figure 6.116) (Topp et al. 1992). There may be a high degree of specificity in the action of these nitrilases, and this may have considerable interest in biotechnology. Two examples may be given as illustration.

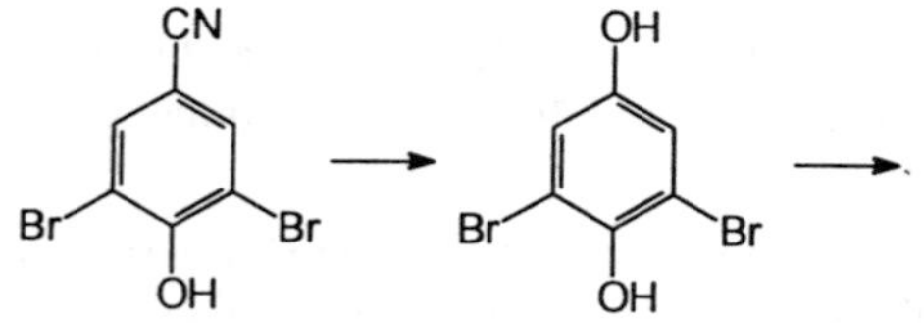

Figure 6.116 Metabolism of bromoxynil.

1. Racemic 2-(4′-*iso*butylphenyl)propionitrile is converted by a strain of *Acinetobacter* sp. to *S*-(+)-2-(4′-*iso*butylphenyl)propionic acid with an optical purity >95% (Yamamoto et al. 1990).
2. The nitrilase from a number of strains of *Pseudomonas* sp. mediated an enantiomerically selective hydrolysis of racemic *O*-acetylmandelonitrile to D-acetylmandelic acid [(*R*(-)-acetylmandelic acid] (Layh et al. 1992).

6.11.2 Hydroxylations, Oxidations, Dehydrogenations, and Reductions

Single-step reactions involving hydroxylation, oxidation, dehydrogenation, or reduction in which the microorganisms function as biocatalysts have been extensively exploited in the synthesis of commercially valuable compounds, particularly sterols. An attempt will be made here to provide a wider perspective and to summarize briefly some other examples of microbial biotransformations which have the potential for application in biotechnology. A valuable compilation of a wide range of oxidations carried out by diverse microorganisms including bacteria, fungi, and yeasts has been provided (Hudlicky 1990) and should be consulted for details and further illustrations. Only a few examples which have rather different motivation will be used to illustrate the extremely wide possibilities.

1. Interest in fungal metabolism as a model for that of higher organisms has been noted briefly in Chapter 4 (Section 4.3), and the use of microorganisms, particularly fungi, to produce hydroxylated sterols of therapeutic interest has been extensively explored. Probably most of these investigations have been directed to the synthesis of 11-, 17-, or 21-hydroxylated compounds, but a recent departure has been motivated by the need for unusual products which may be sterol metabolites. Studies have been directed to the biosynthesis, for example, of the otherwise rare and inaccessible derivatives of progesterone hydroxylated at the 6-, 9-, 14-, or 15-positions, and this could be accomplished by incubating progesterone in a complex medium with the fungus *Apiocrea chrysosperma* (Smith et al. 1988). The spectrum of substrates has been extended to xenobiotics and a single example may suffice: the fungus *Beauveria bassiana* metabolized the hydroxycoumarin rodenticide warfarin not only to the hydroxylated and oxidized metabolites that have already been established in mammalian systems or mediated by fungal cytochrome P450 monooxygenase systems, but also novel products resulting from the reduction of the keto group originating from the 4-hydroxyl group (Griffiths et al. 1992).

2. Haloperoxidase from the fungus *Caldariomyces fumago* has been used to accomplish a number of potentially valuable synthetic reactions including the following:
   - αβ-halohydrins from ethene and propene (Geigert et al. 1983a).
   - Vicinal dihalides from alkenes and alkynes (Geigert et al. 1983b).
   - α-halogenated ketones from alkynes and 3-hydroxyhalides from cyclopropanes (Geigert et al. 1983c).

   In all cases, reactive hypohalous acid is probably the active reagent.

3. Increasing interest has been directed to the microbial synthesis of chiral compounds that would not be readily accessible by chemical synthesis, and a few examples are given to illustrate the diversity of reactions that have been examined; the stereoselective hydrolysis of nitriles to carboxylic acids has already been noted.
   - Attention has already been directed (Section 6.1.3) to the use of a *Xanthobacter* sp. for resolution of 2,3-epoxyalkanes by selective metabolism of one of the enantiomers (Weijers et al. 1988).
   - A number of compounds which would be difficult to synthesize chemically have been produced as pure enantiomers by the oxidation of aromatic compounds by mutant strains of bacteria whereby catechols or dihydrodiols may be produced (Ribbons et al. 1990).
   - A strain of *Rhodococcus* sp. carries out allylic oxidation of α-cedrene to (*R*)-10-hydroxycedrene that undergoes rearrangement to α-curcumene (Figure 6.117) (Takigawa et al. 1993).
   - The unusual reduction of the α,β-double bond in 2-nitro-1-phenyl-1-propene by strains of *Rhodococcus rhodochrous* and several species of *Nocardia* results in a preponderance of one enantiomer (Sakai et al. 1985), and this is of interest in the synthesis of physiologically active norephedrin-type compounds.
   - Reduction of perfluoroalkylated ketones and of the double bond in perfluoroalkenoic acid esters has been accomplished with baker's yeast (*Saccharomyces cerevisiae*), and the products had a high optical purity ranging from 67% to 96% (Kitazume and Ishikawa 1983).
   - The hydration of the double bond in octadec-9-enoate to 10-hydroxyoctadecanoate may be carried out by a strain of *Pseudomonas* sp. with exclusive formation of the 10(*R*) compound (Yang et al. 1993).
   - At low concentrations of 1,3-dichloro-propan-2-ol, cells of a strain of *Corynebacterium* sp. induced with glycerol or a number of chlorinated alcohols were able to convert the substrate into (*R*)-3-chloropropan-1,2-diol in a yield of 97% and with an optical purity exceeding 80% (Nakamura et al. 1993).
   - Enantiomerically pure alkyl aryl sulfoxides have been obtained by microbial oxidation of the corresponding sulfides (Holland 1988); both *Corynebacterium equi* and fungi including *Aspergillus niger*, species of *Helminthosporium*, and *Mortierella isabellina* were effective although the same fungi were not able to carry out enantiomeric-selective oxidation of ethylmethylphenyl phosphine due apparently to the intrusion of non-selective chemical autoxidation (Holland et al. 1993).

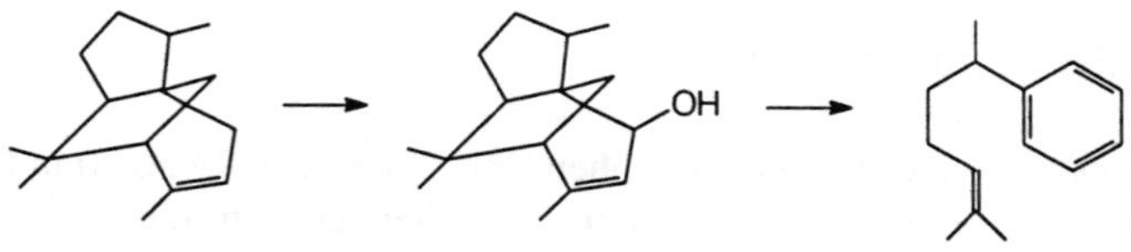

Figure 6.117 Biotransformation of α-cedrene by *Rhodococcus* sp.

6.11.3 The Formation of Dimeric Products from Aromatic Amines

It has been established that aromatic amines may be degraded by bacteria (McClure and Venables 1986; Fuchs et al. 1991), and it has already been noted that amines may be formed by reduction of the corresponding nitro and azo compounds (Sections 6.8.2 and 6.8.3). On the other hand, fungi may oxidize aromatic amines. For example, *Fusarium oxysporum* oxidizes 4-chloroaniline to 4-chlorophenylhydroxylamine, 4-chloronitrosobenzene, and 4-chloronitrobenzene. Condensation of suitable pairs of these intermediates would then reasonably account for the formation of 4,4′-dichloroazoxybenzene and 4,4′-dichloroazobenzene by this organism (Kaufman et al. 1973). The same organism produced 3,3′,4,4′-tetrachloroazoxybenzene from 3,4-dichloroaniline (Kaufman et al. 1972). A number of bacteria growing in the presence of 4-chloroaniline and nitrate produce 4,4′-dichloroazobenzene, and more detailed investigations with *Escherichia coli* and 3,4-dichloroaniline revealed that in addition to the corresponding azo compound and the triazine, the unexpected 3,3′,4,4′-tetrachlorobiphenyl was also formed (Figure 6.118) (Corke et al. 1979). These reactions mediated by bacteria are quite different from the fungal oxidations and plausibly involve the diazonium compound that is produced chemically from 4-chloroaniline and nitrite that is microbiologically formed from nitrate. Although the reactions mediated by fungi may not have a direct major impact on the aquatic environment, their occurrence in landfills with subsequent leaching of the metabolites into water courses cannot be discounted.

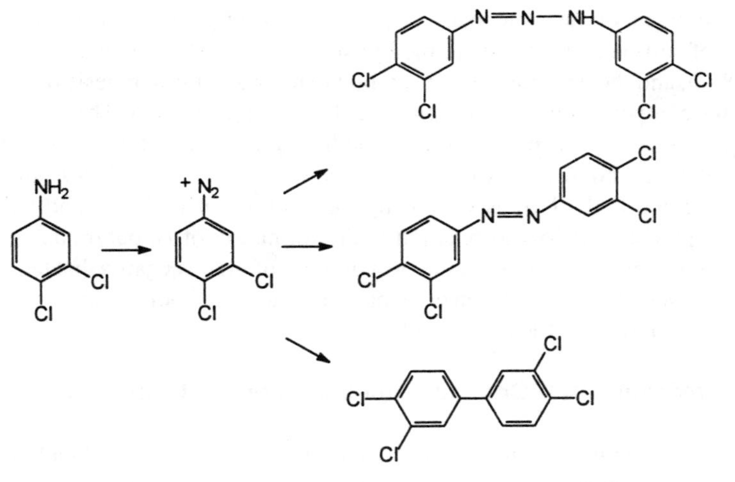

Figure 6.118 Formation of condensation products from 3,4-dichloroaniline via diazo compounds.

6.11.4 Methylation Reactions

The O-methylation of halogenated phenols has been briefly noted in Chapter 4 as an example of a biotransformation reaction, and they have attracted attention in view of the unacceptable flavor that chloroanisoles impart to broiler chickens (Gee and Peel 1974), freshwater fish (Paasivirta et al. 1987), and wine corks (Buser et al. 1982). A few additional comments may usefully be added to place these observations in a wider perspective.

1. The O-methylation of halogenated phenolic compounds by both gram-positive and gram-negative bacteria has been demonstrated (Allard et al. 1987), and in cell extracts of representatives of these bacteria it appears that the methyl group is provided as expected by S-adenosyl methionine (Neilson et al. 1988b).
2. Fungi are able to methylate chloride ion (Harper and Hamilton 1988), and it has been suggested that this could result in a significant biotic input into the environment, independently of anthropogenic discharge. In addition, fungi are able to carry out a transmethylation reaction between methyl chloride and compounds with reactive hydroxyl groups including phenols and both aliphatic and aromatic carboxylic acids (Harper et al. 1989; McNally et al. 1990). Care should therefore be exercised in assigning the ultimate source of halogenated anisoles which are widely distributed in the environment (Wittlinger and Ballschmitter 1990), and which have been identified in environmental samples including the marine atmosphere (Atlas et al. 1986), sediment samples (Tolosa et al. 1992), and biota (Watanabe et al. 1983); they may possibly originate from halogenated phenols that have been identified in automobile exhaust (Müller and Büser 1986) or from brominated phenols that have been used as flame retardants. The origin of some of these anisoles, such as those that have been established as metabolites of agrochemicals incuding 2,4-dichloro- and 2,4,5-trichlorophenoxyacetic acids (McCall et al. 1981; Smith 1985) and 2,4,5-trichloro-2-pyridinyloxyacetate (Lee et al. 1986), is of course unequivocal.
3. Cell extracts of the protozoan *Tetrahymena thermophila* carry out the methylation of sulfide and selenide to methyl sulfide and methyl selenide (Drotar et al. 1987a), and the importance of microbial reactions in the synthesis of methylated sulfur compounds has been examined (Drotar et al. 1987b). The environmental significance of the production of methyl sulfide has already been noted in Chapter 4 (Section 4.2). Aromatic thiols are also S-methylated by *T. thermophila* (Drotar and Fall 1986) as well as by *Euglena gracilis* (Drotar and Fall 1985).
4. An apparently analogous reaction is the formation of *iso*butyraldehyde oxime during the metabolism of valine by a number of gram-negative bacteria (Harper and Nelson 1982), and the enzyme has been purified from a strain of *Pseudomonas* sp. (Harper and Kennedy 1985).

6.11.5 Environmental Consequences of Biotransformation

A number of environmentally relevant biotransformations have already been used as illustration in other contexts, and attention is drawn particularly to two opposing aspects of their significance: their function as a detoxification mechanism towards a potential toxicant and the adverse effect of metabolites on organisms other than those producing them. In addition, these transformation products may have physico-chemical properties very different from those of their precursors so that these metabolites may be accumulated in higher organisms even if they cause no palpably adverse effect.

Biotransformation as an effective mechanism of detoxification — This is discussed more fully in Chapter 7 (Section 7.5) in the wider context of toxicology. Only brief attention will therefore be directed here to the less fully documented role of analogous reactions in microorganisms.

1. One of the most significant illustrations of detoxification is, of course, provided by the mechanisms whereby bacteria develop resistance to antibiotics, and a full discussion of this is available (Franklin and Snow 1981). A variety of different reactions may occur, and whereas all of them leave the structure of the antobiotics largely intact, they effectively destroy the biological activity (toxicity) of the drug. A few of these transformations will be used as illustration.

   • Hydrolysis of the β-lactam ring in penicillins with the formation of inactive penicilloic acid is a serious problem in the development of resistance to penicillin, though this may be at least partially overcome in the semi-synthetic penicillins.

   • Resistance to chloramphenicol is mediated by the synthesis of the 1-acetate and 1,3-diacetate while the corresponding resistance to streptomycin involves phosphorylation.

2. The reductive cleavage of methylmercury to methane (Spangler et al. 1973) and of phenylmercuric acetate to benzene (Furukawa and Tonomura 1971; Nelson et al. 1973) may be considered as mechanisms of detoxification in a wider context since these organomercury compounds are extremely toxic to higher organisms. There are apparently two different enzymes in *Pseudomonas* sp. K-62 that bring about splitting of the phenyl-Hg and methyl-Hg bonds; the S-1 enzymes accept both substrates whereas the S-2 enzyme is specific to substrates with phenyl-Hg bonds (Tezuka and Tonomura 1978).

3. Attention has already been made to the formation of glycoside and sulfate conjugates of phenolic metabolites during the biotransformation of polycyclic aromatic hydrocarbons by fungi (Section 6.2.2). Formation of sulfate esters is not, however, limited to fungal systems since this has been observed, for example, following 4'-hydroxylation of 5-hydroxyflavone in *Streptomyces fulvissimus* (Ibrahim and Abul-Hajj 1989), and the enzyme has been partly purified from a *Eubacterium* sp. (Koizumi et al. 1990).

4. Although the biodegradation of phenols and anilines is well established, these are relatively toxic compounds, and some microorganisms detoxify them by acylation as an alternative to biodegradation. For example, acetylation of pen-tachlorophenol has been observed (Rottt et al. 1979), and acetylation of substi-tuted anilines has been established; 4-chloroaniline is converted into 4-chloroacetanilide by *Fusarium oxysporum* (Kaufman et al. 1973), and 2-nitro-4-aminotoluene that is produced from 2,4-dinitrotoluene by a species of *Mucrosporium* (McCormick et al. 1978) is converted into 3-nitro-4-methylacetanilide. All these neutral compounds may plausibly be assumed to be less toxic than their precursors.

Potentially adverse ecological consequences of biotransformation — A number of different metabolic possibilities that may have extensive — and possibly adverse — environmental repercussions may be recognized; these involve the synthesis of metabolites that are (a) lipophilic and less polar than their precursors, (b) highly water-soluble, or (c) toxic to other biota in the ecosystem. A number of diverse examples will be used to illustrate what is probably a widespread phenomenon.

1. The synthesis of lipophilic metabolites has been demonstrated in a number of studies of the transformation of xenobiotics, and the following are given as illustration:

    • The classic case is the biomethylation of mercury (Jensen and Jernelöv 1969), whereby cationic inorganic Hg is transformed into an organic form that is both lipophilic and more toxic to higher organisms. Both abiotic and biotic processes may be involved, and microbial methylation has been confirmed both in the sediment phase (Furutani and Rudd 1980) and probably even in the intestines of fish (Rudd et al. 1980). A range of microorganisms that are capable of methylating Hg^{2+} has been isolated, and these include aerobic bacteria and fungi (Vonk and Sijpesteijn 1973) and facultatively anaerobic (Hamdy and Noyes 1975) and anaerobic bacteria (Yamada and Tonomura 1972). Methylation of other inorganic compounds has also been observed, although these have been less extensively investigated, and includes compounds of Sn (Jackson et al. 1982), As in which reduction of As(III or IV) is also involved (Cullen et al. 1984), and S and Se; methylation of the last two has been discussed briefly in Section 6.11.4. In all of these cases, the organic forms of these elements should be taken into consideration in constructing global cycles.

    • The O-methylation of halogenated phenolic compounds has been systematically investigated (Allard et al. 1987; Neilson et al. 1988b); these metabolites have a high bioconcentration potential and, therefore, present a potential environmental hazard — quite apart from their demonstrated toxicity to zebra fish larvae (Neilson et al. 1984). These and some related reactions have been discussed in Section 6.11.4. The occurrence of such reactions provides a plausible reason for the occurrence of, for example, the O-methyl ether of the diphenyl ether antibacterial agent triclosan in aquatic biota (Miyazaki et al. 1984).

    • It has already been noted that some microorganisms do not degrade long-chain alkanols completely but synthesize esters by reaction between the carboxylic acid produced and the alkanol (Hankin and Kolattukudy 1968); these metabolites are highly lipophilic, though any adverse effect may be restricted through their subsequent hydrolysis after uptake by higher organisms.

    • Attention has been drawn in Chapter 4 (Section 4.3.1) to a number of decarboxylation reactions carried out by Enterobacteriaceae; the products are much less polar than their precursors and may, therefore, exhibit the potential for concentration in biota. The significance of this reaction is seldom taken into account but merits serious consideration.

2. The synthesis of water-soluble conjugates of metabolites may bring about the widespread dissemination of the metabolites, and this has been briefly noted in Chapter 3 (Section 3.5.2). For example, many xenobiotics are metabolized both by fungi and higher organisms to hydroxylated metabolites which are then conjugated to form sulfate esters, glycosides, or glucuronides. Some examples of the reactions mediated by microorganisms have been discussed in this chapter, and those carried out by higher biota will be discussed in Chapter 7. The significance of these metabolites in a wider context lies in the possibility for the widespread dispersion of these water-soluble compounds in water masses and their subsequent hydrolysis by bacteria to the original phenolic compounds. For example, it is well established that *Escherichia coli* readily hydrolyzes a number of phenolic glucuronides.

3. The synthesis of stable microbial transformation products should be assessed in the wider context of the possible adverse effect of these metabolites on other components of the ecosystem. One of the earliest examples that demonstrated the toxicological significance of metabolites is provided by the drug Prontosil that is noted in Chapter 7 (Section 7.5), and the toxicity of methyl mercury to man and of chlorinated veratroles to zebra fish larvae has been noted above. Another example is the formation of established carcinogenic nitrosamines that may be mediated by microbiologically mediated reactions that are discussed in Chapter 4 (Section 4.1.2).

Clearly, therefore, microbial metabolites may have potentially widespread adverse environmental effects on higher biota, including man.

6.12 SUMMARY OF BASIC MICROBIAL REACTIONS

In the preceding sections, an account has been given of the pathways followed during biodegradation and biotransformation of a range of organic compounds. It may be useful to present this information in an alternative way: in terms of the types of reaction which may be carried out. No references to the literature are given since those that are relevant may be found in the earlier sections. Some examples of typical reactions are given by way of illustration, together with references to the appropriate section in which they have already been discussed. In the case of decarboxylation, reference is made also to Chapter 4.

1. Hydrolysis — Esters, amides, and related compounds, and nitriles (Section 6.11.1)
2. Nucleophilic substitution — Hydrolysis of organohalogen compounds (Sections 6.4.2 and 6.5.1)
3. Addition of water to unsaturated compounds — Alkynes (Sections 6.1.4 and 6.7.1)
4. Elimination — Degradation of DDT, lindane (Section 6.4.1), and aliphatic sulfides (Section 6.9.2)
5. Carboxylation— Aerobic degradation of branched alkanes (Section 6.1.1)
      — Anaerobic degradation of phenols (Section 6.7.3)

6. Decarboxylation (Sections 6.2.1 and 4.3.1)
7. Introduction of oxygen — Hydroxylases, monooxygenases, and dioxygenases
   — Aliphatic and alicyclic compounds: hydroxylation
   (Sections 6.1.1 and 6.1.2)
   — Alkenes: epoxidation (Section 6.1.3)
   — Aromatic compounds: monooxygenation and
   dioxygenation (Sections 6.2 and 6.3)
   — Halogenated aromatic compounds: hydroxylation
   (Section 6.5.1)
   — Aromatic nitro compounds and sulfonates: elimination of nitrite or sulfite by dioxygenation (Sections
   6.8.1 and 6.8.2)
8. Oxidation — Alkanols to aldehydes or ketones
9. Dehydrogenation of alicyclic rings (Section 6.1.2)
10. Reduction — Dehydroxylation (Sections 6.7.2 and 6.7.3)
    — Nitro and azo compounds (Sections 6.8.2 and 6.8.3)
    — Aromatic aldehydes (Section 6.7.3)
    — Dehalogenation (Sections 6.4.4 and 6.6)
11. Reductive cleavage — The C-P bond in organophosphonates (Section 6.9.3)
    — The C-Hg bond in organomercurials (Section 6.11.5)
12. Conjugation — Sulfates and glucuronides of phenols: excretion of water-
    soluble metabolites (Sections 6.2.2 and 6.11.3)
13. Rearrangement — Coenzyme B_{12}-mediated reactions
    · Anaerobic degradation of amino acids (Section 6.7.1)
    · Anaerobic degradation of PEG (Section 6.9.1)
    — The NIH shift (Sections 6.2.2 and 6.5.2)
    — Terpene (Section 6.11.2)
14. Dismutation — The Stickland reaction between pairs of amino acids (Section
    6.7.1)
15. Methylation reactions (Section 6.11.4)

REFERENCES

Abbott, B.J., and C.T. Hou. 1973. Oxidation of 1-alkenes to 1,2-epoxides by *Pseudomonas oleovorans*. *Appl. Microbiol.* 26: 86-91.

Abril, M.-A., C. Michan, K.N. Timmis, and J.L. Ramos. 1989. Regulator and enzyme specificities of the TOL plasmid-encoded upper pathway for degradation of aromatic hydrocarbons and expansion of the substrate range of the pathway. *J. Bacteriol.* 171: 6782-
6790

Adrian, N.R., and J.M. Suflita. 1990. Reductive dehalogenation of a nitrogen heterocyclic
herbicide in anoxic aquifer slurries. *Appl. Environ. Microbiol.* 56: 292-294.

Aeckersberg, F., F. Bak, and F. Widdel. 1991. Anaerobic oxidation of saturated hydrocarbons
to CO_2 by a new type of sulfate-reducing bacterium. *Arch. Microbiol.* 156: 5-14.

Ahmad, D., M. Sylvestre, and M. Sondossi. 1991. Subcloning of *bph* genes from *Pseudomonas
testosteroni* B-356 in *Pseudomonas putida* and *Escherichia coli*: evidence for
dehalogenation during initial attack on chlorobiphenyls. *Appl. Environ. Microbiol.* 57:
2882-2887.

Alexander, M. 1981. Biodegradation of chemicals of environmental concern. *Science* 211: 132-138.

Allard, A.-S., M. Remberger, and A.H. Neilson. 1987. Bacterial O-methylation of halogen-substituted phenols. *Appl. Environ. Microbiol.* 53: 839-845.

Allard, A.-S., C. Lindgren, P.-Å. Hynning, M. Remberger, and A.H. Neilson.1991. Dechlorination of chlorocatechols by stable enrichment cultures of anaerobic bacteria. *Appl. Environ. Microbiol.* 57: 77-84.

Allard, A.-S., P.-Å. Hynning, M. Remberger, and A.H. Neilson.1992. Role of sulfate concentration in dechlorination of 3,4,5-trichlorocatechol by stable enrichment cultures grown with coumarin and flavanone glycones and aglycones *Appl. Environ. Microbiol.* 58: 961-968.

Allison, N., A.J. Skinner, and R.A. Cooper. 1983. The dehalogenases of a 2,2-dichloropropionate-degrading bacterium. *J. Gen. Microbiol.* 129: 1283-1293.

Altenschmidt, U., and G. Fuchs. 1992a. Anaerobic toluene oxidation to benzyl alcohol and benzaldehyde in a denitrifying *Pseudomonas* strain. *J. Bacteriol.* 174: 4860-4862.

Altenschmidt, U., and G. Fuchs. 1992b. Novel aerobic 2-aminobenzoate metabolism. *Eur. J. Biochem.* 205: 721-727.

Altenschmidt, U., B. Oswald, E. Steiner, H. Herrmann, and G. Fuchs. 1993. New aerobic benzoate oxidation pathway via benzoyl-coenzyme A and 3-hydroxybenzoyl-coenzyme A in a denitrifying *Pseudomonas* sp. *J. Bacteriol.* 175: 4851-4858.

Ander, P., K.-E. Eriksson, and H-S. Yu. 1983. Vanillic acid metabolism by *Sporotrichium pulverulentum*: evidence for demethoxylation before ring-cleavage. *Arch. Microbiol.* 136: 1-6.

Anderson, J.J., and S. Dagley. 1980. Catabolism of aromatic acids in *Trichosporon cutaneum*. *J. Bacteriol.* 141: 534-543.

Anderson, R.C., M.A. Rasmussen, and M.J. Allison. 1993. Metabolism of plant toxins nitropropionic acid and nitropropanol by ruminal microorganisms. *Appl. Environ. Microbiol.* 59: 3056-3061.

Anthony, C. 1982. *The biochemistry of methylotrophs*. Academic Press, London.

Apajalahti, J.H.A., and M.S. Salkinoja-Salonen. 1987a. Dechlorination and para-hydroxylation of polychlorinated phenols by *Rhodococcus chlorophenolicus*. *J. Bacteriol.* 169: 675-681.

Apajalahti, J.H.A., and M.S. Salkinoja-Salonen. 1987b. Complete dechlorination of tetrachlorohydroquinone by cell extracts of pentachlorophenol-induced *Rhodococcus chlorophenolicus*. *J. Bacteriol.* 169: 5125-5130.

Arkowitz, R.A., and R.H. Abeles. 1989. Identification of acetyl phosphate as the product of clostridial glycine reductase: evidence for an acyl enzyme intermediate. *Biochemistry* 28: 4639-4644.

Asturias, J.A., and K.N. Timmis. 1993. Three different 2,3-dihydroxybiphenyl-1,2-dioxygenase genes in the gram-positive polychlorobiphenyl-degrading bacterium *Rhodococcus globerulus* P6. *J. Bacteriol.* 175: 4631-4640.

Atlas, E., K. Sullivan, and C.S. Giam. 1986. Widespread occurrence of polyhalogenated aromatic ethers in the marine atmosphere. *Atmos. Environ.* 20: 1217-1220.

Bache, R., and N. Pfennig. 1981. Selective isolation of *Acetobacterium woodii* on methoxylated aromatic acids and determination of growth yields. *Arch. Microbiol.* 130: 255-261.

Baggi, G., P. Barbieri, E. Galli, and S. Tollari. 1987. Isolation of a *Pseudomonas stutzeri* strain that degrades o-xylene. *Appl. Environ. Microbiol.* 53: 2129-2132.

Bak, F., and E. Widdel. 1986. Anaerobic degradation of indolic compounds by sulfate-reducing enrichment cultures, and description of *Desulfobacterium indolicum* gen. nov., sp. nov. *Arch. Microbiol.* 146: 170-176

Bak, F., K. Finster, and F. Rothfuß. 1992. Formation of dimethylsulfide and methanthiol from methoxylated aromatic compounds and inorganic sulfide by newly isolated anaerobic bacteria. *Arch. Microbiol.* 157: 529-534.

Ball, H.A., M. Reinhard, and P.L. McCarty. 1989. Biotransformation of halogenated and nonhalogenated octylphenol polyethoxylate residues under aerobic and anaerobic conditions. *Environ. Sci. Technol.* 23: 951-961.

Barker, H.A. 1961. Fermentations of nitrogenous organic compounds pp.151-207. In *The Bacteria,* Volume 2 (Eds. I.C. Gunsalus and R.Y. Stanier). Academic Press, New York.

Barker, H.A. 1972. Corrinoid-dependent enzymatic reactions. *Ann. Rev. Biochem.* 41: 55-90.

Barker, H.A. 1981. Amino acid degradation by anaerobic bacteria. *Ann. Rev. Biochem.* 50: 23-40.

Barnsley, E.A. 1983. Phthalate pathway of phenanthrene metabolism: formation of 2'-carboxybenzalpyruvate. *J. Bacteriol.* 154: 113-117.

Bartels, I., H.-J. Knackmuss, and W. Reineke. 1984. Suicide inactivation of catechol 2,3-dioxygenase from *Pseudomonas putida* mt-2 by 3-halocatechols. *Appl. Environ. Microbiol.* 47: 500-505.

Bartholomew, B, K.S. Dodgson, G.W.J. Matcham, D.J. Shaw, and G.F. White. 1977. A novel mechanism of enzymatic hydrolysis. Inversion of configuration and carbon-oxygen bond cleavage by secondary alkylsulphohydrolases from detergent-degrading microorganisms. *Biochem. J.* 165: 575-580.

Beadle, C.A., and A.R.W. Smith. 1982. The purification and properties of 2,4-dichlorophenol hydroxylase from a strain of *Acinetobacter* sp. *Eur. J. Biochem.* 123: 323-332.

Bedard, D.L., M.L. Haberl, R.J. May, and M.J. Brennan. 1987. Evidence for novel mechanisms of polychlorinated biphenyl metabolism in *Alcaligenes eutrophus* H 850. *Appl. Environ. Microbiol.* 53: 1103-1112.

Belay, N., and L. Daniels. 1987. Production of ethane, ethylene, and acetylene from halogenated hydrocarbons by methanogenic bacteria. *Appl. Environ. Microbiol.* 53: 1604-1610.

Beller, H.R., M. Reinhard, and D. Grbic-Galic. 1992. Metabolic by-products of anaerobic toluene degradation by sulfate-reducing enrichment cultures. *Appl. Environ. Microbiol.* 58: 3192-3195.

Beunink, J., and H.-J. Rehm. 1990. Coupled reductive and oxidative degradation of 4-chloro-2-nitrophenol by a co-immobilized mixed culture system. *Appl. Microbiol. Biotechnol.* 34: 108-115.

Beuscher, H.U., and J.R. Andressen. 1984. *Eubacterium angustum* sp. nov., a gram-positive anaerobic, non-spore-forming, obligate purine fermenting organism. *Arch. Microbiol.* 140: 2-8.

Bisaillon, J.-G., F. Lépine, R. Beaudet, and M. Sylvestre. 1991. Carboxylation of *o*-cresol by an anaerobic consortium under methanogenic conditions. *Appl. Environ. Microbiol.* 57: 2131-2134.

Blakley, E.R. 1974. The microbial degradation of cyclohexanecarboxylic acid: a pathway involving aromatization to form *p*-hydroxybenzoic acid. *Can. J. Microbiol.* 20: 1297-1306.

Block, D.W., and F. Lingens. 1992. Microbial metabolism of quinoline and related compounds XIV. Purification and properties of 1*H*-3-hydroxy-4-oxoquinoline oxygenase, a new extradiol cleavage enzyme from *Pseudomonas putida* strain 33/1. *Biol. Chem. Hoppe-Seyler* 373: 343-349.

Boethling, R.S., B. Gregg, R. Frederick, N.W. Gabel, S.E. Campbell, and A. Sabljic. 1989. Expert systems survey on biodegradation of xenobiotic chemicals. *Ecotoxicol. Environ. Saf.* 18: 252-267.

Bokkenheuser, V.D., J. Winter, S. O'Rourke, and A.E. Ritchie. 1980. Isolation and characterization of fecal bacteria capable of 16α-dehydroxylating corticoids. *Appl. Environ. Microbiol.* 40: 803-808.

Brackmann, R., and G. Fuchs. 1993. Enzymes of anaerobic metabolism of phenolic compounds. 4-hydroxybenzoyl-CoA reductase (dehydroxylating) from a denitrifying *Pseudomonas* sp. *Eur. J. Biochem.* 213: 563-571.

Branchaud, B.P., and C.T. Walsh. 1985. Functional group diversity in enzymatic oxygenation reactions catalyzed by bacterial flavin-containing cyclohexanone oxygenase. *J. Amer. Chem. Soc.* 107: 2153-2161.

Braun, M., and H. Stolp. 1985. Degradation of methanol by a sulfate reducing bacterium. *Arch. Microbiol.* 142: 77-80.

Britt, A.J., N.C. Bruce, and C.R Lowe. 1992. Identification of a cocaine esterase in a strain of *Pseudomonas maltophilia. J. Bacteriol.* 174: 2087-2094.

Britton, L.N. 1984. Microbial degradation of aliphatic hydrocarbons pp. 89-130. In *Microbial degradation of organic compounds* (Ed. D.T. Gibson). Marcel Dekker Inc., New York.

Broadway, N.M., F.M. Dickinson, and C. Ratledge. 1993. The enzymology of dicarboxylic acid formation by *Corynebacterium* sp. strain 7E1C grown on *n*-alkanes. *J. Gen. Microbiol.* 139: 1337-1344.

Brune, A., and B. Schink. 1990. Pyrogallol-to-phloroglucinol conversion and other hydroxyl-transfer reactions catalyzed by cell extracts of *Pelobacter acidigallici. J. Bacteriol.* 172: 1070-1076.

Brune, A. and B. Schink. 1992. Phloroglucinol pathway in the strictly anaerobic *Pelobacter acidigallici* fermentation of trihydroxybenzenes to acetate via triacetic acid. *Arch. Microbiol.* 157: 417-424.

Bumpus, J.A. 1989. Biodegradation of polycyclic aromatic hydrocarbons by *Phanerochaete chrysosporium. Appl. Environ. Microbiol.* 55: 154-158.

Bumpus, J.A., and S.A. Aust. 1987. Biodegradation of DDT [1,1,1-trichloro-2,3-*bis*(4-chloroohenyl)ethane] by the white rot fungus *Phanerochaete chrysosporium. Appl. Environ. Microbiol.* 53: 2002-2008.

Bünz, P.V., and A.M. Cook. 1993. Dibenzofuran 4,4a-dioxygenase from *Sphingomonas* sp. strain RW1: angular dioxygenation by a three-component system. *J. Bacteriol.* 175: 6467-6475.

Buser, H.-R., C. Zanier, and H. Tanner. 1982. Identification of 2,4,6-trichloroanisole as a potent compound causing cork taint in wine. *J. Agric. Food Chem.* 30: 359-362.

Cain, R.B. 1981. Microbial degradation of surfactants and "builder" components pp. 325-370. In *Microbial degradation of xenobiotics and recalcitrant compounds* (Eds. T. Leisinger, A.M. Cook, R. Hütter, and J. Nüesch). Academic Press, London.

Cain, R.B. 1988. Aromatic metabolism by mycelial organisms: actinomycete and fungal strategies pp. 101-144. In *Microbial metabolism and the carbon cycle* (Eds. S.R. Hagedorn, R.S. Hanson, and D.A. Kunz). Harwood Academic Publishers, Chur, Switzerland.

Cain, R.B., and D.R. Farr. 1968. Metabolism of arylsulphonates by microorganisms. *Biochem. J.* 106: 859-877.

Cain, R.B., R.F. Bilton, and J.A. Darrah. 1968. The metabolism of aromatic acids by micro-organisms. Metabolic pathways in the fungi. *Biochem. J.* 108: 797-828.

Campbell, L.L. 1960. Reductive degradation of pyrimidines V. Enzymatic conversion of *N*-carbamyl-beta-alanine to beta-alanine, carbon dioxide, and ammonia. *J. Biol. Chem.* 235: 2375-2378.

Carey, J.H., and M.E. Cox. 1981. Photodegradation of the lampricide 3-trifluoromethyl-4-nitrophenol (TFM) 1. Pathway of the direct photolysis in solution. *J. Great Lakes Res.* 7: 234-241.

Cartwright, N.J., and R.B. Cain.1959. Bacterial degradation of the nitrobenzoic acids. II. Reduction of the nitro group. *Biochem. J.* 73: 305-314.

Castro, C.E., and N.O. Belser. 1990. Biodehalogenation: oxidative and reductive metabolism of 1,1,2-trichloroethane by *Paseudomonas putida* — biogeneration of vinyl chloride. *Environ. Toxicol. Chem.* 9: 707-714.

Castro, C.E., D.M. Riebeth, and N.O. Belser. 1992a. Biodehalogenation: the metabolism of vinyl chloride by *Methylosinus trichosporium* OB-3b. A sequential oxidative and reductive pathway through chloroethylene oxide. *Environ. Toxicol. Chem.* 11: 749-755.

Castro, C.E., R.S. Wade, D.M. Riebeth, E.W. Bartnicki, and N.O. Belser. 1992b. Biodehalogenation: rapid metabolism of vinyl chloride by a soil *Pseudmonas* sp. Direct hydrolysis of a vinyl C-Cl bond. *Environ. Toxicol. Chem.* 11: 757-764

Cerniglia, C.E., and S.A. Crow. 1981. Metabolism of aromatic hydrocarbons by yeasts. *Arch. Microbiol.* 129: 9-13.

Cerniglia, C.E., and S.K. Yang. 1984. Stereoselective metabolism of anthracene and phenanthrene by the fungus *Cunninghamella elegans*. *Appl. Environ. Microbiol.* 47: 119-124.

Cerniglia, C.E., J.P. Freeman, and R.K. Mitchum. 1982a. Glucuronide and sulfate conjugation in the fungal metabolism of aromatic hydrocarbons. *Appl. Environ. Microbiol.* 43: 1070-1075.

Cerniglia, C.E., P.P. Fu, and S.K. Yang. 1982b. Metabolism of 7-methylbenz[a]anthracene and 7-hydroxymethylbenz[a]anthracene by *Cunninghamella elegans*. *Appl. Environ. Microbiol.* 44: 682-689.

Cerniglia, C.E., J.R. Althaus, F.E. Evans, J.P. Freeman, R.K. Mitchum, and S.K. Yang. 1983. Stereochemistry and evidence for an arene oxide-NIH shift pathway in the fungal metabolism of naphthalene. *Chem.-Biol. Interactions* 44: 119-132.

Cerniglia, C.E., J.P. Freeman, and F.E. Evans. 1984. Evidence for an arene oxide-NIH shift pathway in the transformation of naphthalene to 1-naphthol by *Bacillus cereus*. *Arch. Microbiol.*: 138 283-286.

Cerniglia, C.E., W.L. Campbell, J.P Freeman, and F.E. Evans. 1989. Identification of a novel metabolite in phenanthrene metabolism by the fungus *Cunninghamella elegans*. *Appl. Environ. Microbiol.* 55: 2275-2279.

Charlson, R.J., J.E. Lovelock, M.O. Andreae, and S.G. Warren. 1987. Oceanic phytoplankton, atmospheric sulphur, cloud albedo and climate. *Nature* 326: 655-661.

Chen, C.-M., and P.H. Tomasek. 1991. 3,4-dihydroxyxanthone dioxygenase from *Arthrobacter* sp. strain GFB 100. *Appl. Environ. Microbiol.* 57: 2217-2222.

Cheng, K.-J., H.G. Krishnamurty, G.A. Jones, and F.J. Simpson. 1971. Identification of products produced by the anaerobic degradation of naringin by *Butyrivibrio*. sp. C_3. *Can. J. Microbiol.* 17: 129-131.

Cook, A.M. 1987. Biodegradation of s-triazine xenobiotics. *FEMS Microbiol. Revs.* 46: 93-116.

Cook, A.M., C.G. Daughton, and M. Alexander. 1978a. Phosphonate utilization by bacteria. *J. Bacteriol.* 133: 85-90.

Cook, A.M., C.G. Daughton, and M. Alexander. 1978b. Phosphorus-containing pesticide breakdown products: quantitative utilization as phosphorus sources by bacteria. *Appl. Environ. Microbiol.* 36: 668-672.

Corbett, M.D., and B.R. Corbett. 1981. Metabolism of 4-chloronitrobenzene by the yeast Rhodosporidium sp. *Appl. Environ. Microbiol.* 41: 942-949.

Corke, C.T., N.J. Bunce, A.-L. Beaumont, and R.L. Merrick. 1979. Diazonium cations as intermediates in the microbial transformations of chloroanilines to chlorinated biphenyls, azo compounds and triazenes. *J. Agric. Food Chem.* 27: 644-646.

Crawford, R.L. 1975a. Novel pathways for degradation of protocatechuic acid in *Bacillus* species. *J. Bacteriol.* 121: 531-536.

Crawford, R.L. 1975b. Degradation of 3-hydroxybenzoate by bacteria of the genus *Bacillus*. *Appl. Microbiol.* 30: 439-444.

Crawford, R.L. 1976. Pathways of 4-hydroxybenzoate degradation among species of *Bacillus*. *J. Bacteriol.* 127: 204-290.

Crawford, R.L., and P.P. Olson. 1978. Microbial catabolism of vanillate: decarboxylation to guaiacol. *Appl. Environ. Microbiol.* 36: 539-543.

Crescenzi, A.M.V., K.S. Dodgson, G.F. White, and W.J. Payne. 1985. Initial oxidation and subsequent desulphation of propan-2-yl sulphate by *Pseudomonas syringae* strain GG. *J. Gen. Microbiol.* 131: 469-477.

Cripps, R.E. 1973. The microbial metabolism of thiophen-2-carboxylate. *Biochem. J.* 134: 353-366.

Cripps, R.E., and A.S. Noble. 1973. The metabolism of nitrilotriacetate by a pseudomonad. *Biochem. J.* 136: 1059-1068.

Cripps, R.E., and R.J. Watkinson. 1978. Polycyclic hydrocarbons: metabolism and environmental aspects pp. 113-134. In *Developments in biodegradation of hydrocarbons-1* (Ed. R.J. Watkinson). Applied Science Publishers Ltd., London.

Cripps, R.E., P.W. Trudgill, and J.G. Whateley. 1978. The metabolism of 1-phenylethanol and acetophenone by *Nocardia* T5 and an *Arthrobacter* species. *Eur. J. Biochem.* 86: 175-186.

Cullen, W.R., B.C. McBride, and J. Reglinski. 1984. The reduction of trimethylarsine oxide to trimethylarsine by thiols: a mechanistic model for the biological reduction of arsenicals. *J. Inorg. Biochem.* 21: 45-60.

Dalton, H., B.T. Golding, B.W. Waters, R. Higgins, and J.A. Taylor. 1981. Oxidation of cyclopropane, methylcyclopropane, and arenes with the mono-oxygenase systems from *Methylococcus capsulatus. J. Chem. Soc. Chem. Comm.* 482-483.

Daly, J. W., D.M. Jerina, and B. Witkop. 1972. Arene oxides and the NIH shift: the metabolism, toxicity and carcinogenicity of aromatic compounds. *Experientia* 28: 1129-1149.

Dams, R. 1993. Fluorochemical surfactants for hostile environments. *Speciality Chemicals* 13: 4-6.

Dangel, W., R. Brackmann, A. Lack, M. Mohamed, J. Koch, B. Oswald, B. Seyfried, A. Tschech, and G. Fuchs. 1991. Differential expression of enzyme activities initiating anoxic metabolism of various aromatic compounds via benzoyl-CoA. *Arch. Microbiol.* 155: 256-262.

Darby, J.M., D.G. Taylor, and D.J. Hopper. 1987. Hydroquinone as the ring-fission substrate in the catabolism of 4-ethylphenol and 4-hydroxyacetophenone by *Pseudomonas putida* D1. *J. Gen. Microbiol.* 133: 2137-2146.

D'Ari, L., and W.A. Barker. 1985. *p*-cresol formation by cell-free extracts of *Clostridium difficile. Arch. Microbiol.* 143: 311-312.

Davey, J.F., and D.T. Gibson. 1974. Bacterial metabolism of para- and meta-xylene: oxidation of a methyl substituent. *J. Bacteriol.* 119: 923-929.

Davies, I., G.F. White, and W.J. Payne. 1990. Oxygen-dependent desulphation of monomethyl sulphate by *Agrobacterium* sp. M3C. *Biodegradation* 1: 229-241

de Bont, J.A.M., and M.W. Peck. 1980. Metabolism of acetylene by *Rhodococcus* A1. *Arch. Microbiol.* 127: 99-104.

de Bont, J.A.M., J.P. van Dijken, and C.G. van Ginkel. 1982. The metabolism of 1,2-propanediol by the propylene oxide utilizing bacterium *Nocardia* A60. *Biochim. Biophys. Acta* 714: 465-470.

De Bruin, W.P., M.J.J. Kotterman, M.A. Posthumus, G. Schraa, and A.J.B. Zehnder. 1992. Complete biological reductive transformation of tetrachloroethene to ethane. *Appl. Environ. Microbiol.* 58: 1996-2000.

DeMoll, E., and T. Auffenberg. 1993. Purine metabolism in *Methanococcus vannielii. J. Bacteriol.* 175: 5754-5761.

DeWeerd, K.A., F. Concannon, and J.M. Suflita. 1991. Relationship between hydrogen consumption, dehalogenation, and the reduction of sulfur oxyanions by *Desulfomonile tiedjei. Appl. Environ. Microbiol.* 57: 1929-1934.

DeWeerd, K.A., L. Mandelco, R.S. Tanner, C.R. Woese, and J.M. Suflita. 1990. *Desulfomonile tiedjei* gen. nov. and sp. nov., a novel anaerobic, dehalogenating, sulfate-reducing bacterium. *Arch. Microbiol.* 154: 23-30.

Dickel, O.D., and H.-J. Knackmuss. 1991. Catabolism of 1,3-dinitrobenzene by *Rhodococcus* sp. QT-1. *Arch. Microbiol.* 157: 76-79.

DiStefano, T.D., J.M. Gossett , and S.H. Zinder. 1991. Reductive dechlorination of high concentrations of tetrachloroethene to ethene by an anaerobic enrichment culture in the absence of methanogenesis. *Appl. Environ. Microbiol.* 57: 2287-2292.

Dodson, K.S., and G.F. White. 1983. Some microbial enzymes involved in the biodegradation of sulfated surfactants pp. 90-155. In *Topics in Enzyme and Fermentation Technology*, Vol. 7 (Ed. A. Wiseman). Ellis-Horwood, Chichester.

Dolfing, J. 1990. Reductive dechlorination of 3-chlorobenzoate is coupled to ATP production and growth in an anaerobic bacterium, strain DCB-1. *Arch. Microbiol.* 153: 264-266.

Drotar, A.-M., and R. Fall. 1985. Methylation of xenobiotic thiols by *Euglena gracilis*: characterization of a cytoplasmic thiol methyltransferase. *Plant Cell Physiol.* 26: 847-854.

Drotar, A.-M., and R. Fall. 1986. Characterization of a xenobiotic thiol methyltransferase and its role in detoxication in *Tetrahymena thermophila*. *Pestic. Biochem. Physiol.* 25: 396-406.

Drotar, A.-M., G.A. Burton, J.E. Tavernier, and R. Fall. 1987b. Widespread occurrence of bacterial thiol methyltransferases and the biogenic emission of methylated sulfur gases. *Appl. Environ. Microbiol.* 53: 1626-1631.

Drotar, A.-M., L.R. Fall, E.A. Mishalanie, J.E. Tavernier, and R. Fall. 1987a. Enzymatic methylation of sulfide, selenide, and organic thiols by *Tetrahymena thermophila*. *Appl. Environ. Microbiol.* 53: 2111-2118.

Duque, E., A. Haidour, F. Godoy, and J.L. Ramos. 1993. Construction of a *Pseudomonas* hybrid strain that mineralizes 2,4,6-trinitrotoluene. *J. Bacteriol.* 175: 2278-2283.

Durham, D.R., and D.B. Stewart. 1987. Recruitment of naphthalene dissimilatory enzymes for the oxidation of 1,4-dichloronapththalene to 3,6-dichlorosalicylate, a precursor for the herbicide dicamba. *J. Bacteriol.* 169: 2889-2892.

Dürre, P., and J.R. Andreesen. 1982. Anaerobic degradation of uric acid via pyrimidine derivatives by selenium-starved cells of *Clostridium purinolyticum*. *Arch. Microbiol.* 131: 255-260.

Dürre, P., and J.R. Andreesen. 1983. Purine and glycine metabolism by purinolytic clostridia. *J. Bacteriol.* 154: 192-199.

Dürre, P., W. Andersch, and J.R. Andreesen. 1981. Isolation and characterization of an adenine-utlizing, anaerobic sporeformer, *Clostridium purinolyticum*. sp. nov. Int. *J. Syst. Bacteriol.* 31: 184-194.

Dwyer, D.F., and J.M. Tiedje. 1986. Metabolism of polyethylene glycol by two anaerobic bacteria, *Desulfovibrio desulfuricans* and a *Bacteroides* sp. *Appl. Environ. Microbiol.* 52: 852-856.

Eaton, R.W., and P.J. Chapman. 1992. Bacterial metabolism of naphthalene: construction and use of recombinant bacteria to study ring cleavage of 1,2-dihydroxynaphthalene and subsequent reactions. *J. Bacteriol.* 174: 7542-7554.

Eaton, R.W., and D.W. Ribbons. 1982. Metabolism of dibutylphthalate and phthalate by *Micrococcus* sp. strain 12B. *J. Bacteriol.* 151: 48-57.

Eaton, R.W., and K.N. Timmis. 1986. Characterization of a plasmid-specified pathway for catabolism of isopropylbenzene in *Pseudomonas putida* RE204. *J. Bacteriol.* 168: 123-131

Edwards, E.A., L.E. Wills, M. Reinhard, and D. Grbic-Galic. 1992. Anaerobic degradation of toluene and xylene by aquifer microorganisms under sulfate-reducing conditions. *Appl. Environ. Microbiol.* 58: 794-800.

Egli, C., T. Tschan, R. Scholtz, A.M. Cook, and T. Leisinger. 1988. Transformation of tetrachloromethane to dichloromethane and carbon dioxide by *Acetobacterium woodii*. *Appl. Environ. Microbiol.* 54: 2819-2824.

Eichler, B., and B. Schink. 1984. Oxidation of primary aliphatic alcohols by *Acetobacterium carbinolicum* sp. nov., a homoacetogenic anaerobe. *Arch. Microbiol.* 140: 147-152.

Elder, D.J.E., P. Morgan, and D.J. Kelly. 1992. Anaerobic degradation of trans-cinnamate and ω-phenylalkane carboxylic acids by the photosynthetic bacterium *Rhodopseudomonas palustris*: evidence for a beta-oxidation mechanism. *Arch. Microbiol.* 157: 148-154.

Elmorsi, E.A., and D.J. Hopper. 1979. The catabolism of 5-hydroxyisophthalate by a soil bacterium. *J. Gen. Microbiol.* 111: 145-152.

Elsden, S.R., M.G. Hilton, and D.A. Hopwood. 1976. The end products of the metabolism of aromatic acids by clostridia. *Arch. Microbiol.* 107: 283-288.

Engesser, K.-H., E. Schmidt, and H.-J. Knackmuss. 1980. Adaptation of *Alcaligenes eutrophus* B9 and *Pseudomonas* sp. B13 to 2-fluorobenzoate as growth substrate. *Appl. Environ. Microbiol.* 39: 68-73.

Engesser, K.H., R.B. Cain, and H. J. Knackmuss. 1988b. Bacterial metabolism of side chain fluorinated aromatics: cometabolism of 3-trifluoromethyl (TFM)-benzoate by *Pseudomonas putida* (arvilla) mt-2 and *Rhodococcus rubropertinctus* N657. *Arch. Microbiol.* 149: 188-197.

Engesser, K.H., M.A. Rubio, and D.W. Ribbons. 1988a. Bacterial metabolism of side chain fluorinated aromatics: cometabolism of 4-trifluoromethyl (TFM)-benzoate by 4-isopropylbenzoate grown *Pseudomonas putida* JT strains. *Arch. Microbiol.* 149: 198-206.

Engesser, K.H., V. Strubel, K. Christoglou, P. Fischer, and H.G. Rast. 1989. Dioxygenolytic cleavage of aryl ether bonds: 1,10-dihydro-1,10-dihydroxyfluorene-9-one, a novel arene dihydrodiol as evidence for angular dioxygenation of dibenzofuran. *FEMS Microbiol. Lett.* 65: 205-210.

Engesser, K.H., G. Auling, J. Busse, and H.-J. Knackmuss. 1990a. 3-fluorobenzoate enriched bacterial strain FLB 300 degrades benzoate and all three isomeric monofluorobenzoates. *Arch. Microbiol.* 153: 193-199.

Engesser, K.H., M.A. Rubio, and H.-J. Knackmuss. 1990b. Bacterial metabolism of side-chain-fluorinated aromatics: unproductive meta cleavage of 3-trifluoromethylcatechol. *Appl. Microbiol. Biotechnol.* 32: 600-608.

Ensley, B.D., and D.T. Gibson. 1983. Naphthalene dioxygenase: purification and properties of a terminal oxygenase component. *J. Bacteriol.* 155: 505-511.

Erickson, B.D., and F.J. Mondelo. 1992. Nucleotide sequencing and transcriptional mapping of the genes encoding biphenyl dioxygenase, a multicomponent polychlorinated-biphenyl-degrading enzyme in *Pseudomonas* strain LB 400. *J. Bacteriol.* 174: 2903-2912.

Evans, P.J., W. Ling, B. Goldschmidt, E.R. Ritter, and L.Y. Young. 1992. Metabolites formed during anaerobic transformation of toluene and *o*-xylene and their proposed relationship to the initial steps of toluene mineralization. *Appl. Environ. Microbiol.* 58: 496-501.

Evans, W.C., and G. Fuchs. 1988. Anaerobic degradation of aromatic compounds. *Ann. Rev. Microbiol.* 42: 289-317.

Evans, W.C., H.N. Fernley, and E. Griffiths. 1965. Oxidative metabolism of phenanthrene and anthracene by soil pseudomonads. The ring-fission mechanism. *Biochem. J.* 95: 819-831.

Evans, W.C., B.S.W. Smith, H.N. Zernley, and J.I. Davies. 1971. Bacterial metabolism of 2,4-dichlorophenoxyacetate. *Biochem. J.* 122: 543-551.

Fall, R.R., J.I. Brown, and T.L. Schaeffer. 1979. Enzyme recruitment allows the biodegradation of recalcitrant branched hydrocarbons by *Pseudomonas citronellolis*. *Appl. Environ. Microbiol.* 38: 715-722.

Fathepure, B.Z., and S. A. Boyd. 1988. Dependence of tetrachloroethylene dechlorination on methanogenic substrate consumption by *Methanosarcina* sp. strain DCM. *Appl. Environ. Microbiol.* 54: 2976-2980.

Fathepure, B.Z., J.M. Tiedje, and S. A. Boyd. 1988. Reductive dechlorination of hexachlorobenzene to tri- and dichlorobenzenes in anaerobic sewage sludge. *Appl. Environ. Microbiol.* 54: 327-330.

Faulkner, J.K., and D. Woodcock. 1965. Fungal detoxication. VII. Metabolism of 2,4-dichlorophenoxyacetic and 4-chloro-2-methylphenoxyacetic acids by *Aspergillus niger*. *J. Chem. Soc.* 1187-1191.

Finster, K., and F. Bak. 1993. Complete oxidation of propionate, valerate, succinate, and other organic compounds by newly isolated types of marine, anaerobic, mesophilic, gram-negative sulfur-reducing eubacteria. *Appl. Environ. Microbiol.* 59: 1452-1460.

Firestone, M.K., and J.M. Tiedje. 1978. Pathway of degradation of nitrilotriacetate by a *Pseudomonas* sp. *Appl. Environ. Microbiol.* 35: 955-961.

Fitzgerald, J.W. 1976. Sulfate ester formation and hydrolysis: a potentially important yet often ignored aspect of the sulfur cycle of aerobic soils. *Bacteriol. Revs.* 40: 698-721.

Fitzgerald, J.W., H.W. Maca, and F.A. Rose. 1979. Physiological factors regulating tyrosine-sulphate sulphohydrolase activity in *Comamonas terrigena*: occurrence of constitutive and inducible enzymes. *J. Gen. Microbiol.* 111: 407-415.

Fortnagel, P., H. Harms, R.-M. Wittich, S. Krohn, H. Meter, V. Sinnwell, H. Wilkes, and W. Francke. 1990. Metabolism of dibenzofuran by *Pseudomonas* sp. strain HH69 and the mixed culture HH27. *Appl. Environ. Microbiol.* 56: 1148-1156.

Fraisse, L., and H. Simon. 1988. Observations on the reduction of non-activated carboxylates by *Clostridium formicoaceticum* with carbon monoxide or formate and the influence of various viologens. *Arch. Microbiol.* 150: 381-386.

Franklin, T.J., and G.A. Snow. 1981. *Biochemistry of antimicrobial action*. Chapman and Hall, London.

Freedman, D.L., and J.M. Gossett. 1989. Biological reductive dechlorination of tetrachloroethylene and trichloroethylene under methanogenic conditions. *Appl. Environ. Microbiol.* 55: 2144-2151.

Frimmer, U., and F. Widdel. 1989. Oxidation of ethanol by methanogenic bacteria. Growth experiments and enzymatic studies. *Arch. Microbiol.* 152: 479-483.

Frings, J., E. Schramm, and B. Schink. 1992. Enzymes involved in anaerobic polyethylene glycol degradation by *Pelobacter venetianus* and *Bacteroides* strain PG1. *Appl. Environ. Microbiol.* 58: 2164-2167.

Fuchs, A., W. de Vries, and M.P. Sanz. 1980. The mechanism of pisatin degradation by *Fusarium oxysporum* f. sp. *pisi*. *Physiol. Plant Pathol.* 16: 119-133.

Fuchs, K., A. Schreiner, and F. Lingens. 1991. Degradation of 2-methylaniline and chlorinated isomers of 2-methylaniline by *Rhodococcus rhodochrous* strain CTM. *J. Gen. Microbiol.* 137: 2033-2039.

Fujioka, M., and H. Wada. 1968. The bacterial oxidation of indole. *Biochim. Biophys. Acta* 158: 70-78.

Fukumori, F., and R.P. Hausinger. 1993. *Alcaligenes eutrophus* JMP 134 "2,4-dichlorophenoxyacetate monooxygenase" is an α-ketoglutarate-dependent dioxygenase. *J. Bacteriol.* 175: 2083-2086.

Furukawa, N., and K. Tonomura. 1971. Enzyme system involved in the decomposition of phenyl mercuric acetate by mercury-resistant *Pseudomonas*. *Agric. Biol. Chem.* 35: 604-610.

Furukawa, K., N. Tomizuka, and A. Kamibayashi. 1979. Effect of chlorine substitution on the bacterial metabolism of various polychlorinated biphenyls. *Appl. Environ. Microbiol.* 38: 301-310.

Furutani, A., and J.W.M. Rudd. 1980. Measurement of mercury methylation in lake water and sediment samples. *Appl. Environ. Microbiol.* 40: 770-776.

Gälli, R., and P.L. McCarthy. 1989. Biotransformation of 1,1,1-trichloroethane, trichloromethane, and tetrachloromethane by a *Clostridium* sp. *Appl. Environ. Microbiol.* 55: 837-844.

Gamar, Y., and J.K. Gaunt. 1971. Bacterial metabolism of 4-chloro-2-methylphenoxyacetate (MCPA): formation of glyoxylate by side-chain cleavage. *Biochem. J.* 122: 527-531.

Gauthier, J.J., and S.C. Rittenberg. 1971. The metabolism of nicotinic acid. II. 2,5-dihydroxypyridine oxidation, product formation, and oxygen-18 incorporation. *J. Biol. Chem.* 246: 3743-3748.

Gee, J.M., and J.L. Peel. 1974. Metabolism of 2,3,4,6-tetrachlorophenol by microorganisms from broiler house litter. *J. Gen. Microbiol.* 85: 237-243.

Geigert, J., S.L. Neidleman, D.J. Dalietos, and S.K. DeWitt. 1983a. Haloperoxidases: enzymatic synthesis of α,β-halohydríns from gaseous alkenes. *Appl. Environ. Microbiol.* 45: 366-374.

Geigert, J., S.L. Neidleman, D.J. Dalietos, and S.K. DeWitt. 1983b. Novel haloperoxidase reaction: synthesis of dihalogenated products. *Appl. Environ. Microbiol.* 45: 1575-1578.

Geigert, J., S.L. Neidleman, and D.J. Dalietos. 1983c. Novel haloperoxidase substrates. Alkynes and cyclopropanes. *J. Biol. Chem.* 258: 2273-2277.

Geissler, J.F., C.S. Harwood, and J. Gibson. 1988. Purification and properties of benzoate-coenzyme A ligase, a *Rhodopseudomonas palustris* enzyme involved in the anaerobic degradation of benzoate. *J. Bacteriol.* 170: 1709-1714.

Genther, B.R.S., W.A. Price, and P.H. Pritchard. 1989. Anaerobic degradation of chloroaromatic compounds in aquatic sediments under a variety of enrichment conditions. *Appl. Environ. Microbiol.* 55: 1466-1471.

Genther, B.R.S., G.T. Townsend, and P.J. Chapman. 1990. Effect of fluorinated analogues of phenol and hydroxybenzoates on the anaerobic transformation of phenol to benzoate. *Biodegradation* 1: 65-74.

Gerritse, J., B.J. van der Woude, and J.C. Gottschal. 1992. Specific removal of chlorine from the *ortho*-position of halogenated benzoic acids by reductive dechlorination in anaerobic enrichment cultures. *FEMS Microbiol. Lett.* 100: 273-280.

Gibson, D.T., and V. Subramanian. 1984. Microbial degradation of aromatic hydrocarbons pp. 181-252. In *Microbial degradation of organic compounds* (Ed. D.T. Gibson). Marcel Dekker Inc., New York.

Gibson, D.T., J.R. Koch, and R.E. Kallio. 1968. Oxidative degradation of aromatic hydrocarbons. I. Enzymatic formation of catechol from benzene. *Biochemistry* 9: 2653-2662.

Gibson, D.T., G.E Cardini, F.C. Masales, and R.E. Kallio. 1970. Incorporation of oxygen-18 into benzene by *Pseudomonas putida*. *Biochemistry* 9: 1631-1635.

Gibson, D.T., V. Mahadevan, and J.F. Davey. 1974. Bacterial metabolism of para- and meta-xylene: oxidation of the aromatic ring. *J. Bacteriol.* 119: 930-936.

Gibson, K.J., and J. Gibson. 1992. Potential early intermediates in anaerobic benzoate degradation by *Rhodopseudomonas palustris*. *Appl. Environ. Microbiol.* 58: 696-698.

Gibson, S.A., and G.W. Sewell. 1992. Stimulation of reductive dechlorination of tetrachloroethene in anaerobic aquifer microcosms by addition of short-chain acids or alcohols. *Appl. Environ. Microbiol.* 58: 1392-1393.

Golbeck, J.H., S.A. Albaugh, and R. Radmer. 1983. Metabolism of biphenyl by *Aspergillus toxicarius*: induction of hydroxylating activity and accumulation of water-soluble conjugates. *J. Bacteriol.* 156: 49-57.

Goldman, P. 1965. The enzymatic cleavage of the carbon-fluorine bond in fluoroacetate. *J. Biol. Chem.* 240: 3434-3438.

Gorny, N., and B. Schink. 1994. Hydroquinone degradation via reductive dehydroxylation of gentisyl-CoA by a strictly anaerobic fermenting bacterium. *Arch. Microbiol.* 161: 25-32.

Gorny, N., G. Wahl, A. Brune, and B. Schink. 1992. A strictly anaerobic nitrate-reducing bacterium growing with resorcinol and other aromatic compounds. *Arch. Microobiol.* 158: 48-53.

Gorontzy, T., J. Küver, and K.-H. Blotevogel. 1993. Microbial reduction of nitroaromatic compounds under anaerobic conditions. *J. Gen. Microbiol.* 139: 1331-1336.

Gray, P.H.H., and H.G. Thornton. 1928. Soil bacteria that decompose certain aromatic compounds. *Centralbl. Bakteriol. Parasitenkd. Infektionskr.* (2 Abt.) 73: 74-96.

Grbic-Galic, D., and L.Y. Young. 1985. Methane fermentation of ferulate and benzoate: anaerobic degradation pathways. *Appl. Environ. Microbiol.* 50: 292-297.

Grbic-Galic, D., and T.M. Vogel. 1987. Transformation of toluene and benzene by mixed methanogenic cultures. *Appl. Environ. Microbiol.* 53 254-260.

Grbic-Galic, D., N. Churchman-Eisel, and I. Mrakovic. 1990. Microbial transformation of styrene by anaerobic consortia. *J. Appl. Bacteriol.* 69: 247-260.

Griffin, M., and P.W. Trudgill. 1972. The metabolism of cyclopentanol by *Pseudomonas* N.C.I.B. 9872. *Biochem. J.* 129: 595-603.

Griffiths, D.A., D.E. Brown, and S.G. Jezequel. 1992. Biotransformation of warfarin by the fungus Beauveria bassiana. *Appl. Microbiol. Biotechnol.* 37: 169-175.

Groenewegen, P.E.G., P. Breeuwer, J.M.L.M. van Helvoort, A.A.M. Langenhoff, F.P. de Vries, and J.A.M. de Bont. 1992. Novel degradative pathway of 4-nitrobenzoate in *Comamonas acidovorans* NBA-10. *J. Gen. Microbiol.* 138: 1599-1605.

Grund, E., B. Denecke, and R. Eichenlaub. 1992. Naphthalene degradation via salicylate and gentisate by *Rhodococcus* sp. strain B4. *Appl. Environ. Microbiol.* 58: 1874-1877.

Gurarini, A., G. Guglielmetti, M. Vincenti, P. Guarda, and G. Marchionni. 1993. Characterization of perfluoropolyethers by desorption chemical ionization and tandem mass spectrometry. *Anal.Chem.* 65: 970-975.

Haddock, J.D., and J.G. Ferry. 1993. Initial steps in the anaerobic degradation of 3,4,5-trihydroxybenzoate by *Eubacterium oxidoreducens*: characterization of mutamts and role of 1,2,3,5-tetrahydroxybenzene. *J. Bacteriol.* 175: 669-673.

Haemmerli, S.D., M.S.A. Leisola, D. Sanglard, and A. Fiechter. 1986. Oxidation of benzo[a]pyrene by extracellular ligninases of *Phanerochaete chrysosporium. J. Biol. Chem.* 261: 6900-6903.

Hagedorn, S.R., R.S. Hanson, and D.A. Kunz (Eds.).1988. *Microbial metabolism and the carbon cycle.* Harwood Academic Publishers, Chur, Switzerland.

Häggblom, M.M., and L.Y. Young. 1990. Chlorophenol degradation coupled to sulfate reduction. *Appl. Environ. Microbiol.* 56: 3255-3260.

Haigler, B.E., and J.C. Spain. 1993. Biodegradation of 4-nitrotoluene by *Pseudomonas* sp. strain 4NT. *Appl. Environ. Microbiol.* 59: 2239-2243.

Hales, S.G., G.K. Watson, K.S. Dodson, and G.F. White. 1986. A comparative study of the biodegradation of the surfactant sodium dodecyltriethoxy sulphate by four detergent-degrading bacteria. *J. Gen. Microbiol.* 132: 953-961.

Hallas, L.E., and M. Alexander. 1983. Microbial transformation of nitroaromatic compounds in sewage effluent. *Appl. Environ. Microbiol.* 45: 1234-1241.

Hamdy, M.K., and O.R. Noyes. 1975. Formation of methyl mercury by bacteria. *Appl. Microbiol.* 30: 424-432.

Hammel, K.E., W.Z. Gai, B. Green, and M.A. Moen. 1992. Oxidative degradation of phenanthrene by the lignoolytic fungus *Phanerochaete chrysosporium. Appl. Environ. Microbiol.* 58: 1832-1838.

Hankin, L., and P.E. Kolattukudy. 1968 Metabolism of a plant wax paraffin (*n*-nonacosane) by a soil bacterium (*Micrococcus cerificans*). *J. Gen. Microbiol.* 51: 457-463.

Happe, B., L.D. Eltis, H. Poth, R. Hedderich, and K.N. Timmis. 1993. Characterization of 2,2',3-trihydroxybiphenyl dioxygenase, an extradiol dioxygenase from the dibenzofuran- and dibenzo-*p*-dioxin-degrading bacterium *Sphingomonsas* sp. strain RW1. *J. Bacteriol.* 175: 7313-7320.

Hareland, W.A., R.L. Crawford, P.J. Chapman, and S. Dagley. 1975. Metabolic function and properties of 4-hydroxyphenylacetic acid 1-hydrolase from *Pseudomonas acidovorans. J. Bacteriol.* 121: 272-285.

Harper, D.B. 1977. Microbial metabolism of aromatic nitriles. Enzymology of C-N cleavage by *Nocardia* sp. (Rhodochrous group) N.C.I.B. 11216. *Biochem. J.* 165: 309-319.

Harper, D.B., and J. Nelson. 1982. The bacterial biogenesis of isobutyraldoxine O-methyl ether, a novel volatile secondary metabolite. *J. Gen. Microbiol.* 128: 1667-1678.

Harper, D.B., and J.T. Kennedy. 1985. Purification and properties of S-adenosylmethionine: aldoxime O-methyltransferase from *Pseudomonas* sp. N.C.I.B. 11652. *Biochem. J.* 226: 147-153.

Harper, D.B., and J.T.G. Hamilton. 1988. Biosynthesis of chloromethane in *Phellinus pomaceus*. *J. Gen. Microbiol.* 13 4: 2831-2839.

Harper, D.B., J.T.G. Hamilton, J.T. Kennedy, and K. J. McNally. 1989. Chloromethane, a novel methyl donor for biosynthesis of esters and anisoles in *Phellinus pomaceus*. *Appl. Environ. Microbiol.* 55: 1981-1989.

Hartmans, S., and J.A.M. de Bont. 1992. Aerobic vinyl chloride metabolism in *Mycobacterium aurum* L1. *Appl. Environ. Microbiol.* 58: 1220-1226.

Hartmans, S., J.P. Smits, M.J. van der Werf, F. Volkering, and J.A.M. de Bont. 1989. Metabolism of styrene oxide and 2-phenylethanol in the styrene-degrading *Xanthobacter* strain 124X. *Appl. Environ. Microbiol.* 55: 2850-2855.

Hartmans, S., F.J. Weber, B.P.M. Somhorst, and J.A.M. de Bont. 1991. Alkene monooxygenase from *Mycobacterium* E3: a multicomponent enzyme. *J. Gen. Microbiol.* 137: 2555-2560.

Haug, W., A. Schmidt, B. Nörtemann, D.C. Hempel, A. Stolz, and H.-J. Knackmuss. 1991. Mineralization of the sulfonated azo dye mordant Yellow 3 by a 6-aminonaphthalene-2-sulfonate-degrading bacterial consortium. *Appl. Environ. Microbiol.* 57: 3144-3149.

Haugland, R.A., D.J. Schlemm, R.P. Lyons, P.R. Sferra, and A. M. Chakrabarty. 1990. Degradation of the chlorinated phenoxyacetate herbicides 2,4-dichlorophenoxyacetic acid and 2,4,5-trichlorophenoxyacetic acid by pure and mixed bacterial cultures. *Appl. Environ. Microbiol.* 56: 1357-1362.

Havel, J., and W. Reineke. 1993. Microbial degradation of chlorinated acetophenones. *Appl. Environ. Microbiol.* 59: 2706-2712.

Hegeman, G.D. 1966. Synthesis of the enzymes of the mandelate pathway by *Pseudomonas putida*. I. Synthesis of enzymes of the wild type. *J. Bacteriol.* 91: 1140-1154.

Heijthuijsen, J.H.F.G., and T.A. Hansen. 1989a. Anaerobic degradation of betaine by marine *Desulfobacterium* strains. *Arch. Microbiol.* 152: 393-396.

Heijthuijsen, J.H.F.G., and T.A. Hansen. 1989b. Betaine fermentation and oxidation by marine *Desulfuromonas* strains. *Appl. Environ. Microbiol.* 55: 965-969.

Heitkamp, M.A., J.P. Freeman, D.W. Miller, and C.E. Cerniglia. 1988. Pyrene degradation by a *Mycobacterium* sp.: identification of ring oxidation and ring fission products. *Appl. Environ. Microbiol.* 54: 2556-2565.

Hensgens, C.M.H., J. Vonck, J. Van Beeumen, E.F.J. van Bruggen, and T.A. Hansen. 1993. Purification and characterization of an oxygen-labile, NAD-dependent alcohol dehydrogenase from *Desulfovibrio gigas*. *J. Bacteriol.* 175: 2859-2863.

Hickey, W.J., and D.D. Focht. 1990. Degradation of mono-, di, and trihalogenated benzoic acids by *Pseudomonas aeruginosa* JB2. *Appl. Environ. Microbiol.* 56: 3842-3850.

Higson, F.K., and D.D. Focht. 1990. Degradation of 2-bromobenzoate by a strain of *Pseudomonas aeruginosa*. *Appl. Environ. Microbiol.* 56: 1615-1619.

Hilton, M.D., and W.J. Cain. 1990. Bioconversion of cinnamic acid to acetophenone by a pseudomonad: microbial production of a natural flavor compound. *Appl. Environ. Microbiol.* 56: 623-627.

Hippe, H., D. Caspari, K. Fiebig, and G. Gottschalk. 1979. Utilization of trimethylamine and other N-methyl compounds for growth and methane formation by *Methanosarcina barkeri*. *Proc. Natl. Acad. Sci. U.S.A.* 76: 494-498.

Hirschberg, R., and J.C. Ensign. 1971. Oxidation of nicotinic acid by a *Bacillus* species: source of oxygen atoms for the hydroxylation of nicotinic acid and 6-hydroxynicotinic acid. *J. Bacteriol.* 108: 757-759.

Högn, T., and L. Jaenicke. 1972. Benzene metabolism of *Moraxella* species. *Eur. J. Biochem.* 30: 369-375.

Holland, H.L. 1988. Chiral sulfoxidation by biotransformation of organic sulfides. *Chem. Revs.* 88: 473-485.

Holland, H.L., M. Carey, and S. Kumaresan. 1993. Fungal biotransformation of organophosphines. *Xenobiotica* 23: 519-524.

Holliger, C., G. Schraa, E. Stuperich, A.J.M. Stams, and A.J.B. Zehnder. 1992a. Evidence for the involvement of corrinoids and factor F_{430} in the reductive dechlorination of 1,2-dichloroethane by *Methanosarcina barkeri*. *J. Bacteriol.* 174: 4427-4434.

Holliger, C., S.W.M. Kengen, G. Schraa, A.J.M. Stams, and AJ.B. Zehnder. 1992b. Methyl-coenzyme M reductase of *Methanobacterium thermoautrotrophicum* delta H catalyzes the reductive dechlorination of 1,2-dichloroethane to ethylene and chloroethane. *J. Bacteriol.* 174: 4435-4443.

Hopper, D.J. 1978. Microbial degradation of aromatic hydrocarbons pp. 85-112. In *Developments in Biodegradation of hydrocarbons-1* (Ed. R.J. Watkinson). Applied Science Publishers Ltd., London.

Hopper, D.J. 1988. Properties of *p*-cresol methylhydroxylases. In *Microbial metabolism and the carbon cycle* pp. 247-258 (Eds. S.R. Hagedorn, R.S. Hanson and D.A. Kunz). Harwood Academic Publishers, Chur, Switzerland.

Hopper, D.J., H.G. Jones, E.A. Elmorisi, and M.E. Rhodes-Roberts. 1985. The catabolism of 4-hydroxyacetophenone by an *Alcaligenes* sp. *J. Gen. Microbiol.* 131: 1807-1814.

Hormann, K., and J.R. Andreesen. 1991. A flavin-dependent oxygenase reaction initiates the degradation of pyrrole-2-carboxylate in *Arthrobacter* strain Py1 (DSM 6386). *Arch. Microbiol.* 157: 43-48.

Horowitz, A., J.M. Suflita, and J.M. Tiedje. 1983. Reductive dehalogenations of halobenzoates by anaerobic lake sediment microorganisms. *Appl. Environ. Microbiol.* 45: 1459-1465.

Hou, C.T., R. Patel, A.I. Laskin, N. Barnabe, and I. Barist. 1983. Epoxidation of short-chain alkenes by resting-cell suspensions of propane-grown bacteria. *Appl. Environ. Microbiol.* 46: 171-177.

Howell, L.G., T. Spector, and V. Massey. 1972. Purification and properties of p-hydroxybenzoate hydroxylase from *Pseudomonas fluorescens*. *J. Biol. Chem.* 247: 4340-4350.

Hsu, T., S.L. Daniel, M.F. Lux, and H.L. Drake. 1990. Biotransformation of carboxylated aromatic compounds by the acetogen *Clostridium thermoaceticum*: generation of growth-supportive CO_2 equivalents under CO_2-limited conditions. *J. Bacteriol.* 172: 212-217.

Hudlicky, M. 1990. *Oxidations in organic chemistry*. ACS Monograph 186. American Chemical Society, Washington, D.C.

Husain, M., B. Entsch, D.P. Ballou, V. Massey, and P.J Chapman. 1980. Fluoride elimination from substrates in hydroxylation reactions catalyzed by *p*-hydroxybenzoate hydroxylase. *J. Biol. Chem.* 255: 4189-4197.

Ibrahim, A.-R., and Y.J. Abul-Hajj. 1989. Aromatic hydroxylation and sulfation of 5-hydroxyflavone by *Streptomyces fulvissimus*. *Appl. Environ. Microbiol.* 55: 3140-3142.

Ibrahim, A.-R. S., and Y. J. Abul-Hajj. 1990. Microbiological transformation of flavone and isoflavone. *Xenobiotica* 20: 363-373.

Imhoff, D., and J.R. Andreesen. 1979. Nicotinic acid hydroxylase from *Clostridium barkeri*: selenium-dependent formation of active enzyme. *FEMS Microbiol. Lett.* 5: 155-158.

Imhoff-Stuckle, D., and N. Pfennig. 1983. Isolation and characterization of a nicotinic acid-degrading sulfate-reducing bacterium, *Desulfococcus niacini* sp. nov. *Arch. Microbiol.* 136: 194-198.

Ince, J.E., and C.J. Knowles. 1986. Ethylene formation by cell-free extracts of *Escherichia coli*. *Arch. Microbiol.* 146: 151-158

Ingledew, W.M., M.E.F. Tresguerres, and J.L. Cánovas. 1971. Regulation of the enzymes of the hydroaromatic pathway in *Acinetobacter calco-aceticus*. *J. Gen. Microbiol.* 68: 273-282.

Jackson, J.-A., W.R. Blair, F.E. Brinckman, and W.P. Iverson. 1982. Gas-chromatographic speciation of methylstannanes in the Chesapeake bay using purge and trap sampling with a tin-selective detector. *Environ. Sci. Technol.* 16: 110-119.

Jacob, G.S., J.R. Garbow, L.E. Hallas, N.M. Kimack, G.N. Kishore, and J. Schaefer. 1988. Metabolism of glyphosate in *Pseudomonas* sp. strain LBr. *Appl. Environ. Microbiol.* 54: 2953-2958.

Jagnow, G., K. Haider, and P.-C. Ellwardt. 1977. Anaerobic dechlorination and degradation of hexachlorocyclohexane by anaerobic and facultatively anaerobic bacteria. *Arch. Microbiol.* 115: 285-292.

Jeffrey, A.M., H.J.C. Yeh, D.M. Jerina, T.R. Patel, J.F. Davey, and D.T. Gibson. 1975. Initial reactions in the oxidation of naphthalene by *Pseudomonas putida*. *Biochemistry* 14: 575-584.

Jensen, S., and A. Jernelöv. 1969. Biological methylation of mercury in aquatic organisms. *Nature* (London) 233: 753-754.

Jones, K.H., R.T. Smith, and P.W. Trudgill. 1993. Diketocamphane enantiomer-specific 'Bayer-Villiger' monooxygenases from camphor-grown *Pseudomonas putida* ATCC 17453. *J. Gen. Microbiol.* 139: 797-805.

Joshi, D.K., and M.H. Gold. 1993. Degradation of 2,4,5-trichlorophenol by the lignin-degrading basidiomycete *Phanerochaete chrysosporium*. *Appl. Environ. Microbiol.* 59: 1779-1785.

Jutzi, K., A.M. Cook, and R. Hütter. 1982. The degradative pathway of the *s*-triazine melamine. *Biochem. J.* 208: 679-684.

Kaufman, D.D., J.. Plimmer, J. Iwan, and U.I. Klingebiel. 1972. 3,3',4,4'-tetrachloroazoxybenzene from 3,4-dichloroaniline. *J. Agric. Food Chem.* 20: 916-919.

Kaufman, D.D., J.. Plimmer, and U.I. Klingebiel. 1973. Microbial oxidation of 4-chloroaniline. *J. Agric. Food Chem.* 21: 127-132.

Kelley, I., J.P. Freeman, F.E. Evans, and C.E. Cerniglia. 1993. Identification of metabolites from the degradation of fluoranthene by *Mycobacterium* sp. strain PYR-1. *Appl. Environ. Microbiol.* 59: 800-806.

Kende, H. 1989. Enzymes of ethylene biosynthesis. *Plant Physiol.* 91: 1-4.

Kersten, P.J., S. Dagley, J.W. Whittaker, D.M. Arciero, and J.D. Lipscomb. 1982. 2-pyrone-4,6-dicarboxylic acid, a catabolite of gallic acids in *Pseudomonas* species. *J. Bacteriol.* 152: 1154-1162.

Kido, T., and K. Soda. 1976. A new oxygenase, 2-nitropropane dioxygenase of *Hansenula mrakii*. Enzymologic and spectrophotometric properties. *J. Biol. Chem.* 251: 6994-7000.

Kiene, R.P., and B.F. Taylor. 1988. Demethylation of dimethylsulfoniopropionate and production of thiols in anoxic marine sediments. *Appl. Environ. Microbiol.* 54: 2208-2212.

King, G.M. 1984. Metabolism of trimethylamine, choline and glycine betaine by sulfate-reducing and methanogenic bacteria in marine sediments. *Appl. Environ. Microbiol.* 48: 719-725.

King, G.M. 1988. Dehalogenation in marine sediments containing natural sources of halophenols. *Appl. Environ. Microbiol.* 54: 3079-3085.

Kitazume, T., and N. Ishikawa. 1983. Asymmetrical reduction of perfluoroalkylated ketones, ketoesters and vinyl compounds with baker's yeast. *Chemistry Lett.* 237-238.

Kiyohara, H., and K. Nagao. 1978. The catabolism of phenanthrene and naphthalene by bacteria. *J. Gen. Microbiol.* 105: 69-75

Kiyohara, H., K. Nagao, and R. Nomi. 1976. Degradation of phenanthrene through *o*-phthalate by an *Aeromonas* sp. *Agric. Biol. Chem.* 40: 1075-1082.

Klecka, G.M., and D.T. Gibson. 1981. Inhibition of catechol 2,3-dioxygenase from *Pseudomonas putida* by 3-chlorocatechol. *Appl. Environ. Microbiol.* 41: 1159-1165.

Knoll, G., and J. Winter. 1989. Degradation of phenol via carboxylation to benzoate by a defined, obligate syntrophic consortium of anaerobic bacteria. *Appl. Microbiol. Biotechnol.* 30: 318-324.

Koch, J., and G. Fuchs. 1992. Enzymatic reduction of benzoyl-CoA to alicyclic compounds, a key reaction in anaerobic aromatic metabolism. *Eur. J. Biochem.* 205: 195-202.

Kodama, K., K. Umehara, K. Shikmizu, S. Nakatani, Y. Minoda, and K. Yamada. 1973. Identification of microbial products from dibenzothiophene and its proposed oxidation pathway. *Agric. Biol. Chem.* 37: 45-50.

Kohler-Staub, D., and H.-P. E. Kohler. 1989. Microbial degradation of beta-chlorinated four-carbon aliphatic acids. *J. Bacteriol.* 171: 1428-1434.

Kohring, G.-W., X. Zhang, and J. Wiegel. 1989 Anaerobic dechlorination of 2,4-dichlorophenol in freshwater sediments in the presence of sulfate. *Appl. Environ. Microbiol.* 55: 2735-2737.

Koizumi, M., M. Shimizu, and K. Kobashi. 1990. Enzymatic sulfation of quercitin by arylsulfotransferase from a human intestinal bacterium. *Chem. Pharm. Bull.* 38: 794-796.

Kolattukudy, P.E., and L. Hankin. 1968. Production of ω-haloesters from alkyl halides by *Micrococcus cerificans*. *J. Gen. Microbiol.* 54: 333-336.

Kondo, R., H. Yamagami, and K. Sakai. 1993. Xylosation of phenolic hydroxyl groups of the monomeric lignin model compounds 4-methylguaiacol and vanillyl alcohol by *Coriolus versicolor*. *Appl. Environ. Microbiol.* 59: 438-441.

Konig, H., and K.O. Stetter. 1982. Isolation and characterization of *Methanolobus tindarius*, sp. nov., a coccoid methanogen growing only on methanol, and methylamines. *Zentralbl. Bakteriol. Parasitenkd. Infektionskr. Hyg.* Abt 1 Orig. Reihe C 3: 478-490.

Kozarich, J.W. 1988. Enzyme chemistry and evolution in the β-ketoadipate pathway. In *Microbial metabolism and the carbon cycle* pp. 283-302. (Eds. S.R. Hagedorn, R.S. Hanson and D.A. Kunz). Harwood Academic Publishers, Chur, Switzerland.

Kredich, N.M., L.J. Foote, and B.S. Keenan. 1973. The stoichiometry and kinetics of the inducible cysteine desulfhydrase from *Salmonella typhimurium*. *J. Biol. Chem.* 248: 6187-6196.

Kreft, J.-U., and B. Schink. 1993. Demethyation and degradation of phenylmethylethers by the sulfide-methyliationg homoacetogenic bacterium strain TMBS 4. *Arch. Microbiol.* 159: 308-315.

Kretzer, A., and J.R. Andreesen. 1991. A new pathway for isonicotinate degradation by *Mycobacterium* sp. INA1. *J. Gen. Microbiol.* 137: 1073-1080.

Krishnamurty, H.G., and F.J. Simpson. 1970. Degradation of rutin by *Aspergillus flavus*. Studies with oxygen 18 on the action of a dioxygenase on quercitin. *J. Biol. Chem.* 245: 1467-1471.

Krishnamurty, HG., K.-J. Cheng, G.A. Jones, F.J. Simpson, and J.E. Watkin. 1970. Identification of products produced by the anaerobic degradation of rutin and related flavonoids by *Butyrivibrio* sp C_3. *Can. J. Microbiol.* 16: 759-767.

Krumholz, L.R., and M.P. Bryant. 1985. *Clostridium pfennigi* sp. nov. uses methoxyl groups of monobenzenoids and produces butyrate. *Appl. Environ. Microbiol.* 35: 454-456.

Krumholz, L.R., and M.P. Bryant. 1988. Characterization of the pyrogallol-phloroglucinol isomerase of *Eubacterium oxidoreducens*. *J. Bacteriol.* 170: 2472-2479.

Krumholz, L.R., R.L. Crawford, M.E. Hemling, and M.P. Bryant. 1987. Metabolism of gallate and phloroglucinol in *Eubacterium oxidoreducens* via 3-hydroxy-5-oxohexanoate. *J. Bacteriol.* 169: 1886-1890.

Kuever, J., J. Kulmer, S. Janssen, U. Fischer, and K.-H. Blotevogel. 1993. Isolation and characterization of a new spore-forming sulfate-reducing bacterium growing by complete oxidation of catechol. *Arch. Microbiol.* 159: 282-288.

Kuhm, A.E., A. Stolz, K.-L. Ngai, and H.-J. Knackmuss. 1991. Purification and characterization of a 1,2-dihydroxynaphthalene dioxygenase from a bacterium that degrades naphthalenesulfonic acids. *J. Bacteriol.* 173: 3795-3802.

Kuhn, E.P., G.T. Townsend, and J.M. Suflita. 1990. Effect of sulfate and organic carbon supplements on reductive dehalogenation of chloroanilines in anaerobic aquifer slurries. *Appl. Environ. Microbiol.* 56: 2630-2637.

Kung, H.-F., and T.C. Stadtman. 1971. Nicotinic acid metabolism. VI. Purification and properties of alpha-methyleneglutarate mutase (B_{12}-dependent) and methylitaconate isomerase. *J. Biol. Chem.* 246: 3378-3388.

Kung, H.-F., and L. Tsai. 1971. Nicotinic acid metabolism. VII. Mechanism of action of clostridial alpha-methyleneglutarate mutase (B_{12}-dependent) and methylitaconate isomerase. *J. Biol. Chem.* 246: 6436-6443.

Kunz, D.A., and P.J. Weimer. 1983. Bacterial formation and metabolism of 6-hydroxyhexanoate: evidence of a potential role for ω-oxidation. *J. Bacteriol.* 156: 567-575.

La Nauze, J.M., and H. Rosenberg. 1968. The identification of 2-phosphonoacetaldehyde as an intermediate in the degradation of 2-aminoethylphosphonate by *Bacillus cereus*. *Biochim. Biophys. Acta* 165: 438-447.

La Nauze, J.M., H. Rosenberg, and D.C. Shaw. 1970. The enzymatic cleavage of the carbon-phosphorus bond: purification and properties of phosphonatase. *Biochim. Biophys. Acta* 212: 332-350.

La Roche, S.D., and T. Leisinger. 1991. Identification of *dcmR*, the regulatory gene governing expression of dichloromethane dehalogenase in *Methylobacterium* sp. strain DM4. *J. Bacteriol.* 173: 6714-6721.

Laanio, T.L, P.C. Kearney, and D.D. Kaufman. 1973. Microbial metabolism of dinitramine. *Pest. Biochem. Physiol.* 3: 271-277.

Lack, A., and G. Fuchs. 1992. Carboxylation of phenylphosphate by phenol carboxylase, an enzyme system of anaerobic phenol metabolism. *J. Bacteriol.* 174: 3629-3636.

Layh, N., A. Stolz, S. Förster, F. Effenberger, and H.-J. Knackmuss. 1992. Enantioselective hydrolysis of *O*-acetylmandelonitrile to O-acetylmandelic acid by bacterial nitrilases. *Arch. Microbiol.* 158: 405-411.

Lee, C.H., P.C. Oloffs, and S.Y. Szeto. 1986. Persistence, degradation, and movement of triclopyr and its ethylene glycol butyl ether ester in a forest soil. *J. Agric. Food Chem.* 34: 1075-1079.

Lenke, H., and H.-J. Knackmuss. 1992. Initial hydrogenation during catabolism of picric acid by *Rhodococcus erythropolis* HL 24-2. *Appl. Environ. Microbiol.* 58: 2933-2937.

Lenke, H., D.H. Pieper, C. Bruhn, and H.-J. Knackmuss. 1991. Degradation of 2,4-dinitrophenol by two *Rhodococcus erythropolis* strains, HL 24-1 and HL 24-2. *Appl. Environ. Microbiol.* 58: 2928-2932.

Leppik, R.A. 1989. Steroid catechol degradation: disecoandrostane intermediates accumulated by *Pseudomonas transposon* mutant strains. *J. Gen. Microbiol.* 135: 1979-1988.

Li, D.-Y., J. Eberspächer, B. Wagner, J. Kuntzer, and F. Lingens. 1991. Degradation of 2,4,6-trichlorophenol by Azotobacter sp. strain GP1. *Appl. Environ. Microbiol.* 57: 1920-1928.

Lieberman, I., and A. Kornberg. 1955. Enzymatic synthesis and breakdown of a pyrimidine, orotic acid III. Ureidosuccinase. *J. Biol. Chem.* 212: 909-920.

Liebert, F. 1909. The decomposition of uric acid by bacteria. *Proc. K. Acad. Ned. Wetensch.* 12: 54-64.

Little, P.J., M.O. James, J.B. Pritchard, and J.R. Bend. 1984. Benzo(a)pyrene metabolism in hepatic microsomes from feral and 3-methylcholanthrene-treated southern flounder, *Paralichthys lethostigma*. *J. Environ. Pathol. Toxicol. Oncol.* 5: 309-320.

Little, C.D, A.V. Palumbo, S.E. Herbes, M.E. Lidstrom, R.L. Tyndall, and P.J. Gilmer. 1988. Trichloroethylene biodegradation by a methane-oxidizing bacterium. *Appl. Environ. Microbiol.* 54: 951-956.

Liu, C.-M., P.A. McLean, C.C. Sookdeo, and F.C. Cannon. 1991. Degradation of the herbicide glyphosate by members of the family Rhizobiaceae. *Appl. Environ. Microbiol.* 57: 1799-1804.

Liu, S., and J.M. Suflita. 1993. H_2-CO_2-dependent anaerobic *O*-demethylation activity in subsurface sediments and by an isolated bacterium. *Appl. Environ. Microbiol.* 59: 1325-1331.

Lobos, J.H., T.K. Leib, and T.-M. Su. 1992. Biodegradation of bisphenol A and other bisphenols by a gram-negative aerobic bacterium. *Appl. Environ. Microbiol.* 58: 1823-1831.

Locher, H.H., T. Leisinger, and A.M. Cook. 1991. 4-toluene sulfonate methyl-monooxygenase from *Comamonas testosteroni*: purification and some properties of the oxygenase component. *J. Bacteriol.* 173: 3741-3748.

Lochmeyer, C., J. Koch, and G. Fuchs. 1992. Anaerobic degradation of 2-aminobenzoic acid (anthranilic acid) via benzoyl-coenzyme A (CoA) and cyclohex-1-enecarboxyl-CoA in a denitrifying bacterium. *J. Bacteriol.* 174: 3621-3628.

Londry, K.L., and P.M. Fedorak. 1993. Use of fluorinated compounds to detect aromatic metabolites from *m*-cresol in a methanogenic consortium: evidence for a demethylation reaction. *Appl. Environ. Microbiol.* 59: 2229-2238.

Mackie, R.I., B.A. White, and M.P. Bryant. 1991. Lipid metabolism in anaerobic ecosystems. *Crit. Revs. Microbiol.* 17: 449-479.

Madsen, T., and J. Aamand. 1991. Effects of sulfuroxy anions on degradation of pentachlorophenol by a methanogenic enrichment culture. *Appl. Environ. Microbiol.* 57: 2453-2458.

Madsen, T., and D. Licht. 1992. Isolation and characterization of an anaerobic chlorophenol-transforming bacterium. *Appl. Environ. Microbiol.* 58: 2874-2878.

Mahaffey, W. R., D.T. Gibson, and C.E. Cerniglia. 1988. Bacterial oxidation of chemical carcinogens: formation of polycyclic aromatic acids from benz[a]anthracene. *Appl. Environ. Microbiol.* 54: 2415-2423.

Mandelstam, J., K. McQuillen, and I. Dawes. 1982. *Biochemistry of bacterial growth*. Blackwell Scientific Publications, Oxford.

Mansouri, S., and A.W. Bunch. 1989. Bacterial synthesis from 2-oxo-4-thiobutyric acid and from methionine. *J. Gen. Microbiol.* 135: 2819-2827.

Massé, R., D. Lalanne, F. Mssier, and M. Sylvestre. 1989. Characterization of new bacterial transformation products of 1,1,1-trichloro-2,2-bis-(4-chlorophenyl)ethane (DDT) by gas chromatography/mass spectrometry. *Biomed. Environ. Mass Spectrom.* 18: 741-752.

Masuda, N., H. Oda, S. Hirano, M. Masuda, and H. Tanaka. 1984. 7α-dehydroxylation of bile acids by resting cells of a *Eubacterium lentum*-like intestinal anaerobe, strain c-25. *Appl. Environ. Microbiol.* 47: 735-739.

May, S.W., and B.J. Abbott. 1973. Enzymatic epoxidation. II. Comparison between the epoxidation and hydroxylation reactions catalyzed by the omega-hydroxylation system of *Pseudomonas oleovorans*. *J. Biol. Chem.* 248: 1725-1730.

McBride, K.E., J.W. Kenny, and D.M. Stalker. 1986. Metabolism of the herbicide bromoxynil by *Klebsiella pneumoniae* subspecies *ozaenae*. *Appl. Environ. Microbiol.* 52: 325-330.

McCall, P.J., S.A. Vrona, and S.S. Kelly. 1981. Fate of uniformly carbon-14 ring labeled 2,4,5-trichlorophenoxyacetic acid and 2,4-dichlorophenoxyacetic acid. *J. Agric. Food Chem.* 29: 100-107.

McClure, N.C., and W.A Venables. 1986. Adaptation of *Pseudomonas putida* mt-2 to growth on aromatic amines. *J. Gen. Microbiol.* 132: 2209-2218.

McCormick, N.G., F.E. Feeherry, and H.S. Levinson. 1976. Microbial transformation of 2,4,6-trinitrotoluene and other nitroaromatic compounds. *Appl. Environ. Microbiol.* 31: 949-958.

McCormick, N.G., J.H. Cornell, and A.M. Kaplan. 1978. Identification of biotransformation products from 2,4-dinitrotoluene. *Appl. Environ. Microbiol.* 35: 945-948.

McInerney, M.J., M.P. Bryant, R.B. Hespell, and J.W. Costerton. 1981. *Syntrophomonas wolfei* gen. nov. sp. nov., an anaerobic, syntrophic, fatty acid-oxidizing bacterium. *Appl. Environ. Microbiol.* 41: 1029-1039.

McKenna, E.J., and R.E. Kallio. 1971. Microbial metabolism of the isoprenoid alkane pristane. *Proc. Natl. Acad. Sci. U.S.A.* 68: 1552-1554.

McNally, K.J., J.T.G. Hamilton, and D.B. Harper. 1990. The methylation of benzoic and n-butyric acids by chloromethane in *Phellinus pomaceus*. *J. Gen. Microbiol.* 136: 1509-1515.

Meßmer, M., G. Wohlfarth, and G. Diekert. 1993. Methyl chloride metabolism of the strictly anaerobic, methyl chloride-utilizing homoacetogen strain MC. *Arch. Microbiol.* 160: 383-387.

Metcalf, W.W., and B.L. Wanner. 1991. Involvement of the *Escherichia coli phn (psiD)* gene cluster in assimilation of phosphorus in the form of phosphonates, phosphite, P_i esters, and P_i. *J. Bacteriol.* 173: 587-600.

Meyer, J.J.M., N. Grobbelaar, and P.L. Steyn. 1990. Fluoroacetate-metabolizing pseudomonad isolated from *Dichapetalum cymosum*. *Appl. Environ. Microbiol.* 56: 2152-2155.

Michaelsen, M., R. Hulsch, T. Höpner, and L. Berthe-Corti. 1992. Hexadecane mineralization in oxygen-controlled sediment-seawater cultivations with autochthonous microorganisms. *Appl. Environ. Microbiol.* 58: 3072-3077.

Mikesell, M.D., and S.A. Boyd. 1986. Complete reductive dechlorination and mineralization of pentachlorophenol by anaerobic microorganisms. *Appl. Environ. Microbiol.* 52: 861-865.

Miller, J.M., and D.O. Gray. 1982. The utilization of nitriles and amides by a *Rhodococcus* species. *J. Gen. Microbiol.* 128: 1803-1809.

Miller, K.W. 1992. Reductive desulfurization of dibenzyldisulfide. *Appl. Environ. Microbiol.* 58: 2176-2179.

Miyazaki, T., T. Yamagishi, and M. Matsumoto. 1984. Residues of 4-chloro-1-(2,4-dichlorophenoxy)-2-methoxybenzene (triclosan methyl) in aquatic biota. *Bull. Environ. Contam. Toxicol.* 32: 227-232.

Modrzakowski, M.C., and W.R. Finnerty. 1980. Metabolism of symmetrical dialkyl ethers by Acinetobacter sp. HO1-N. *Arch. Microbiol.* 126: 285-290.

Mohamed, M. E.-S., B. Seyfried, A. Tschech, and G. Fuchs. 1993. Anaerobic oxidation of phenylacetate and 4-hydroxyphenylacetate to benzoyl-coenzyme A and CO_2 in denitrifying *Pseudomonas* sp. *Arch. Microbiol.* 159: 563-573.

Mohn, W.W, and K.J. Kennedy. 1992. Reductive dehalogenation of chlorophenols by *Desulfomonile tiedjei* DCB-1. *Appl. Environ. Microbiol.* 58: 1367-1370.

Mohn, W.W., and J.M. Tiedje. 1991. Evidence for chemiosmotic coupling of reductive dechlorination and ATP synthesis in *Desulfomonile tiedjei*. *Arch. Microbiol.* 157: 1-6.

Morris, P.J., J.F. Quensen III, J.M. Tiedje, and S.A. Boyd. 1992. Reductive debromination of the commercial polybrominated biphenyl mixture Firemaster BP6 by anaerobic microorganisms from sediments. *Appl. Environ. Microbiol.* 58: 3249-3256.

Mott, G.E., A.W. Brinkley, and C.L. Mersinger. 1980. Biochemical characterization of cholesterol-reducing *Eubacterium*. *Appl. Environ. Microbiol.* 40: 1017-1022.

Mountfort, D.O. 1987. The rumen anaerobic fungi. *FEMS Microbiol. Revs.* 46: 401-408.

Mulbry, W.W. 1994. Purification and characterization of an inducible *s*-triazine hydrolase from *Rhodococcus corallinus* NRRL B-15444R. *Appl. Environ. Microbiol.* 60: 613-618.

Müller, E., K. Fahlbusch, R. Walther, and G. Gottschalk. 1981. Formation of *N,N*-dimethylglycine, acetic acid and butyric acid from betaine by *Eubacterium limosum*. *Appl. Environ. Microbiol.* 42: 439-445.

Müller, M.D., and H.-R. Büser. 1986. Halogenated aromatic compounds in autmotive emissions from leaded gasoline additives. *Environ. Sci. Technol.* 20: 1151-1157.

Munnecke, D.M., L.M. Johnson, H.W. Talbot, and S. Barik. 1982. Microbial metabolism and enzymology of selected pesticides pp.1-32. In *Biodegradation and detoxification of environmental pollutants* (Ed. A.M. Chrakrabarty). CRC Press, Boca Raton, Florida.

Nagasawa, S., R. Kikuchi, Y. Nagata, M. Takagi, and M. Matsuo. 1993a. Stereochemical analysis of γ-HCH degradation by *Pseudomonas paucimobilis* UT26. *Chemosphere* 26: 1187-1201.

Nagasawa, S., R. Kikuchi, Y. Nagata, M. Takagi, and M. Matsuo. 1993b. Aerobic mineralization of γ-HCH by *Pseudomonas paucimobilis* UT26. *Chemosphere* 26: 1719-1728.

Nagata, Y., T. Nariya, R. Ohtomo, M. Fukuda, K. Yano, and M. Takagi. 1993. Cloning and sequencing of a dehalogenase gene encoding an enzyme with hydrolase activity involved in the degradation of γ-hexachlorocyclohexane in *Pseudomonas paucimobilis*. *J. Bacteriol.* 175: 6403-6410.

Nakamura, T., F. Yu, W. Mizunashi, and I. Watanabe. 1993. Production of (*R*)-3-chloro-1,2-propandiol from prochiral 1,3-dichloro-2-propanol by *Corynebacterium* sp. strain N-1074. *Appl. Environ. Microbiol.* 59: 227-230.

Nanninga, H.J., and J.C. Gottschal. 1987. Properties of *Desulfovibrio carbinolicus* sp. nov. and other sulfate-reducing bacteria isolated from an anaerobic-purification plant. *Appl. Environ. Microbiol.* 53: 802-809.

Narro, M.L., C.E. Cerniglia, C. Van Baalen, and D.T. Gibson. 1992. Evidence for an NIH shift in oxidation of naphthalene by the marine cyanobacterium *Oscillatoria* sp. strain JCM. *Appl. Environ. Microbiol.* 58: 1360-1363.

Naumann, E., H. Hippe, and G. Gottschalk. 1983. Betaine: new oxidant in the Stickland reaction and methanogenesis from betaine and L-alanine by a *Clostridium sporogenes-Methanosarcina barkeri* coculture. *Appl. Environ. Microbiol.* 45: 474-483.

Nawaz, M.S., T.M. Heinze, and C.E. Cerniglia. 1992. Metabolism of benzonitrile and butyronitrile by *Klebsiella pneumoniae*. *Appl. Environ. Microbiol.* 58: 27-31.

Neilson, A. H., A.-S. Allard, P.-Å. Hynning, and M. Remberger. 1988a. Transformations of halogenated aromatic aldehydes by metabolically stable anaerobic enrichment cultures. *Appl. Environ. Microbiol.* 54: 2226-2236.

Neilson, A. H., C. Lindgren, P.-Å. Hynning, and M. Remberger. 1988b. Methylation of halogenated phenols and thiophenols by cell extracts of gram-positive and gram-negative bacteria. *Appl. Environ. Microbiol.* 54: 524-530.

Neilson, A. H., A.-S. Allard, P.-Å. Hynning, and M. Remberger. 1991. Distribution, fate and persistence of organochlorine compounds formed during production of bleached pulp. *Toxicol. Environ. Chem.* 30: 3-41.

Nelson, J.D., W. Blair, F.E. Brinckman, R.R. Colwell, and W.P. Iverson. 1973. Biodegradation of phenylmercuric acetate by mercury-resistant bacteria. *Appl. Microbiol.* 26: 321-326.

Newman, L.M., and L.P. Wackett. 1991. Fate of 2,2,2-trichloroacetaldehyde (chloral hydrate) produced during trichloroethylene oxidation by methanotrophs. *Appl. Environ. Microbiol.* 57: 2399-2402.

Nishino, S.F., and J.C. Spain. 1993. Degradation of nitrobenzene by a *Pseudomonas pseudoalcaligenes*. *Appl. Environ. Microbiol.* 59: 2520-2525.

Ohisa, N., M. Yamaguchi, and N. Kurihara. 1980. Lindane degradation by cell-free extracts of *Clostridium rectum*. *Arch. Microbiol.* 125: 221-225.

Ohisa, N., N. Kurihara, and M. Nakajima. 1982. ATP synthesis associated with the conversion of hexachlorocyclohexane related compounds. *Arch. Microbiol.* 131: 330-333.

Oldenhuis, R., R.L.J.M. Vink, D.B. Janssen, and B. Witholt. 1989. Degradation of chlorinated aliphatic hydrocarbons by *Methylosinus trichosporium* OB3b expressing soluble methane monooxygenase. *Appl. Environ. Microbiol.* 55: 2819-2816.

Oltmanns, R.H., R. Müller, M.K. Otto, and F. Lingens. 1989. Evidence for a new pathway in the bacterial degradation of 4-fluorobenzoate. *Appl. Environ. Microbiol.* 55: 2499-2504.

Omori, T., L. Monna, Y. Saiki, and T. Kodama. 1992. Desulfurization of dibenzothiophene by *Corynebacterium* sp. strain SY1. *Appl. Environ. Microbiol.* 58: 911-915.

Oremland, R.S., R.S. Kiene, I. Mathrani, M.J. Whiticar, and D.R. Boone. 1989. Description of an estuarine methylotrophic methanogen which grows on dimethyl sulfide. *Appl. Environ. Microbiol.* 55: 994-1002.

Ornston, L.N. 1966. The conversion of catechol and protocatechuate to beta-ketoadipate by *Pseudomonas putida.* IV. Regulation. *J. Biol. Chem.* 241: 3800-3810.

Ougham, H.J., and P.W. Trudgill. 1982. Metabolism of cyclohexaneacetic acid and cyclohexanebutyric acid by *Arthrobacter* sp. strain CA1. *J. Bacteriol.* 150: 1172-1182.

Ougham, H.J., D.G. Taylor, and P.W. Trudgill. 1983. Camphor revisited: involvement of a unique monooxygenase in metabolism of 2-oxo-Δ^3-4,5,5-trimethylcyclopentenylacetic acid by *Pseudomonas putida. J. Bacteriol.* 153: 140-152.

Paasivirta, J., P. Klein, M. Knuutila, J. Knuutinen, M. Lahtiperä, R. Paukku, A. Veijanen, L. Welling, M. Vuorinen, and P.J. Vuorinen. 1987. Chlorinated anisoles and veratroles in fish. Model compounds. Instrumental and sensory determinations. *Chemosphere* 16: 1231-1241.

Parales, R.E., and C.S. Harwood. 1993. Regulation of the *pcaIJ* genes for aromatic acid degradation in *Pseudomonas putida. J. Bacteriol.* 175: 5829-5838.

Parekh, V.R., R.W. Traxler, and J.M. Sobek. 1977. *n*-alkane oxidation enzymes of a pseudomonad. *Appl. Environ. Micbrobiol.* 33: 881-884.

Parisot, D., M.C. Malet-Martino, R. Martino, and P. Crasnier. 1991. 19$_F$ nuclear magnetic resonance analysis of 5-fluorouracil metabolism in four differently pigmented strains of *Nectria haematococca. Appl. Environ. Microbiol.* 57: 3605-3612.

Patel, R.N., C.T. Hou, A.I. Laskin, and A. Felix. 1982. Microbial oxidation of hydrocarbons: properties of a soluble methane monooxygenase from a facultative methane-utilizing organism, *Methylobacterium* sp. strain CRL-26. *Appl. Environ. Microbiol.* 44: 1130-1137.

Patel, T.R., and E.A. Barnsley. 1980. Naphthalene metabolism by pseudomonads: purification and properties of 1,2-dihydroxynaphthalene oxygenase. *J. Bacteriol.* 143: 668-673.

Patel, T.R., and D.T. Gibson. 1974. Purification and properties of (+)-cis-naphthalene dihydrodiol dehydrogenase of *Pseudomonas putida. J. Bacteriol.* 119: 879-888.

Patterson, J.A., and R.B. Hespell. 1979. Trimethylamine and methylamine as growth substrates for rumen bacteria and *Methanosarcina barkeri. Curr. Microbiol.* 3: 79-83.

Pettigrew, C.A., B.E. Haigler, and J.C. Spain. 1991. Simultaneous biodegradation of chlorobenzene and toluene by a *Pseudomonas* strain. *Appl. Environ. Microbiol.* 57: 157-162.

Pfeifer, F., S. Schacht, J. Klein, and H.G. Trüper. 1989. Degradation of diphenyl ether by *Pseudomonas cepacia. Arch. Microbiol.* 152: 515-519.

Pfeifer, F., H.G. Trüper, J. Klein, and S. Schacht. 1993. Degradation of diphenylether by *Pseudomonas cepacia* Et4: enzymatic release of phenol from 2,3-dihydroxydiphenylether. *Arch. Microbiol.* 159: 323-329.

Pieper, D.H., W. Reineke, K.-H. Engesser, and H.-J. Knackmuss. 1988. Metabolism of 2,4-dichlorophenoxyacetic acid, 4-chloro-2-methylphenoxyacetic acid and 2-methylphenoxyacetic acid by *Alcaliges eutrophus* JMP 134. *Arch. Microbiol.* 150: 95-102.

Pipke, R., and N. Amrhein. 1988. Degradation of the phosphonate herbicide glyphosate by *Arthrobacter atrocyaneus* ATCC 13752. *Appl. Environ. Microbiol.* 54: 1293-1296.

Pipke, R., N. Amrhein, G.S. Jacob, J. Schaefer, and G.M. Kishore. 1987. Metabolism of glyphosate in an *Arthrobacter* sp. GLP-1. *Eur. J. Biochem.* 165: 267-273.

Pirnik, M.P., R.M. Atlas, and R. Bartha. 1974. Hydrocarbon metabolism by *Brevibacterium erythrogenes*: normal and branched alkanes. *J. Bacteriol.* 119: 868-878.

Plugge, C.M., C. Dijkema, and A.J.M. Stams. 1993. Acetyl-CoA cleavage patyhways in a syntrophic propionate oxidizing bacterium growing on fumarate in the absence of methanogens. *FEMS Microbiol. Lett.* 110: 71-76.

Poh, C.L., and R.C. Bayly. 1980. Evidence for isofunctional enzymes used in *meta*-cresol and 2,4-xylenol degradation via the gentisate pathway in *Pseudomonas alcaligenes*. *J. Bacteriol.* 143: 59-69.

Pons, J.-L., A. Rimbault, J.C. Darbord, and G. Leluan. 1984. Biosynthèse de toluène chez *Clostridium aerofoetidum* souche WS. *Ann. Microbiol.* (Inst. Pasteur) 135B: 219-222.

Poth, M., and D.D. Focht. 1985. [15]N kinetic analysis of N_2O production by *Nitrosomonas europaea*: an examination of nitrifier denitrification. *Appl. Environ. Microbiol.* 49: 1134-1141.

Pothuluri, J.V., J.P. Freeman, F.E. Evans, and C.E. Cerniglia. 1990. Fungal transformation of fluoranthene. *Appl. Environ. Microbiol.* 56: 2974-2983.

Pujar, B.G., and D.W. Ribbons. 1985. Phthalate metabolism in *Pseudomonas florescens* PHK: purification and properties of 4,5-dihydroxyphthalate decarboxylase. *Appl. Environ. Microbiol.* 49: 374-376.

Qatibi, A.I., V. Niviére, and J.L. Garcia. 1991. *Desulfovibrio alcoholovorans* sp. nov., a sulfate-reducing bacterium able to grow on glycerol, 1,2- and 1,3-propanediol. *Arch. Microbiol.* 155: 143-148.

Rabus, R., R. Nordhaus, W. Ludwig, and F. Widdel. 1993. Complete oxidation of toluene under strictly anoxic conditions by a new sulfate-reducing bacterium. *Appl. Environ. Microbiol.* 59: 1444-1451.

Rafii, F., and C.E. Cerniglia. 1993. Comparison of the azoreductase and nitroreductase from *Clostridium perfringens*. *Appl. Environ. Microbiol.* 59: 1731-1734.

Rafii, F., W. Franklin, R.H. Heflich, and C.E. Cerniglia. 1991. Reduction of nitroaromatic compounds by anaerobic bacteria isolated from the human gastrointestinal tract. *Appl. Environ. Microbiol.* 57: 962-968.

Ramanand, K., and J.M. Suflita. 1991. Anaerobic degradation of *m*-cresol in anoxic aquifer slurries: carboxylation reactions in a sulfate-reducing bacterial enrichment. *Appl. Environ. Microbiol.* 57: 1689-1695.

Rao, J.R., N.D. Sharma, J.T.G. Hamilton, D.R. Boyd, and J.E. Cooper. 1991. Biotransformation of the pentahydroxy flavone quercitin by *Rhizobium loti* and *Bradyrhizobium* strains (Lotus). *Appl. Environ. Microbiol.* 57: 1563-1565.

Ratledge, C. 1978. Degradation of aliphatic hydrocarbons pp. 1-46. In *Developments in biodegradation of hydrocarbons-1* (Ed. R.J. Watkinson), Applied Science Publishers Ltd., London.

Reineke, W. 1984. Microbial degradation of halogenated aromatic compounds. pp. 319-360. In *Microbial degradation of organic compounds* (Ed. D.T. Gibson). Marcel Dekker Inc., New York.

Reineke, W., and H.-J. Knackmuss. 1988. Microbial degradation of haloaromatics. *Ann. Rev. Microbiol.* 42: 263-287.

Renganathan, V. 1989. Possible involvement of toluene-2,3-dioxygenase in defluorination of 3-fluoro-substituted benzenes by toluene-degrading *Pseudomonas* sp. strain T-12. *Appl. Environ. Microbiol.* 55: 330-334.

Rhys-Williams, W., S.C. Taylor, and P.A. Williams. 1993. A novel pathway for the catabolism of 4-nitrotoluene by *Pseudomonas*. *J. Gen. Microbiol.* 139: 1967-1972.

Ribbons, D.W., and R.W. Eaton. 1982. Chemical transformations of aromatic hydrocarbons that support the growth of microorganisms pp. 59-84. In *Biodegradation and detoxification of environmental pollutants* (Ed. A.M. Chakrabarty). CRC Press, Boca Raton.

Ribbons, D.W., S.J.C. Taylor, C.T. Evans, S.T. Thomas, J.T. Rossiter, D.A. Widdowson, and D.J. Williams. 1990. Biodegradations yield novel intermediates for chemical synthesis pp. 213-245. In *Biotechnology and Biodegradation* (Eds. D. Kamely, A. Chakrabarty, and G.S. Omenn). Portfolio Publishing Company, Texas.

Roberts, D.J., P.M. Fedorak, and S.E. Hrudey. 1990. CO_2 incorporation and 4-hydroxy-2-methylbenzoic acid formation during anaerobic metabolism of *m*-cresol by a methanogenic consortium. *Appl. Environ. Microbiol.* 56: 472-478.

Robertson, J.B., J.C. Spain, J.D. Haddock, and D.T. GIbson. 1992. Oxidation of nitrotoluenes by toluene dioxygenase: evidence for a monooxygenase reaction. *Appl. Environ. Microbiol.* 58: 2643-2648.

Rossol, I., and A. Pühler. 1992. The *Corynebacterium glutamicum aecD* gene encodes a C-S lyase with alpha-beta-elimination activity that degrades aminoethylcysteine. *J. Bacteriol.* 174: 2968-2977.

Rothmel, R.K., D.L. Shinbarger, M.R. Parsek, T.L. Aldrich, and A.M. Chakrabarty. 1991. Functional analysis of the *Pseudomonas putida* regulatory protein CatR: transcriptional studies and determination of the CatR DNA-binding site by hydroxyl-radical footprinting. *J. Bacteriol.* 173: 4717-4724.

Rott, B., S. Nitz, and F. Korte. 1979. Microbial decomposition of pentachlorophenolate. *J. Agric. Food Chem.* 27: 306-310.

Rudd, J.W.M., A. Furutani, and M.A. Turner. 1980. Mercury methylation by fish intestinal contents. *Appl. Environ. Microbiol.* 40: 777-782.

Rudolphi, A., A. Tschech, and G. Fuchs. 1991. Anaerobic degradation of cresols by denitrifying bacteria. *Arch. Microbiol.* 155: 238-248.

Sakai, K., A. Nazakawa, K. Kondo, and H. Ohta. 1985. Microbial reduction of nitroolefins. *Agric. Biol. Chem.* 49: 2331-2335.

Sallis, P.J., S.J. Armfield, A.T. Bull, and D.J. Hardman. 1990. Isolation and characterization of a haloalkane halidohydrolase from *Rhodococcus erythropolis* Y2. *J. Gen. Microbiol.* 136: 115-120.

Samain, E., H.C. Dubourguier, and G. Albagnac. 1984. Isolation and characterization of *Desulfobulbus elongatus* sp. nov. from a mesophilic industrial digester. *Syst. Appl. Microbiol.* 5: 391-401.

Sander, P., R.-M. Wittich, P. Fortnagel, H. Wilkes, and W. Francke. 1991. Degradation of 1,2,4-trichloro- and 1,2,4,5-tetrachlorobenzene by *Pseudomonas* strains. *Appl. Environ. Microbiol.* 57: 1430-1440.

Sayama, M., M. Inoue, M.-A. Mori, Y. Maruyama, and H. Kozuka. 1992. Bacterial metabolism of 2,6-dinitrotoluene with *Salmonella typhimurium* and mutagenicity of the metabolites of 2,6-dinitrotoluene and related compounds. *Xenobiotica* 22: 633-640.

Schackmann, A., and R. Müller. 1991. Reduction of nitroaromatic compounds by different *Pseudomonas* species under aerobic conditions. *Appl. Microbiol. Biotechnol.* 34: 809-813.

Schauder, R., and B. Schink. 1989. *Anaerovibrio glycerini* sp. nov., an anaerobic bacterium fermenting glycerol to propionate, cell matter and hydrogen. *Arch. Microbiol.* 152: 473-478.

Schiefer-Ullrich, H., and J.R. Andreesen. 1985. *Peptostreptoccus barnesae* sp. nov., a gram-positive, anaerobic, obligately purine utilizing coccus from chicken feces. *Arch. Microbiol.* 143: 26-31.

Schieffer-Ullrich, H., R. Wagner, P. Dürre, and J.R. Andreesen. 1984. Comparative studies on physiology and taxonomy of obligately purinolytic clostridia. *Arch. Microbiol.* 138: 345-353.

Schink, B. 1985a. Degradation of unsaturated hydrocarbons by methanogenic enrichments cultures. *FEMS Microbiol. Ecol.* 31: 69-77.

Schink, B. 1985b. Fermentation of acetylene by an obligate anaerobe, *Pelobacter acetylenicus* sp. nov. *Arch. Microbiol.* 142: 295-301.

Schink, B., and N. Pfennig. 1982. Fermentation of trihydroxybenzenes by *Pelobacter acidigallici* gen. nov. sp. nov., a new strictly anaerobic, non-spore-forming bacterium. *Arch. Microbiol.* 133: 195-201.

Schink, B., and M. Stieb. 1983. Fermentative degradation of polyethylene glycol by a strictly anaerobic, gram-negative, nonsporeforming bacterium, *Pelobacter venetianus.* sp. nov. *Appl. Environ. Microbiol.* 45: 1905-1913.

Schmidt, S., R.-M. Wittich, D. Erdmann, H. Wilkes, W. Francke, and P. Fortnagel. 1992. Biodegradation of diphenyl ether and its monohalogenated derivatives by *Sphingomonas* sp. strain SS3. *Appl. Environ. Microbiol.* 58: 2744-2750.

Schmidt, S., P. Fortnagel, and R.-M. Wittich. 1993. Biodegradation and transformation of 4,4'- and 2,4-dihalodiphenyl ethers by *Sphingomonas* sp. strain SS33. *Appl. Environ. Microbiol.* 59: 3931-3933.

Schnell, S., and B. Schink. 1991. Anaerobic aniline degradation via reductive deamination of 4-aminobenzoyl-CoA in Desulfobacterium anilini. *Arch. Microbiol.* 155: 183-190.

Schnell, S., F. Bak, and N. Pfennig. 1989. Anaerobic degradation of aniline and dihydroxybenzenes by newly isolated sulfate-reducing bacteria and description of *Desulfobacterium anilini. Arch. Microbiol.* 152: 556-563.

Schocher, R.J., B. Seyfried, F. Vazquez, and J. Zeyer. 1991. Anaerobic degradation of toluene by pure cultures of denitrifying bacteria. *Arch. Microbiol.* 157: 7-12.

Scholtz, R., F. Messi, T. Leisinger, and A.M. Cook. 1988. Three dehalogenases and physiological restraints in the biodegradation of haloalkanes by *Arthrobacter* sp. strain HA1. *Appl. Environ. Microbiol.* 54: 3034-3038.

Schowanek, D., and W. Verstraete. 1990. Phosphonate utilization by bacterial cultures and enrichments from environmental samples. *Appl. Environ. Microbiol.* 56: 895-903.

Schramm, E., and B. Schink. 1991. Ether-cleaving enzyme and diol dehydratase involved in anaerobic polyethylene glycol degradation by a new *Acetobacterium* sp. *Biodegradation* 2: 71-79.

Schwarz, G., R. Bauder, M. Speer, T.O. Rommel, and F. Lingens. 1989. Microbial metabolism of quinoline and related compounds II. Degradation of quinoline by *Pseudomonas fluorescens* 3, *Pseudomonas putida* 86 and *Rhodococcus* spec. B1. *Biol. Chem. Hoppe-Seyler* 370: 1183-1189.

Sendoo, K., and H. Wada. 1989. Isolation and identification of an aerobic γ-HCH-decomposing bacterium from soil. *Soil Sci. Plant Nutr.* 35: 79-87.

Shields, M.S., S.O. Montgomery, P.J. Chapman, S.M. Cuskey, and P.H. Pritchard.1989. Novel pathway of toluene catabolism in the trichloroethylene-degrading bacterium G4. *Appl. Environ. Microbiol.* 55: 1624-1629.

Shimizu, S., S. Jareonkitmongkol, H. Kawashima, K. Akimoto, and H. Yamada. 1991. Production of a novel ω1-eicosapentaenoic acid by *Mortierella alpina* 1S-4 grown on 1-hexadecene. *Arch. Microbiol.* 156: 163-166.

Shochat, E., I. Hermoni, Z. Cohen, A. Abeliovich, and S. Belkin. 1993. Bromoalkane-degrading *Pseudomonas* strains. *Appl. Environ. Microbiol.* 59: 1403-1409.

Simpson, F.J., N. Narasimhachari, and D.W.S. Westlake. 1963. Degradation of rutin by *Aspergillus flavus*. The carbon monoxide producing system. *Can. J. Microbiol.* 9: 15-25.

Singer, M.E., and W.R. Finnerty. 1985a. Fatty aldehyde dehydrogenases in *Acinetobacter* sp. strain HO1-N: role in hexadecane and hexadecanol metabolism. *J. Bacteriol.* 164: 1011-1016.

Singer, M.E., and W.R. Finnerty. 1985b. Alcohol dehydrogenases in *Acinetobacter* sp. strain HO1-N: role in hexadecane and hexadecanol metabolism. *J. Bacteriol.* 164: 1017-1024.

Smith, A.E. 1985. Identification of 2,4-dichloroanisole and 2,4-dichlorophenol as soil degradation products of ring-labeled [^{14}C] 2,4-D. *Bull. Environ. Contam. Toxicol.* 34: 150-157.

Smith, N.A., and D.P. Kelly. 1988. Isolation and physiological characterization of autotrophic sulfur bacteria oxidizing dimethyl disulphide as sole source of energy. *J. Gen. Microbiol.* 134: 1407-1417.

Smith, R.V., and J.P. Rosazza. 1983. Microbial models of mammalian metabolism. *J. Nat. Prod.* 46: 79-91.

Smith, R.V., P.J. Davies, A.M. Clark, and S.K. Prasatik. 1981. Mechanism of hydroxylation of biphenyl by *Cunninghamella echinulata. Biochem. J.* 196: 369-371.

Smith, K.E., S. Latif, D.N. Kirk, and K.A. White. 1988. Microbial transformations of steroids. I. Rare transformations of progesterone by *Apiocrea chrysosperma*. *J. Steroid Biochem.* 31: 83-89.

Söhngen, N.L. 1913. Benzin, Petroleum, Paraffinöl und Paraffin als Kohlenstoff- und Energiequelle für Mikroben. *Centralbl. Bakteriol. Parasitenkd. Infektionskr.* (2 Abt.) 37: 595-609.

Sowers, K.R., and J.G. Ferry. 1983. Isolation and characterization of a methylotrophic marine methanogen, *Methanococcoides methylutens* gen. nov., sp. nov. *Appl. Environ. Microbiol.* 45: 684-690.

Spain, J.C., and D.T. Gibson. 1991. Pathway for biodegradation of *p*-nitrophenol in a *Moraxella sp. Appl. Environ. Microbiol.* 57: 812-819.

Spain, J.C., G.J. Zylstra, C.K. Blake, and D.T. Gibson. 1989. Monohydroxylation of phenol and 2,5-dichlorophenol by toluene dioxygenase in *Pseudomonas putida* F1. *Appl. Environ. Microbiol.* 55: 2648-2652.

Spanggord, R.J., J.C. Spain, S.F. Nishino, and K. E. Mortelmans. 1991. Biodegradation of 2,4-dinitrotoluene by a *Pseudomonas* sp. *Appl. Environ. Microbiol.* 57: 3200-3205.

Spangler, W.J., J.L. Spigarelli, J.M. Rose, R.S. Flipin, and H.H. Miller. 1973. Degradation of methylmercury by bacteria isolated from environmental samples. *Appl. Microbiol.* 25: 488-493.

Sparnins, V.L., and P.J. Chapman. 1976. Catabolism of L-tyrosine by the homoprotocatechuate pathway in gram-positive bacteria. *J. Bacteriol.* 127: 363-366.

Sparnins, V.L., P.J. Chapman, and S. Dagley. 1974. Bacterial degradation of 4-hydroxyphenylacetic acid and homoprotocatechuic acid. *J. Bacteriol.* 120: 159-167.

Stams, A.J.M., D.R. Kremer, K. Nicolay, G.H. Weenk, and T.A. Hansen. 1984. Pathway of propionate formation in *Desulfobulbus propionicus*. *Arch. Microbiol.* 139: 167-173.

Stanier, R.Y. 1968. Biochemical and immunological studies on the evolution of a metabolic pathway in bacteria pp. 201-225. In *Chemotaxonomy and Serotaxonomy* (Ed. J.G. Hawkes). Systematics Association Special Volume No. 2. Academic Press, London.

Steiert, J.G., and R.L. Crawford. 1986. Catabolism of pentachlorophenol by a *Flavobacterium* sp. *Biochem. Biophys. Res. Commun.* 141: 1421-1427.

Stickland, L.H. 1934. Studies in the metabolism of the strict anaerobes (Genus *Clostridium*). I. The chemical reactions by which *Cl. sporogenes* obtains its energy. *Biochem. J.* 28: 1746-1759.

Stickland, L.H. 1935a. Studies in the metabolism of the strict anaerobes (Genus *Clostridium*). II. The reduction of proline by *Cl. sporogenes*. *Biochem. J.* 29: 288-290.

Stickland, L.H. 1935b. Studies in the metabolism of the strict anaerobes (Genus *Clostridium*). III. The oxidation of alanine by *Cl. sporogenes*. IV. The reduction of glycine by Cl. sporogenes. *Biochem. J.* 29: 898.

Stieb, M., and B. Schink. 1985. Anaerobic oxidation of fatty acids by *Clostridium bryantii* sp. nov., a sporeforming, obligately syntrophic bacterium. *Arch. Microbiol.* 140: 387-390.

Stirling, L.A., R.J. Watkinson, and I.J. Higgins. 1977. Microbial metabolism of alicyclic hydrocarbons: isolation and properties of a cyclohexane-degrading bacterium. *J. Gen. Microbiol.* 99: 119-125.

Strubel, V., K.-H. Engesser, P. Fischer, and H.-J. Knackmuss. 1991. 3-(2-hydroxy-phenyl)catechol as substrate for proximal meta ring cleavage in dibenzofuran degradation by *Brevibacterium* sp. strain DPO 1361. *J. Bacteriol.* 173: 1932-1937.

Stucki, G., R. Gälli, H.R. Ebersold, and T. Leisinger. 1981. Dehalogenation of dichloromethane by cell extracts of *Hyphomicrobium* DM2. *Arch. Microbiol.* 130: 366-371.

Sundström, G., Å. Larsson, and M. Tarkpea. 1986. Creosote pp. 159-205. In *Handbook of Environmental Chemistry* (Ed. O. Hutzinger) Vol. 3/Part D. Springer-Verlag, Berlin.

Sutherland, J.B., A.L. Selby, J.P. Freeman, F.E. Evans, and C.E. Cerniglia. 1991. Metabolism of phenanthrene by *Phanerochaete chrysosporium*. *Appl. Environ. Microbiol.* 57: 3310-3316

Sutherland, J.B., P.P. Fu, S.K. Yang, L.S. von Tungelnm, R.P. Casillas, S.A. Crow, and C.E. Cerniglia. 1993. Enantiomeric composition of the *trans*-dihydrodiols produced from phenanthrene by fungi. *Appl. Environ. Microbiol.* 59: 2145-2149.

Suylen, G.M.H., G.C. Stefess, and J.G. Kuenen. 1986. Chemolithotrophic potential of a *Hyphomicrobium* species, capable of growth on methylated sulfur compounds. *Arch. Microbiol.* 146: 192-198.

Suylen, G.M.H., P.J. Large, J.P. van Dijken, and J.G. Kuenen. 1987. Methyl mercaptan oxidase, a key enzyme in the metabolism of methylated sulphur compounds by *Hyphomicrobium* EG. *J. Gen. Microbiol.* 133: 2989-2997.

Swarts, F. 1922. Sur l'acide trifluoracétique. *Bull. Acad. Roy. Belg.* 8: 343-370.

Sylvestre, M., R. Massé, F. Messier, J. Fauteux, J.-G. Bisaillon, and R. Beaudet. 1982. Bacterial nitration of 4-chlorobiphenyl. *Appl. Environ. Microbiol.* 44: 871-877.

Szewzyk, R., and N. Pfennig. 1987. Complete oxidation of catechol by the strictly anaerobic sulfate-reducing *Desulfobacterium catecholicum* sp. nov. *Arch. Microbiol.* 147: 163-168.

Taeger, K., H.-J. Knackmuss, and E. Schmidt. 1988. Biodegradability of mixtures of chloro- and methylsubstituted aromatics: simultaneous degradation of 3-chlorobenzoate and 3-methylbenzoate. *Appl. Microbiol. Biotechnol.* 28: 603-608.

Takigawa, H., H. Kubota, H. Sonohara, M. Okuda, S. Tanaka, Y. Fujikura, and S. Ito. 1993. Novel allylic oxidation of α-cedrene to *sec*-cedrenol by a *Rhodococcus* strain. *Appl. Environ. Microbiol.* 59: 1336-1341.

Tasaki, M., Y. Kamagata, K. Nakamura, and E. Mikami. 1992. Utilization of methoxylated benzoates and formation of intermediates by *Desulfotomaculum thermobenzoicum* in the presence or absence of sulfate. *Arch. Microbiol.* 157: 209-212.

Tausson, W.O. 1927. Naphthalin als Kohlenstoffquelle für Bakterien. *Planta* 4: 214-256.

Taylor, B.F., J.A. Amador, and H.S. Levinson. 1993. Degradation of *meta*-trifluoro-methylbenzoate by sequential microbial and photochemical treatments. *FEMS Microbiol. Lett.* 110: 213-216.

Taylor, D.G., and P.W. Trudgill. 1978. Metabolism of cyclohexane carboxylic acid by *Alcaligenes* strain W1. *J. Bacteriol.* 134: 401-411.

Taylor, S.J.C., D.W. Ribbons, A.M.Z. Slawin, D.A. Widdowson, and D.J. Williams. 1987. Biochemically generated chiral intermediates for organic synthesis: the absolute stereochemistry of 4-bromo-*cis*-2,3-dihydroxycyclohexa-4,6-diene-1-carboxylic acid formed from 4-bromobenzoic acid by a mutant of *Pseudomonas putida*. *Tetrahedron Lett.* 28: 6391-6392.

Tezuka, T., and K. Tonomura. 1978. Purification and properties of a second enzyme catalyzing the splitting of carbon-mercury linkages from mercury-resistant *Pseudomonas* K-62. *J. Bacteriol.* 135: 138-143.

Thauer, R.K., D. Möller-Zinkhan, and A.M Spormann. 1989. *Biochemistry* of acetate catabolism in anaerobic chemotrophic bacteria. *Ann. Rev. Microbiol.* 43: 43-67.

Thomas, O.R.T., and G.F. White. 1989. Metabolic pathway for the biodegradation of sodium dodecyl sulfate by *Pseudomonas* sp.C12B. *Biotechnol. Appl. Biochem.* 11: 318-327.

Tipton, C.L., and N.M. Al-Shathir. 1974. The metabolism of cyclopropane fatty acids by *Tetrahymena pyriformis*. *J. Biol. Chem.* 249: 886-889.

Tolosa, I., J.M. Bayona, and J. Albaigés. 1992. Identification and occurrence of brominated and nitrated phenols in estuarine sediments. *Mar. Pollut. Bull.* 22: 603-607.

Tomasek, P.H., and R.L. Crawford. 1986. Initial reactions of xanthone biodegradation by an *Arthrobacter* sp. *J. Bacteriol.* 167: 818-827.

Topp, E., L. Xun, and C.S. Orser. 1992. Biodegradation of the herbicide bromoxynil (3,5-dibromo-4-hydroxybenzonitrile) by purified pentachlorophenol hydroxylase and whole cells of *Flavobacterium* sp. strain ATCC 39723 is accompanied by cyanogenesis. *Appl. Environ. Microbiol.* 58: 502-506.

Traunecker, J., A. Preuß, and G. Diekert. 1991. Isolation and characterization of a methyl chloride utilizing, strictly anaerobic bacterium. *Arch. Microbiol.* 156: 416-421.

Trower, M.K., R.M. Buckland, R. Higgins, and M. Griffin. 1985. Isolation and characterization of a cyclohexane-metabolizing *Xanthobacter* sp. *Appl. Environ. Microbiol.* 49: 1282-1289.

Trudgill, P.W. 1969. The metabolism of 2-furoic acid by *Pseudomonas* F2. *Biochem. J.* 113: 577-587.

Trudgill, P.W. 1978. Microbial degradation of alicyclic hydrocarbons pp. 47-84. In *Developments in Biodegradation of hydrocarbons-1* (Ed. R.J. Watkinson). Applied Science Publishers Ltd., London.

Trudgill P.W. 1984. Microbial degradation of the alicyclic ring: structural relationships and metabolic pathways pp. 131-180. In *Microbial degradation of organic compounds* (Ed. D.T. Gibson). Marcel Dekker Inc., New York.

Tschech, A., and B. Schink. 1985. Fermentative degradation of resorcinol and resorcylic acids. *Arch. Microbiol.* 143: 52-59.

Tu, C.M. 1976. Utilization and degradation of lindane by soil microorganisms. *Arch. Microbiol.* 108: 259-263.

Uetz, T., R. Schneider, M. Snozzi, and T. Egli. 1992. Purification and characterization of a two-component monooxygenase that hydroxylates nitrilotriacetate from "*Chelatobacter*" strain ATCC 29600. *J. Bacteriol.* 174: 1179-1188.

Uotila, J.S., V.H. Kitunen, T. Saastamoinen, T. Coote, M.M. Häggblom, and M.S. Salkinoja-Salonen. 1992. Characterization of aromatic dehalogenases of *Mycobacterium fortuitum* CG-2. *J. Bacteriol.* 174: 5669-5675.

Urakami, T., H. Araki, H. Oyanagi, K.-I. Suzuki, and K. Komagata. 1992. Transfer of *Pseudomonas aminovorans* (den Dooren de Jong 1926) to *Aminobacter* gen. nov. as *Aminobacter aminovorans* comb. nov. and description of *Aminobacter aganoensis* sp. nov. and *Aminobacter niigataensis* sp. nov. *Int. J. Syst. Bacteriol.* 42: 84-92.

Valli, K., and M.H. Gold. 1991. Degradation of 2,4-dichlorophenol by the lignin-degrading fungus *Phanerochaete chrysosporium*. *J. Bacteriol.* 173: 345-352.

Valli, K., H. Wariishi, and M.H. Gold. 1992a. Degradation of 2,7-dichlorodibenzo-*p*-dioxin by the lignin-degrading basidiomycete *Phanerochaete chrysosporium*. *J. Bacteriol.* 174: 2131-2137.

Valli, K., B.J. Brock, D.K. Joshi, and M.H. Gold. 1992b. Degradation of 2,4-dinitrotoluene by the lignin-degrading fungus *Phanerochaete chrysosporium*. *Appl. Environ. Microbiol.* 58: 221-228.

van Afferden, M., S. Schacht, J. Klein, and H.G. Trüper. 1990. Degradation of dibenzothiophene by *Brevibacterium* sp. DO. *Arch. Microbiol.* 153: 324-328.

Van Alfen, N.K., and T. Kosuge. 1974. Microbial metabolism of the fungicide 2,6-dichloro-4-nitroaniline. *J. Agric. Food Chem.* 22: 221-224.

van den Tweel, W.J. J., J.B. Kok, and J.A.M. de Bont. 1987. Reductive dechlorination of 2,4-dichlorobenzoate to 4-chlorobenzoate and hydrolytic dehalogenation of 4-chloro, 4-bromo-, and 4-iodobenzoate by *Alcaligenes denitrificans* NTB-1. *Appl. Environ. Microbiol.* 53: 810-815.

van den Tweel, W.J.J., and J.A.M. de Bont. 1985. Metabolism of 3-butyl-1-ol by *Pseudomonas* BB1. *J. Gen. Microbiol.* 131: 3155-3162.

van den Wijngaard, A.J., K.W.H.J. van der Kamp, J. van der Ploeg, F. Pries, B. Kazemier, and D.B. Janssen. 1992. Degradation of 1,2-dichloroethane by *Ancylobacter aquaticus* and other facultative methylotrophs. *Appl. Environ. Microbiol.* 58: 976-983.

Van der Maarel, P. Quist, L. Dijkhuizen, and T.A. Hansen. 1993. Anaerobic degradation of dimethylsulfoniopropionate to 3-S-methylmercaptopropionate by a marine *Desulfobacterium* strain. *Arch. Microbiol.* 411-412.

van der Meer, W.M. de Vos, S. Harayama, and A.J.B. Zehnder. 1992. Molecular mechanisms of genetic adaptation to xenobiotic compounds. *Microbiol. Revs.* 56: 677-694.

Van Dort, H.M., and D.L. Bedard. 1991. Reductive ortho and meta dechlorination of a polychlorinated biphenyl congener by anaerobic microorganisms. *Appl. Environ. Microbiol.* 57: 1576-1578.

van Ginkel, C.G., and J.A.M. de Bont. 1986. Isolation and characterization of alkene-utilizing Xanthobacter spp. *Arch. Microbiol.* 145: 403-407.

Visscher, P.T., and B.F. Taylor. 1993. A new mechanism for the aerobic catabolism of dimethyl sulfide. *Appl. Environ. Microbiol.* 59: 3784-3789.

Vogel, T.M., and D. Grbic-Galic. 1986. Incorporation of oxygen from water into toluene and benzene during anaerobic fermentative transformation. *Appl. Environ. Microbiol.* 52: 200-202.

Vogel, T.M., and P.L. McCarty. 1985. Biotransformation of tetrachloroethylene to trichloro-ethylene, dichloroethylene, vinyl chloride, and carbon dioxide under methanogenic conditions. *Appl. Environ. Microbiol.* 49: 1080-1083.

Vonk, J.W., and A.K. Sijpesteijn. 1973. Studies on the methylation of mercuric chloride by pure cultures of bacteria and fungi. *Antonie van Leeuwenhoek* 39: 505-513.

Vorbeck, C., H. Lenke, P. Fischer, and H.-J. Knackmuss. 1994. Identification of a hydride-Meisenheimer complex as a metabolite of 2,4,6-trinitrotoluene by a *Mycobacterium* strain. *J. Bacteriol.* 176: 932-934.

Wackett, L.P., S.L. Shames, C.P. Venditti, and C.T. Walsh. 1987. Bacterial carbon-phosphorus lyase: products, rates and regulation of phosphonic and phosphinic acid metabolism. *J. Bacteriol.* 169: 710-717.

Wagner, C., and E.R. Stadtman. 1962. Bacterial fermentation of dimethyl-β-propiothetin. *Arch. Biochem. Biophys.* 98: 331-336.

Watanabe, I., T. Kashimoto, and R. Tatsukawa. 1983. Polybromianted anisoles in marine fish, shellfish, and sediments in Japan. *Arch. Environ. Contam. Toxicol.* 12: 615-620.

Watson, G.K., and R.B. Cain. 1975. Microbial metabolism of the pyridine ring. Metabolic pathways of pyridine biodegradation by soil bacteria. *Biochem. J.* 146: 157-172.

Watson, G.K., C. Houghton, and R.B. Cain. 1974. Microbial metabolism of the pyridine ring. The metabolism of pyridine-3,4-diol (3,4-dihydroxypyridine) by *Agrobacterium* sp. *Biochem. J.* 140: 277-292.

Wayne, L.G. 1961. Recognition of Mycobacterium fortuitum by means of the three-day phenolphthalein sulfatase test. *Amer. J. Clin. Pathol.* 36: 185-187.

Wayne, L.G., and 17 coauthors.1991. Fourth report of the cooperative, open-ended study of slowly growing mycobacteria by the international working group on mycobacterial taxonomy. *Int. J. Syst. Bacteriol.* 41: 463-472.

Wedemeyer, G. 1967. Dechlorination of 1,1,1-trichloro-2,2-bis[p-chlorophenyl]ethane by Aerobacter aerogenes. *Appl. Microbiol.* 15: 569-574.

Wegener, W.S., H.C. Reeves, R. Rabin, and S.J. Ajl. 1968. Alternate pathways of metabolism of short-chain fatty acids. *Bacteriol. Revs.* 32: 1-26.

Weijers, C.A.G.M., A. de Haan, and J.A.M. de Bont. 1988. Chiral resolution of 2,3-epoxyalkanes by *Xanthobacter* Py2. *Appl. Microbiol. Biotechnol.* 27: 337-340.

Wheelis, M., N.J. Palleroni, and R.Y. Stanier. 1967. The metabolism of aromatic acids by *Pseudomonas testosteroni* and *P. acidovorans. Arch. Microbiol.* 59: 302-314.

White, G.F., and J.R. Snape. 1993. Microbial cleavage of nitrate esters: defusing the environment. *J. Gen. Microbiol.* 139: 1947-1957.

White, G.F., K.S. Dodson, I. Davies, P.J. Matts, J.P. Shapleigh, and W.J. Payne. 1987. Bacterial utilisation of short-chain primary alkyl sulphate esters. *FEMS Microbiol. Lett.* 40: 173-177.

White, H., R. Feicht, C. Huber, F. Lottspeich, and H. Simon. 1991. Purification and some properties of the tungsten-containing carboxylic acid reductase from *Clostridium formicoaceticum. Biol. Chem. Hoppe-Seyler.* 372: 999-1005.

White, H., C. Huber, R. Feicht, and H. Simon. 1993. On a reversible molybdenum-containing aldehyde oxidoreductase from *Clostridium formicoaceticum*. *Arch. Microbiol.* 159: 244-249.

White-Stevens, R.H., H. Kamin, and Q.H. Gibson. 1972. Studies of a flavoprotein, salicylate hydroxylase. II. Enzyme mechanism. *J. Biol. Chem.* 247: 2371-2381.

Whited, G.M., and D.T. Gibson.1991. Separation and partial characterization of the enzymes of the toluene-4-monooxygenase catabolic pathway in *Pseudomonas mendocina* KR1. *J. Bacteriol.* 173: 3017-3020.

Whited, G.M., W.R. McCombie, L.D. Kwart, and D.T. Gibson. 1986. Identificaton of *cis*-diols as intermediates in the oxidation of aromatic acids by a strain of *Pseudomonas putida* that contains a TOL plasmid. *J. Bacteriol.* 166: 1028-1039.

Widdel, F. 1986. Growth of methanogenic bacteria in pure culture with 2-propanol and other alcohols as hydrogen donors. *Appl. Environ. Microbiol.* 51: 1056-1062.

Widdel, F. 1987. New types of acetate-oxidizing, sulfate-reducing *Desulfobacter* species, *D. hydrogenophilus* sp. nov., *D. latus* sp. nov., and *D. curvatus* sp. nov. *Arch. Microbiol.* 148: 286-291.

Widdel, F., and N. Pfennig. 1981. Sporulation and further nutritional characteristics of *Desulfotomaculum acetoxidans* (emend.). *Arch. Microbiol.* 112: 119-122.

Widdel, F., and N. Pfennig. 1982. Studies on dissimilatory sulfate-reducing bacteria that decompose fatty acids. II. Incomplete opxidation of propionate by *Desulfobulbus propionicus* gen. no., sp. nov. *Arch. Microbiol.* 131: 360-365.

Widdel, F., P.E. Rouvière, and R.S. Wolfe. 1988. Classification of secondary alcohol-utilizing methanogens including a new thermophilic isolate. *Arch. Microbiol.* 150: 477-481.

Willeke, U., and W. Barz. 1982. Catabolism of 5-hydroxyisoflavones by fungi of the genus *Fusarium*. *Arch. Microbiol.* 132: 266-269.

Winter, J., L.H. Moore, V.R. Dowell, and V.D. Bokkenheuser. 1989. C-ring cleavage of flavonoids by human intestinal bacteria. *Appl. Environ. Microbiol.* 55: 1203-1208.

Wittich, R.M., H.G. Rast, and H.-J. Knackmuss. 1988. Degradation of naphthalene-2,6- and naphthalene-1,6-disulfonic acid by a *Moraxella* sp. *Appl. Environ. Microbiol.* 54: 1842-1847.

Wittich, R.-M., H. Wilkes, V. Sinnwell, W. Francke, and P. Fortnagel. 1992. Metabolism of dibenzo-*p*-dioxin by *Sphingomonas* sp. strain RW1. *Appl. Environ. Microbiol.* 58: 1005-1010.

Wittlinger, R., and K. Ballschmitter. 1990. Studies of the global baseline pollution XIII. C6-C14 organohalogens (a and g [sic]-HCH, HCB, PCB, 4,4'-DDT, 4,4-DDE, cis- and trans-chlordane, trans-nonachlor, anisols) in the lower troposphere of the southern Indian Ocean. *Fresenius J. Anal. Chem.* 336: 193-200.

Wolgel, S.A., J.E. Dege, P.E. Perkins-Olson, C.H. Juarez-Garcia, R.L. Crawford, E. Münck, and J.D. Lipscomb. 1993. Purification and characterization of protocatechuate 2,3-dioxygenase from *Bacillus macerans*: a new extradiol catecholic dioxygenase. *J. Bacteriol.* 175: 4414-4426.

Xu, G., and T.P. West. 1992. Reductive catabolism of pyrimidine bases by *Pseudomonas stutzeri*. *J. Gen. Microbiol.* 138: 2459-2463.

Xun, L., E. Topp, and C.S. Orser. 1992a. Confirmation of oxidative dehalogenation of pentachlorophenol by a *Flavobacterium* pentachlorophenol hydroxylase. *J. Bacteriol.* 174: 5745-5747.

Xun, L., E. Topp, and C.S. Orser. 1992b. Diverse substrate range of a *Flavobacterium* pentachlorophenol hydroxylase and reaction stoichiometries. *J. Bacteriol.* 174: 2898-2902.

Xun, L., E. Topp, and C.S. Orser. 1992c. Purification and characterization of a tetrachloro-*p*-hydroquinone reductive dehalogenase from a *Flavobacterium*. sp. *J. Bacteriol.* 174: 8003-8007.

Yamada, E.W., and W.B. Jakoby. 1959. Enzymatic utilization of acetylenic compounds II. Acetylenemonocarboxylic acid hydrase. *J. Biol. Chem.* 234: 941-945.

Yamada, M., and K. Tonomura. 1972. Further study of formation of methylmercury from inorganic mercury by *Clostridium cochlearium* T-2. *J. Ferment. Technol.* 50: 893-900.

Yamamoto, K., Y. Ueno, K. Otsubo, K. Kawakami, and K.-I. Komatsu. 1990. Production of *S*-(+)-ibuprofen from a nitrile compound by *Acinetobacter* sp. strain AK 226. *Appl. Environ. Microbiol.* 56: 3125-3129.

Yang, W., L. Dostal, and J.P.N. Rosazza. 1993. Stereospecificity of microbial hydrations of oleic acid to 10-hydroxystearic acid. *Appl. Environ. Microbiol.* 59: 281-284.

Zeikus, J.G. 1980. Chemical and fuel production by anaerobic bacteria. *Ann. Rev. Microbiol.* 34: 423-464.

Zellner, G., H. Kneifel, and J. Winter. 1990. Oxidation of benzaldehydes to benzoic acid derivatives by three *Desulfovibrio* strains. *Appl. Environ. Microbiol.* 56: 2228-2233.

Zeyer, J., and P.C. Kearney. 1984. Degradation of *o*-nitrophenol and *m*-nitrophenol by a *Pseudomonas putida*. *J. Agric. Food Chem.* 32: 238-242.

Zeyer, J., and H.P. Kocher. 1988. Purification and characterization of a bacterial nitrophenol oxygenase which converts ortho-nitrophenol to catechol and nitrite. *J. Bacteriol.* 170: 1789-1794.

Zhang, X., and J. Wiegel. 1992. The anaerobic degradation of 3-chloro-4-hydroxybenzoate in freshwater sediment proceeds via either chlorophenol or hydroxybenzoate to phenol and subsequently to benzoate. *Appl. Environ. Microbiol.* 58: 3580-3585.

Zimmermann, T., F. Gasser, H.G. Kulla, and T.Leisinger. 1984. Comparison of two bacterial azoreductases acquired during adaptation to growth on azo dyes. *Arch. Microbiol.* 138: 37-43.

CHAPTER 7

Ecotoxicology

SYNOPSIS

The basic input required for assessing the toxicity of xenobiotics is summarized and includes data both on the exposure to the toxicant and an evaluation of its biological effect in terms of numerically determined end-points. A brief discussion is directed to the choice of test species, to the range of acceptable end-points, and to the choice of media for laboratory tests. Some comments are provided on the commonly used tests using single organisms that include representatives of algae, crustaceans, and fish with emphasis on assays for sub-lethal effects. Assays directed to evaluating sediment toxicity and to detecting genotoxic effects are noted. A discussion is presented on multi-component test systems including different types of mesocosms. Attention is directed to the important question of metabolism by the test organisms with emphasis on fish. The use of biomarkers is briefly discussed and includes application of both biochemical and physiological parameters. Some cautionary comments are given on the application of these to feral fish. Attention is briefly drawn to the relation between chemical structure and toxicity. It is suggested that a hierarchical approach to evaluating toxicity could be used, and that assays at the higher levels are justified if no effects are observed at the lower ones. The system should be flexible and should be able to incorporate studies of partition and additional factors relevant to a particular environment.

INTRODUCTION

Previous chapters have been devoted to the distribution and persistence of xenobiotics after discharge into the aquatic environment. This chapter is devoted to the effect of xenobiotics on aquatic organisms. Its depth and orientation should be clearly recognized: like Chapter 2 on analytical procedures, this chapter is not directed to the professional ecotoxicologist. The aim has been to provide an overview of the kinds of bioassays that are being used in environmental research and to indicate a few of the areas to which further attention might profitably be directed. No attempt has been made to provide protocols for standardized procedures, nor to indicate which of the many possible assay procedures are acceptable to the administrative authorities that issue discharge permits.

Toxicology may be defined as the science of poisons and has traditionally been devoted to their effect on higher organisms including man and domestic animals. The term ecotoxicology has been coined to include the effect of toxicants on biota, both in natural ecosystems and in laboratory test systems; these may evaluate the effect on a wide spectrum of organisms ranging from bacteria through algae and crustaceans to vertebrates. In addition, attention has increasingly been directed to the application of a number of parameters in routine clinical use and to their adaption for use both in laboratory experiments and in evaluations using feral fish; for example, the levels of specific enzymes, morphological changes in organs such as liver, and blood parameters have been successfully used. The two approaches are not, of course, mutually exclusive.

Man is a predator of organisms at higher trophic levels so that, for example, the consumption of fish provides a mechanism whereby man may be indirectly exposed to xenobiotics. In this case, the critical question, therefore, is the degree of contamination of fish by toxicants; the mechanisms whereby xenobiotics may be accumulated in biota have been discussed in Chapter 3 (Section 3.1), and the metabolism of these by higher aquatic biota will be reviewed briefly later in this chapter (Section 7.5). There is an enormous literature on human toxicology from which many useful ideas applicable to ecotoxicology may be gleaned, and these can profitably be adapted with only minor modification. In human toxicology, a number of basic problems have been extensively explored and these include the following:

1. The fundamental issue of the relation between the dose of a toxicant and the response elicited.
2. The vexatious question of the existence or otherwise of threshold concentrations below which toxicity is not displayed.
3. The statistical design of experiments.

It is appropriate, therefore, to make brief reference to some essentially popular accounts in which these are illustrated by readily understood examples (Ottoboni 1984; Rodricks 1992) in addition to the substantial discussions presented by Casarett and Doull (1975).

Most of the present discussion will be illustrated by examples from experiments with pure organic compounds, but it should be appreciated that the discharge of single compounds into the aquatic environment is exceptional; almost invariably the effluents consist of a complex mixture of compounds, and these are generally evaluated as unfractionated effluents or occasionally on the basis of the effect of their major components. The question of synergistic or antagonistic effects, therefore, remains essentially unresolved. Toxicity equivalent factors (TEFs) have been used to provide an overall estimate of toxicity in situations where mixtures of compounds are present, and this is discussed again in Chapter 8 (Section 8.3.1). Briefly, the toxicity of each component is evaluated using a given test system, and this value is multiplied by the concentration of that component in the mixture; these values are then summed to provide an estimate of the toxicity of the mixture.

The problem of assessing effects on natural ecosystems is so complex that it is generally simplified to a greater or lesser extent; experiments may be conducted in the laboratory or outdoors in model ecosystems, and a plethora of single species have been used for assaying biological effects. These assays attempt to encompass various trophic levels and different physiologies and metabolisms. For example, representatives of algae are generally used to assess effects on photosynthesis and on primary production; crustaceans and fish may be used to evaluate effects on secondary producers, while communities may be used to explore interactions among components of natural ecosystems. In the final analysis, however, there are three well-defined stages in all test systems:

1. Exposure of the test organism(s) to the toxicant.
2. Evaluation of the effect(s) in terms of numerically accessible end-points.
3. Analysis of the data to provide a single value representing the biological effect (toxicity).

In the past, many industrial effluents were significantly toxic so that tests relying on acute toxicity to fish were routinely used; these usually involved exposure to the toxicant for a maximum of 96h. Rainbow trout were traditionally used, and the results were reported as LD50 values. With increased demand for less toxic effluents before discharge into aquatic systems and increased appreciation of the complexity of ecosystem effects (Rosenthal and Alderdice 1976), assays for acute toxicity have gradually been replaced by considerably more sophisticated test systems.

It is desirable clearly to distinguish and appreciate the differences between the various terms; in this account the following usage has been adopted.

Acute implies that the organism does not survive the exposure and often — though not necessarily — implies a short-term exposure. It should be appreciated that organisms at various stages of development may be used and that earlier stages will generally display greater sensitivity to the xenobiotic.

Sub-acute or sub-lethal implies that the test organism survives exposure but is, nonetheless, impaired in some specific way; a test may, for example, examine the effect of a toxicant on growth or reproduction. An old, though stimulating, review with valuable references to the basic literature is available (Sprague 1971).

Chronic tests aim at examining the effect of prolonged exposure and will, therefore, of necessity examine sub-acute effects. Considerable ambiguity surrounds the application of the term chronic, and this has been carefully analyzed by Suter et al. 1987; the length of the exposure is not rigorously defined but should probably be related to both the growth rate and the life expectation of the test organism. Life-cycle tests using, for example, fish clearly represent a truly chronic assay. The introduction of the term sub-chronic and its use in the context of relatively short-term tests lasting 7d (Norberg and Mount 1985; Norberg-King 1989) seems, therefore, regrettable.

Reproduction tests may be directed to a single generation or, especially for fish, to a complete life-cycle: growth of fish from spawn to maturity followed by growth of the next generation. The useful term *fecundity* has been used for tests that examine early stages in the development of the test organism.

The exclusive use of single-species test systems has been the subject of justified criticism (Cairns 1984), and it should be pointed out that fundamental objections have been raised against the application of conventional bioassays to predictive assessment. A valuable critique has been provided (Maltby and Calow 1989) in which the limitations of conventional approaches were carefully delineated; it was suggested that the intrinsic limitations of inductive procedures make these of low predictive value except in restricted circumstances. This issue is discussed again in Chapter 8 (Section 8.1.3).

7.1 CHOICE OF SPECIES IN LABORATORY TESTS

Broadly three different philosophies may be adopted, although these should not be regarded as mutually exclusive.

1. Internationally recognized organisms using standardized protocols may be used and, for judicial purposes, may be obligatory. These procedures have the advantage of enabling comparison with extensive published data, although their relevance to a specific ecosystem may be restricted.
2. Alternatively, use may be made of indigenous species in which case there can be no doubt of their relevance to the ecosystem from which they were isolated; standardized protocols must exist for these organisms also, and cloned cultures of taxonomically defined individuals should be used. An interesting example is afforded by the widespread European use of zebra fish as a test organism. This is, of course, a tropical fish, and it may reasonably be questioned whether this is appropriate for application to the cold-water environment of northern Europe. A study using 3,4-dichloroaniline has revealed, however, that in early life-stage studies, there was no significant difference between the results from tests using zebra fish and those using perch (*Perca fluviatilis*) that is widely distributed in Europe (Schäfers and Nagel 1991). This investigation alone clearly does not provide a *nihil obstat* to the use of zebra fish but provides valuable support for its widespread application in ecotoxicology.
3. Wild organisms have sometimes been used for each test series, and stocks of these have not been maintained in the laboratory. There are, therefore unavoidable limitations in their affiliation:

   - Variations in the natural population may remain unnoticed.
   - Experiments may be restricted to particular seasons of the year.
   - In view of the virtual global dissemination of xenobiotics, including potentially toxic organochlorines and PAHs, the test organisms may have already been exposed to background levels of such toxicants.

It should be emphasized that the choice of test organism is not primarily determined by the requirement for maximum sensitivity to toxicants; indeed, extreme sensitivity may be a disadvantage if it results in problems of repeatability or reproducibility. On the other hand, it should be clearly appreciated that some groups of organisms may be significantly more sensitive than others to a given class of toxicant — that may have a common mode of action. For example, the substituted diphenyl ether pyrethroids permethrin, fenvalerate, and cypermethrin were generally more toxic to marine invertebrates than to marine fish, and among the former the mysid *Mysidopsis bahia* was the most sensitive (Clark et al. 1989).

There exist also technical issues of considerable importance.

1. Some organisms, such as strains of algae or cultures of crustaceans, may be maintained in the laboratory as stocks; this has the advantage that putatively unaltered test strains are always available, but it is important that additional stress or selection is not imposed during maintenance.
2. Many species of fish may be available from commercial breeders, and this avoids extensive labor in keeping such stocks. On the other hand, absolute uniformity cannot be guaranteed, and genetic variations are clearly possible. One possible danger results from the use of antibiotics by fish breeders to maintain healthy stocks free from microbial infection.
3. Relatively little attention has been directed to genetic variation within specific taxa. Two examples of the care which should be exercised are provided by studies with the midge *Chironomus tentans* and the water-flea *Daphnia magna*.

   - Analysis of gel electrophoresis patterns of a number of glycolytic enzymes in different strains of *C. tentans* was used to assess a number of relevant genetic parameters including heterozygosity, percent polymorphic loci, and genetic distance within the populations (Woods et al. 1989). The observed variations between strains of the same organism from different sources were considerable, and the results strongly underscore the critical importance of taking this into consideration in assessing the effects of toxicants on different populations of the same organism.
   - The study of clonal variations in various strains of *D. magna* to cadmium chloride and 3,4-dichloroaniline was examined (Baird et al. 1990) in both actute and in 21d life-cycle (chronic) tests. There were wide differences in results from the acute tests, particularly for cadmium chloride, though these were relatively small for the life-cycle tests; it was, therefore, concluded that different mechanisms of toxicity operate in the two test systems, and this illustrates, in addition, the advantage of including both kinds of assay.

These results not only reinforce the conclusions on the possible significance of genetic variations in test species within the same taxon but also indicate the subtleties in expression of toxicity that may be revealed under different exposure conditions. A commendable development is, therefore, the assessment of the genetic structure in the mayfly *Cloeon triangulifer* that has been proposed as a suitable assay organism (Sweeney et al. 1993). Since the analysis of alloenzyme composition for enzymes representing different genetic foci is well developed and generally straightforward, more widespread application of this technique could profitably be made to other populations of organisms already used for bioassay.

7.2 EXPERIMENTAL DETERMINANTS

There are a number of important experimental considerations which affect both the design of the experiments and the interpretation of the data, and some of the most important will, therefore, be briefly summarized. It should be noted that in order to display its effect, the potential toxicant must be transported into the cells and that the compound may then be metabolized. In assays for toxicity, no distinction can therefore be made between transport into the cell, toxicity of the compound supplied, or that of potential metabolites; these effects are assayed collectively and indiscriminately. The situation has, therefore, certain features in common with that of bioconcentration that are discussed in Chapter 3 (Section 3.1.3).

7.2.1 Exposure

The most widely used exposure of the test organisms is to aqueous solutions of the toxicant prepared in media that are suitable for their growth and reproduction; this represents the situation for many potential toxicants, although attention is briefly drawn later to alternative procedures that have been applied to compounds with poor water solubility.

Exposure may take advantage of three procedures:

1. Static systems without renewal of the toxicant during exposure.
2. Semi-static systems in which the medium containing the toxicant is renewed periodically during the test.
3. Continuous flow-through systems in which the concentration of the toxicant is essentially constant.

In the first procedure — and to a lesser extent, in the second — toxicant concentrations will not remain constant during exposure, and the range of exposure concentrations during the test could advantageously be reported. In any case, analytically controlled concentrations of pure compounds should be provided, since some of them may be unstable under the conditions used for testing. Whereas exposure for a predetermined time is acceptable for experiments using crustaceans and fish, this cannot be done for growth tests using algae, since an unpredictable lag-phase may exist before growth commences. The length of the lag-phase may provide useful information on adaptation to the toxicant, but growth must be measured during the whole of the test and into the stationary phase.

A particularly troublesome problem arises with compounds having only low solubility in water, and this is particularly acute if the compound is only slightly toxic since high concentrations are then necessary to elicit a response from the test organism. Solutions of such compounds have been prepared in water-miscible organic solvents, such as acetone, ethanol, dimethyl sulfoxide, or dimethylformamide, which are added to the test medium. Although the toxicity of these solvents can readily be evaluated, a much greater uncertainty surrounds the true state of the toxicant: Is a true solution attained or merely a suspension of finely divided particles? It has also been shown that the acute toxicity of three xenobiotics towards a number of crustacean and rotifers was influenced by the organic solvent independently of the possible effect of the solvent alone (Calleja and Persoone 1993). It is, therefore, preferable to use saturated solutions of the toxicant in water; these may conveniently be prepared by passing the test medium through a column containing glass beads or other sorbents coated with the toxicant (Veith and Comstock 1975; Billington et al. 1988). Although this is the method of choice, preparation of large volumes for testing may be impractical though a design incorporating continuous flow has been developed and is clearly attractive (Veith and Comstock 1975). There is an additional problem that may be encountered with compounds that have extremely low water solubility; in an investigation on the uptake of highly chlorinated dibenzo-1,4-dioxins by fish, it was found that the concentrations in saturated solutions prepared by this procedure exceeded the established water solubility of the octachloro congener (Muir et al. 1986). It was hypothesized that the compound was associated with low concentrations of dissolved organic carbon in the water and that the very low BCF values that were measured could be the result of the poor bioavailability of the compound; this is equally relevant to the toxicity and is an issue to which further attention should be devoted.

For compounds with extremely low water solubility, such as 2,3,7,8-tetra-chlorodibenzo-1,4-dioxin, two other exposure regimes have been used: (1) egg injection using a single dose for rainbow trout (Walker and Peterson 1991; Walker et al. 1992) and (2) intraperitoneal injection into fingerling (Spitsbergen et al. 1988) or juvenile (van der Welden et al. 1992) rainbow trout. Although these provide valuable data and provide a solution to a technical problem, the extent to which they simulate environmental exposure appears questionable in most circumstances.

There are, however, situations in which organisms in natural systems may not be exposed to the essentially constant concentrations of the toxicant used in flow-through laboratory systems. This may be the case in quite different field situations:

1. Accidental spills that result in sudden and temporarily high concentrations of the toxicant.
2. Anadromous fish that may be exposed temporarily to a plume of toxicant on their way to the spawning ground.
3. Non-stationary fish that are, therefore, exposed to varying concentrations of a toxicant.

The question then arises of the extent to which any non-lethal effect is reversible after removal of the toxicant. The answer seems to be that in the few cases which have been examined, this may indeed be the case: all of them have examined phenolic compounds, one (McCahon et al. 1990) using the crustacean *Asellus aquaticus*, one assessing respiratory/cardiovascular effects on rainbow trout (Bradbury et al. 1989), and the third (Neilson et al. 1990) using the embryo/larvae assay with zebra fish (*Brachydanio rerio*). The last of these has been developed into a protocol that is modeled on the standard bioconcentration procedure in which a period of depuration is included after exposure to the toxicant.

7.2.2 End-Points

Any parameter which can be assessed numerically may be used, and these are generally adapted to the test organism and to experimental accessibility. An interesting survey (Maltby and Calow 1989) of papers during the periods pre-1979 and 1979 to 1987 using single-species laboratory tests revealed that survival was by far the most common end-point, followed by growth and reproduction; physiological/biochemical, behavioral, and morphological end-points were much less common. The chosen end-points differ widely. For example, the growth of algae may be estimated using cell number or turbidity or chlorophyll concentration — though care must be exercised in using the last, since the toxicant may affect chlorophyll levels without significantly affecting growth. Although growth has been proposed (Norberg-King 1989) as a simple end-point in tests using fathead minnows (*Pimephales promelas*), growth has not generally been used for larger fish since it is a relatively insensitive parameter. For reproduction tests using crustaceans, the number of offspring is unambiguous, though for life-cycle tests using fish, a number of different end-points have been suggested including the number of eggs produced after spawning. The design of the test may also determine the end-point; for example, in one version of the zebra fish embryo/larvae test, food is not provided during exposure, and the test uses the median survival time of the larvae which ingest the egg-yoke until starvation (Landner et al. 1985). Tests using physiological or biochemical or histological parameters have access to a much wider array of end-points which are discussed in more detail in Section 7.6.

Three examples are given as illustration of other less common parameters that have been used as end-points.

1. The scope for growth (SfG) which is the difference between the energy absorbed and that metabolized indicates how much energy is available for growth and reproduction. This has been used with marine invertebrates, particularly the mussel *Mytilus edulis*, and has also been examined (Maltby et al. 1990a) in the amphipod *Gammarus pulex*. Concentrations of 0.5 mg/L 3,4-dichloroaniline significantly reduced the SfG, and this concentration may be compared with the LC50 (48h) value of 7.9 mg/L. The test has also been evaluated in a field bioassay (Maltby et al. 1990b) and in mesocosms (Maltby 1992).

2. Behavioral responses have been used since these may be important determinants of the ability for a species to survive and reproduce under natural conditions. This has been examined in the amphipod *Pontoporeia affinis* that is an important food for fish in the Baltic Sea. In organisms exposed to sub-lethal concentrations of phenols and styrene in a flow-though system (Lindström and Lindström 1980), there was an initial stimulation of motility, although the long-term effect was an impairment of swimming at toxicant concentrations that had no effect on mortality. The response of fish to toxicants has been examined in a rotary-flow system (Lindahl et al. 1976), and this has been incorporated into chronic exposure tests (Bengtsson 1980). The impairment of swimming could be highly significant in determining the survival and reproduction of feral fish exposed to toxicants in natural ecosystems.

3. A single example is given to illustrate the fact that virtually any quantifiable effect may be used as end-point. The net spinning behavior of the caddis-fly *Hydropsyche angustipennis* has been used to evaluate the aquatic toxicity of 4,5,6-trichloroguaiacol; both the increased frequency of different types of net distortions and the time necessary for pupilation were examined (Petersen and Petersen 1984). Since these insects may be common in ecosystems and are themselves an important source of food for fish, the effect of toxicants on them may have widespread repercussions.

The results from such experiments will provide a relation between the exposure concentration of the toxicant and the observed biological effect — a dose-response relationship. From this data, assessments of toxicity can be calculated in terms of the relative effect, for example, EC50, EC20, or EC10 — the concentrations eliciting 50%, 20%, or 10% of the maximally observed effect — or used to calculate threshold toxic concentrations — the lowest concentrations causing an observed effect (LOEC) that are significant compared with those causing no observed effect (NOEC). It should be carefully noted that the term "observed" is used; use of "observable" is clearly misleading since effects might be registered under other more sensitive test conditions. The existence of threshold concentrations below which biological effects are not manifested has been extensively discussed in connection with exposure of humans to carcinogens but has been less fully explored in the context of ecotoxicology (Cairns 1992). The existence or otherwise of such thresholds is extremely important and seems not unreasonable in view of the concentration dependence of mechanisms whereby toxicants are transported into cells. It should be appreciated that the range of concentrations of toxicants that are examined determines the accuracy of the value assigned to LOEC; there may, therefore, be a considerable gap between the values of NOEC and LOEC, and this can only be diminished at considerable expense by repeating the experiment with a more appropriate range of concentrations.

The assessment of effects on community structure which are applicable to micro-cosm and mesocosm experiments is very much more complex and will generally necessitate careful statistical analysis. A summary of the most commonly used indices is given by Maltby and Calow (1989).

7.2.3 Test Conditions

One basic consideration is the choice of the test medium — which may conven-iently be termed the dilution medium for the toxicant. It is desirable that this medium is defined as closely as possible, and in an ideal situation a completely defined medium using laboratory distilled water supplemented with inorganic nutrients and adjusted to a suitable degree of hardness and salinity would be used. This is possible for many algae that are used as test organisms since these organisms may generally be grown in defined mineral media, supplemented if necessary with trace elements and vitamins. Even here, however, problems may arise. A good illustration is provided by the variation in the sensitivity of a range of marine algae towards tetrabromobisphenol-A which depended critically on the test medium and that could not be rationalized on a simple basis (Walsh et al. 1987).

This apparently ideal situation may prove, however, to be impossible for many other groups of test organisms: on the one hand, some are highly sensitive to the impurities unavoidably introduced into such media, while, on the other hand, many have obligate nutritional requirements that are accessible at low concentrations in natural waters but are not provided by synthetic media. If the laboratory is situated close to a clean and unpolluted site, suitable natural water may readily be obtained, but unfortunately, those who decide laboratory locations do not always consider such important matters. Most higher organisms have complex nutritional requirements which have seldom been defined, so that food is supplied, for example, to fish as commercial preparations or to crustaceans in cookbook-type preparations. Addition of food during the test period also inevitably introduces several complications through sorption of the toxicant, so that a multiple exposure route may exist: this has already been discussed in Chapter 3 (Section 3.1.1) in the particular context of hydrophobic compounds. Also, the organically rich test medium may favor the development of microbial populations of bacteria or fungi which are pathogenic to the test organism: this problem can be especially severe when industrial effluents which have been subjected to biological treatment are tested.

Test media are generally chosen to be as nearly as possible optimal for growth and reproduction of the test organism which should not be subjected to any stress additional to that imposed by the presence of the toxicant. This means, however, that dilution media for different organisms may differ appreciably in pH, salinity, or degree of hardness. Two examples may be used to illustrate these problems.

1. For ionizable organic compounds, such as carboxylic acids and phenols or amines, it is generally assumed that it is the free acid (or free base) that is transported into the cells and which mediates the toxicity. The relative proportion of the free acid is a function of the degree of its dissociation and of the pH of the medium. If test media have different pH, the effect of the toxicant may be compounded by the pH of the medium as well as by the sensitivity of the test organism. The problem probably becomes significant for compounds with pK_a values below ca. 6 and for organisms grown in media with pH > 7.5 which is the case for most fish and crustaceans. Most test media are only weakly buffered, generally with bicarbonate, and only for algal systems, and the MICROTOX system has the use of buffered media been consistently investigated (Neilson and Larsson 1980; Neilson et al. 1990). Even for neutral compounds, the pH of the test medium may significantly affect toxicity. For example, whereas assays for the acute toxicity of 4-nitrophenol and 2,4-dinitrophenol using rainbow trout and the amphipod *Gammarus pseudolimnaeus* showed the expected decrease in toxicity between pH 6.5 and pH 9.5, the toxicity of the neutral phosphate triester trichlorophon increased with increasing pH in the same range (Howe et al. 1994).
2. The presence of Ca^{2+} or Mg^{2+} cations in test media of appreciable hardness may introduce complications due to complex formation with the toxicant or, in extreme cases, formation of precipitates which make it impossible to carry out the test. One example of this is provided by 2,5-dichloro-3,6-dihydroxybenzo-1,4-quinone which could be examined in the zebra fish system although not in the *Ceriodaphnia dubia* system using dilution media of greater hardness (Remberger et al. 1991). The additional sigificance of ionic strength is noted in Section 7.3.1.

7.2.4 Evaluation of Variability in the Sensitivity of Test Organism

It is obligatory to assess periodically the sensitivity of the organisms used in laboratory tests, and there are some basic requirements that have been generally accepted:

1. That the standard compound is available in high purity and is readily soluble in the test medium.
2. That it is stable under the test conditions.
3. That it is sufficiently toxic to provide a reasonable response in the test organism.

It is also desirable that the test compound can be analyzed to provide data on the true exposure concentration. The choice of the compounds fulfilling these specifications is less easy. It seems clearly unsatisfactory to use a compound such as potassium dichromate to control the sensitivity of organisms used for evaluating organic toxicants. The problem with organic compounds is that of choosing one or a few from the huge structural range that is available. As a general rule, it might be stated that the compound should ideally bear some structural resemblance to the compounds which are to be evaluated and, if possible, related to the mechanism of their toxicity: it should be a surrogate. Halogenated phenols have been quite widely used for one important class of toxicants, but for neutral compounds, the choice is more difficult primarily due to volatility on the one hand (e.g., naphthalene) or poor water solubility (e.g., anthracene and PCBs) on the other. In the last analysis, probably an unsatisfactory compromise must be accepted and advantage be taken of the established sensitivity of the organism to as wide a range of toxicants as is realistic. The systematic collection and availability of this basic data for all test organisms would be extremely valuable.

7.3 TEST SYSTEMS: SINGLE ORGANISMS

7.3.1 Aquatic Organisms

A wide range of organisms is available, and many of them have been extensively evaluated with structurally diverse organic compounds. The choice of organism and the design of the study will depend critically on its objective. If, for example, the aim is the elicidation of the mode of action of a toxicant, detailed investigations may be devoted to one or only a few taxa. If, on the other hand, the aim is to evaluate the effect of a toxicant on an ecosystem, several taxonomically and physiologically diverse taxa should be included. The data from these may then be incorporated into a hierarchical system for environmental impact assessment of the toxicant (Section 7.8). In the following paragraphs, an attempt is made to describe briefly the major groups of organisms which have been most widely used in ecotoxicological studies.

The MICROTOX system — This is really a surrogate test for toxicity and uses a luminescent strain of a marine bacterium; the end-point is the inhibition of luminescence measured after 5 or 15 min exposure to the toxicant. The rapidity and low cost of this assay make this an attractive screening system, though care should be exercised in drawing conclusions of environmental relevance from the data. Many correlations between the acute toxicity measured in this system with values using higher organisms such as crustaceans and fish have been made, but it should be noted that these are made on a log-log basis so that agreement may really be no better than within a power of ten. Attempts have been made to use buffered media, but even in this case, correlation with the acute toxicity of a range of phenolic compounds to other organisms was singularly unimpressive (Neilson et al. 1990), although within a structural class of compounds, this may be a valuable test system. The general issue of multicomponent microbial test systems has been discussed (Cairns et al. 1992), although it should be clearly appreciated that, from a biochemical and evolutionary point of view, there are very substantial and important differences between bacteria and higher organisms.

Algae — Algae have been used for toxic assessment over many years since, as primary producers, they represent an important element in aquatic food chains. In addition, unicellular strains may generally be isolated, maintained, and grown using standard bacteriological techniques. In spite of this, however, there are a number of important issues to which attention should be directed, since these influence the conclusions that may be drawn from the results of a given test.

1. Use of axenic cultures is attractive in view of the high degree of reproducibility and avoids potentially serious problems due to the possible occurrence of bacterial transformation of the toxicant during tests with unialgal cultures containing bacteria.

2. The algal kingdom includes a very wide range of biochemically, physiologically, and morphologically distinct organisms of which only a very few have been used in ecotoxicology. The choice is naturally dictated largely by their ease of cultivation, and representatives of Chlorophyceae, Xanthophyceae, and Bacillariophyceae have been used; some Cyanobacteria (Cyanophyta) have been used less frequently. Widely differing sensitivities of representatives of these groups towards a range of toxicants have been observed (Blanck et al. 1984; Neilson et al. 1990), so that no single organism can reflect the spectrum of responses observed. Use of a group of taxonomically diverse strains is, therefore, to be preferred to the use of a single standardized species such as *Selenastrum capricornutum*.

3. Two essentially different end-points have been widely used in experiments with algae: growth rate or biomass, and estimates of toxicity using these may not be comparable (Nyholm 1985). Measurements of $H^{14}CO_3^-$ uptake include both the increase in the biomass and growth rate and have generally been applied to undefined populations of algae taken from receiving waters. Clearly, comparison between the results of such experiments carried out at different times is not possible, though the results are valuable in assessing the effects of a toxicant to organisms in a given ecosystem. Attention has been drawn to possible ambiguities resulting from measurements of chlorophyll concentrations for growth measurements (Neilson and Larsson 1980).

4. The laminarian marine alga *Macrocystis pyrifera* has been used in a laboratory assay (Anderson and Hunt 1988) and takes advantage of the different stages in the life-cycle of these algae; suitable end-points are the germination and growth of the spores which have been incorporated into a short-term (48h) assay, or the production of sporophytes by fertilization of female gametophytes that may be used in a 16d assay. The possibility of using this or other algae with a comparable life-cycle clearly merits further attention.

5. An interesting assay has been developed that uses communities of naturally occurring algae. The assay is based on the concept of pollution-induced community tolerance (PICT) (Blanck et al. 1988), whereby exposure to a toxicant results in the elimination of sensitive species and the dominance of tolerant ones; quantification is achieved by using short-term assays for the inhibition of photosynthesis under laboratory conditions (Blanck and Wänberg 1988). One of the attractive features of the system is that it can be used in widely different situations ranging from micocosms and mesocosms to natural ecosystems.

Higher plants — Growth of higher plants has been used for assessing toxicity, although these assays have not apparently been widely adopted. The use of higher plants may be particularly advantageous when discharge of toxicants is made to specific ecosystems where reliance solely on evaluating the effects on algae may be considered insufficient or where water is used for irrigation of agricultural crops; the possible bioconcentration of the xenobiotic (Chapter 3, Section 3.1.1) may also be conveniently evaluated. A range of different plants including both monocotyledons, such as oats and wheat, and dicotyledons, such as beans, carrots, and cucumbers, has been used. Suitable end-points are the frequency of seed germination and the extent of root elongation, and the tests are readily carried out even though systematic evaluation has apparently been restricted to only a few species (Wang 1991); inhibition of root development has been used to assay a few phenols (Wang 1987), and millet (*Panicum miliaceum*) was the most sensitive compared with cucumber (*Cucumis sativus*) and lettuce (*Lactuca sativa*). The toxicity of a range of compounds including substituted phenols, anilines, chlorobenzenes, aromatic hydrocarbons, and pesticides to lettuce (*L. sativa*) has been examined both in a semi-static assay in nutrient solutions and in soil (Hulzebos et al. 1993), and the toxicity correlated with values of P_{ow}. The duckweed *Lemna minor* has attracted greater attention and been proposed as a suitable representative organism (Taraldsen and Norberg-King 1990); after 96 h exposure to the toxicant, the biomass is assessed from the number of fronds or from the chlorophyll concentration. Presumably, these experiments could also be adapted to measure growth in which, for example, root length is measured. In many ways, the experimental conditions are similar to those used for algae with the added advantage that the medium can be renewed during the test. In experiments using a few substituted phenols, *L. minor* was comparable in sensitivity to the algae tested (Neilson et al.1990), though this should not be regarded as the determining advantage of the test. Monocotyledons have also been used to a limited extent; for example, the onion *Allium cepa* has been used and offers the additional advantage that the analysis of chromosome breaks and c-mitotic events in the root cells may be used to provide a further toxicological parameter. A micronucleus assay which detects chromosome breaks in *Tradescantia* sp. planted in areas suspected of pollution has been used (Sandhu et al. 1991) and suggests the possible value of such assays in the terrestrial environment or as an indicator of atmospheric pollution.

Crustaceans — Several freshwater daphnids have been extensively used, including *Daphnia magna*, *Daphnia pulex,* and *Ceriodaphnia dubia*, and standardized protocols have been developed. It has already been noted that only a single stress induced by the toxicant should be imposed during exposure. The importance of using the same media for the cultivation of *Daphnia magna* and in toxicity tests has emerged together with the possible influence not only of the hardness of the water but of its specific ionic composition (Buhl et al. 1993). All of these organisms have been used both in acute tests and in reproduction tests, and a life-cycle test for *D. magna* has been proposed (Meyerhoff et al. 1985) though it has been less extensively applied. There has been an increased tendency to use *C. dubia* which is apparently less susceptible to some of the cultivation problems associated especially with *D. magna*, and it seems (Winner 1988) that *C. dubia* is at least as sensitive as — or even more sensitive than — *D. magna*. Quite extensive effort has been devoted to development of the reproduction test using *C. dubia*, although some of the experimental problems appear to remain incompletely resolved. For the sake of completeness, it may be noted that a cloned strain of *D. pulicaria* from Lake Erie has been used to assess the toxicity of 2,2'-dichlorobiphenyl in full life-cycle tests and that this experiment demonstrated effects at the extremely low concentrations in the range of 50-100 ng/L (Bridgham 1988).

The harpacticoid copepod *Nitocra spinipes* which is common in the brackish waters of the Baltic Sea has been cloned and used extensively for toxic evaluation in Sweden (Bengtsson 1978), and a reproduction test in a flow-through system has been developed (Bengtsson and Bergström 1987). This is a valuable organism since it has an appreciable tolerance of the varying salinities which exist in the Gulf of Bothnia and the Baltic Sea, though it has possibly a more restricted geographical application than other test organisms.

Fish — Fish have traditionally been used for assessing the toxicity of effluents, and the 96h acute test with rainbow trout belongs to the classic era of ecotoxicology. There have been at least three significant developments over the intervening years: (1) use of a wider range of test fish, (2) development of tests aimed at evaluating reproduction efficiency and long-term (chronic) exposure, and (3) the use of bio-chemical and physiological parameters to assess the effect of toxicants.

Probably most investigations have used one or more of the following fish: rainbow trout (*Oncorhynchus mykiss*, syn. *Salmo gairdneri*), guppy (*Poecilia reticulata*), fathead minnow (*Pimephales promelas*), sheepshead minnow (*Cyprinidon variegatus*), channel catfish (*Ictalurus punctatus*), bluegill sunfish (*Lepomis macrochirus*), zebra fish (*Brachydanio rerio*), and flagfish (*Jordanella floridae*). Efforts have been de-voted to developing standardized protocols for all of these, and for many of them accessible tests for assessing sub-lethal or chronic effects have been developed. Less attention has been directed to the use of marine fish, although limited application has been made (Shenker and Cherr 1990) of assays using larvae of English sole (*Parophrys vetulus*) and topsmelt (*Atherinops affinis*).

The sensitivity of the test species — It may be valuable to attempt an assessment of the relative sensitivities of different fish towards organic toxicants. Short-term LC50 values for several compounds including pentachlorophenol and picloram (4-amino-3,4,5-trichloropicolinic acid) have been compared using rainbow trout, zebra fish, and flagfish. The results showed considerable variation in sensitivity among the test fish, although these were not judged to be significantly greater than those encountered using the same fish in the same laboratory during repeated testing (Fogels and Sprague 1977). In a similar way, comparison of the data for the sensitivity to pentachlorophenol of zebra fish, rainbow trout, and fathead minnows in early life-stage tests showed that these fish were quite similar (Neilson et al. 1990). A much more extensive correlation of acute toxicity has been carried out (Doherty 1983) using data for rainbow trout, bluegill sunfish, and fathead minnow. Although reasonable correlation was obtained between pairs of data, it seems perilous to evaluate applications for a discharge permit on the basis of data for a single organism instead of relying on the established practice of using the results for one cold-water fish, one warm-water fish, and an invertebrate.

It is critically important to assess the significance of the differences in the response of various fish to the same toxicant; for example, how significant is the difference between the 96 h acute toxicity of chlorothalonil (2,4,5,6-tetrachloro-1,3-dicyanobenzene) to channel catfish (52 µg/L) and to rainbow trout (18 µg/L) (Gallagher et al. 1992)? Such differences would probably not emerge as significant in a correlation study that plots data on a log-log basis — though it could be highly relevant for a given receiving water. In addition, the conclusion that the use of a single test organism is acceptable would not be supported by either the increased appreciation of the limitations of using only data for acute toxicity in environmental hazard assessments or the philosophy of implementing a hierarchical evaluation system (Svanberg and Renberg 1989).

Experimental determinants — In attempting to provide an overview of the effect of toxicants on fish, a number of significant determinative parameters should be taken into consideration.

1. Marked differences in the effect of temperature on the toxicity of pentachlorophenol to rainbow trout have been observed (Hodson and Blunt 1981), and early life-cycle stages were more adversely affected in fish exposed to a cold-water regime (6°C) than with those exposed to a warm-water regime (10°C). These results could have serious implications for natural populations exposed to pentachlorophenol during low temperatures when spring egg development occurs.

2. In most laboratory tests, fish are exposed to an essentially continuous concentration of the toxicant during the exposure regime. It has been shown, however, that even brief exposure of fathead minnows to pesticides such as chloropyrifos, endrin, or fenvalerate at high concentration may induce chronic effects including deformation and reduced growth (Jarvinen et al. 1988). A test in which adult zebra fish are pre-exposed to a toxicant before examining the sensitivity of the offspring to the same toxicant has been used to simulate, in principle, the exposure of anadromous or non-stationary fish to toxicants in natural ecosystems; increased sensitivity by a factor of ca. 5 was observed with pure test compounds, but this ratio appears to be related to the magnitude of the toxic effect; i.e., it decreases with decreasing toxicity (Landner et al. 1985; Neilson et al. 1990). These represent ecologically important considerations that could readily be incorporated into standardized protocols.

3. It has already been briefly pointed out in Chapter 3 (Section 3.1.3) that probably most xenobiotics can be metabolized by the test organisms, albeit to varying extents, and that this may compromise estimates of bioconcentration potential. The same is also true for toxicity, since whereas metabolism may, in many cases, serve as a detoxification mechanism, it may also result in the synthesis of toxic metabolites (Section 7.5); the question of enzyme induction and its implication for toxicity has been discussed in detail (Kleinow et al. 1987), and some examples of the metabolism of xenobiotics by fish are given in Section 7.5.1.

Sub-acute and chronic tests — The classic experiments of McKim et al. (1976) on the chronic effects of toxicants in which three generations of brook trout (*Salvelinus fontinalis*) were exposed to methyl mercuric chloride led to a scientifically based appreciation of the fact that the early life stages could be used for assessing chronic toxicity (McKim 1977). On account of the cost of complete life-cycle tests and the time required for their completion, increasing attention has been directed to truncated tests for chronic toxicity and to the effects of toxicants on sensitive life stages. Nonetheless, there is substantial evidence that clearly exposes the limitations in the conclusions which can be drawn if *only* early life-cycle stages are examined.

Attention is directed both to two end-points which have been examined and to the different concentrations of the toxicant at which the same end-point is affected in successive generations. There is clear evidence for a number of fish that toxicants may have significant effects on their fecundity, even though the effects on early life stages are marginal (Suter et al. 1987). A dramatic example is provided by the differences in the effect of chloropyrifos that were observed during a chronic test using fathead minnows lasting 200 d (Jarvinen et al. 1983); growth of first generation fish was reduced at a concentration of 2.68 µg/L within 30 d, whereas the comparable value for second generation fish was 0.12 µg/L. The results of such studies have revealed their value and justifiably led to a revival of interest in full life-cycle tests.

The use of the term sub-chronic for a test with fathead minnows lasting 7d seems regrettable, and the reporting of the results in terms of a chronic value which is the geometric mean of the LOEC and NOEC values (Norberg-King 1989) appears potentially misleading; such data clearly do not represent the effect of long-term exposure and cannot, therefore, be considered chronic in any etymologically acceptable sense of the word. This does not mean, however, that the results of such evaluations are not valuable as a basis for further examination, and some descriptive term such as early life-cycle test would seem much appropriate for such experiments.

It is important to appreciate methodological differences in the procedures by which tests for sub-acute toxicity are carried out and differences in the results that may be obtained with fish having different reproduction strategies. The two test protocols for zebra fish may be used as illustration of the first and the divergent results for a single substance using zebra fish and guppy used as illustration of the second.

In one protocol using zebra fish, the spawn from unexposed adults are collected, the fertilization rate determined after 24 h, and the test continued with fertilized ova; the survival rate and body length of the larvae are then determined after 6 weeks, during which time the larvae are exposed to the toxicant and provided with food (Nagel et al. 1991).

In an alternative procedure also using zebra fish, the end-point is the median survival time of the larvae which are exposed to the toxicant but are not provided with food; the termination of the test is determined by the time of survival that may be achieved by ingestion of the yoke sac (Landner et al. 1985). In these experiments, food is withheld for two specific reasons: to inhibit growth of microorganisms, particularly in industrial effluents which have been treated biologically, and to circumvent the possibility of sorption of the toxicant to the food.

Guppy and zebra fish are both warm-water fish with short generation times, and both of them reproduce throughout the year; the guppy is, however, viviparous with a high energy cost per larva (K-strategy) compared with the low energy cost per larva (r-strategy) of the zebra fish. The acute toxicity of 3,4-dichlorophenol is similar in both fish, with values of 8.4 mg/L (zebra fish) and 8.7-9.0 mg/L (guppy). For zebra fish, however, survival of the larvae is the most sensitive parameter, whereas reproduction is the most sensitive for guppy. Reproduction of guppy was reduced by 35% at a concentration of 2µg/L which does not affect zebra fish. On the other hand, at a concentration of 200µg/L, zebra fish populations will be eliminated, whereas in guppies this results only in a reduction of 40% in the number of offspring (Schäfers and Nagel 1991). Whether this conclusion may be extended to other cold-water fish with an r-strategy — which probably represent most European fish — is presently unknown.

By way of offering a wider perspective on the range of compounds that have been evaluated and the different fish that have been used in early life-stage evaluations, it may be mentioned that Californian grunion (*Leuresthes tenuis*) have been employed for assessing the toxicity of chlorpyrifos (*O,O*-diethyl-*O*-[3,5,6-trichloro-2-pyridyl] phosphorothioate) (Goodman et al. 1985) and pike (*Esox lucius*) for 2,3,7,8-tetrachlorodibenzo-1,4-dioxin (Helder 1980).

A truly chronic test using zebra fish and extending over three generations has been described and evaluated using 4-chloroaniline (Bresch et al. 1990). Reduction in egg release was the most sensitive parameter, and this was affected at a concentration of 40 µg/L which is some ten times lower than the threshold toxic concentration for growth. These results illustrate the valuable information that may be gained from such long-term toxicity tests; their only serious limitations are the time required for carrying them out and the expense involved. Possibly some compromise such as that outlined above employing pre-exposure of adults before assessing the sensitivity of their offspring to the toxicant might be acceptable, though there clearly remain serious difficulties in estimating concentrations which are environmentally innocuous — if indeed this value is scientifically accessible.

Integration with other criteria — In all these experiments using relatively long-term exposure, advantage may usefully be taken of the opportunity to examine behavioral, morphological, metabolic, and biochemical effects. Some of these are discussed in greater detail in Section 7.5 and Section 7.6 so that only three examples will be used to illustrate the possibilities.

1. During experiments investigating the toxicity and bioconcentration of chlorinated veratroles in zebra fish, the metabolism of these compounds was also examined. Successive *O*-demethylation of 3,4,5-trichloroveratrole to 3,4,5- and 4,5,6-trichloroguaiacol and 3,4,5-trichlorocatechol occurred, and these were then further metabolized to form sulfate and glucuronic acid conjugates (Neilson et al. 1989).

2. Exposure of zebra fish or rainbow trout to 4-chloroaniline resulted in numerous morphological alterations in the ultrastructure of the liver (Braunbeck et al. 1990a). In zebra fish, the effects were observed at concentrations as low as 40 µg/L and included hepatic compartmentation, invasion of macrophages, effects on the rough endoplasmic reticulum, and increase in the number of lysosomes, of autophagosomes, and of myelinated bodies. These changes apparently indicate response to stress and induction of biotransformation (detoxification) processes. Morphological changes including deformation of the spine have also been observed in the F-1 generation after 8 months exposure to 4-chloroaniline (Bresch et al. 1990). Long-term exposure of zebra fish to γ-hexachlorocyclohexane (lindane) during a full life-cycle test resulted in a number of pathological effects including hepatic steatosis, glycogen depletion, and occurrence of deformed mitochondria (Braunbeck et al. 1990b). Exposure of zebra fish larvae to chloroveratroles resulted in curvature of the larvae and deformation of the notochord (Neilson et al. 1984), and this is at least formally comparable to the serious skeletal deformations that have been observed after exposure to a number

of other neutral xenobiotics (refs. in Van den Avyle et al. 1989). The observation of such pathologies is an important supplement to the diagnosis of adverse effects in feral fish.

3. The behavior of fish in a rotary-flow system (Lindahl et al. 1976) has been incorporated into chronic exposure tests (Bengtsson 1980), and such behavioral alterations could be particularly important to fish which are exposed to toxicants in natural habitats.

Other aquatic organisms — The preceding organisms represent groups of organisms that have been most widely used in ecotoxicology. Increased appreciation of the need for evaluating the effect of a toxicant on a spectrum of organisms has directed attention to other groups of organisms, and these developments have encompassed important biotic components of both freshwater and marine ecosystems. A brief discussion of some of these seems, therefore, appropriate.

1. Rotifers are major components of the zooplankton and may graze on phytoplankton and themselves serve as food for larval fish. A 2 d life-cycle test using *Brachionus calyciflorus* has been developed (Snell and Moffat 1992) and is unusual in several respects; the rate of *population* growth is used as the endpoint, and all life stages are present during the test including embyros, juveniles, and adults. For the relatively few organic compounds which have so far been examined, the sensitivity appears to be comparable to that of *Ceriodaphnia dubia*.

2. Aquatic insects are an important component of stream and river ecosystems and may be particularly vulnerable to the adverse effect of xenobiotics. An assay has been developed using the mayfly *Cloeon triangulifer* that exhibits many features that distinguish it from other insects and make it accessible to laboratory bioassays; it has a relatively short life cycle; egg and larval stages can be reared under laboratory conditions, and important parameters in the life cycle can be quantified. This organism has been used to evaluate the effect of a commercial mixture of chlorinated camphenes on all stages of the life cycle (Sweeney et al. 1993), although certain critical aspects of the design including the exposure regime merit further development.

3. In general, less attention has been given to the development of test systems using marine organisms although assays using larvae of marine fish have been used (Shemkev and Cherr 1990). Two others are briefly noted even though they do not appear to have been extensively evaluated with organic compounds, and in the two examples cited, azide has been used as a reference toxicant.

   • A study based on assessing fertilization and embryo/larval development in the mussel *Mytilus californianus* (Cherr et al. 1990) has addressed important factors in the design of the test protocol including the use of the alga *Isochrysis galbana* to induce spawning, determination of the optimal sperm/ egg ratio, and the use of small chambers to avoid selective adherance of different development stages to the walls of the containing vessel. The endpoint was the frequency of development of veliger larvae with a complete shell, and this was assessed using polarization microscopy.

- Sea urchin (*Strongylocentrotus purpuratus*) sperm has been used to assess the toxicity of some components of bleachery effluents (Cherr et al.1987); the sperm is preincubated with the toxicant, and then both the sperm and eggs are exposed to the toxicant. The end-points are the inhibition of fertilization and the effect on the motility of the sperm; as in the assay using *Mytilus californiamus*, the sperm/egg ratio is critical. One of the important results from this study was the difference in the toxicity of some of the compounds compared with that which had been estimated from the effects on juvenile salmonid fish.

7.3.2 Organisms for Evaluating the Toxicity of Sediments

The association between xenobiotics and sediments has been discussed in Chapter 3 (Section 3.2), and it is now clearly established that many xenobiotics can be recovered from contaminated sediments which may act as a "sink" for these compounds. Attention has, therefore, been increasingly directed to assessing the toxicity of such sediments (Burton and Scott 1992), and assays have used either of three different exposure strategies: (1) to whole sediments or (2) to elutriates or (3) to interstitial water. Indigenous organisms have been extensively used and bred in the laboratory; in general, however, these organisms have complex nutritional requirements, and the degree and route of exposure to the toxicant are significantly determined by the affinity of the toxicant for the food supplied as well as by the organic carbon present in the system. The results of sediment assays using elutriates and typical aquatic organisms such as daphnids should, therefore, be interpreted with caution unless there is free desorption of the xenobiotics from the sediments; aspects of this have been discussed in Chapter 3 (Sections 3.2.2 and 3.3.2). A number of examples of these assays is given, though since the tests have not generally been standardized, it is not justifiable to compare relative sensitivity.

1. Numerous studies have used freshwater amphipods. For example, *Diporeia* sp.(syn. *Pontoporeia hoyi*) — which is one of the important benthic invertebrates in the Great Lakes — has been used to study basic determinants of both toxicity and bioavailability (Landrum et al. 1991), while *Hyalella azteca* has been quite extensively used for assessing toxicity (Nebeker et al. 1989; Ankley et al. 1991), and *Gammarus pulex* has been proposed as a suitable organism for routine toxicity evaluation (McCahon and Pascoe 1988; Taylor et al. 1991), although the nutritional demands of the organism must be taken into account. The marine amphipod *Rhepoxynius abronius* collected from an unpolluted site in Yaquina Bay (Oregon) has been used to assess the toxicity of creosote-contaminated sediments in Eagle Harbor, Washington (Swartz et al. 1989), and the estuarine amphipod *Leptocheirus plumulosus* has also been examined (McGee et al. 1993).
2. A life-cycle test using interstitial water and the polychaete *Dinophilus gyrociliatus* has been proposed for incorporation into a national screening program (Carr et al. 1989), and a whole-sediment assay using the marine polychaete *Nereis arenaceodentata* has been developed although important issues including the feeding regime should be critically evaluated (Dillon et al. 1993).

3. An extensive study has been devoted to examining the use of freshwater oligochaetes as test organisms for whole sediments (Wiederholm et al. 1987), and
of those tested including *Tubifex tubifex*, three species of *Limnodrilus*, and
Potamothrix hammoniensis, the first was the preferred organism due to the ease
of manipulation. Possible complications arising from the provision of food
during the test, and from differences in the nutritional status of sediments from
different lake environments were addressed and the importance of the relevant
controls underscored. On the basis of further evidence and development, *Tabifex
tabifex* has been recommended as one component in a group of tests for assessing sediment toxicity (Reynoldson et al. 1991). Protocols have also been developed for bioassays and estimation of bioaccumulation using *Lumbriculus
variegatus* (Phipps et al. 1993).

4. Midge larvae of several species of *Chironomus* including *C. tentans, C. decorus,*
and *C. riparius* have been quite extensively used (Nebeker et al. 1984), and these
organisms have been incorporated into microcosms (Fisher and Lohner 1987).
Two end-points have been effectively used: adult emergence and growth of
larvae (Nebeker et al. 1988) using *C. tentans*. Both acute toxicity and effects on
reproduction (Kosalwat and Knight 1986a,b) have been examined with cupric
sulfate using *C. decorus*, and as expected, larval stages were the more sensitive.
It is clearly important to define the development stage, and for practical reasons
the second larval instar of *C. riparius* has been used (Taylor et al. 1991): it was
also shown that the lethal toxicity varied during exposure during 240 h, and
comparable test protocols could be more generally exploited with advantage.
The effect of lindane (γ-hexachlorocyclohexane) on *C. riparius* has also been
examined during the complete life-cycle (Taylor et al. 1993): the LOEC of 9.9
μg/L was less than that found in assays that focused on growth or emergence,
although the authors point out that this could also be the result of experimental
factors in the life-cycle assay. The possibility of examining larval deformation
has also been examined using mentum deformities along a gradient of contamination (van Urk et al. 1992), though further laboratory investigations would be
needed to reveal the causative agent. The important issue of the genetic variation
in laboratory populations of *C. tentans* (Woods et al. 1989) has already been
noted in Section 7.1.

7.3.3 Assays for Genotoxic Effects

There are a large number of compounds which may induce carcinomas in man.
Very substantial effort has, therefore, been devoted to detecting these compounds so
as to eliminate — or at least to reduce — human exposure to them as far as possible.
Tests for induction of cancer in laboratory animals are extremely time-consuming
and costly and require strict attention to the design of the experimental protocols.
Attention has, therefore, been directed to developing sceening tests which are rapid
and reproducible and whose results display a satisfying correlation with cancer
induction. They make an attempt to take into account the exposure routes. Two
widely different organisms have been used for such assays — microorganisms and
fish. Attention should, however, also be directed to the use of morphological alterations of the polytene chromosomes in the salivary glands of larval *Chironomus
tentans* exposed to established carcinogens (Bentivegna and Cooper 1993); attractive

features are the use of whole organisms and the possible direct application to field samples. The use of higher plants for assessing toxicity including chromosomal aberrations has been noted in Section 7.3.1

Assays using microorganisms — Numerous tests for mutagenesis using micro-organisms have been developed and of these, the Ames test using mutant strains of *Salmonella typhimurium* has been widely used. Care should be exercised in correlating mutagenicity with carcinogenicity, and this has been consistently pointed out by Ames. The Ames test has been extensively evaluated and has therefore become the most commonly applied routine test. It is normally combined with a microsomal activation system prepared from rat liver, but two interesting developments are worth noting. One of these uses a preparation from the liver of rainbow trout for activation of promutagens; an important modification of the test protocol is necessitated by the temperature sensitivity of the metabolizing system which is optimal at 10-15°C (Johnson 1992). Although the second has probably greatest relevance to the terrestrial environment, it is clearly applicable to aquatic plants that have generally received somewhat scant attention. Activation of aromatic amines by plant-cell cultures has been examined, and a provisional model incorporating the formation of mutagenic complexes with macromolecules has been proposed (Plewa et al. 1993).

A revised protocol for the Ames test with experimental details and including new tester strains has been issued (Maron and Ames 1983), and a modification has been proposed that uses higher cell densities and diminishes the addition of microsomes to increase the sensitivity (Kado et al. 1986). A good account of the results using a wide variety of test organisms — including conventional assays using *S. typhimurium* — and their relative response to a selection of compounds has been given (De Serres and Ashby 1981).

The Ames test uses a set of histidine-requiring mutants of *S. typhimurium* and assesses the number of revertants after exposure to various concentrations of the test compound. The test can readily be carried out by an experienced bacteriologist and offers the following advantages:

1. The test strains include those that are able to detect both frameshift and base-pair mutations.
2. A metabolizing preparation of microsomes can be included to evaluate the synthesis of mutagenic metabolites from the test compound (promatagen).
3. Solutions in solvents such as dimethyl sulfoxide can be used so that problems of poor water solubility are minimized.
4. The test can be carried out within 2 d, and the results are generally readily interpreted; since a large number of cells are used, the test is extremely sensitive.
5. The test may be carried out either on plates of agar medium or as a fluctuation test in liquid medium; the latter is probably preferable in terms of sensitivity.
6. There is a reasonable correlation between mutagenic effects revealed by the test and the development of cancer in higher animals.

Naturally, care should be exercised in using the results, and they should be complemented, at least, by a cell transformation test using a suitable cell line. Difficulties may emerge in the form of complex or non-linear dose-response curves, but in any case, quantitative comparison of numerical values should not normally be attempted. The greatest ambiguity probably arises from the microsomal metabolizing system. Most often, a crude preparation termed S-9 prepared from rats after induction with a PCB mixture is used. It must, however, be appreciated that other animals may have widely different metabolizing systems so that both care and judgment should be exercised in interpreting the results. The attractive alternative using microsomes from fish has already been noted.

Other microbial systems for evaluating genotoxic effects have been developed. These include a forward-mutation assay using *S. typhimurium*, and assays using the yeast *Saccharomyces cerevisiae* or *Bacillus subtilis* although these have not been as extensively used as the Ames test. An ingenious assay is based on induction of the prophage λ in *Escherichia coli* (Moreau et al. 1976), is as readily carried out as the Ames test, and has the advantage of greater sensitivity towards specific groups of compounds such as chlorinated pesticides (Houk and DiMarini 1987) and chlorophenols (DeMarini et al. 1990). There is, however, no question of the superiority of one of these systems over the other; they are complementary, and both may advantageously be employed.

Probably any compound suspected of entering higher trophic levels and which may, therefore, be consumed by man is worth evaluating for its mutagenic potential before carrying out assessment of its carcinogenicity in higher animals.

Assays using fish embryos and fish — These tests have been directed primarily to the possibility of inducing cancer in man. It has, however, become increasingly clear that many biota in aquatic systems are exposed to potentially carcinogenic compounds and that feral fish may display pathological evidence of disease in certain organs (Malins et al. 1985, 1987); this is noted again in Chapter 8 (Section 8.2). Both of these objectives may be combined in tests in which fish embryos are injected with the toxicant dissolved in dimethyl sulfoxide; the embryos are allowed to develop into fry and, 8 to 9 months after exposure, samples are examined for the development of neoplasms (Black et al. 1985). Both rainbow trout (*Oncorhynchus mykiss*) and coho salmon (*Oncorhynchus kisutch*) were examined, and although a comprehensive evaluation of the test does not appear to be available, this is an attractive test which is suitable for screening large numbers of samples. Initiatory investigations (Hawkins et al. 1988) have been directed to the use of small laboratory fish, such as sheepshead minnow and fathead minnow, to detect carcinogenic effects. From these investigations, it is clear that the susceptibility of fish to induction of cancer varies widely among the species evaluated and that the influence of, for example, nutrition should be evaluated before a standardized protocol can be presented. Subsequent investigations (Hawkins et al. 1990) have demonstrated the potential value of both Japanese medaka (*Oryzias latipes*) and guppy (*Poecilia reticulata*), and the differential sensitivity of these to the known human carcinogens benzo[a]pyrene and 7,12-dimethylbenz[a]anthracene.

7.4 TEST SYSTEMS: SEVERAL ORGANISMS

7.4.1 Introduction

The relevance of tests using single species for evaluating the effects of xenobiotics on natural ecosystems has been the subject of considerable discussion (Cairns 1984), and no final resolution of the conflicting views is to be expected. On the one hand, it is obvious that few natural ecosystems consist of only one taxon and that the effects of a toxicant should be evaluated in the context of communities; on the other hand, it is clear that the effect will ultimately be displayed in individual organisms as a result of reactions at the molecular level. Neither approach is, therefore, exclusive, and evaluations of the biological effect of xenobiotics may profitably be carried out at three levels: (1) the individual, (2) the population, and (3) the community. Detailed discussions of all three approaches have been given (Landner et al. 1989).

The design of the experiments to address these questions is necessarily complex, and complementary monitoring to reveal adverse effects of xenobiotics on wild populations is probably necessary. It should, however, be emphasized that extreme care must be exercised in attributing the observed effect to the exposure of biota to a given toxicant; this problem is discussed in Chapter 8 (Section 8.2). There is, therefore, a clear need for test systems that lie between the two extremes of single species and natural ecosystems; these test systems should attempt to combine the reproducibility of single-species tests with the ecological relevance of field experiments. It is hardly realistic to expect a non-specialist like the author to evaluate cardinal issues such as the design of the systems, the end-points used, or the relevance of the results to natural ecosystems. With inevitable shortcomings, some attempt is, however, made to provide at least an overview of what appear to be the critical issues. It should also be appreciated that it may not be possible to present the data collected from such experiments in formats which are as accessible as those from singe-species testing; this fact reflects, in essence, the complexity of the natural systems whose simulation is the primary objective of the experiments.

Two broad types of multi-species test systems have been developed, differing essentially in their dimensions: laboratory microcosms having a volume of a few liters and mesocosm systems with volumes of cubic meters. There is, of course, no absolute distinction between these, and some systems have volumes between the two extremes.

7.4.2 Microcosms

The communities that are used generally consist of organisms at different trophic levels, and two rather different approaches have been used.

1. Taube microcosms (Larsen et al. 1986) may include representatives of the following groups of organisms which are added to a synthetic medium: heterotrophic bacteria and fungi, algae, protozoans, amphipods, and ostracods. A sediment component is used generally supplemented with insoluble organic material such as cellulose or chitin.
2. Alternatively, organisms such as protozoans may be collected on polyurethane foam blocks suspended in a stream which the system is designed to simulate and macroinvertebrate communities assembled on rocks held in plastic containers (Pontasch et al. 1989).

In these experiments, a range of end-points has been used such as the population densities of individual taxa (Pontasch et al. 1989) or appropriate measures of primary and secondary production including $^{14}CO_2$ uptake, the concentration of chlorophyll, the oxygen concentration, and the rate of respiration (Larsen et al. 1986; Stay et al. 1988). Use of the rate of respiration of glucose to measure heterotrophic activity (Stay et al. 1988) seems, however, unduly restrictive since many bacteria and even heterotrophic algae are unable to mineralize glucose as a substrate even though they may be able to incorporate it into cell material.

Those studies which have examined single compounds (atrazine or fluorene) or complex effluents were designed to assess the extent to which the results of these test systems could be used to predict effects in specific ecosystems. The overall agreement between the two was good, and the results complemented those using single-species assays. The results also clearly supported the view that the ecological effects of a toxicant should be assessed on the basis of the results from a range of organisms, even within the same taxonomic group. On the other hand, they indicated that significant differences could occur between the effects at population and at ecosystem levels, and between results obtained from microcosms using biota from different environments (Stay et al. 1988). The suggestion (Niederlehner et al. 1990) that such disagreement with predictions implies the need for a review both of the design of the experiments and their objective is highly commendable. It, therefore, seems unlikely that a single test system will adequately represent a range of natural ecosystems, and the assay system that is used should reflect as far as possible the structure of specific ecosystems. In addition, it is entirely conceivable that some significant variables in all ecosystems remain to be evaluated.

Systems using organisms at other trophic levels have also been used, and two may be used as illustration.

1. The use of indigenous microbial communities has been suggested not as surrogates for macroscopic communities but because they provide communities of comparable complexity and offer the possibility of studying the dynamics of their susceptibility to the effect of xenobiotics. It should be noted that the term *microbial* is generally used in a restricted sense to include only protozoa and algae and that emphasis has, hitherto, been placed on the use of the distribution of taxa in the system as an end-point (Cairns et al. 1992).
2. An assay system has been developed that used sediment containing micro-, meio-, and macrofauna. The sediment was sieved, and the amphipod *Pontoporeia affinis* and the mussel *Macoma baltica* were used as test organisms in a flow-through system (Landner et al. 1989). A number of end-points are possible, including the survival and reproduction of *P. affinis* during chronic exposure (Sundelin 1983) and population changes in various groups of meiofauna. Apparently, this system has not, however, been systematically evaluated with a range of compounds so that its potential remains unknown.

7.4.3. Mesocosms

A valuable review of the widely different experimental systems which have been used has been provided (Lundgren 1985). Model systems have been developed to simulate at least three different types of natural ecosystems: (1) littoral zones using pools or ponds; (2) benthic zones using tanks; and (3) riverine systems using model streams. These have been discussed in Chapter 5 (Section 5.3.3) in the context of biodegradation and biotransformation.

The objective and design of model ecosystems — It is necessary to appreciate at the start that model ecosystems may aim at answering quite diverse questions:

1. The effect on biota of chronic exposure to low concentrations of a toxicant.
2. The relative significance of various routes of exposure of biota to the toxicant.
3. The fate of toxicants and their distribution among the environmental compartments.

One of the very substantial attractions of model ecosystems is the possibility of attaining all three objectives in a single set of experiments. On the other hand, one of the inherent limitations is the virtual impossibility of accurately reproducing the same conditions, so that particular care in the design of experiments should be directed to providing a sufficient number of replicates and an unequivocal control. Climatic variations during the year may be an advantage in providing a degree of realism not readily accomplished in laboratory experiments, though in more northern latitudes, ice formation may exclude the possibility of continuing the experiments during the winter months.

Attention should also be drawn to some fairly obvious experimental issues. Most of these apply equally to tests using single species, but they may be particularly important in the more complex mesocosm systems.

1. Particular care is needed in testing compounds with poor water solubility since concomitant application of dispersion agents or organic solvents may lead to overgrowth of the systems with bacteria or algae.
2. In systems with fish, care should be exercised in their selection, e.g., use of both largemouth bass (*Micropterus salmoides*) and bluegills (*Lepomis macrochirus*) to maintain both of them in good condition (Deutsch et al. 1992).
3. If physiological or biochemical parameters on fish are assessed, particular care should be taken in the handling of the fish to avoid stress and its interference with the measurements (Munkittrick et al. 1991; Hontela et al. 1992).
4. Analytical control of toxicant concentration in the system should be carried out since some compounds may partition rapidly into the sediment phase; although this partition reflects an important aspect of natural ecosystems, it may clearly influence the exposure concentration to aquatic biota.

A large number of different end-points may be used in experiments using model ecosystems and will generally include those used for single species, e.g., growth and reproduction of fish or planktonic algae. In addition, assessment of the composition of populations of invertebrates in sediments or of algae attached to surfaces (periphyton) may be made, as well as measurements of any of the physiological and biochemical parameters in fish which are discussed in more detail in Section 7.6. An interesting development is the use of physiological energetics in the amphipod *Gammarus pulex* (Maltby 1992) that has the attractive advantage that this may also be applied to laboratory experiments.

A striking example of the toxic effect of a single compound which was initially revealed in experiments with model mecosystems is that of chlorate towards bladder wrack (*Fucus vesiculosus*) which is a component of the Baltic Sea littoral zone (Rosemarin et al. 1986); in this case, it emerged that, compared with unicellular algae that are generally used for toxicity testing, *F. vesiculosus* was exquisitely sensitive to chlorate. In general, however, model ecosystems provide the possibility of revealing interactions between the various components, and they are less fully exploited if they merely measure isolated effects on single species. The total effect of a toxicant may be greater than the sum of these single-species effects if the interactions between them are significant.

Illustrative examples — It has been noted that mesocosm experiments have the potential for revealing both the toxicity and the fate of a xenobiotic; results from investigations that illustrate some aspects of these are provided in the following examples.

1. Fate and distribution

- The fate of 2,3,7,8-tetrachlorobenzo-1,4-dioxin (TCDD) was examined in freshwater ponds using ^{14}C-labeled substrate over a period of up to 2 years. Equilibrium concentrations in pondweeds (*Elodea nuttali* and *Cerotophyllon demersum*), in fathead minnows (*Pimephales promelas*), and in sediment were attained after 1, 2, and 6 months, respectively. After 1 year, most of the remaining TCDD was found in the pondweed, and after 2 years, through death of the plants, almost all the TCDD was found in the sediment phase that included plant detritus. Unidentified metabolites were confined to the aqueous phase and the plants (Tsushimoto et al. 1982).

- The fate and distribution of the pyrethroid insecticide deltamethrin was studied in small outdoor ponds over a period of 306d (Muir et al. 1985). The compound rapidly partitioned into suspended solids, plants, sediment, and the atmosphere, and at the termination of the experiment, the major sink for the intact compound was the sediment phase. Although fathead minnows (*Pimephales promelas*) concentrated the compound from the aquatic phase, no mortality was observed.

- The persistence of benzo[a]pyrene was investigated in a marine model ecosystem with planktonic primary producers and a heterotrophic benthos. The experiment was conducted over 202 d and showed that the initial substrate was substantially degraded into polar products, but that after a period of ca. 2 months, both the remaining substrate and its metabolites were contained in the sediment and were apparently then protected from further degradation or metabolism (Hinga and Pilson 1987).

- Brackish-water systems simulating the littoral zone of the Baltic Sea were used to examine the fate and distribution of 4,5,6-trichloroguaiacol and its metabolites. The known metabolites including 3,4,5-trichloroveratrole, 3,4,5-trichlorocatechol, and dechlorination products were widely distributed in the biota, and a mass balance showed that the bulk of them were partitioned into the sediment phase. The results effectively confirmed those from previous laboratory experiments (Neilson et al. 1989).

The results of all these controlled experiments are consistent with the principles of partitioning that have been discussed in Chapter 3 (Section 3.2) and with the established recovery of many xenobiotics from sediments.

2. Effect on biota

Two kinds of experiments have been carried out. In one, only a single organism such as the water-flea *Daphnia longispina* (Crossland and Hillaby 1985) has been the primary organism of interest, but attention should be drawn to the much more sophisticated full life-cycle tests with rainbow trout carried out in model stream systems subjected to constant input of diluted bleached kraft mill effluent during 3.5 years (Hall et al. 1991). In the second type of mesocosm experiment, advantage is taken of the possibility of examining the population changes in the major components of the biota. In this respect, the procedures are similar to those used in microcosm studies.

- A study of the effect of 4-nitrophenol used a divided pond and contained algae, macrophytes, and a range of aquatic fauna. There was an alteration in the structure of the algal population in which cryptophytes replaced chlorophytes, and there were reduced populations of the aquatic fauna, particularly the crustaceans belonging to the orders Cladocera and Cyclopida. With the exception of the alga *Chara hispida*, the population of macrophytes was eliminated 1 year after application of the toxicant (Zieris et al. 1988). It is important to point out that the results of this study differed significantly from those obtained using single-species tests: the effect on algae was less, and on Cladocera greater. In this case, false predictions would therefore have resulted from extrapolation of the results of assays using single species.

- The effect of single applications of esfenvalerate [*S*-alpha-cyano-3-phenoxybenzyl (*S*)-2-(4-chlorophenyl)-3-methylbutyrate] was examined (Lozano et al 1992) in a pond system containing plant and animal communities including the following: macrophytes and phytoplankton, micro- and macroinvertebrates and fish (fathead minnows, bluegills, and northern redbelly dace [*Phoxinus eos*]) which were placed in enclosures. A range of toxicant exposure concentrations was employed, but at the lowest that were used, the concentrations of the toxicant were unmeasurable after 24 h. At the higher concentrations tested (1 µg/L and 5 µg/L), drastic reduction or elimination of populations was observed, though some invertebrate communities recovered at the lower concentrations used. It should be noted that this experiment was designed to assess the effect of low concentrations and not chronic exposure which is the more common objective.

The same compound was examined during application every two weeks to a mesocosm system, and a number of dynamic parameters were evaluated including growth of bluegill, primary production, community respiration, and enumeration of benthic invertebrates and zooplankton (Fairchild et al. 1992). Again, the important conclusion was drawn that although effects on individual organisms such as fish or crustaceans may be assessed from single-species tests, alterations in the composition of the ecosystem as a whole may prove both more subtle and, in the long run, more significant.

These results clearly illustrate the limitations in the conclusions that may be drawn from conventional single-species assays and support the views of Maltby and Calow (1989) that are discussed in greater detail in Chapter 8 (Section 8.1.3).

7.5 METABOLISM OF XENOBIOTICS BY HIGHER ORGANISMS

It will have become clear from arguments presented in previous chapters that very few xenobiotics remain unaltered in the environment for any length of time after their release. Although metabolism primarily by microorganisms has been discussed in detail in Chapter 6, some brief comments on metabolism — particularly by fish — are briefly summarized here. The biotransformation of xenobiotics in many higher organisms is mediated by the cytochrome P-450 monooxygenase system, and the complexity of factors that regulate the synthesis of this in fish has been reviewed (Andersson and Förlin 1992).

Possibly the motivation for discussing metabolism in a chapter dealing with toxicology should be briefly addressed. The impact of metabolism on experiments on bioconcentration has been outlined briefly in Chapter 3 (Section 3.1.3), and similar principles apply equally to toxicity; the kinetics and products of metabolism critically influence the nature and the concentrations of the xenobiotic and its transformation products to which the cells are exposed. Increasing evidence from different sources has shown that the effective toxicant may, indeed, be a metabolite synthesized from the compound originally supplied and not the xenobiotic itself (Buhler and Williams 1988). Please refer to additional list of references. It is important to appreciate that — as with toxicity — the extent of metabolism will generally depend on the nature and position of substituents on aromatic rings as well as on their number. For example, although 2,3,4- and 3,4,5-trichloroaniline were N-acetylated in guppy, this did not occur with 2,4,5-trichloroaniline (de Wolf et al. 1993). A few diverse examples may be given to illustrate the important toxicological consequences of metabolism.

1. The classic case is that of Prontosil (Figure 7.1) in which the compound is active against bacterial infection in animals though inactive against the bacteria in pure culture. The toxicity in animals is the result of reduction to the sulfanilamide (4-aminobenzenesulfonamide) that competitively blocks the incorporation of 4-aminobenzoate into the vitamin folic acid.
2. The case of fluoroacetate in which the toxic substance is fluorocitrate synthesized by incomplete operation of the tricarboxylic acid cycle (Peters 1952) has already been noted in Chapter 4 (Section 4.7).
3. Considerable attention has been directed to the synthesis of the epoxides of polynuclear hydrocarbons mediated by the action of cytochrome P-450 systems and their role in inducing carcinogenesis in fish (Varanasi et al. 1987). Tumors observed in feral fish exposed to PAHs may plausibly — though not necessarily exclusively — be the result of this transformation. Even though an apparently causal relationship between exposure of fish to PAHs and disease may be established (Malins et al 1984, 1985), caution should be exercised due to the possibility that other — and unknown — substances may have induced carcinogenesis. Further discussion of this is presented in Chapter 8 (Section 8.2). It is also important to appreciate that induction of the metabolic system for PAHs may be induced by other compounds. For example, exposure of rainbow trout to PCBs increases the effectiveness of liver enzymes to transform benzo[a]pyrene to carcinogenic intermediates (Egaas and Varanasi 1982).

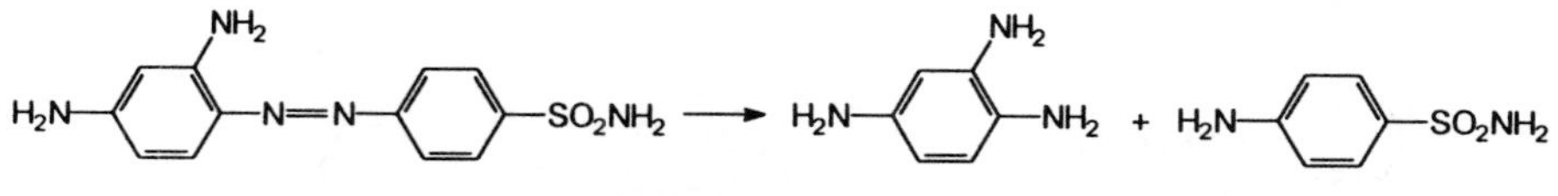

Figure 7.1 Metabolism of Prontosil in animals.

Figure 7.2 Metabolism of aldicarb by rainbow trout.

4. The carbamate insecticide aldicarb (Figure 7.2) that exerts its effect by inactivating acetylcholinesterase is metabolized by a flavin monooxygenase from rainbow trout to the sulfoxide which is a more effective inhibitor (Schlenk and Buhler 1991).

5. Pre-exposure to the organophosphate diazinon at exposures half the LC_{50} values increased the LC_{50} value by a factor of about five for guppy (*Poecilia reticulata*) but had no effect on the value for zebra fish (*Brachydanio rerio*). This was consistent with the observation that during pre-exposure of guppy, there was a marked inhibition in the synthesis of the toxic metabolites diazoxon and pyrimidinol, whereas this did not occur with zebra fish in which the toxicity was mediated primarily by the parent compound (Keizer et al. 1993).

At the other extreme, if metabolism of the xenobiotic by the organism does not occur at all — or at insignificant rates — after exposure, the compound will be persistent in the organism and may therefore be consumed by predators. This has been briefly discussed in the context of biomagnification in Chapter 3 (Section 3.5.4).

Most of the reactions carried out by fish and by higher aquatic organisms are relatively limited transformations in which the skeletal structure of the compounds remains intact. Three widely distributed reactions are probably of greatest significance:

1. Cytochrome P-450 type monooxygenase systems which have a generally low substrate specificity are widely distributed in the species of fish used for toxicity testing (Funari et al. 1987).

2. Glutathione *S*-transferases (Donnarumma et al. 1988; Nimmo 1987) which are important in the metabolism of highly reactive compounds containing electrophilic groups, such as epoxides, and aromatic rings with several strongly electron-attracting substituents such as halogen, cyano, or nitro groups.

3. Conjugation of polar groups, such as amines, carboxylic acids, and phenolic hydroxyl groups, produce water-soluble compounds that are excreted, and these reactions, therefore, function as a detoxification mechanism.

7.5.1 Metabolism by Fish

The metabolic potential of fish may appear restricted compared to that of microorganisms, but it may have been considerably underestimated. For example,

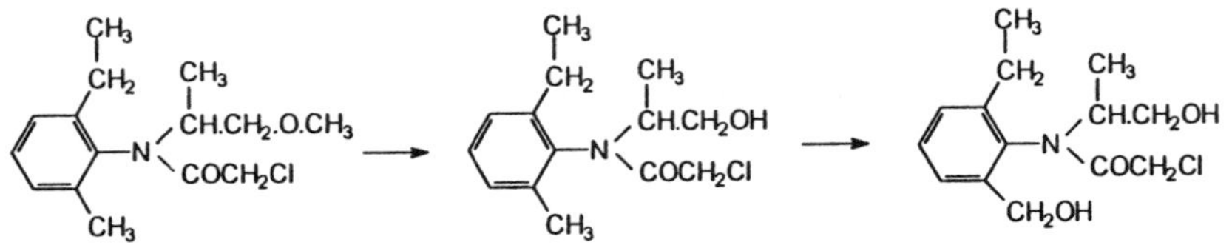

Figure 7.3 Metabolism of metolachlor by bluegill sunfish.

metolachlor (2-chloro-N-[2-ethyl-6-methylphenyl]-N-[2-methoxy-1-methylethyl]-acetamide) is metabolized by bluegill sunfish (*Lepomis macrochirus*) by reactions involving initially O-demethylation and hydroxylation (Cruz et al. 1993) (Figure 7.3) that are comparable to those carried out by an actinomycete (Krause et al. 1985), while the benzylic hydroxylation is analogous to that involved in the biotransformation of the structurally similar alachlor by the fungus *Cunninghamella elegans* to which reference has been made in Chapter 4 (Section 4.3.5). The metabolism of PAHs has attracted considerable attention both in the context of assays for genotoxic effects (Section 7.3.3) and on account of the incidence of tumors in natural fish populations exposed to PAHs that has been noted in the introduction to Section 7.5 and briefly again in Chapter 8 (Section 8.2). A few examples of limited transformation reactions carried out by fish are summarized by way of illustration:

1. N-dealkylation of dinitramine to 1,3-diamino-2,4-dinitro-6-trifluoromethyl-benzene (Olson et al. 1977) by carp (*Cyprinus carpio*) (Figure 7.4).
2. O-demethylation of pentachloroanisole in rainbow trout (Glickman et al. 1977) and of chlorinated veratroles by zebra fish (Neilson et al.1989).
3. Acetylation of 3-amino ethylbenzoate in rainbow trout (Hunn et al. 1968).
4. Displacement of the nitro group in pentachloronitrobenzene by hydroxy- and thiol groups (Figure 7.5) (Bahig et al. 1981) in golden orfs (*Idus idus*).
5. Oxidation of a number of PAHs has been demonstrated in a variety of fish. For example, coho salmon (*Oncorhynchus kisutch*) metabolized naphthalene to a number of compounds consistent with oxidation to the epoxide, hydrolysis to the dihydrodiol, and dehydration of the dihydrodiol to naphth-1-ol (Figure 7.6) (Collier et al. 1978). For the carcinogen benzo[a]pyrene, a much wider range of metabolites has been identified in southern flounder (*Paralichthys lethostigma*) including the 4,5-, 7,8-, and 9,10-diols, the 1,6-, 3,6-, and 6,12-quinones as well as the 1-, 3-, and 9-benzopyreneols (Figure 7.7) (Little et al. 1984).
6. N-hydroxylation of aniline and 4-chloroaniline by rainbow trout to hydroxyl-amines that could plausibly account for the sub-chronic toxicity of the original compounds (Dady et al. 1991).

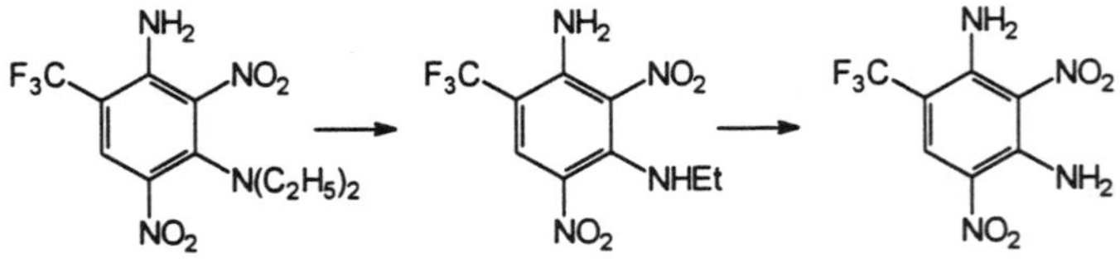

Figure 7.4 Metabolism of dinitramine by carp.

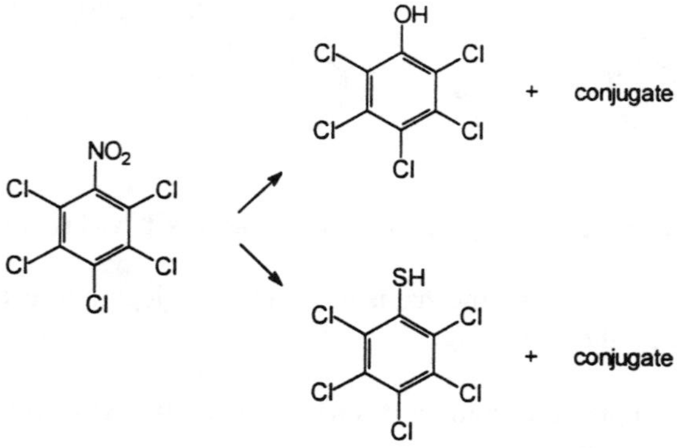

Figure 7.5 Metabolism of pentachloronitrobenzene by golden orfs.

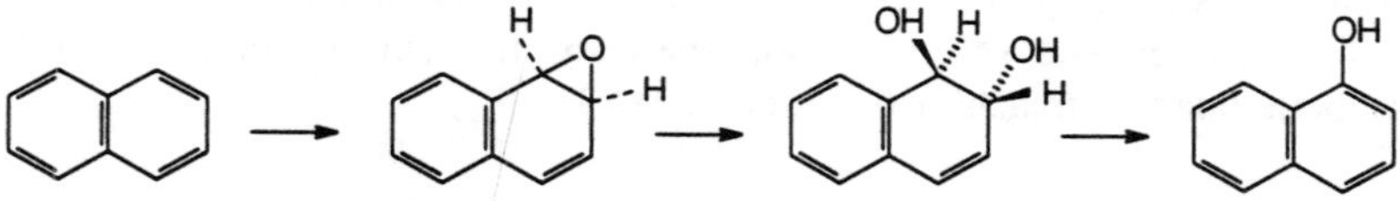

Figure 7.6 Metabolism of naphthalene by coho salmon.

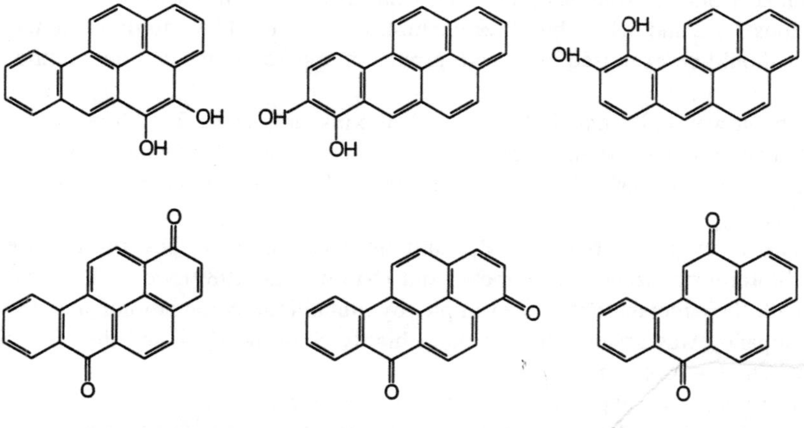

Figure 7.7 Metabolites produced from benzo[a]pyrene by southern flounder.

Initially formed polar metabolites such as phenols and amines may then be conjugated to produce terminal metabolites that are excreted into the medium. For example, in the illustrations given above, pentachlorophenol and pentachlorothiophenol produced from pentachloronitrobenzene were conjugated to yield the major metabolites; although the naphthalene dihydrodiol was the major metabolite produced from naphthalene, the further transformation product naphth-1-ol was also isolated as the sulfate, glucuronic acid, and glucose conjugates. Conjugation of xenobiotics is extremely important since this reaction results in the production of highly water-soluble products that are then excreted from the fish; this reaction, therefore, functions

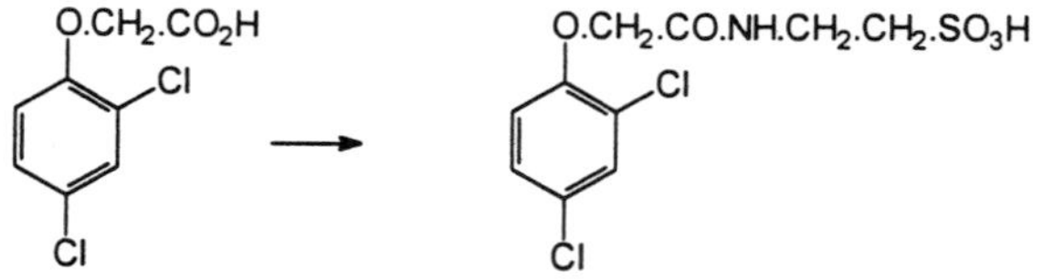

Figure 7.8 Conjugation of carboxylic acids with amino acids.

as an effective detoxificiation mechanism. Diverse conjugation reactions have been described including the following:

1. Conjugation of phenolic compounds with formation of glucuronides, sulfates, or glucosides as noted above.
2. Reaction of carboxylic acids with the amino groups of glycine (Huang and Collins 1962) or taurine (Figure 7.8) (James and Bend 1976) to form the amides.
3. Reaction between glutathione and reactive chloro compounds such as 1-chloro-2,4-dinitrobenzene (Niimi et al. 1989) or the chloroacetamide group in demethylated metolachlor (Cruz et al. 1993).

One additional consequence of such reactions is that experiments designed to measure bioconcentration may be seriously compromised, and this will be the case particularly with compounds which are metabolized at rates comparable with the rate of uptake from the aqueous phase. This has already been discussed in Chapter 3 (Sections 3.1.2 and 3.1.3), but three examples will be used here to illustrate the care which should be exercised even with groups of apparently recalcitrant compounds.

1. The lower chlorinated dibenzo-1,4-dioxins and even 2,3,7,8-tetrachlorodibenzofuran (Opperhuizen and Sijm 1990) were apparently metabolized by aquatic organisms so that values of their BCFs could not be predicted on the basis of their P_{ow} values.
2. The discrepancy between P_{ow} and observed BCF values for a series of chloronitrobenzenes in rainbow trout (Niimi et al. 1989) would be consistent with their metabolism, for example, by glutathione conjugation, since the discrepancy was greatest for the more highly chlorinated — and therefore more reactive — congeners.
3. The results of experiments on the bioconcentration of azaarenes in fathead minnows showed significant differences between the observed bioconcentration factors for some azaarenes and the values predicted from their P_{ow} values; this was attributed to metabolism of the test substances that was significant for benz[a]acridine and dibenz[a,h]acridine but not for acridine itself or for quinoline (Southworth et al. 1989). This could plausibly be correlated with the higher π-electron density of the rings in the benzo compounds.

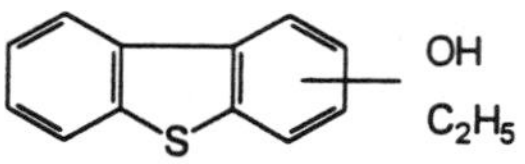

Figure 7.9 Alkylated dibenzothiophenes suggested as markers.

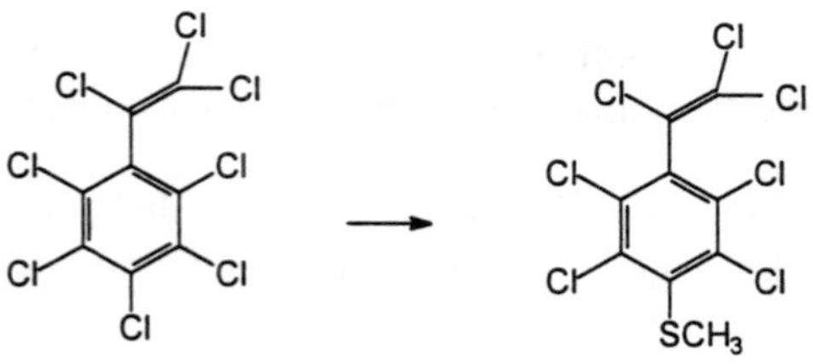

Figure 7.10 Metabolism of octachlorostyrene by the blue mussel.

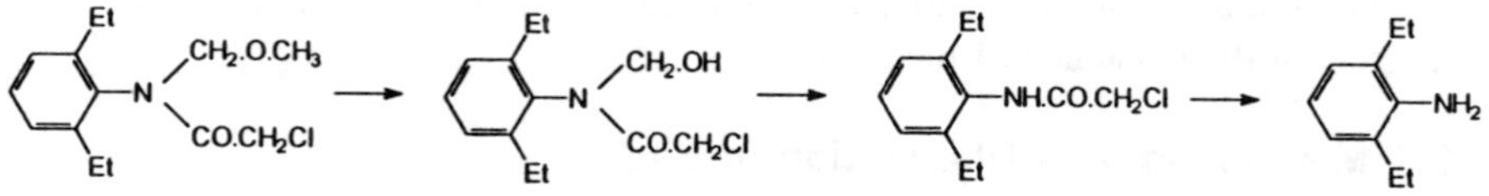

Figure 7.11 Metabolism of alachlor by chironomid larvae.

The metabolic capability of fish has found two complementary applications in monitoring studies.

1. Analysis of xenobiotic conjugates in fish bile has been used to demonstrate the exposure of fish to chlorophenolic compounds — though not to quantify it (Oikari and Kunnamo-Ojala 1987; Wachtmeister et al. 1991). Analysis of the conjugates of metabolites of aromatic components of oil after a spill has been used to identify the origin of the oil (Krahn et al. 1992), and specific use of the metabolites of alkylated dibenzothiophenes (Figure 7.9) was suggested as a valuable indicator of the source of the oil.
2. Biomarkers including the level of monooxygenase enzymes involved in metabolizing xenobiotics have been used to demonstrate the exposure of biota to xenobiotics, and this application is discussed in greater detail in Section 7.6.

7.5.2 Metabolism by Other Higher Aquatic Organisms

1. It has been generally assumed that mussels do not carry out more than the limited reactions of oxidation and conjugation, although variations between summer and winter levels for both cytochrome P-450 and NADPH-independent 7-ethoxycoumarin O-deethylase have been found in the common mussel *Mytilus edulis* (Kirchin et al. 1992). Levels of cytochrome P-450 and the rates of metabolism of PAHs are apparently low compared with those found in fish, and an investigation using subcellular extracts of the digestive glands from the mussel *M. galloprovincialis* showed that although the formation of diols and phenols from benzo[a]pyrene was dependent on NADPH, the quinones that were the major metabolites were produced in the absence of NADPH apparently by radical-mediated reactions involving lipid peroxidase systems (Michel et al. 1992).
2. A reaction presumably mediated by glutathione *S*-transferase is the replacement of the 4-chloro substitutent in octachlorostyrene in the blue mussel (*Mytilus edulis*) by a thiomethyl group (Figure 7.10) (Bauer et al. 1989). A similar

reaction of glutathione with arene oxides produced by aquatic mammals from PCBs and DDT results ultimately in the production of dimethyl sulfones (Bergman et al. 1994).
3. The herbicide alachlor is transformed by chironomid larvae and proceeds by *O*-demethylation followed by loss of the chloroacetyl group to produce ultimately 2,6-diethylaniline (Figure 7.11) (Wei and Vossbrinck 1992).

The results of these investigations suggest that caution should be exercised in interpreting not only the results of toxicity assays in which such organisms are employed but also data accumulated in monitoring studies that may not have taken into account the existence of metabolites.

7.5.3 Metabolism and Disposition in Birds

It seems appropriate to add some brief comments on birds since it was the attention directed to the effect of agrochemicals on these which effectively mobilized public awareness of the environmental hazard of chemicals. Birds may play a role in the dissemination of xenobiotics initially discharged into the aquatic environment, and in some ecosystems fish-eating birds may be the top predator and may thereby be exposed to a variety of xenobiotics. For example, there is continued concern that the existence of a variety of deformities among fish-eating water birds in the Great Lakes is causally associated with exposure to xenobiotics (Giesy et al. 1994). In addition, extensive data suggest that some agrochemicals are more toxic to birds than to mammals. Although birds may possess lower levels of microsomal monooxygenase enzymes which are involved in the metabolism of xenobiotics than fish (Walker 1983; Ade et al. 1984), like mammals they seem to have phenobarbital-inducible systems that mediate the metabolism of important PCB congeners. Attention may briefly be directed to the results of different kinds of investigation.

1. The existence of residues of PCBs in cormorants (*Phalacrocorax carbo sinensis*) captured on the German North Sea coast (Scharenberg 1991), of PCBs and DDT in samples of muscle, and in eggs of various birds from Sweden (Andersson et al. 1988), and of chlorinated benzenes in herring gulls from the Great Lakes (Hallett et al. 1982) supports the widespread dissemination of these compounds to which the birds must have been exposed through consumption of contaminated food. This has been discussed in the context of biomagnification in Chapter 3 (Section 3.5.4).
2. Evidence of porphyria in herring gulls (*Larus argentatus*) in the Great Lakes (Fox et al. 1988) has been plausibly associated with exposure to organochlorine compounds that may plausibly be assumed to have originated in fish.
3. By comparing rates of elimination of 2,3,7,8-tetrachlorodibenzo-1,4-dioxin in egg-laying and non-laying hens of ring-necked pheasants (*Phasianus colchicus*), it was shown that egg laying was an important route of elimination of the toxicant and could contribute up to ca. 35% of the total amount of the toxicant that was administered. There was no evidence of metabolism during the experiment, and the toxicant was found only in the yoke of the eggs (Nosek et al. 1992). The consumption of eggs by predators, therefore, would expose them to the xenobiotic.

In addition, migratory birds may be exposed via their food to diverse xenobiotics from widely separated geographical areas. Although numerous questions, therefore, remain to be answered, sufficient evidence for concern clearly remains.

7.6 BIOMARKERS:
BIOCHEMICAL AND PHYSIOLOGICAL END-POINTS

Concomitant with an appreciation of the importance of sub-lethal effects of xenobiotics on natural populations, there has been increased interest in alternative procedures for assessing these effects. Considerable interest has centered on the application to fish of procedures developed in clinical medicine; these have used biochemical assays for specific enzyme activities and physiological parameters in blood and serum samples. The term "biomarker" has been used, but it should be pointed out that this has also been applied in a completely different context to compounds isolated from samples of sediment, coal, and oil and which plausibly have a biological origin. There seems, however, little reason for confusion in the application of the same term in these widely different contexts.

7.6.1 Biochemical Parameters

The previous discussion has illustrated the role of monooxygenase systems in fish for the metabolism of xenobiotics. Measurements of this activity have, therefore, been used as a measure of the extent to which fish have been exposed to xenobiotics; at the same time, of course, increased levels enable the fish to metabolize xenobiotics effectively (Kleinow et al. 1987). Although specific assays of cytochtome P-450 activity may be made by immunoblot methods (Monosson and Stegeman 1991), it may be expedient to measure specific enzyme activity using defined substrates. Two assays have been widely used: (1) aryl hydrocarbon hydroxylase activity that may be assayed using benzo[a]pyrene as substrate, although this substrate has been replaced recently by the less hazardous 2,5-diphenyloxazole, and (2) activity for O-deethylation of 7-ethoxyresorufin (EROD) that has been extensively used and is a simple and convenient assay.

In interpreting the results of such assays, however, a number of important factors should be considered. Induction by exposure to a given xenobiotic may be strongly influenced by previous exposure to other environmental toxicants (Monosson and Stegeman 1991). Environmental factors such as the temperature and the feeding regime may influence EROD activity (Jimenez et al. 1988), and both the rate of induction and the levels of activity that are attained may be influenced by the temperature at which fish are maintained (Andersson and Koivusaari 1985). This is consistent with the role of temperature on toxicity that has already been noted (Hodson and Blunt 1981).

Other metabolic enzymes have also been used in assessing exposure to toxicants. These include conjugation enzymes that are important in the metabolism and depuration of xenobiotics; for example, an assay for uridine diphosphoglucuronosyl transferase (UDP glucuronosyl transferase) (Andersson et al. 1985) has been used quite extensively in Sweden in conjunction with assays for EROD activity. Two important factors in interpreting the results of such assays are worth pointing out.

1. The pattern of induction of metabolizing and of conjugating enzymes may be quite different, even though, for example, cytochrome P-450 inducers are capable of inducing both glutathione transferase and UDP glucuronosyl transferase activities (Andersson et al. 1985).
2. Levels of UDP glucuronosyl transferase activity after induction with phenolic compounds differed considerably among various fish (Förlin et al. 1989).

An assay for plasma leucine aminonaphthylamidase has been examined as a measure of stress in rainbow trout, but variability in the levels due, for example, to diet combined with the rather low sensitivity of the assay do not seem encouraging for its widespread application especially to field samples (Dixon et al. 1985).

Metallothionein is a metal-binding protein whose level generally increases in response to exposure to metals (Hennig 1986), although the demonstration that its level is also increased by exposure to hydrocarbon fuel oil even in the absence of metals illustrates the care that must continuously be exercised in interpreting data on enzyme levels (Steadman et al. 1991).

A study in which biochemical indicators were used to assess the effect of hydrocarbon fuel oil on rainbow trout clearly revealed a number of factors which could compromise interpretation of the results (Steadman et al. 1991), and these seem of sufficient general importance to merit brief summary.

1. Enzyme levels may not exhibit a dose-response due to the effect of toxicity of components contained in the mixture of toxicants.
2. Enzyme levels may be affected by the exposure route and by chronic exposure to toxicants, and these factors have not generally been systematically examined.
3. At high levels of exposure to a toxicant, fish may no longer synthesize increased levels of detoxification enzymes but respond by displaying liver hypertrophy and a reduction in spleen size.
4. The increased levels of metallothein in the putative absence of metals to which reference has already been made.

7.6.2 Physiological Parameters

These have been chosen to reflect disturbances in function, and some have been based on those which are routinely used in clinical medicine. Some examples of those which have been most widely used are the following:

1. Effect on liver function: liver somatic index, level of ascorbic acid, together with enzyme levels, e.g., EROD noted above.
2. Gonad growth: gonad somatic index.
3. Carbohydrate metabolism: glycogen content of liver and muscle, glucose and lactate levels in blood.
4. Osmotic and ion regulation: concentrations of Na^+, K^+, Ca^{2+}, Mg^{2+}, and Cl^- in blood plasma.
5. Status of blood cells: red and white blood cell count, hematocrit, methemoglobin.

Many of these parameters have been used to assess the effect of industrial effluents on both laboratory-reared and feral fish (Andersson et al. 1987; Härdig et al. 1988; Larsson et al. 1988).

Respiratory-cardiovascular responses in rainbow trout have been used to assess the effects of toxicants, and the results with two phenols and three anilines indicated different narcosis syndromes which, in the case of phenols, could be reversed by exposure of the fish to uncontaminated medium (Bradbury et al. 1989).

In addition, analysis of hormone levels in serum samples of white sucker (*Catostomus commersoni*) in Lake Superior has been used to complement these measurements of physiological response; significant differences in the levels of testosterone and estradiol in females and of testosterone levels in males have been associated with pollution by bleachery effluents (Munkittrick et al. 1991). It has been observed that feral fish captured from areas exposed to established contamination by polycyclic aromatic hydrocarbons, polychlorinated biphenyls, and mercury did not exhibit the increased levels of hydrocortisone normally resulting from capture (Hontela et al. 1992). These results were interpreted as showing the adverse effect on sterol metabolism in fish chronically exposed to such pollutants. Collectively, such data draw attention to the more subtle effects of exposure to xenobiotics.

A word of caution should be inserted on the difficulties of drawing unequivocal correlations between effects observed in feral fish and the levels of their exposure to putative toxicants. The data are often interpreted in terms of distance from an established point discharge — sometimes supplemented with analyses for specific toxicants (Goksøyr et al. 1991). Feral fish may, however, be exposed — simultaneously or consecutively — to a number of potential stresses which may be reflected in their physiological and biochemical response; the significance of these different factors may not always be sufficiently carefully evaluated. A good illustration of the difficulties is afforded by observations on the occurrence of skeletal deformities in smallmouth bass (*Micropterus dolomieui*) which could not be correlated with exposure to any of the toxicants known to induce such symptoms (Van Den Avyle et al. 1989); the cause of the effect was not established, and the results clearly underscore the need for a deeper understanding of fish physiology. The whole issue of associating observed biological effects with presumptive exposure to toxicants is discussed more fully in Chapter 8 (Section 8.2).

7.6.3 Chromosomal Alterations

Attention has already been briefly drawn to the use of morphological alterations of the polytene chromosomes in the salivary glands of larval *Chironomus tentans* (Bentivegna and Cooper 1993) as an assay for genotoxicity. This could be applied equally to field samples although care would have to be exercised concerning the variability in natural populations which may contain several species of *Chironomus*.

7.7 THE STRUCTURE OF THE TOXICANT AND ITS BIOLOGICAL EFFECT

There has been substantial interest in correlating the toxicity of xenobiotics with their chemical structure. Whereas it is unrealistic to expect that widely applicable principles can be formulated, it has been convincingly demonstrated that within groups of structurally related compounds, valuable correlations exist between their structure and their biological activity. This approach was pioneered by Hansch and his coworkers (Hansch and Leo 1979), and attention is drawn to its extensive application to ecotoxicology (Kaiser 1987; Hermens 1989) while brief mention has been made in Chapter 4 (Section 4.6.1) of its application to problems in biodegradability and biotransformation. Advantage may be taken not only of electronic and steric factors but also of physico-chemical parameters that may be considered significant in determining both the bioavailability of the toxicant and its mode of action. The procedure is especially effective when data for large numbers of compounds differing only in minor structural features are available, although it is clearly more limited in its application to completely novel structures. Only brief note will be made here of a few salient issues on the relation between the chemical structure of a compound and its toxicity.

1. Attention has already been drawn (Section 7.5) to the fact that a number of compounds exert their biological effect through metabolites that may be considerably more toxic than their precursors. This clearly cannot be taken into account by such procedures.
2. The toxicity may depend critically not only on the number of substituents on aromatic rings but also on their orientation. This is illustrated by data for mammalian toxicity of chlorinated dibenzo-1,4-dioxins and dibenzofurans (Kutz et al. 1990) and for PCBs (Safe 1987). Although this aspect of toxicity has been much less extensively explored for aquatic organisms, there is certainly no reason to doubt its general relevance.

7.8 A HIERARCHICAL SYSTEM FOR EVALUATING THE BIOLOGICAL EFFECTS OF TOXICANTS

It will have become clear that attempts to assess the environmental impact of a toxicant are fraught with difficulties. No single system is sufficiently embracing; no single organism is uniquely sensitive; life-cycle tests with fish and multi-species systems are complex, time-consuming, and expensive, and inherent ambiguities exist in interpreting measurements of biochemical and physiological parameters. The simplest solution is to cut the Gordian knot and take maximum advantage of each system while recognizing its limitations. The various assay procedures could be built into a hierarchical system such as the following:

Stage 1: Acute tests using, for example, algae, crustaceans, and fish.
Stage 2: Subacute tests including, for example,

- Reproduction tests using crustaceans and fish.
- Physiological/biochemical indices in fish.
- Community-induced tolerance in algae.

Stage 3: Life-cycle tests with fish including examination of pathologies.
Stage 4: Multi-component systems

- Microcosms.
- Mesocosms.

The four stages could be regarded as constituting an algorithm: for screening purposes, if a toxicant fails at any level, testing at a higher level is not justified. Conversely, if no effect is observed at a lower level, testing should be continued to the next higher level. Since, however, the cost of pre-exposure tests and multi-component systems is relatively high compared with those for other tests, at least some of these other single-species tests should be carried out simultaneously as a routine complement. For completeness, assays for carcinogenicity using either fish (Black et al. 1985) or surrogate assays using microorganisms (De Serres and Ashby 1981) may be included.

It should also be emphasized that difficulties may remain in correlating results from tests at the different stages (Stay et al. 1988). In addition, the boundaries between them should not be sharply drawn so that at Stage 2, for example, a high degree of flexibility should exist, and the various alternatives regarded as complementary rather than mutually exclusive. Application of sophisticated biochemical/physiological indices in fish, for example, might justifiably be considered an element in Stage 3 systems. It should also be pointed out that a number of additional features might be incorporated in order to provide a more comprehensive environmental hazard assessment in a given situation. Among these features, the following might, for example, be considered for inclusion:

1. Use of species relevant to a specific ecosystem.
2. Use of media with varying salinity, hardness, or pH simulating those of the receiving system.
3. Assessment of bioconcentration potential either directly, or by use of surrogate systems, which have been discussed in Chapter 3 (Section 3.1).

7.9 CONCLUSIONS

Opinions may differ widely on matters of detail, but it may be valuable to provide some general conclusions in an attempt to provide a wider perspective on this complex area.

1. Increasing attention is being directed to the sub-lethal and chronic effects of xenobiotics, and this means that assay systems are becoming increasingly sophisticated.
2. A wide range of end-points is available, and attempts should be made to extract the maximal information from the results of a toxicological assay.
3. No single assay will provide data that make it possible to predict the effect of a xenobiotic on a complex ecosystem, and serious attempts should be made to design assays that will reflect important effects of xenobiotics at the molecular level.
4. It is not practical to examine the toxic effects of all compounds in highly sophisticated assays so that a hierarchical approach may be used; the results of the simpler tests together with an assessment of the magnitude of the problem will then determine the extent to which the application of more complex assay systems is justified.

5. The toxicity of a xenobiotic cannot be dissociated from its transport into the organism and its subsequent metabolism; it is experimentally attractive to combine assays of toxicity with assessments of bioconcentration potential and metabolism.

6. The application of biochemical and physiological assays to feral fish is a valuable complement to laboratory tests for toxicity, though care should be exercised in making correlations between the observed responses and exposure to toxicants: a number of compromising factors may be involved.

REFERENCES

Ade, P., and 16 coauthors, 1984. Biochemical and morphological comparison of microsomal preparations from rat, quail, trout, mussel, and water flea. *Ecotoxicol. Environ. Saf.* 8: 423-446.

Anderson, B.S., and J.W. Hunt. 1988. Bioassay methods for evaluating the toxicity of heavy metals, biocides and sewage effluents using microscopic stages of giant kelp *Macrocystis pyrifera* (Agardh): a preliminary report. *Mar. Environ. Res.* 26: 113-134.

Andersson, T., and L. Förlin. 1992. Regulation of the cytochrome P450 enzyme system in fish. *Aquat. Toxicol.* 24: 1-20.

Andersson, T., and U. Koivusaari. 1985. Influence of environmental temperature on the induction of xenobiotic metabolism by β-naphthoflavone in rainbow trout, *Salmo gairdneri*. *Toxicol. Appl. Pharmacol.* 80: 43-50.

Andersson, T., B.-E. Bengtsson, L. Förlin, J. Härdig, and Å. Larsson. 1987. Long-term effects of bleached kraft mill effluents on carbohydrate metabolism and hepatic xenobiotic biotransformation enzymes in fish. *Ecotoxicol. Environ. Saf.* 13: 53-60.

Andersson, T., M. Pesonen, and C. Johansson. 1985. Differential induction of cytochrome P-450-dependent monoxygenase, epoxide hydrolase, glutathione transferase and UDP glucuronyltransferase activities in the liver of the rainbow trout by β-naphthoflavone or Clophen A50. *Biochem. Pharmacol.* 34: 3309-3314.

Andersson, Ö., C.-E. Linder, M. Olsson, L. Reutergård, U.-B. Uvemo, and U. Wideqvist. 1988. Spatial differences and temporal trends of organochlorine compounds in biota from the northwestern hemisphere. *Arch. Environ. Contam. Toxicol.* 17: 755-765.

Ankley, G.T., M.K. Schubauer-Berigan, and J.R. Dierkes. 1991. Predicting the toxicity of bulk sediments to aquatic organisms with aqueous test fractions: pore water vs. elutriate. *Environ. Toxicol. Chem.* 10: 1359-1366.

Bahig, M.E., A. Kraus, W. Klein, and F. Korte. 1981. Metabolism of pentachloronitrobenzene-[14]C (quintozene) in fish. *Chemosphere* 10: 319-322.

Baird, D.J., I. Barber, and P. Calow. 1990. Clonal variation in general responses of *Daphnia magna* Straus to toxic stress. I. Chronic life-history effects. *Funct. Ecol.* 4: 399-407.

Bauer, I., K. Weber, and W. Ernst. 1989. Metabolism of octachlorostyrene in the blue mussel (*Mytilus edulis*). *Chemosphere* 18: 1573-1579.

Bengtsson, B.-E. 1978. Use of a harpacticoid copepod in toxicity tests. *Mar. Pollut. Bull.* 9: 238-241.

Bengtsson, B.-E. 1980. Long-term effects of PCB (Clophen A 50) on growth, reproduction and swimming performance in the minnow *Phoxinus phoxinus*. *Water Res.* 14: 681-687.

Bengtsson, B.-E., and B. Bergström. 1987. A flowthrough fecundity test with *Nitocra spinipes* (*Harpacticoidea Crustacea*) for aquatic toxicity. *Ecotoxicol. Environ. Saf.* 14: 260-268.

Bentivegna, C.S., and K.R. Cooper. 1993. Reduced chromosomal puffing in *Chironomus tentans* as a biomarker for potentially genotoxic substances. *Environ. Toxicol. Chem.* 12: 1001-1011.

Bergman, Å., R.J. Norstrom, K. Haraguchi, H. Kuroki, and P. Béland. 1994. PCB and DDE methyl sulfones in mammals from Canada and Sweden. *Environ. Toxicol. Chem.* 13: 121-128.

Billington, J.W., G.-L. Huang, F. Szeto, W.Y. Shiu, and D. Mackay. 1988. Preparation of aqueous solutions of sparingly soluble organic substances. I. Single component systems. *Environ. Toxicol. Chem.* 7: 117-124.

Black, J.J., A. E. Maccubbin, and M. Schiffert. 1985. A reliable, efficient, microinjection apparatus and methodology for the in vivo exposure of rainbow trout and salmon embryos to chemical carcinogens. *J. Natl. Cancer Inst.* 75: 1123-1128.

Blanck, H., and S.Å. Wänberg. 1988. The validity of an ecotoxicological test system. Short-term and long-term effects of arsenate on marine phytoplankton communities in laboratory systems. *Can. J. Fish. Aquat. Sci.* 45: 1807-1815.

Blanck, H., G. Wallin, and S.-Å. Wänberg. 1984. Species-dependent variation in algal sensitivity to chemical compounds. *Ecotoxicol. Environ. Saf.* 8: 339-351.

Blanck, H., S.-Å. Wänberg, and S. Molander. 1988. Pollution-induced community tolerance — a new ecotoxicological tool. In *Functional testing of aquatic biota for estimating hazards of chemicals* pp. 219-230 (Eds. J. Cairns and J.P. Pratt). ASTM STP 988. American Society for Testing and Materials, Philadelphia.

Bradbury, S.P., T.R. Henry, G.J. Niemi, R.W. Carlson, and V.M. Snarski. 1989. Use of respiratory-cardiovascular responses of rainbow trout (*Salmo gairdneri*) in identifying acute toxicity syndromes in fish: Part 3. Polar narcotics. *Environ. Toxicol. Chem.* 8: 247-261.

Braunbeck, T., V. Storch, and H. Bresch. 1990a. Species-specific reaction of liver ultrastructure in zebrafish (*Brachydanio rerio*) and trout (*Salmo gairdneri*) after prolonged exposure to 4-chloroaniline. *Arch. Environ. Contam. Toxicol.* 19: 405-418.

Braunbeck, T., G. Görge, V. Storch, and R. Nagel. 1990b. Hepatoc steatosis in zebra fish (*Brachydanio rerio*) induced by long-term exposure to gamma-hexchlorocyclohexane. *Environ. Toxicol. Chem.* 19: 355-374.

Bresch, H., H. Beck, D. Ehlermann, H. Schlaszus, and M. Urbanek. 1990. A long-term toxicity test comprising reproduction and growth of zebrafish with 4-chloroaniline. *Arch. Environ. Contam. Toxicol.* 19: 419-427.

Bridgham, S.D. 1988. Chronic effects of 2,2'-dichlorobiphenyl on reproduction, mortality, growth, and respiration of *Daphnia pulicaria*. *Arch. Environ. Contam. Toxicol.* 17: 731-740.

Buhl, K.J., S.J. Hamilton, and J.C. Schmulbach. 1993. Acute toxicity of the herbicide bromoxynil to *Daphnia magna*. *Environ. Toxicol. Chem.* 12: 1455-1468.

Buhler, D.R., and D.E. Williams, 1988. The role of biotransformation in the toxicity of chemicals. *Aquat. Toxicol.* 11: 19-28.

Burton, G.A., and K. J. Scott. 1992. Sediment toxicity evaluations. Their niche in ecological assessments. *Environ. Sci. Technol.* 26: 2068-2075.

Cairns, J. 1984. Are single species toxicity tests alone adequate for estimating environmental hazard? *Environ. Monitor. Assess.* 4: 259-273.

Cairns, J. 1992. The threshold problem in ecotoxicology. *Ecotoxicology* 1: 3-16.

Cairns, J., P.V. McCormick, and B.R. Niederlehner. 1992. Estimating ecotoxicological risk and impact using indigenous aquatic microbial communities. *Hydrobiologia* 237: 131-145.

Calleja, M.C., and G. Persoone. 1993.The influence of solvents on the acute toxicity of some lipophilic chemicals to aquatic invertebrates. *Chemosphere* 26: 2007-2022.

Carr, R.S., J.W. Williams, and C. T.B. Ffragata. 1989. Development and evaluation of a novel marine sediment pore water toxicity test with the polychaete *Dinophilus gyrociliatus*. *Environ. Toxicol. Chem.* 8: 533-543.

Casarett, L.J., and J. Doull. 1975. *Toxicology. The basic science of poisons.* Macmillan Publishing Co., Inc., New York.

Cherr, G.N., J.M. Shenker, C. Lundmark, and K.O. Turner. 1987. Toxic effects of selected bleached kraft mill effluent constituents on sea urchin sperm cell. *Environ. Toxicol. Chem.* 6: 561-569.

Cherr, G.N., J. Shoffner-McGee, and J.M. Shenker. 1990. Method for assessing fertilization and embryonic/larval development in toxicity tests using the California mussel (*Mytilus californianus*). *Environ. Toxicol. Chem.* 9: 1137-1145.

Clark, J.R., L.R. Goodman, P.W. Borthwick, J.M. Patrick, G.M. Cripe, P.M. Moody, J.C. Moore, and E.M. Lores. 1989. Toxicity of pyrethroids to marine invertebrates and fish: a literature review and test results with sediment-sorbed chemicals. *Environ. Toxicol. Chem.* 8: 393-401.

Collier, T.K., L.C. Thomas, and D.C. Malins.1978. Influence of environmental temprature on disposition of dietary naphthalene in coho salmon (*Oncorhynchus kisutch*): isolation and identification of individual metabolites. *Comp. Biochem. Physiol.* 61C: 23-28.

Crossland, N.O., and J.M. Hillaby. 1985. Fate and effects of 3,4-dichloroaniline in the laboratory and in outdoor ponds. II. Chronic toxicity to Daphnia spp. and other invertebrates. *Environ. Toxicol. Chem.* 4: 489-499.

Cruz, S.M., M.N. Scott, and A.K. Merritt. 1993. Metabolism of [^{14}C]metolachlor in blueguill sunfish. *J. Agric. Food Chem.* 41: 662-668.

Dady, J.M., S.P. Bradbury, A.D. Hoffman, M.M. Voit, and D.L. Olson. 1991. Hepatic microsomal N-hydroxylation of aniline and 4-chloroaniline by rainbow trout (*Oncorhynchus mykiss*). *Xenobiotica* 21: 1605-1620.

DeMarini, D.M., H.G. Brooks, and D.G. Parkes. 1990. Induction of prophage lambda by chlorophenols *Environ. Mol. Mutagen.* 15: 1-9.

De Serres, F.J., and J. Ashby (Eds.). 1981. *Evaluation of short-term tests for carcinogenesis.* Elsevier/North Holland, New York.

Deutsch, W.G., E.C. Webber, D.R. Bayne, and C.W. Reed. 1992. Effects of largemouth bass stocking rate on fish populations in aquatic mesocosms used for pesticide research. *Environ. Toxicol. Chem.* 11: 5-10.

De Wolf, W., W. Seinen, and J.L.M. Hermens. 1993. Biotransformation and toxicokinetics of trichloroanilines in fish in relation to their hydrophobicity. *Arch. Environ. Contam. Toxicol.* 25: 110-117.

Dillon, T.M., D.W. Moore, and A.B. Gibson. 1993. Development of a chronic sublethal bioassay for evaluating contaminated sediment with the marine polychaete worm *Nereis* (*Neanthes*) *arenaceodentata*. *Environ. Toxicol. Chem.* 12: 589-605.

Dixon, D.G., C.E.A. Hill, PV. Hodson, E.J. Kempe, and K.L.E. Kaiser. 1985. Plasma leucine aminonaphthylamidase as an indicator of acute sublethal toxicant stress in rainbow trout. *Environ. Toxicol. Chem.* 4: 789-796.

Doherty, F.G. 1983. Interspecies correlations of acute aquatic median lethal concentration for four standard testing species. *Environ. Sci. Technol.* 17: 661-665.

Donnarumma, L., G. de Angelis, F. Gramenzi, and L. Vittozzi. 1988. Xenobiotic metabolizing enzyme systems in test fish. III. Comparative studies of liver cytosolic glutathione S-transferases. *Ecotoxicol. Environ. Saf.* 16: 180-186.

Egaas, E., and U. Varanasi. 1982. Effects of polychlorinated biphenyls and environmental temperature on in vitro formation of benzo[a]pyrene metabolites by liver of trout (*Salmo gairdneri*). *Biochem. Pharmacol.* 31: 561-566.

Fairchild, J.F., T.W. La Point, J.L. Zajicek, M.K. Nelson, F.J. Dwyer, and P.A. Lovely. 1992. Population-, community- and ecosystem-level responses of aquatic mesocosms to pulsed doses of a pyrethroid insecticide. *Environ. Toxicol. Chem.* 11: 115-129.

Fisher, S. W., and T.W. Lohner. 1987. Changes in the aqueous behavior of parathion under varying conditions of pH. *Arch. Environ. Contam. Toxicol.* 16: 79-84.

Fogels, A., and J.B. Sprague. 1977. Comparative short-term tolerance of zebrafish, flagfish, and rainbow trout to five poisons including potential reference toxicants. *Water Res.* 11: 811-817.

Förlin, L., T. Andersson, and C.A. Wachtmeister. 1989. Hepatic microsomal 4,5,6-trichloroguaiacol glucuronidation in five species of fish. *Comp. Biochem. Physiol.* 93B: 653-656.

Fox, G.A., S.W. Kennedy, RJ. Norstrom, and D.C. Wigfield. 1988. Porphyria in herring gulls: a biochemical response to chemical contamination of Great Lakes food chains. *Environ. Toxicol. Chem.* 7: 831-839.

Funari, E., A. Zoppinki, A. Verdina, G. de Angelis, and L. Vittozzi. 1987. Xenobiotic metabolizing enzyme systems in test fish. I. Comparative studies of liver microsomal monooxygenases. *Ecotoxicol. Environ. Saf.* 13: 24-31.

Gallagher, E.P., G.L. Kedders, and R.T. di Giulio. 1991. Glutathione S-transferase-mediated chlorothalonil metabolism in liver and gill subcellular fractions of channel catfish. *Biochem. Pharmacol.* 42: 139-145.

Gallagher, E.P., R.C. Cattley, and R.T. di Giulio. 1992. The acute toxicity and sublethal effects of chlorothalonil in channel catfish (*Ictalurus punctatus*). *Chemosphere* 24: 3-10.

Giesy, J.P., J.P. Ludwig, and D.E. Tillitt. 1994. Deformities in birds of the Great Lakes region. Assigning causality. *Environ. Sci. Technol.* 28: 128A-135A.

Glickman, A.H., C.N. Statham, A. Wu, and J.J. Lech. 1977. Studies on the uptake, metabolism, and disposition of pentachlorophenol and pentachloroanisole in rainbow trout. *Toxicol. Appl. Pharmacol.* 41: 649-658.

Goksöyr, A.-M Husöy, H. E. Larsen, J. Klungsöyr, S. Wilhelmsen, A. Maage, E.M. Brevik, T. Andersson, M. Celander, M. Pesonen, and L. Förlin. 1991. Environmental contaminants and biochemical responses in flatfish from the Hvaler archipelago in Norway. *Arch. Environ. Contam. Toxicol.* 21: 486-496.

Goodman, L.R., D.J. Hansen, G.M. Cripe, D.P. Middaugh, and J.C. Moore. 1985. A new early life-stage toxicity test using the Californian grunion (*Leuresthes tenuis*) and results with chlorpyrifos. *Ecotoxicol. Environ. Saf.* 10: 12-21.

Hall, T.J., R.K. Haley, and L.E. LaFleur. 1991. Effects of biologically treated bleached kraft mill effluent on cold water stream productivity in experimental stream channels. *Environ. Toxicol. Chem.* 10: 1051-1060.

Hallett, D.J., R.J. Norstrom, F.I. Onuska, and M.E. Comba. 1982. Chlorinated benzenes in Great Lakes herring gulls. *Chemosphere* 11: 277-285.

Hansch, C., and A. Leo. 1979. *Substituent constants for correlation analysis in chemistry and biology*. John Wiley and Sons Inc., New York.

Härdig J., T. Andersson, B.-E. Bengtsson, L. Förlin, and Å. Larsson.1988. Long-term effects of bleached kraft mill effluents on red and white blood cell status, ion balance, and vertebral structure in fish. *Ecotoxicol. Environ. Saf.* 15: 96-106.

Hawkins, W.E., R.M. Overstreet, and W.W. Walker. 1988. Carcinogenicity tests with small fish species. *Aquat. Toxicol.* 11: 113-128.

Hawkins, W.E., W.W. Walker, R.M. Overstreet, J.S. Lytle, and T.F. Lytle. 1990. Carcinogenic effects of some polycyclic aromatic hydrocarbons on the Japanese medaka and guppy in waterborne exposures. *Sci. Total Environ.* 94: 155-167.

Helder, T. 1980. Effects of 2,3,7,8-tetrachlorodibenzo-p-dioxin (TCDD) on early life stages of the pike (*Esox lucius L.*). *Sci. Tot. Environ.* 14: 255-264.

Hennig, H. F.-K. O. 1986. Metal-binding proteins as metal pollution indicators. *Environ. Health Perspect.* 65: 175-187.

Hermens, J.L.M. 1989. Quantitative structure-activity relationships of environmental pollutants. pp. 111-162. In *Handbook of Environmental Chemistry* (Ed. O. Hutzinger) Vol. 2E. Springer-Verlag, Berlin.

Hinga, K.R., and M.E.Q. Pilson. 1987. Persistence of benz[a]anthracene degradation products in an enclosed marine ecosystem. *Environ. Sci. Technol.* 21: 648-653.

Hodson, P.V., and B.R. Blunt. 1981. Temperature-induced changes in pentachlorophenol chronic toxicity to early life stages of rainbow trout. *Aquat. Toxicol.* 1: 113-127.

Hontela, A., J.B. Rasmussen, C. Audet, and G. Chevalier. 1992. Impaired cortisol stress response in fish from environments polluted by PAHs, PCBs, and mercury. *Arch. Environ. Contam. Toxicol.* 22: 278-283.

Houk, V.S., and D.M. DeMarini. 1987. Induction of prophage lambda by chlorinated pesticides. *Mutation Res.* 182: 193-201.

Howe, G.E., L.L. Marking, T.D. Bills, J.J. Rach, and F.L. Mayer. 1994. Effects of water temperature and pH on toxicity of terbufos, trichlorofon, 4-nitrophenol and 2,4-dinitrophenol to the amphipod *Gammarus pseudolimnaeus* and rainbow trout (*Oncorhynchus mykiss*). *Environ. Toxicol. Chem.* 13: 51-66.

Huang, K.C., and S.F. Collins. 1962. Conjugation and excretion of aminobenzoic acid isomers in marine fishes. *J. Cell. Comp. Physiol.* 60: 49-52.

Hulzebos, E.M., D.M.M. Adema, E.M. Dirven-van Breemen, L. Henzen, W.A. van Dis, H.A. Herbold, J.A. Hoekstra, R. Baerselman, and C.A.M. van Gestel. 1993. Phytotoxicity studies with *Lactuca sativa* in soil and nutrient solution. *Environ. Toxicol. Chem.* 12: 1079-1094.

Hunn, J.B., R.A. Schoettger, and W.A. Willford. 1968. Turnover and urinary excretion of free and acetylated M.S. 222 by rainbow trout, *Salmo gairdneri. J. Fish. Res. Bd. Can.* 25: 215-31.

James, M.O., and J.R. Bend. 1976. Taurine conjugation of 2,4-dichlorophenoxyacetic acid and phenylacetic acid in two marine species. *Xenobiotica* 6: 393-398.

Jarvinen, A.Q.W., B.R. Nordling, and M.E. Henry. 1983. Chronic toxicity of Dursban (Chloropyrifos) to the fathead minnow (*Pimephales promelas*) and the resultant acetylcholinesterase inhibition. *Ecotoxicol. Environ. Saf.* 7: 423-434.

Jarvinen, A.W., D.K. Tanner, and E.R. Kline. 1988. Toxicity of chlorpyrifos, endrin, or fenvalerate to fathead minnows following episodic or continuous exposure. *Ecotoxicol. Environ. Saf.* 15: 78-95.

Jimenez, B.D., L.S. Burtis, G.H. Ezell, B.Z. Egan, N.E. Lee, J.J. Beauchamp, and J.F. McCarthy. 1988. The mixed function oxidase system of bluegill sunfish, *Lepomis macrochirus*: correlation of activities in experimental and wild fish. *Environ. Toxicol. Chem.* 7: 623-634.

Johnson, B.T. 1992. Rainbow trout liver activation systems with the Ames mutagenicity test. *Environ. Toxicol. Chem.* 9: 1183-1192.

Kado, N.Y., G.N. Guirguis, C.P. Flessel, R.C. Chan, K.-I. Chang, and J.J. Wesolowski. 1986. Mutagenicity of fine (< 2.4µm) airborne particles: diurnal variation in community air determined by a Salmonella micro preincubation (microsuspension) procedure. *Environ. Mutagen.* 8: 53-66.

Kaiser, K.L. (Ed.). 1987. *QSAR in Environmental Toxicology*. Reidel Publishers, Dordrecht.

Keizer, J., G. d 'Agostino, R. Nagel, F. Gramenzi, and L. Vittozzi. 1993. Comparative diazinon toxicity in guppy and zebra fish: different role of oxidative metabolism. *Environ. Toxicol. Chem.* 12: 1243-1250.

Kirchin, M.A., A. Wiseman, and D.R. Livingstone. 1992. Seasonal and sex variation in the mixed-function oxygenase system of digestive gland microsomes of the common mussel, *Mytilus edulis* L. *Comp. Biochem. Physiol.* 101C: 81-91.

Kleinow, K.M., M.J. Melancon, and J.J. Lech. 1987. Biotransformation and induction: implications for toxicity, bioaccumulation and monitoring of environmental xenobiotics in fish. *Environ. Health Perspec.* 71: 105-119.

Kosalwat, P., and A.W. Knight. 1986a. Acute toxicity of aqueous and substrate-bound copper to the midge, *Chironomus decorus. Arch. Environ. Contam. Toxicol.* 16: 275-282.

Kosalwat, P., and A.W. Knight. 1986b. Chronic toxicity of copper to a partial life stage of the midge, *Chironomus decorus*. *Arch. Environ. Contam. Toxicol.* 16: 283-290.

Krahn, M.M., D.G. Burrows, G.M. Ylitalo, D.W. Brown, C.A. Wigren, T.K. Collier, S.-L. Chan, and U. Varanasi. 1992. Mass spectrometric analysis for aromatic compounds in bile of fish sampled after the Exxon Valdez oil spill. *Environ. Sci. Technol.* 26: 116-126.

Krause, A., W.G. Hancock, R.D. Minard, A.J. Freyer, R.C. Honeycutt, H.M. LeBaron, D.L. Paulson, S.Y. Liu, and J.M. Bollag. 1985. Microbial transformation of the herbicide metolachlor by a soil actinomycete. *J. Agric. Food Chem.* 33: 584-589.

Kutz, F.W., D.G. Barnes, E.W. Bretthauer, D.P. Bottimore, and H. Greim. 1990. The international toxicity equivalency factor (I-TEF) method for estimating risks associated with exposures to complex mixtures of dioxins and related compounds. *Toxicol. Environ. Chem.* 26: 99-109.

Landner, L., A.H. Neilson, L. Sörensen, A. Tärnholm, and T. Viktor. 1985. Short-term test for predicting the potential of xenobiotics to impair reproductive success in fish. *Ecotoxicol. Environ. Saf.* 9: 282-293.

Landner, L., H. Blanck, U. Heyman, A. Lundgren, M. Notini, A. Rosemarin, and B. Sundelin. 1989. Community testing, microcosm and mesocosm experiments: ecotoxicological tools with high ecological realism pp. 216-254. In *Chemicals in the Aquatic Environment* (Ed. L. Landner). Springer-Verlag, Berlin.

Landrum, P.F., B.J. Eadie, and W.R. Faust. 1991. Toxicokinetics and toxicity of a mixture of sediment-associated polycyclic aromatic hydrocarbons to the amphipod *Diporeia* sp. *Environ. Toxicol. Chem.* 10: 35-46.

Larsen, D.P., F. deNoyelles, F. Stay, and T. Shiroyama. 1986. Comparisons of singe-species, microcosm and experimental pond responses to atrzine exposure. *Environ. Toxicol. Chem.* 5: 179-190.

Larsson, Å., T. Andersson, L. Förlin, and J. Härdig. 1988. Physiological disturbances in fish exposed to bleached kraft mill effluents. *Water Sci. Technol.* 20 (2): 67-76.

Lindahl, P.E., S. Olofsson, and E. Schwanbom. 1976. Improved rotary-flow technique applied to cod (*Gadus morrhua* L). *Water Res.* 10: 833-845.

Lindström, M., and A. Lindström. 1980. Changes in the swimming activity of *Pontoporeia affinis* (Crustacea, Amphipoda) after exposure to sublethal concentratins of phenol, chlorophenol and styrene. *Ann. Zool. Fennici* 17: 221-231.

Little, P.J., M.O. James, J.B. Pritchard, and J.R. Bend. 1984. Benzo(a)pyrene metabolism in hepatic microsomes from feral and 3-methylcholanthrene-treated southern flounder, *Paralichthys lethostigma*. *J. Environ. Pathol. Toxicol. Oncol.* 5: 309-320.

Lozano, S.L., S.L. O'Halloran, K.W. Sargent, and J.C. Brazner. 1992. Effects of esfenvalerate on aquatic organisms in littoral enclosures. *Environ. Toxicol. Chem.* 11: 35-47.

Lundgren, A. 1985. Model ecosystems as a tool in freshwater and marine research. *Arch. Hydrobiol. Suppl. Bd.* 70: 157-196.

Malins, D.C., M.M. Krahn, M.S. Myers, L.D. Rhodes, D.W. Brown, C.A. Krone, B.B. McCain, and S.-L. Chan. 1985. Toxic chemicals in sediments and biota from a creosote-polluted harbor: relationships with hepatic neoplasms and other hepatic lesions in English sole (*Parophrys vetulus*). *Carcinogenesis* 6: 1463-1469.

Malins, D.C., B.B. McCain, D.W. Brown, M.S. Myers, M.M. Krahn, and S.-L. Chan. 1987. Toxic chemicals, including aromatic and chlorinated hydrocarbons and their derivatives, and liver lesions in white croaker (*Genyonemus lineatus*) from the vicinity of Los Angeles. *Environ. Sci. Technol.* 21: 765-770.

Maltby, L. 1992. The use of the physiological energetics of *Gammarus pulex* to assess toxicity: a study using artificial streams. *Environ. Toxicol. Chem.* 11: 79-85.

Maltby, L., and P. Calow. 1989. The application of bioassays in the resolution of environmental problems: past, present and future. *Hydrobiologia* 188/189: 65-76.

Maltby, L., C. Naylor, and P. Calow. 1990a. Effect of stress on a freshwater benthic detrivore: scope for growth in *Gammarus pulex*. *Ecotoxicol. Environ. Saf.* 19: 285-291.

Maltby, L., C. Naylor, and P. Calow. 1990b. Field deployment of a scope for growth assay involving *Gammarus pulex*, a freshwater benthic invertebrate. *Ecotoxicol. Environ. Saf.* 19: 292-300.

Maron, D.M., and B.N. Ames. 1983. Revised methods for the *Salmonella* mutagenicity test. *Mutat. Res.* 113: 173-215.

McCahon, C.P., and D. Pascoe. 1988. Use of *Gammarus pulex* (L.) in safety evaluation tests: culture and selection of a sensitive life stage. *Ecotoxicol. Environ. Saf.* 15: 245-252.

McCahon, C.P, S.F. Barton, and D. Pascoe. 1990. The toxicity of phenol to the freshwater crustacean *Aselus aquaticus* (L.) during periodic exposure — relationship between sublethal responses and body phenol concentrations. *Arch. Environ. Contam. Toxicol.* 19: 926-929.

McGee, B.L., C.E. Schlekat, and E. Reinharz. 1993. Assessing sublethal levels of sediment contamination using the estuarine amphipod *Leptocheirus plumulosus*. *Environ. Toxicol. Chem.* 12: 577-587.

McKim, J.M. 1977. Evaluation of tests with early life stages of fish for predicting long-term toxicity. *J. Fish. Res. Bd. Can.* 34: 1148-1154.

McKim, J.M., G.F. Olson, G.W. Holcombe, and E.P. Hunt. 1976. Long-term effects of methylmercuric chloride on three generations of brook trout (*Salvelinus fontinalis*): toxicity, accumulation, distribution and elimination. *J. Fish. Res. Bd. Can.* 33: 1726-2739.

Meyerhoff, R.D., D.W. Grothe, S. Sauer, and G.K. Dorulla. 1985. Chronic toxicity of tebuthiuron to an alga (*Selenastrum capricornutum*), a cladoceran (*Daphnia magna*), and the fathead minnow (*Pimephales promelas*). *Environ. Toxicol. Chem.* 4: 695-701.

Michel, X.R., P.M. Cassand, D.G. Ribera, and J.-F. Narbonne. 1992. Metabolism and mutagenic activation of benzo(a)pyrene by subcellular fractions from mussel (*Mytilus galloprovincialis*) digestive gland and sea bass (*Discenthrarcus labrax*) liver. *Comp. Biochem. Physiol.* 103C: 43-51.

Monosson, E., and J.J. Stegeman. 1991. Cytochrome P450E (P4501A) induction and inhibition in winter flounder by 3,3',4,4'-tetrachlorobiphenyl: comparison of response in fish from Georges Bank and Narragansett Bay. *Environ. Toxicol. Chem.* 10: 765-774.

Moreau, P., A. Bailone, and R. Devoret. 1976. Prophage λ induction in *Escherichia coli* K12 *env A urv B*: a highly sensitive test for potential carcinogens. *Proc. Natl. Acad. Sci. U.S.A.* 73: 3700-3704.

Muir, D.C.G., G.P. Rawn, and N.P. Grift. 1985. Fate of the pyrethroid insecticide deltamethrin in small ponds: a mass balance study. *J. Agric. Food Chem.* 33: 603-609.

Muir, D.C.G., A.L. Yarechewski, and A. Knoll. 1986. Bioconcentration and disposition of 1,3,6,8-tetrachlorodibenzo-*p*-dioxin and octachlorodibenzo-*p*-dioxin by rainbow trout and fathead minnows. *Environ. Toxicol. Chem.* 5: 261-272.

Munkittrick, K.R., C.B. Portt, G.J. Van Der Kraak, I.R. Smith, and D.A. Rokosh. 1991. Impact of bleached kraft mill effluent on population characteristics, liver MFO activity, and serum steroid levels of a Lake Superior white sucker (*Catostomus commersoni*) population. *Can. J. Fish. Aquat. Sci.* 48: 1371-1380.

Nagel, R., H. Bresch, N. Caspers, P.D. Hansen, M. Markert, R. Munk, N. Scnholz, and B.B. ter Höfte. 1991. Effect of 3,4-dichloroaniline on the early life stages of the zebrafish (*Brachydanio rerio*): results of a comparative laboratory study. *Ecotoxicol. Environ. Saf.* 21: 157-164.

Nebeker, A.V., M.A. Cairns, J.H. Gakstatter, K.W. Malueg, G.S. Schuytema, and D.F. Krawczyk. 1984. Biological methods for determining toxicity of contaminated freshwater sediments to invertebrates. 3: 617-630.

Nebeker, A.V., S.T. Onjukka, and M.A. Cairns. 1988. Chronic effects of contaminated sediment on *Daphnia magna* and *Chironomus tentans*. *Bull. Environ. Contam. Toxicol.* 41: 574-581.

Nebeker, A.V., G.S. Schuytema, W.L. Griffis, J.A. Barbitta, and L.A. Carey. 1989. Effect of sediment organic carbon on survival of *Hyalella azteca* exposed to DDT and endrin. *Environ. Toxicol. Chem.* 705-718.

Neilson, A.H., and T. Larsson. 1980. The utilization of organic nitrogen for growth of algae: physiological aspects. *Physiol. Plant.* 48: 542-553.

Neilson, A.H., A.-S. Allard, S. Reiland, M. Remberger, A. Tärnholm, T. Viktor, and L. Landner. 1984. Tri- and tetra-chloroveratrole, metabolites produced by bacterial O-methylation of tri- and tetra-chloroguaiacol: an assessment of their bioconcentration potential and their effects on fish reproduction. *Can. J. Fish. Aquat. Sci.* 41: 1502-1512.

Neilson, A.H., H. Blanck, L. Förlin, L. Landner, P. Pärt, A. Rosemarin, and M. Söderström. 1989. Advanced hazard assessment of 4,5,6-trichloroguaiacol in the Swedish Environment. In *Chemicals in the Aquatic Environment*, pp. 329-374. Ed. L. Landner, Springer-Verlag, Berlin.

Neilson, A.H., A.-S. Allard, S. Fischer, M. Malmberg, and T. Viktor. 1990. Incorporation of a subacute test with zebra fish into a hieracrchical system for evaluating the effect of toxicants in the aquatic environment. *Ecotoxicol. Environ. Saf.* 20: 82-97.

Niederlehner, B,.R., K.W. Pontasch, J.R. Pratt, and J. Cairns. 1990. Field evaluation of predictions of environmental effects from a multispecies-microcosm toxicity test. *Arch. Environ. Contam. Toxicol.* 19: 62-71.

Niimi, A.J., H.B. Lee, and G.P. Kissoon. 1989. Octanol/water partition coefficients and bioconcentration factors of chloronitrobenzenes in rainbow trout (*Salmo gairdneri*). *Environ. Toxicol. Chem.* 8: 817-823.

Nimmo, I.A. 1987. The glutathione S-transferases of fish. *Fish Physiol. Biochem.* 3: 163-172.

Norberg, T.J., and D.I. Mount. 1985. A new fathead minnow (*Pimephales promelas*) subchronic toxicity test. *Environ. Toxicol. Chem.* 4: 711-718.

Norberg-King, T.J. 1989. An evaluation of the fathead minnow seven-day subchronic test for estimating chronic toxicity. *Environ. Toxicol. Saf.* 8: 1075-1089.

Nosek, J.A., S.R. Craven, J.R. Sullivan, J.R. Olsen, and R.E. Peterson. 1992. Metabolism and disposition of 2,3,7,8-tetrachlorodibenzo-*p*-dioxin in ring-necked pheasant hens, chicks and eggs. *J. Toxicol. Environ. Health* 35: 153-164.

Nyholm, N. 1985. Response variable in algal growth inhibition tests — biomass or growth rate? *Water Res.* 19: 273-279.

Oikari, A., and T. Kunnamo-Ojala. 1987. Tracing of xenobiotic contamination in water with the aid of fish bile metabolites: a field study with caged rainbow trout (*Salmo gairdneri*). *Aquat. Pollut.* 9: 327-341.

Olson, L.E., J.L. Allen, and J.W. Hogan. 1977. Biotransformation and elimination of the herbicide dinitramine in carp. *J. Agric. Food Chem.* 25: 554-556.

Opperhuizen, A., and D.T.H.M. Sijm. 1990. Bioaccumulation and biotransformation of polychlorinated dibenzo-*p*-dioxins and dibenzofurans in fish. *Environ. Toxicol. Chem.* 9: 175-186.

Ottoboni, M.A. 1984. *The dose makes the poison.* Vincente Books, Berkeley, California.

Peters, R. 1952. Lethal synthesis. *Proc. Roy. Soc.* (London) B 139: 143-170.

Petersen, L. B.-M., and R.C. Petersen. 1984. Effect of kraft pulp mill effluent and 4,5,6-trichloroguaiacol on the net spinning behavior of *Hydropsyche angustipennis* (Trichoptera). *Ecol. Bull.* 36: 68-74.

Phipps, G.L., G.T. Ankley, D.A. Benoit, and V.R. Mattson. 1993. Use of the aquatic oligochaete Lumbriculus variegatus for assessing the toxicitiy and bioaccumulation of sediment-associated contaminants. *Environ. Toxicol. Chem.* 12: 269-279.

Plewa, M.J., T. Gichner, H. Xin, K.-Y. Seo, S.R. Smith, and E.D. Wagner. 1993. Biochemical and mutagenic characterization of plant-activated aromatic amines. *Environ. Toxicol. Chem.* 12: 1353-1363.

Pontasch, K.W., B.R. Niederlehner, and J. Cairns. 1989. Comparison of singe-species microcosm and field responses to a complex effluent. *Environ. Toxicol. Chem.* 8: 521-532.

Remberger, M., P.-Å. Hynning, and A.H. Neilson. 1991. 2,5-dichloro-3,6-dihydroxybenzo-1,4-quinone: identification of a new organochlorine compound in kraft mill bleachery effluents. *Environ. Sci. Technol.* 25: 1903-1907.

Reynoldson, T.B., S.P. Thompson, and J.L. Bamsey. 1991. A sediment bioassay using the tubificid oligochaete worm *Tubifex tubifex*. *Environ. Toxicol. Chem.* 10: 1061-1072.

Rodricks, J.V. 1992. *Calculated risks.* Cambridge University Press, Cambridge.

Rosemarin, A., J. Mattsson, K.-J. Lehtinen, M. Notini, and E. Nylén. 1986. Effects of pulp mill chlorate (ClO_3^-) on *Fucus vesiculosus* — a summary of projects. *Ophelia Suppl.* 4: 219-224.

Rosenthal, H., and D.F. Alderdice. 1976. Sublethal effects of environmental stressors, natural and pollutional, on marine fish eggs and larvae. *J. Fish Res. Bd. Can.* 33: 2047-2065.

Safe, S. (Ed.) 1987. *Mammalian Biologic and Toxic Effects of PCBs.* Springer-Verlag, Berlin.

Sandhu, S.S., B.S. Gill, B.C. Casto, and J.W. Rice. 1991. Application of *Tradescantia micronucleus* assay for in situ evaluation of potential genetic hazards from exposure to chemicals at a wood-preserving site. *Hazardous Waste Hazardous Materials.* 8: 257-262.

Schäfers, C., and R. Nagel. 1991. Effects of 3,4-dichloroaniline on fish populations. Comparison between r- and K-strategists: a complete life cycle test with the guppy (*Poecilia reticulata*). *Arch. Environ. Contam. Toxicol.* 21: 297-302.

Schäfers, C., and R. Nagel. 1993. Toxicity of 3,4-dichloroaniline to perch (*Perca fluviatilis*) in acute and early stage life stage exposures. *Chemosphere* 26: 1641-1651.

Scharenberg, W. 1991. Cormorants (*Phalacrocorax carbo sinensis*) as bioindicators for polychlorinated biphenyls. *Arch. Environ. Contam. Toxicol.* 21: 536-540.

Schlenk, D., and D.R. Buhler. 1991. Role of flavin-containing monooxygenase in the in vitro biotransformation of aldicarb in rainbow trout (*Oncorhynchus mykiss*). *Xenobiotica* 21: 1583-1589.

Shenker, J.M., and G.N. Cherr. 1990. Toxicity of zinc and bleached kraft mill effluent to larval English sole (*Parophrys vetulus*) and topsmelt (*Atherinops affinis*). *Arch. Environ. Contam. Toxicol.* 19: 680-685.

Snell, T.W., and B.D. Moffat. 1992. A 2-d life cycle test with the rotifer *Brachionus calyciflorus*. *Environ. Toxicol. Chem.* 11: 1249-1257.

Southworth, G.R., C.C. Keffer, and J.J. Beauchamp. 1980. Potential and realized bioconcentration. A comparison of observed and predicted bioconcentration of azaarenes in the fathead minnow (*Pimephales promelas*). *Environ. Sci. Technol.* 14: 1529-1531.

Spitsbergen, J.M., J.M. Kleeman, and R.E. Peterson. 1988. Morphological lesions and acute toxicity in rainbow trout (*Salmo gairdneri*) treated with 2,3,7,8-tetrachlorodibenzo-p-dioxin. *J. Toxicol. Environ. Health* 23: 333-358.

Sprague, J.B. 1971. Measurement of pollutant toxicity to fish — III. Sublethal effects and "safe" concentrations. *Water Res.* 5: 245-266.

Stay, F.S., A. Kateko, C.M. Rohm, M.A. Fix, and D.P. Larsen. 1988. Effects of fluorene on microcosms developed from four natural communities. *Environ. Toxicol. Chem.* 7: 635-644.

Steadman, B.L., A.M. Farag, and H.L. Bergman. 1991. Exposure-related patterns of biochemical indicators in rainbow trout exposed to no. 2 fuel oil. *Environ. Toxicol. Chem.* 10: 365-374.

Sundelin, B. 1983. Effects of cadmium on *Pontoporeia affinis* (Crustacea: Amphipoda) in laboratory soft-bottom microcosms. *Mar. Biol.* 74: 203-212.

Suter, G.W., A. E. Rosen, E. Linder, and D.F. Parkhurst. 1987. End-points for responses of fish to chronic toxic exposures. *Environ. Toxicol. Chem.* 6: 793-809.

Svanberg, O., and L. Renberg. 1988. Biological-chemical characterization of effluents for the evaluation of the potential impact on the aquatic environment pp. 244-255. In *Organic micropollutants in the aquatic environment* (Eds. G. Angeletti and A. Björseth). Kluwer Academic Publishers, Dordrecht.

Swartz, R.C., P.F. Kemp, D.W. Schults, G.R. Ditsworth, and R.J Ozretich. 1989. Acute toxicity of sediment from Eagle Harbor, Washington, to the infaunal amphipod *Rhepoxynius abronius. Environ. Toxicol. Chem.* 8: 215-222.

Sweeney, B.W., D.H. Funk, and L.J. Standley. 1993. Use of the stream mayfly *Cloeon triangulifer* as a bioasssy organism: life history response and body burden following exposure to technical chlordane. *Environ. Toxicol. Chem.* 12: 115-125.

Taraldsen, J.E., and T.J. Norberg-King. 1990. New method for determining effluent toxicity using duckweed (*Lemna minor*). *Environ. Toxicol. Chem.* 9: 761-767.

Taylor, E.J., S.J. Maund, and D. Pascoe. 1991. Toxicity of four common pollutants to the freshwater macroinvertebrates *Chironomus riparius* (Meigen) (Insecta: diptera) and *Gammarus Pulex* (L.) (Crustacea: amphipoda). *Arch. Environ. Contam. Toxicol.* 21: 371-376.

Taylor, E.J., S.J. Blockwell, S.J. Maund, and D. Pascoe. 1993. Effects of lindane on the life-cycle of a freshwater macroinvertebrate *Chironomus riparius Meigen* (Insecta: diptera). *Arch. Environ. Contam. Toxicol.* 24: 145-150.

Tsushimoto, G., F. Matsumara, and R. Sago. 1982. Fate of 2,3,7,8-tetrachlorodibenzo-*p*-dioxin (TCDD) in an outdoor pond and in model aquatic ecosystems. *Environ. Toxicol. Chem.* 1: 61-68.

Van Den Avyle, M.J., S.J. Garvick, V.S. Blazer, S.J. Hamilton, and W.G. Brumbaugh. 1989. Skeletal deformities in smallmouth bass, (*Micropterus dolomieui*) from southern Appalachian reservoirs. *Arch. Environ. Contam. Toxicol.* 18: 688-696.

Van der Welden, M.E.J., J. van der Kolk, R. Bleuminck, W. Seinen, and M. van der Berg. 1992. Concurrence of P450 1A1 induction and toxic effects after administration of a low dose of 2,3,7,8-tetrachlorodibenzo-*p*-dioxin (TCDD) to the rainbow trout (*Oncorhynchus mykiss*). *Aquat. Toxicol.* 24: 123-142.

Van Urk, F.C.M. Kerkum, and H. Smit. 1992. Lide cycle patterns, density, and frequency of deformities in *Chironomus* larvae (Diptera: Chironomidae) over a contaminated sediment gradient. *Can. J. Fish. Aquat. Sci.* 49: 2291-2299.

Varanasi, U., J.E. Stein, M. Nishimoto, W.L. Reichert, and T.K. Collier. 1987. Chemical carcinogenesis in feral fish: uptake, activation, and detoxication of organic xenobiotics. *Environ. Health Perspect.* 71: 155-170.

Veith, G.D., and V.M. Comstock. 1975. Apparatus for continuously saturating water with hydrophobic organic chemicals. *J. Fish Res. Bd. Can.* 32: 1849-1851.

Wachtmeister, C.A., L. Förlin, K.C. Arnoldsson, and J. Larsson. 1991. Fish bile as a tool for monitoring aquatic pollutants: studies with radioactively labeled 4,5,6-trichloroguaiacol. *Chemosphere* 22: 39-46.

Walker, C.H. 1983. Pesticides and birds — mechanisms of selective toxicity. *Agric. Ecosyst. Environ.* 9: 211-226.

Walker, M.K., and R.E. Peterson. 1991. Potencies of polychlorinated dibenzo-p-dioxin, dibenzofuran, and biphenyl congeners, relative to 2,3,7,8-tetrachlorodibenzo-p-dioxin, for producing early life stage mortality in rainbow trout (*Oncorhynchus mykiss*). *Aquat. Toxicol.* 21: 219-238.

Walker, M.K., L.C. Hufnagle, M.K. Clayton, and R.E. Peterson. 1992. An egg injection method for assessing early life stage mortality of polychlorinated dibenzo-*p*-dioxins, dibenzofurans, and biphenyls in rainbow trout (*Oncorhynchus mykiss*). *Aquat. Toxicol.* 22: 15-38.

Walsh, G.E., M.J. Yoder, L.L. McLaughlin, and E.M. Lores. 1987. Responses of marine unicellular algae to brominated organic compounds in six growth media. *Ecotoxicol. Environ. Saf.* 14: 215-222.

Wang, W. 1987. Root elongation method for toxicity testing of organic and inorganic pollutants. *Environ. Toxicol. Chem.* 6: 409-414.

Wang, W. 1991. Literature review on higher plants for toxicity testing. *Water, Air, Soil Pollut.* 59: 381-400.

Wei, L.Y., and C.R. Vossbrinck. 1992. Degradation of alachlor in chironomid larvae (Diptera: Chironomidae). *J. Agric. Food Chem.* 40: 1695-1699.

Wiederholm, T., A.-M. Wiederholm, and G. Milbrink. 1987. Bulk sediment bioassays with five species of fresh-water oligochaetes. *Water, Air, Soil Pollut.* 36: 131-154.

Winner, R.W. 1988. Evaluation of the relative sensitivities of a 7-d *Daphnia magna* and *Ceriodaphnia dubia* toxicity tests for cadmium and sodium pentachlorophenolate. *Environ. Toxicol. Chem.* 7: 153-159.

Woods, P.E., J.D. Paulauskis, L.A. Weight, M.A. Romano, and S.I. Guttman. 1989. Genetic variation in laboratory and field populations of the midge, *Chironomus tentans* Fab.: Implications for toxicology. *Environ. Toxicol. Chem.* 8: 1067-1074.

Zieris, F.-J., D. Feind, and W. Huber 1988. Long-term effects of 4-nitrophenol in an outdoor synthetic aquatic system. *Arch. Environ. Contam. Toxicol.* 17: 165-175.

CHAPTER 8

Environmental Hazard Assessment

SYNOPSIS

A number of alternative procedures for carrying out hazard assessment of organic compounds discharged into the aquatic environment are outlined, and inherent limitations due to the fact that some questions inevitably belong to the realm of trans-science are briefly noted. A degree of acceptable risk is, therefore, probably unavoidable. Two essentially different procedures are outlined — retroactive and predictive. In predictive systems, account should be taken of basic data on persistence, partition, and toxicity. Some of the necessary components are outlined, and examples are given of evaluations using various strategies and having different objectives. An attempt is made to illustrate the problem of achieving a numerically satisfying overall assessment; it is, therefore, emphasized that such evaluations are relative rather than absolute. Retroactive procedures based on the principles of epidemiology are briefly outlined, and the inherent limitations are pointed out. The purpose of these procedures is to explore — as far as is realistic — without prejudice all the possibilities and to produce a balanced judgment. The cardinal significance of appreciating that it is nearly always impossible to establish a single cause for perturbations in natural populations is underscored. The valuable concept of toxicity equivalent factors for application to mixtures of toxicants is examined, and the application of critical loading to organic xenobiotics is analyzed. Attention is drawn to its apparent neglect of imperfections in current understanding of the regulation of natural ecosystems and of the cardinal importance of chronic and sub-lethal effects. The limitations of different approaches to environmental hazard evaluation are illustrated with brief accounts of four case studies. Effort should be made to take into account as many as possible of the relevant factors and to evaluate possible alternatives in ecoepidemiological investigations.

INTRODUCTION

The material presented in the preceding chapters is not merely a series of disconnected essays on diverse topics: collectively, the content of these chapters should be regarded as providing the basic understanding required to provide a rational and comprehensive evaluation of the hazard associated with the discharge of an organic compound into the aquatic environment.

It is desirable at the beginning to clarify some terms to avoid possible misunderstanding. There is very little real difference in meaning between the terms "risk" and "hazard": the former deriving possibly from the Greek for a "cliff", and the readily perceived danger of venturing too near the edge appears to be less problematic than the latter with its overtone of tendency or potential. The term "hazard" will be used here. The term "impact" seems to convey the correct impression of an effect that is not necessarily adverse, whereas the term "insult", sometimes used in clinical terminology, seems overloaded with negative implications and is probably best avoided in the present context. The same principle applies to the use of the word "pollutant" that has a definitely negative connotation; the neutral term "xenobiotic" has, therefore, been used throughout this book.

There are two broad strategies for assessing environmental hazard, and these differ in important respects. One attempts a predictive evaluation, whereas the other is essentially retroactive; the basic construction of these procedures is discussed in the next two sections. There are several books (Fischhoff et al.1981; Ottoboni 1984; Lewis 1990; Rodricks 1992) devoted to the general area of hazard assessment, and although these are generally directed to human exposure and vary considerably both in subject matter, in perspective, and in depth, they offer valuable background to the area. They provide many illustrative examples that will be familiar to most readers and that can readily be appreciated without expert knowledge.

8.1 HAZARD EVALUATION AS A PREDICTIVE EXERCISE

The primary objective may be to assess the effects resulting from the discharge of a given compound (or group of compounds) into the environment, and it is clearly desirable that this evaluation be carried out before discharge begins; this is, therefore, an exercise in prediction. If the magnitude of possible adverse effects is deemed significantly great, the use and discharge of the compounds may be limited or even totally prohibited. It should be emphasized at the outset, however, that virtually no prediction can be made with absolute certainty; evaluations will therefore be *relative* rather than *absolute,* and it may be necessary to adopt the concept of "acceptable (or unacceptable) hazard".

In the following paragraphs, specific procedures for predictive assessment will be outlined. Since most of the details have been discussed in previous chapters, reference will be made to the appropriate sections of these rather than to the literature. It should be appreciated, however, that the discussion takes into account only strictly scientific issues, although some more philosophical ones are briefly discussed in Section 8.4. This statement should not, however, be taken to imply that legislative or socio-political, or even ethical and moral aspects may not be of cardinal importance, although such considerations lie beyond the scope of the present study.

In predictive environmental hazard evaluations, it is customary to make use of data on the three basic properties of a substance: (1) its persistence, (2) its toxicity, and (3) its distribution among environmental matrices. These questions may be addressed by providing the results of experiments on ready biodegradability using conventional procedures that have been outlined in Chapter 5 (Section 5.1), on acute toxicity by methods that have been given in Chapter 7 (Sections 7.2 and 7.3), and of estimates of bioconcentration potential often using surrogate procedures such as those that have been discussed in Chapter 3 (Section 3.1.2). Collectively, this would provide the basis for what may conveniently be termed an *initial hazard assessment*. The limitations of such evaluations are, however, being increasingly realized, and a suite of data from progressively sophisticated experiments is being proposed and increasingly accepted; this exercise constitutes what may be termed an *advanced hazard assessment,* and it is to these procedures that this book has been largely devoted. The concept of a hierarchical system that was suggested for toxicity testing in Chapter 7 (Section 7.7) could be usefully applied to all the individual components that will be discussed in Sections 8.1.1, 8.1.2, and 8.1.3. Varying levels of sophistication may be involved, though no attempt is made here to define in detail how comprehensive a given investigation should be; this must be determined by the specific objectives and will often of necessity take into account economic and socio-political constraints.

In order to accomplish these objectives effectively, an integrated strategy is possible since several of the necessary investigations interface closely with others, for example, partition with both degradability and toxicity. Most important, it cannot be emphasized too strongly that all the investigations must have access to the necessary analytical support and that this should be incorporated into the overall strategy during initial planning of the exercise.

It is appropriate to examine in some detail each of the components outlined above. In doing so, it has been assumed that the compounds — as is the case for most of current interest — have passed the initial screening designated as an initial hazard assessment, so that attention will be directed only to those additional considerations that should be incorporated into an advanced hazard assessment. Much fuller details of all aspects of these procedures have been given in previous chapters.

8.1.1 Persistence

The single most important feature that should be incorporated into the design of experiments on microbial degradation and transformation is a maximal degree of environmental realism. For example, inocula for laboratory studies of biodegradation and biotransformation should be taken from the environment to which the results are to be applied. Either selective enrichment that has traditionally been applied to investigating metabolic pathways, or multi-component microcosms, or mesocosms may be used. A number of important determinants should be incorporated into all of these experiments, and these need only be summarized briefly since a more detailed discussion has been given in Chapter 5 that provides details of the relevant experimental procedures.

Oxygen concentration — Experiments should encompass both aerobic and anaerobic conditions, since the latter prevail in many sediments contaminated with xenobiotics, many of which are partitioned into the sediment phase from the aquatic phase into which they were initially discharged. In addition, attention should be devoted to some important details in the design of experiments under anaerobic conditions.

1. If the environment is contaminated with appreciable concentrations of nitrate, experiments under anaerobic conditions using this as electron acceptor would be appropriate. As has already been noted in Chapter 4 (Section 4.3.3), however, attention should also be directed to obligately nitrate-dependent anaerobic organisms.
2. If the environment is a brackish water or marine ecosystem, the role of sulfate concentration should be evaluated, and appropriate attention should be directed to the role of anaerobic sulfate-reducing bacteria whose metabolic versatility has been discussed in various parts of Chapters 5 and 6.

Substrate concentration — Experiments should be conducted at substrate concentrations approaching as closely as possible those in the system to which discharge is made. The lower limit will, of course, be determined by the sensitivity of the analytical procedures adopted, and if possible three concentrations such as 10, 100, and 1000 µg/L may be examined with advantage. If radiolabelled substrates are available, considerably lower concentrations may be accessible and should be employed.

Biotransformation — Effort should be devoted to exploring the possible synthesis of metabolites which may or may not be persistent during the experiment since these transformation products may display toxicity both to the organism producing them as well as to other organisms in ecosystems; some examples have been given in Chapter 6 (Section 6.11.5). The mechanistic arguments set forth in Chapter 6 may be used as guidelines to predict possible metabolites from compounds with structures which have been previously examined; otherwise, an understanding of the basic biochemical mechanisms supplemented with the ingenuity of the investigator will have to be relied upon. Both the identification and the quantification of metabolites is strongly recommended even though this may represent a substantial increase in the level of necessary analytical support.

Bioavailability — It is important to appreciate an intrinsic limitation in the design of many laboratory experiments; in these, the substrate will almost always be dissolved in the test medium or at least accessible as a suspension. As has been emphasized repeatedly in Chapters 3 and 4, however, this situation may not prevail in natural ecosystems where xenobiotics may be associated with organic or mineral components of the water and sediment phases. This has been discussed in some detail in Chapter 3 (Section 3.2) and in Chapter 4 (Section 4.5.3). As a result of this restricted bioavailability, laboratory experiments may, therefore, seriously underestimate the persistence of xenobiotics in natural environments.

8.1.2 Partition

Traditionally, attention has been directed almost exclusively to the uptake and concentration of xenobiotics in biota — particularly fish. The discussions in Chapter 3 on partitioning justify, however, a wider circumscription of partitioning to include (1) concentration from the aquatic phase into biota, (2) partitioning on to particulate matter in the water phase and into the sediment phase, and (3) partitioning from the aquatic phase into the atmosphere. Thereby, partitioning between all the relevant phases would be taken into account, and this extension of the partition concept appears to be justified for a number of reasons:

1. The uptake of xenobiotics by biota occurs both via the water phase (bioconcentration) and by ingestion of particulate matter (bioaccumulation), and for some compounds the latter is the more important (Chapter 3, Section 3.1.1).
2. The toxicity of xenobiotics is determined by the extent to which the compound is biologically accessible, and this includes both "free" and "bound" forms (Chapter 3, Section 3.2.2).
3. The recalcitrance of xenobiotics is affected by the degree of their accessibility to the relevant microorganisms, and this should address the role of organic matter which may influence their bioavailability (Chapter 4, Section 4.5.3).
4. The transport of xenobiotics originally discharged into the aquatic phase may take place via the atmosphere and result in their dissemination into areas remote from the point of original discharge (Chapter 3, Section 3.5.3).

Even in monitoring programs, assessment of the concentration of xenobiotics in environmental matrices should take into account both "free" and "bound" residues and include transformation products formed by both abiotic (Chapter 4, Section 4.1) and biotic reactions (Chapter 6).

Assessment of partitioning between environmental phases may be carried out at different levels of sophistication. The assembly of basic physico-chemical data for a compound is invaluable as a general guide to partitioning and may comprise, for example, data on water solubility, vapor pressure, boiling point, or melting point. As an alternative to direct assessment of partitioning, surrogate procedures that have been outlined in Chapter 3 (Section 3.1.2 and Section 3.2.1) have been widely used in initial hazard evaluations. For an advanced assessment, however, they should be supplemented by direct measurements of water/sediment partition using samples from the area under investigation and direct evaluation of bioconcentration and (bioaccumulation) potential into relevant biota such as fish. It should also be underscored that the appropriate measurements should not be restricted to the initial substances under evaluation but should also include metabolites revealed by studies on microbial degradation and transformation.

Experiments on water/sediment partition are readily carried out, and it is valuable to supplement these with data on the degree of reversibility of this process over an extended period of time even though this may involve entry into the controversial area of "free" and "bound" or "associated" xenobiotics. There can, however, be little doubt either that associations of different degrees of reversibility do, in fact, exist or of their critical relevance to bioavailability and, hence, both to persistence and to toxicity.

The design of experiments on bioconcentration presents a number of technical issues upon which universal agreement is unlikely to be reached. Among these, at least two are of cardinal significance: (1) the choice of organism and (2) the degree to which metabolic studies should be incorporated. The choice of fish is probably best left to the laboratory carrying out the study; they have most experience of the test organism, its feeding habits, its relevance to a given environment, and how to avoid compromising stress during the experiments. In practice, the following fish have probably been most widely used: rainbow trout (*Oncorhynchus mykiss*), zebra fish (*Brachydanio rerio*), guppy (*Poecilia reticulata*), and fathead minnow (*Pimephales promelas*); but any of those used for toxicity testing and which have been noted in Chapter 7 (Section 7.3.1) are clearly applicable, and all of them have been used.

Experiments should be carried out for a sufficient length of time so that an apparent equilibrium is reached, and a post-exposure period in the absence of the toxicant to study elimination is a valuable supplement. For compounds with adequate solubility in water, aqueous solutions should be used since the use of organic solvents probably does not generally result in true solution, and they may introduce additional complications that have been noted in Chapter 7 (Section 7.2.1). If the ecosystem to which the results are to be applied is rich in organic matter, either suspended or dissolved, it may be valuable to carry out comparable experiments under these conditions in order to evaluate the significance of bioaccumulation. In some cases elimination of a xenobiotic is accomplished almost exclusively by metabolism (Welling and de Vrise 1992), so that it is highly desirable to examine both the medium and the fish for the presence of metabolites. These matrices will, in any case, be analyzed for the test substance although differences in the properties of the metabolites — and particularly water-soluble conjugates — will almost always be different from those of the initial substrate, so that specific analytical procedures will generally have to be developed.

8.1.3 Toxicity

It is generally recognized that the toxicity of a xenobiotic should be evaluated using a spectrum of organisms which comprises different trophic levels: for example, representatives of algae, crustaceans, and fish. In addition, interest in sediment toxicity suggests the inclusion of any of a wide range of sediment-dwelling organisms including insect larvae. It should always be borne in mind that the development of a new test system may reveal hitherto unexpected types and degrees of toxicity, so that current assessment norms should be regarded as provisional and possibly conservative.

Strenuous attempts have been made to introduce sets of established test organisms and standardized procedures: this is justified by the desire to facilitate the comparison of results obtained under essentially identical conditions. The underlying assumption that differences in the sensitivity to a given toxicant within a given taxonomic group, for example, fish or crustaceans, are not generally very great is not necessarily true; attention has already been drawn in Chapter 7 (Section 7.3.1) to the case of algae which differ widely not only morphologically and biochemically but also in their response towards the same toxicant. A number of important considerations should, therefore, be clearly kept in mind in choosing and evaluating the results from the assay system.

1. The use of assays for acute toxicity using fish belongs to the historical past; the concentration of a toxicant that kills 50% of the population is no longer an environmentally acceptable measurement of toxic effects on natural ecosystems. Interest has, therefore, been increasingly directed to tests for non-lethal toxicity and to the effects of chronic exposure. This has been discussed in greater detail in Chapter 7 (Section 7.3.1), and the only limitation to their wider application lies in the experimental complexities of some of the test systems. From a long-term ecological point of view, the value of tests for assessing success in reproduction — including full life-cycle tests — can hardly be exaggerated.

2. The effect of pre-exposure of the test organism to low levels of the toxicant before its evaluation in a standard test has been examined, and as in the case for experiments on bioconcentration, the possibility that detoxification of the test substance occurs after transfer to toxicant-free medium is a valuable adjunct and is readily incorporated into the test protocol. This is pragmatically — though not strictly accurately — covered by the term "reversibility" of toxicity; "recovery" would be a more appropriate term.

3. Attention is directed to the application of end-points other than those that have traditionally been used. Their choice is dependent both on the test organism and on the test protocol; the application of end-points as diverse as scope-for-growth, energy metabolism, and behavior such as swimming in fish has been noted in Chapter 7 (Sections 7.2.2 and 7.3.1).

4. The importance of carrying out analysis for the concentration of the toxicant during the test period seems elementary but does not appear always to have been carried out. Such analyses reveal possible instability of the test substance and may readily be extended to identifying and quantifying the transformation products; an illustrative example has been given in which the test substance (tetrachlorobenzo-1,2-quinone) had a half-life in the assay medium of < 0.5 h and where the toxicity could be attributed to one of the transformation products (Remberger et al. 1991).

All of these are fairly self-evident factors that can, at least in principle, be taken into account in laboratory-based systems. There are, however, at least four others which are less readily incorporated into laboratory assay systems but which may, nonetheless, be of considerable ecological importance:

1. The route by which biota are exposed to the xenobiotic; this has been discussed in Chapter 3 (Section 3.1.1) and in Chapter 7 (Section 7.2.1).

2. The possible adverse effect of a xenobiotic on non-target organisms.

3. The possibility that effects not observable at concentrations used in current assays may be of environmental significance.

4. The unresolved issue of the existence of threshold concentrations below which toxic effects are not manifested, and the more controversial issue of "sufficient challenge" (Smyth 1967). This is the phenomenon whereby a *positive* effect of a toxicant is compared with the control group observed at low doses of the toxicant and which might plausibly be attributed to the stimulation of defense mechanisms.

These factors are difficult to evaluate so that they are not always taken into account in environmental hazard evaluations; indeed, some of them illustrate the limitations of conventional toxicological procedures. In addition, fundamental objections have been raised against the application of conventional bioassays to predictive assessment. A valuable critique has been provided (Maltby and Calow 1989) in which the limitations of conventional approaches were carefully delineated. Briefly, it was suggested that the intrinsic limitations of inductive procedures make these of low predictive value except in restricted circumstances. A plea was therefore made for the adoption of a mechanistic view for assessing the effects of perturbations on biological systems at increasingly higher trophic levels. Three important considerations were put forward for examination:

1. The construction of test systems of increasing complexity so that they more nearly resemble natural systems although the limitations of this procedure are fully appreciated. This is consistent with Stage 4 systems described in Chapter 7 (Section 7.7).
2. No matter how complex the test system is, there is no way of assessing *in general* how similar the response of *different* ecosystems to the same perturbation will be.
3. Bioassays with predictive power should be directed to identifying the impact of perturbations at a reproducible level independent of the system; this would support a molecular approach to ecotoxicology which seems highly desirable even though this represents a long-term research effort.

Clearly then, a good deal of conceptual development in the design of bioassays remains to be done before an acceptably rational basis for predicting the biological effect of xenobiotics discharged into the environment can be achieved.

Finally, it should be pointed out that integration of some of the components of Sections 8.1.2 and 8.1.3 may be possible and is indeed highly desirable; this will provide both rationalization of the experiments and at the same time increase the level of sophistication of the investigation. For example, experiments on bioconcentration, metabolism, and the application of biomarkers in fish could readily and advantageously be interfaced.

8.1.4 Coordinating the Results

When the data from the experiments outlined in the preceding sections have been assembled, probably the most difficult part still remains — its interpretation and evaluation. This will involve subjective assessment and will inevitably be biased by the experience and prejudice of the investigator. The evaluation of the data is a complex process, and no strictly logical procedure is available; a certain amount of personal judgment will therefore have to be accepted. In addition, there is a conceptual difficulty: most traditional procedures for hazard assessment produce evaluations in numerical terms, but this may be a misleading approach in the present case. A brief analysis may be used to illustrate the problem.

1. Each test result could be scored on a numerical basis, for example, from 1 to 5, and the sum of the the scores for all the tests that were applied could then be used to make a decision — either by comparison with reference compounds scored using the same tests or by setting an arbitrary limit.
2. Alternatively, one could set the limit first and if the score exceeded this — regardless of the number of tests incorporated — the result could be assessed as acceptable or unacceptable. If the result were considered acceptable by either procedure, the substance could then, if desired, be subjected to a more rigorous set of assays. Thereby, an increasing approximation to the ideal could be reached. There are, however, at least two potentially serious difficulties in the adoption of these approaches:

- In the first procedure, judgment has still to be exercised on the *quantitiative* significance of the numerical differences — and this is a subjective evaluation.
- A substance may score well on all of the tests except one — yet still attain an acceptable total. Personal judgment may, however, strongly suggest a negative ruling on the basis of the use of the compound, the exposure route, or on previous experience with compounds possessing the doubtful attribute that was revealed in the testing.

In the last analysis, therefore, a rather high degree of subjective judgment seems unavoidable in spite of the fact that the procedures appear objective and are numerically based. The situation is indeed reminiscent of the application of Bayesian statistics which is hardly beloved of most professional statisticians. A light-hearted though good illustration of its application is given in the book by Lewis (1990): the essential issue is that information additional to that provided by the data is, in fact, frequently needed to accomplish the final evaluation.

As is the case for pharmaceutical products which undergo a much more rigorous safety evaluation than compounds which may present an environmental hazard, there is, however, a mechanism for providing a safeguard and a warning in advance of disaster: monitoring for hitherto unperceived perturbations in natural ecosystems.

In many cases, due to a limited understanding of the mechanism of chronic toxicity and the effects on feral populations, some uncertainty may necessarily remain on the biological consequences resulting from the discharge of a compound into the environment. Comparable unresolved issues may surround questions of persistence, bioconcentration, or accumulation in the sediment phase. In such circumstances, a monitoring program should be initiated at an early stage; indeed, this is generally desirable and could valuably include a wide range of compounds. The whole issue of monitoring is discussed in Chapter 3 (Section 3.5.6), and examples of representative programs have been provided. Although these are directed to the analysis of tissue residues, the interpretation of their significance leads naturally to the role of epidemiology that is discussed in more detail in Section 8.2.

8.2 RETROACTIVE ANALYSIS: ECOEPIDEMIOLOGY

Although it is clearly desirable to recognize a potential hazard in advance and thereby prevent the occurrence of seriously adverse effects on ecosystems, it is also necessary in some circumstances to ascertain the probable cause or causes of environmental perturbations that have already been observed. The biological effects may be highly visible and often quite dramatic. The following examples illustrate the range of biota that may be affected:

1. The extensive deaths of sea-mammals including bottlenose dolphins (Sarokin and Schulkin 1992) and several species of seals (Visser et al. 1990).
2. The decline in the population of rockhopper penguins (*Eudyptes chrysocome*) on Campbell Island, New Zealand (Moors 1986).
3. The incidence of tumors in feral fish populations putatively exposed to polycyclic aromatic hydrocarbons (Harshbarger and Clark 1990).

4. The blooming of algae which may produce toxins (Underdal et al. 1989).
5. The incidence of porphyria in fish-eating herring gulls (*Larus argentatus*) in the Great Lakes (Fox et al. 1988).

Some examples of the difficulties of correlating effects with exposure to specific toxicants have already been given in Chapter 7 (Sections 7.5 and 7.6), and attention is directed particularly to a careful review that examines the evidence for the adverse effect of exposure to organochlorine compounds on the reproductive success of various populations of different marine mammals (Addison 1989). This activity may conveniently be termed ecoepidemiology by analogy with the precedures used in human disease control and prevention. It should be clearly appreciated that there are inherent difficulties in carrying out such an analysis, with the result that it may be impossible to establish an absolute correlation between the observed effect and the putative cause. This is clearly illustrated by investigations into the possible adverse effect on humans who consume fish from the Great Lakes that is contaminated with organochlorine compounds (Swain 1991). Although levels of Pb, Cd, PCB, and DDT in humans are higher in fish-consuming groups, the increased levels of Pb and Cd were apparently influenced primarily by differences in life-style rather than by fish consumption (Hovinga et al. 1993). Some of these complicating issues are explored in the next few paragraphs.

Epidemiology is concerned with the health of human populations and may be defined (Last 1983) as "the study of the distribution and determinants of disease and health-related status in populations and related phenomena in human population groups". This is readily translated into the corresponding activity in ecosystems and is then designated ecoepidemiology. It is, however, valuable to make a few brief comments on epidemiology itself since this is a highly developed discipline that has been conveniently formalized (Susser 1986).

1. Provided that the correlation between the exposure and the observed effect is sufficiently strong, valuable — though temporary — measures to alleviate a problem may be possible without a mechanistic or rational understanding of the cause. A good example quoted by Fox (1991) was the recognition by Snow that the outbreak of cholera in London during 1848-1854 was the result of drinking sewage-polluted water, even though the organism responsible was not isolated until many years later by Koch. Similarly, although the investigations of Potts in 1775 demonstrated the association between the frequency of scrotal cancer in chimney sweeps and their occupation, the underlying association with exposure to PAHs was not revealed until the early years of this century. In both of these cases, however, the exposed population was either confined to a restricted geographical locality or to a specific occupation. This is, however, seldom the case with environmental pollutants, many of which now have a global distribution; the difficulty of establishing a correlation between exposure and observed effect is, therefore, greatly exacerbated.

2. Investigations of human disease may reveal major difficulties due to the emergence of hitherto undiscovered pathogenic agents. One good example is the discovery of the causative agent of legionnaires' disease whose outbreak in 1976 stimulated a massive investigation as a result of which the organism responsible

(*Legionella pneumophila*) was ultimately isolated. In this case, further progress could not have been made without isolation of the organism which seems to be thermally resistant and able to survive and multiply under various conditions particularly in the presence of other microorganisms including amoebae (Wadowsky et al. 1988); it may apparently even survive effectively within organisms such as *Tetrahymena vorax* (Smith-Somerville et al. 1991). Even though many serotypes of *Legionella pneumoniae* have since been identified and other species of the genus described, there remain important unresolved issues on the ecology and pathogenicity of this group of organisms. In a comparable way, significant gaps in evaluating important determinants of degradability or toxicity may await the emergence of results from experiments whose rationale has not so far been motivated.

The ecoepidemiological approach does, in fact, accept that it may be impossible to establish a strictly causal relationship between the observed effect and the compound putatively responsible. The objective is to establish how persuasive the correlation is and the degree to which it is legitimate to exclude alternative explanations. A few specific examples of the difficulties may be used as illustration at this stage; these are taken from a review by Sarokin and Schulkin (1992) which includes other interesting examples.

At least three situations may be distinguished.

1. A causal relation between field observations and exposure to a xenobiotic may be supported by laboratory studies that show the induction of the symptoms on exposure of the appropriate organisms to the supposed toxicant. An example of this is provided by the occurrence of liver tumors in feral fish (Harshbarger et al. 1990) which has been associated with exposure to polycyclic aromatic hydrocarbons (PAHs) (Malins et al. 1985; Myers et al. 1991). This correlation is strongly supported by extensive information on the mechanism of carcinogenesis induced by these compounds both in fish (Hawkins et al. 1990) and in other organisms as well as by the specific nature of the tumors. On the other hand, an extensive statistical study (Johnson et al. 1993) of hepatic lesions in winter flounder (*Pleuronectes americanus*) captured from 22 sites on the northeast coast of the United States illustrates the complexities that may be encountered. The single most important biological factor was age, while PAHs, DDT, and chlordanes — but not PCBs — emerged as important risk factors, although it was not possible to evaluate the quantitative contribution of each of them since they were generally present simultaneously. The practical limitation to the further application of this approach is that whereas it is readily carried out using, for example, fish, it cannot be extended to the large sea mammals that have understandably attracted so much concern.

2. The immediate causative agent may have been identified, but the effect may be suspected of being exacerbated by other simultaneous exposures. A good illustration of this is provided by the extensive death of harbor seals (*Phoca vitulina*) in the North Sea. This has been attributed to infection by a phocine distemper virus belonging to a new member of the genus *Morbillivirus* (Visser et al. 1990), although susceptibility to infection may, in fact, be the ultimate link in a chain

of events triggered by exposure of the seals to pollutants such as PCBs. The possible significance of this factor is supported by the observations that seals fed with PCB-contaminated fish had decreased levels of plasma retinol and thyroxins compared with seals fed with fish containing only low levels of PCB (Brouwer et al. 1989). By analogy with the pathogenic microorganisms that have serious consequences for compromised human beings, these organisms may be termed opportunistic pathogens. Attention has already been drawn in Chapter 7 (Section 7.5.1) to the increase in the level of metabolites of benzo[a]pyrene in the liver of rainbow trout exposed to PCBs. A critical review of these and other associations has been given (Addison 1989).

3. It may be impossible to establish a correlation between the observed effect and a single cause; there may be several interacting factors, none of which alone is sufficient. Two examples are given to illustrate a situation that is probably widespread as a result of the innate complexity of biological control mechanisms and of the ubiquity of many xenobiotics in several environmental compartments and in geographically separated areas.

   • The occurrence of toxic blooms of algae may be due to a number of interacting factors including, for example, pollution, input of humus, acid rain, or temperature, and it is very difficult to establish a correlation with any single one of these to the exclusion of the others: introduction of preventive measures to prevent their occurrence is, therefore, extremely difficult.
   • As a result of the investigations into the cause of the massive deaths of bottle-nose dolphins (Sarokin and Schulkin 1992), at least three possible — and widely different — explanations have been advanced: (1) massive systemic bacterial infection by opportunistic pathogens; (2) ingestion of the toxin produced by the alga *Ptychodis brevis*; or (3) impaired function of immune systems following exposure to toxicants and to PCBs in particular.

This indeterminacy will probably in most cases have to be accepted since there are too many interacting factors, and the determination of a single factor may simply lie beyond the resolution of current experimental science.

An additional complicating factor arises from the background level of some apparently ubiquitous contaminants whose partitioning and dissemination have been discussed in Chapter 3 (Section 3.5). For example, there is a global — though by no means uniform — distribution of a range of organochlorine compounds including hexachlorocyclohexanes, chlordanes, PCBs, and DDT (Iwata et al. 1993), and the same is also true for polycyclic aromatic hydrocarbons. Whereas the ultimate source of many toxicants can be established with a high degree of certainty, this may not always be the case; two examples of compounds that are apparently widely distributed in the environment may be used as illustration.

1. Although tris(4-chlorophenyl)methanol has been identified in samples of marine mammals and bird eggs from different areas throughout the world (Jarman et al. 1992), its source has not been conclusively identified though a number are under suspicion.

2. Analysis of sediment cores from Japan that have been dated to ca. 8000 y revealed the presence of the more persistent heptachloro and octachloro congeners of dibenzo-1,4-dioxins (Hashimoto et al. 1990), and at present there appears to be a background level of chlorinated dibenzo-1,4-dioxins and dibenzofurans that is not attributable to the production of pulp using molecular chlorine (Berry et al. 1993). Their production may possibly be associated with incineration processes or with industrial activity such as the manfacture of pentachlorophenol. Extensive data suggests that it is only the most highly chlorinated congeners that are persistent in the atmosphere (Hites 1990), and these are, therefore, the most likely to be recovered from environmental samples except those of recent origin. This study also showed that, in apparent contradition of the Japanese study, the concentrations of octachlorodibenzo-1,4-dioxin in samples from a remote area (Siskiwit Lake, Isle Royale, Lake Superior) were extremely low before 1930.

Care should therefore be exercised in drawing conclusions on putative causes of biological effects against the background levels of these virtually ubiquitous compounds.

Certain basic criteria in epidemiology (Susser 1986) have been widely accepted and appear to be equally applicable to ecoepidemiology (Fox 1991). It should, however, be clearly appreciated that these techniques are not designed to establish unambiguous causal relationships but rather to indicate which criteria may most usefully be used to provide a balanced evaluation in support — or in contradiction — of a given hypothesis; it should be noted that the relatively loose term "association" is consistently used. These five criteria are (1) consistency, (2) strength, (3) specificity, (4) temporal relationship, and (5) coherence.

The application of these principles has been explored more extensively by Fox (1991) in his review, and in specific application to the reproductive success of lake trout (*Salvelinus namaycush*) in the Great Lakes (Mac and Edsall 1991). In that investigation, although the evidence suggested that the survival of fry was a significant factor in reproductive impairment, no specific suggestion for the involvement of a significant pollutant was put forward. Even when the *source* of the perturbation has been plausibly identified, the problem of ameliorative measures is by no means simple. For example, there is extensive and fairly compulsive evidence that impaired reproduction in white sucker (*Catostomus commersoni*) may be associated with the exposure of adults to bleachery effluents (McMaster et al. 1992; Van der Kraak et al. 1992); until, however, the responsible component(s) have been identified in the effluents, it is difficult to determine the causatvive agent and to implement preventive measures. This example provides, incidentally, a good illustration of the central role of chemical analysis in ecoepidemiology.

The examples noted at the beginning of this section on highly visible perturbations of natural populations and the lack of consensus on their underlying causes illustrate the tentative nature of ecoepidemiology. In spite of this apparent limitation, there are, however, some very attractive features of ecoepidemiology which may be summarized:

1. It will bring to light often diverging — and even contradictory — views, and this is entirely consistent with the inherent complexity of almost all natural situations.

2. It will encourage the careful exploration of many different aspects of the problem and reveal the possibility that several factors may be complementary rather than mutually exclusive.
3. It will focus attention on the mechanisms that regulate natural populations and the interactions whereby populations react to alterations in their environment — and hopefully attract research effort to this important area.

Examination of all of these aspects has been profitable and — possibly even more valuable — may suggest strategies relevant to other environmental issues.

8.3 ADDITIONAL PROCEDURES

8.3.1 Toxicity Equivalent Factors

Reference has already been made to the fact that many industrial effluents and unfortunately some commercial products such as PCBs, PCCs, or chlordane contain several components which have widely differing toxicity. The toxicity of the effluent or commercial product is therefore a function of the relative concentrations of the components together with their toxicities. To meet such situations, the concept of toxicity equivalent factors has been introduced. Briefly, each component is assigned a relative toxicity using a given organism or test system; the concentrations of the components in the mixture are then multiplied by this factor, and the sum of the products for all the components is used to evaluate the toxicity of the effluent or product. This approach has been evaluated extensively for PCBs, PCDDs, and PCDFs (Safe 1990; Kutz et al. 1990) and has been widely used. There is, however, at least one important issue that should be clearly appreciated: the numerical values assigned are valid only for the organism used for assessment. For example, if the assessment of toxicity to man is the objective, toxicity factors should be obtained from values for human toxicity; in this case, of course, the use of surrogate organisms is acceptable provided that the inherent limitations are clearly appreciated. As was wisely stated (Pochin 1983) in the context of the risk to humans of exposure to radionuclides: "I don't know whether we are closer to the dog, the mouse, or the rat in terms of lymph node behavior". The same limitation surrounds the use of such equivalency factors in environmental hazard assessment, and attention should be directed to the differing morphologies, physiologies, and biochemistry of the test organisms; at the very least, toxicity equivalent factors in environmental studies should be assigned on the basis of toxicity results for algae, crustaceans, and fish and take into account possible differences in the sensitivity towards individual representatives of these groups. A commendable study (Kovacs et al. 1993) has addressed the application of the concept to a number of chlorophenolic compounds and has taken advantage of sub-acute data from various organisms including different species of fish and from the sea-urchin sperm assay. The authors carefully point out that the values proposed should be regarded as provisional and that further refinements should take into account the position of substituents and of possible differences in the sensitivity of the test organisms.

There are both merits and demerits in the application of this concept: on the positive side, undeserved attention would not be directed to components existing in low concentration unless they were highly toxic, whereas on the negative side components of low toxicity — though present in relatively high concentration — might not receive the attention they rightly deserved. In addition, an apparent contradiction may result from the use of different determinants of environmental impact. A good illustration is provided by the more highly chlorinated (6 or more chlorine substituents) PCBs and PCDDs which are present in high concentration in some kinds of environmental samples; although these are relatively less toxic than congeners with lower degrees of chlorination, they are apparently highly persistent to microbial attack and to metabolism in higher biota.

These remarks should not be taken to imply that a critical evaluation of the concept is not a worthwhile exercise providing that a sense of balance is preserved; it may indeed be the only rational approach for assessing the toxicity of complex effluents. It is, however, imperative to be aware of the fact that toxicity equivalent factors are critical functions of the organisms used for their assessments and that the values obtained using one organism will not necessarily — or even reasonably — be applicable to any other. Compared even with organisms within the same group, a single taxon may be highly sensitive to a particular toxicant.

8.3.2 Critical Loading

The term critical loading has been used to imply that ecosystems are able to survive environmental impacts providing these are lower than a certain critical threshold level. Although the concept may appear to be consistent with the view that environmental hazard assessments are relative rather than absolute, and that a degree of risk may have to be accepted, this seems a somewhat presumptious and even dangerous concept. A number of reasons which support this critical view are therefore briefly summarized.

1. It assumes that the degree of impact does not exceed that which may be mitigated by homeostatic processes but does not provide procedures for evaluating this pre-set level. It assumes the existence of a buffer capacity for an ecosystem even though, in practice, this is difficult to evaluate in view of the varying sensitivity of different biota and the long-term significance of compounding biochemical and geochemical factors.
2. The concept appears to have been applied to atmospheric deposition of oxidized sulfur and nitrogen compounds (NO_x and SO_x). It may indeed be applicable to this case since these are components of relatively well-established geochemical cycles. This is not, however, the case for organic compounds, many of which are apparently persistent in the environment and whose ultimate fate often remains unresolved.
3. The history of environmental research has repeatedly illustrated the complexity of natural ecosystems; current concern is therefore directed to sub-lethal and chronic effects of xenobiotics, and these more subtle effects could very readily be neglected or insufficiently evaluated if the concept were generally and uncritically adopted.

4. The assumption would appear to be made that current understanding of ecosystems, their populations, and their regulation is adequate to make the appropriate
judgment. Although it might be comforting to accept this complacent point of
view, the application of increased sophistication in assessing the biological
effects of xenobiotics — or indeed the development of completely new concepts
— make this a highly questionable assumption.

8.4 A PHILOSOPHICAL INTERLUDE

8.4.1 Polarization of Viewpoints

It is necessary to appreciate clearly the object of the hazard assessment procedures, since partly conflicting viewpoints may arise; for example, one legitimate aim
is to provide decision makers with clear-cut guidelines based on available data;
another is to understand and evaluate the complexities of problems which may not
necessarily be amenable to simplistic analysis and that may require additional
research effort. The conflict is essentially one of time scale. In addition, there has —
unfortunately, but possibly inevitably — developed a polarization of viewpoints on
environmental issues: extreme and opposing viewpoints have been adopted by their
proponents sometimes without adequate appreciation that the complexity of the
issues almost certainly validates some of the arguments from both sides. The problem
is complicated by the inescapable fact that even the same data may not necessarily
be interpreted in the same way by different experts: for example, by authorities
granting discharge permits and by consulting companies engaged to provide the basic
data. This may present a dilemma which underscores the absolute necessity for the
involvement of non-partisan bodies; possibly all that can be achieved realistically is
an understanding of the magnitude of the difficulties and the hope that scientifically
acceptable decisions will be made on the basis of existing knowledge with the
possibility for incorporating new data as it becomes available.

There is, however, the unescapable though controversial fact that many important
issues belong to the realm of "trans-science" (Weinberg 1972): "In principle they
refer to questions of fact and can be formulated in terms of existing scientific
methods and procedures, but they are in practice unresolvable in these terms alone"
(Crouch 1986). This places the administrator in a difficult and vulnerable position,
and possibly the best solution — indeed, maybe the only one — is to accept the
inherent complexity of the problem and admit the major uncertainties in advance
rather than retrospectively. The statement of Dunster (1984) seems to express nicely
the desired balance: "We must be neither blandly reassuring nor deliberately alarm-
ist".

A good example of the convergence of scientific, anthropological, and moral
dilemmas is provided by the sometimes acrimonious debate on the resumption of
whaling: a representative selection of the various views has been presented in *Arctic*
(1993:46[2]), and this includes an evaluation of the impact of international regulation
and the imposition of sanctions. There seem to be at least three basic issues that are
involved:

1. The size of sustainable populations of different species of whales: this is essentially a matter that should be amenable to scientific resolution using established procedures that have been used in fishery management.
2. The nutritional value of the catch to circumpolar aboriginal populations as an alternative to expensive and poor-quality imported food that is discussed again in Section 8.4.3, as well as the maintenance of a cultural heritage. It is important to appreciate that this level of hunting is of an entirely different magnitude from that exercised indiscriminately in the Antarctic during the 1930s and that this activity may legitimately be regarded as artisanal fishing.
3. The moral issue of hunting and the totemic role of whales to man. The "precautionary principle" emerges as the only one acceptable in view of the understandably conflicting — and sometimes inflammatory — stances that have been taken on both sides.

8.4.2 Anthropogenic and Natural Compounds

A tendency has emerged to evaluate synthetic compounds and anthropogenic discharges against the occurrence of what are generally termed "natural products" synthesized by animals, plants, algae, fungi, and bacteria. This has arisen through a perhaps understandable distrust of synthetic chemicals by the general public and from the feeling that the consumption of naturally occurring compounds is less fraught with danger. The potential dangers behind this argumentation have been carefully pointed out (Ames 1991; Ames and Gold 1991). It is important to appreciate also that many naturally occurring compounds are far from innocuous to species other than those producing them; many examples may be cited including bacterial, fungal, and algal toxins, plant alkaloids, sterol saponins, and toxins and carcinogens produced by higher plants.

Attention has already been drawn to naturally occurring halogenated organic compounds in Chapter 1 (Section 1.3) and to the occurrence of polycyclic aromatic compounds as a result of diagenesis of di- and triterpenes in Chapter 2 (Section 2.4.2). It may even emerge that toxic compounds are synthesized *in vivo* by organisms other than those conventionally accepted as the primary sources; a striking example is the synthesis of alkaloid-like compounds in man from a variety of precursors including natural vitamin B_6 and therapeutically administered chloral hydrate (Bringmann et al. 1991). In addition, industrial processes such as the production of pulp may result in the release of toxic compounds such as terpenes, even though these are in no way directly related to the final product. The apparently sharp line between natural and anthropogenic products is, therefore, becoming increasingly blurred.

8.4.3 Decision-Making: Procedures and Ambiguities

In the final analysis, a decision has to be made on the basis of all the available data. Decision-making dendrograms can, of course, be readily constructed and may incorporate not only dichotomous steps but weighting of the input at each successive step. Eventually, however, subjective judgment must be used and may indeed be inevitable. One might draw the analogy with taxonomic keys which generally work well for higher organisms such as plants; the real difficulty in using these emerges when hybrids are encountered or when species are plastic so that a more or less continuous range of individual morphologies separates that of the defined taxa. The taxonomic assignment will therefore be subjective and will be strongly influenced by experience with the given group of organisms. Judgment may, in fact, surprisingly often be subjective and relative rather than absolute.

Rather different examples may be used to illustrate the difficulties in choosing between alternatives and the importance of accepting relative evaluations:

1. If a chemical is considered environmentally unacceptable, it may be necessary from its recognized value in modern society to develop a replacement. It is essential, however, that the replacement is environmentally less hazardous than its predecessor although this apparently obvious criterion may be very difficult to establish in advance. Three examples may be used to illustrate this issue.

   - Replacement of PCBs that are generally considered to be environmentally unacceptable with non-chlorinated aromatic compounds cannot necessarily be judged as a satisfactory solution to the problem, since some of these newer compounds seem to be both persistent in the environment and partition into the sediment phase (Peterman and Delphino 1990).
   - Replacement of tetraethyl lead in gasoline with additives containing quaternary carbon groups such as t-butanol and t-butyl-methyl ether seems a doubtful improvement in view of the persistence of these compounds to degradation under methanogenic conditions that could prevail in contaminated groundwater (Suflita and Mormile 1993).
   - The interest in chlorinated carbohydrates such as sucralose (4-chloro-4-deoxy-α-D-galactopyranosly-1,6-dichloro-1,6-dideoxy-β-D-fructofuranoside) as artificial sweeteners to replace natural compounds such as sucrose seems remarkable in the light of general concern about organochlorine compounds in general, and evidence suggesting that sucralose is not readily degradable in a number of environments (Labare and Alexander 1993).

2. There has been extensive exchange of views on the role of agrochemicals in inducing cancer in man (Ames and Gold, and other contributors 1991), and it has been convincingly pointed out that many naturally occurring compounds in fruits and vegetables are at least as carcinogenic as synthetic compounds used as agrochemicals and that the former may, in fact, be ingested in substantially higher concentrations. The conclusion to be drawn is not that unrestricted use should be made of agrochemicals but that a balance should be struck between their advantages and the conceivable dangers from alternative sources of carcinogens.

3. It has been clearly shown that high levels of organochlorines including PCBs and PCCs exist in the traditional food of Inuit. From the health point of view, these levels may be unacceptably high, but on the other hand the consumption of meat and fat from marine mammals and fish provides not only protein and lipids but also polyunsaturated fatty acids and retinol together with necessary minerals including Fe, Zn, Ca, and P. It is not therefore obvious that it would be

advantageous to replace this consumption with expensive store-purchased food of questionable quality; this presents a dilemma that should be faced with a sense of balance and in the hope that the levels of the contaminants will be reduced with time (Kinloch et al. 1992).

4. Even a superficial perusal of the literature clearly reveals the massive attention that has been directed to chlorinated dibenzo-1,4-dioxins and chlorinated dibenzofurans. Although there can be little doubt that these compounds are toxic to certain species, there must emerge more than a lingering doubt that the focus on these compounds — in particular, in view of the cost of carrying out the relevant analyses — has inevitably distracted research effort from other environmental issues. This has been discussed in the specific context of cancer induction in man (Ames 1991).

8.5 EXEMPLIFICATION OF HAZARD ASSESSMENT EVALUATIONS

By way of illustration, a brief discussion is devoted to five studies differing both in their objectives and in their methodologies; these exercises did not attempt to fulfill all of the criteria outlined above but are used to illustrate various approaches to the problem. In addition, they show that procedures for predictive and retroactive analysis are not mutually exclusive and that some features of both can be advantageously combined. The first four have been described in great detail in publications to which reference is made, so that only an outline of the salient features of their design and the highlights of their conclusions will be attempted here.

The first project involved a comprehensive series of laboratory studies and mesocosm experiments, whereas the second and third were field studies on true marine ecosystems which have possibly not received the attention that they merit. The fourth was a combination of both field and laboratory investigations, while the fifth has been chosen to illustrate how procedures developed in an entirely different context may provide valuable insights into possible methodologies. An attempt has been made to combine the features of all of these studies into a hierarchical protocol.

8.5.1 The ESTHER Project

A comprehensive study in Sweden (Landner 1989) had three principal objectives:

1. To provide a comprehensive overview of the environmental situation in a restricted ecosystem (the Baltic Sea and the Gulf of Bothnia).
2. To examine the various procedures that have been used for evaluating the effect of xenobiotics on biota.
3. To develop experimental procedures for environmental hazard assessment in the aquatic environment and to apply these to a few structurally diverse compounds.

The project was designed to assemble a range of different disciplines, each contributing a different facet of the project. Emphasis was placed upon understanding the effects of xenobiotics at different organizational levels: individuals, communities, and populations. Both laboratory and mesocosm studies were applied to three diverse compounds of environmental interest, arsenate, 4,5,6-trichloroguaiacol, and hexachlorobenzene — although the latter was not subjected to the same intensity of investigation as the two other compounds. Some of the results from this project have already been discussed in Chapter 5 (Section 5.3.3) and in Chapter 7 (Section 7.4.3). No attempt will be made to summarize the detailed conclusions of this comprehensive study which have been presented in a book (Landner 1989), but some of the conclusions with a wider relevance seem particularly worth emphasizing:

1. The complexity of the factors that determine the fate of a compound after its discharge into the environment was clearly demonstrated, and details of the transformations which may take place were elucidated in considerable detail for arsenate and 4,5,6-trichloroguaiacol. The results of these experiments illustrated the need to take into account the persistence, the toxicity, and the partition among the various phases not only of the initial compounds but also of their metabolites.
2. The possibility of designing laboratory experiments with a high degree of environmental realism emerged. This applies both to the evaluation of persistence and to the determination of toxicity. Particular attention was consistently directed to sub-lethal effects, to the effect of chronic exposure, and to the effect of pre-exposure to xenobiotics. As a result, a number of new methodologies emerged during the study.
3. It is most convincing to use organisms from the area under investigation rather than rely exclusively upon the use of other standardized test organisms, and this applies both to the assessment of persistence and toxicity. This supports and confirms important issues of principle which have emerged consistently throughout this account.

It should be added that an extensive monitoring program was not part of the study which was designed primarily to evaluate possible strategies and evolve the basic principles.

8.5.2 The Loch Eil Project

The Loch Eil project in Scotland was concerned with the environmental impact created by discharge into a sea loch of effluents from a combined pulp and paper mill which used the sulfite process. The effluent consisted of readily degraded compounds together with suspended cellulose fiber. The investigations were carried out over a number of years, and their scope was admittedly determined by the available expertise so that they were possibly neither optimally designed nor executed (Pearson 1981). This resulted in an inevitable imbalance in various aspects of the investigations; for example, both the chemical analysis and the microbiological investigations were quite restricted in scope, while the biological examination of field material was by comparison quite comprehensive. A number of conclusions were drawn (Pearson 1982), of which a few of the most interesting are summarized:

1. There was no accumulation of organic carbon in the sediments, and the increased input into the ecosystem resulted primarily in a substantial increase in the protozoan population, whereas the bacterial biomass was scarcely affected. The last conclusion is unexpected in view of the fact that there is a considerable input of organic carbon into the system without an apparent increase in the organic carbon in the sediment; a more detailed investigation into microbiological prcesses in the sediment phase might possibly have resolved this important issue. Conclusions based on estimation of microbial biomass may also have been too restrictive.

2. The ecosystem was inherently stable in a long-term perspective, and this could plausibly be attributed to the effective tidal flushing of the area and to the procedures used for discharge of the effluent which was retained in holding basins for 1.5 h during slack water.

Whereas the important conclusions from this study are clearly supported by the data presented, the occurrence of possibly persistent compounds such as biocides or other chemical additives in the effluents was apparently not examined; at least retrospectively, the study could clearly have profited from a more extensive analytical program on the composition of the effluents and the distribution and dissemination of specific compounds after discharge. The conclusions briefly summarized above are, however, both valuable and provocative.

8.5.3 The Baffin Island Oil Spill Project (BIOS)

This was a unique investigation involving the deliberate discharge of oil and oil dispersants into a geographically remote and restricted environment in the Canadian Arctic. The reasons for adopting this strategy were themselves interesting; it was felt that, in spite of the effort already devoted to the persistence and toxicity of hydrocarbons in the marine environment, much of this knowledge could not validly be extrapolated to an arctic environment with low water temperatures, and a restricted — and possibly sensitive — pristine biota. It is not possible to review in detail the conclusions from this study which have been published in a series of comprehensive articles in a supplement to *Arctic* (1987: 40) (Sergy 1987); this idea alone should be warmly welcomed as an alternative to the publication of details in miscellaneous reports that may not be readily accessible to interested members of the scientific community. This study is also important since, although it was designed to provide a predictive basis for future activity in the event of an oil spill, it also clearly illustrated the kinds of data that provide the basis for epidemiological studies. Possibly of most direct relevance in the present context are those aspects of its design that could profitably be incorporated into a wider range of hazard assessment procedures, and an attempt is therefore made to summarize these features.

1. Comprehensive chemical analyses of samples of water, sediment, and of biota were carried out both *before* and *after* the spill; this cannot of course be carried out in most cases, and this illustrates a serious limitation in field studies where lack of background data or difficulty in finding an uncontaminated control

locality is frequently encountered. Sum parameters were sparingly employed in BIOS, and emphasis was placed on the analysis of specific compounds; attention was directed not only to polycyclic aromatic hydrocarbons but also to azaarenes, dibenzothiophenes, and hopanes. Thereby, a clear distinction could be made between the input from the oil deliberately discharged and that arising from natural biological reactions or mediated by atmospheric transport. The importance of both these issues has been noted earlier in appropriate parts of this book.

2. Ice scouring of the intertidal zone in arctic waters makes this virtually sterile — a fact that was noted more than 160 years ago by Keilhau (1831) — so that attention was directed to components of the sub-tidal zone to which little attention had previously been directed and which was expected to be particularly sensitive to oil spills. Changes in the components of the macrobenthos including infauna, epibenthos, and macroalgae were examined, and attention was also directed to the histopathological and biochemical responses of bivalve molluscs that were affected in different ways by exposure to the dispersed and the undispersed oil.

The original reports should be consulted for details of the conclusions drawn from this comprehensive investigation, but the following seem of particular importance for this specific class of pollutant. The study provided the following:

1. A basis for determining remedial action in the event of an oil spill.
2. An evaluation of the advantage or disadvantage of using oil dispersants.
3. An estimate of the persistence of oil washed ashore.
4. An assessment of the effectiveness of procedures for removing stranded oil from beaches.

8.5.4 The Fox River–Green Bay Project

This project was directed to providing a basis for determining remedial action of highly contaminated sediments in the Fox River and Green Bay, Wisconsin (Ankley et al. 1992). The investigation encompassed assessments of the abundance of major groups of benthic invertebrates, assays for the toxicity of interstitial water towards a number of test organisms including fish (*Pimephales promelas*), a crustacean (*Ceriodaphnia dubia*) and an alga (*Selenastrum capricornutum*), and congener-specific analysis for PCBs, PCDDs, and PCDFs. There were a number of interesting conclusions, some of which were unexpected:

1. Although the diversity and abundance of benthic organism indicated an adverse impact, its nature could not be unequivocally determined, and it was suggested that ammonia generated by microbial action such as deamination of amino acids or proteins could be an important toxicant.
2. In biota, the concentrations of planar PCBs — rather than PCDDs or PCDFs — probably contributed significantly to toxicity. The concentrations of 1,2,3,4,6,7,8-heptachloro- and octachlorodibenzo-1,4-dioxins exceeded those of any other congener, and this finding is consistent with the persistence of these compounds that has been revealed in other studies that have been referred to in Chapter 3 (Section 3.5.2).

3. Attention was drawn to the emergence of an important and interesting dilemma in respect of remedial strategies. Since removal of PCBs by dredging, for example, may be unrealistic, dilution by unpolluted sediments might be an effective alternative. On the one hand, the maintenance of eutrophic conditions with a highly particulate water mass could possibly help to accomplish this — provided, of course, that organic toxicants which would simultaneously be accumulated were absent — but, on the other hand, this strategy would result in continued poor water quality. The decision between these two alternatives is not readily made, and the adoption of either strategy would require careful evaluation of its consequences.

8.5.5 The Hazard From Release of Genetically Modified Microorganisms

There are many potential appications for genetically modified microorganisms in agriculture although there is a concomitant awareness of possibly unacceptable hazards as a result of their application to terrestrial systems. This subject may not at first sight seem germane to the present discussion — except possibly for the fact that such a substantial part of this book is devoted to microorganisms. It is included primarily for two reasons:

1. Closer examination of the problem reveals substantial similarities to the procedures — and even unresolved issues — in evaluating the hazard following the discharge of an organic compound into the aquatic environment.
2. The problem interfaces rather neatly with some of the critical issues in determining the hazard resulting from the release of organic compounds into the terrestrial environment and thereby provides a convenient elaboration of principles outlined for the aquatic environment.

A brief discussion is, therefore, presented, but before doing so, it may be valuable to spell out the specific analogies.

1. In the hazard assessment of xenobiotics discharged into the aquatic environment — or of that arising from the release of genetically manipulated microorganisms into the terrestrial environment — extreme situations can readily be identified. The real difficulties, however, arise in those situations which lie in the grey zone between these extremes; in both cases, any adverse effect has to be balanced against the expected benefit, so that the question of "acceptable risk" again seems inevitable.
2. The questions of survival and gene transfer and translocation of genetically modified microorganisms may be broadly equated with the concepts of persistence and partition of organic compounds — and in both cases, the key issue is the extent to which ecosystems are adversely affected. Indeed, the application of biological constraints to set a limit on the survival of genetically modified organisms is formally similar to that requiring limited persistence of an organic compound in the environment after discharge.
3. Many of the parameters that have been suggested for monitoring the effects on terrestrial ecosystems (Smit et al. 1992) are identical to those that would be

applied to the release of an organic compound into the same environment; for example, the effect on microbial populations, on key microbial processes and enzymatic activity, and on effects on the components of higher trophic levels are common to both.

4. The most difficult and least accessible problem is the evaluation of the impact at the ecosystem level, and both issues converge closely at this point.

The problem of assessing the potential hazard resulting from the release of genetically manipulated microorganisms into the terrestrial environment has been reviewed (Smit et al. 1992), and various possible strategies for assessing the impact were assembled. The corresponding situation for the aquatic environment presents many important, conflicting, and unresolved issues, most of which have been approached from a strictly microbiological point of view and with special emphasis on the problem of the stability of such organisms in the natural environment. A summary of some of the cardinal issues may be found in Sobecky et al. 1992.

Ideally, components of all of these studies could usefully be combined into a hierarchical system that takes optimal advantage of the strengths of each of them; these are indicated in parentheses. The protocol could incorporate the following stages:

Stage 1: Laboratory studies of the basic issues of persistence, partition, and toxicity (ESTHER).
Stage 2: Verification of principles in model ecosystems (ESTHER).
Stage 3: Monitoring of the area receiving the discharge.
- Chemical analysis (BIOS).
- Estimates of population and distribution of higher biota (Loch Eil).
- Evaluation of toxicity of field samples in laboratory assays (Fox River).
- Assessment of the effect of the xenobiotic on non-target organisms including microbial populations (Genetic hazard).

It should be emphasized, however, that in many situations where the relevant pollutant has already been established, a substantial amount of relevant data will already have been accumulated; this is clearly the case with the second and third studies, in contrast to the first that was devoted primarily to establishing and evaluating the necessary principles of operation. One important feature of all these studies is that they have drawn attention to hitherto unresolved issues and, at the same time, provided novel strategies and methodological procedures; for details of these, reference should be made to the original publications.

8.6 CONCLUDING REMARKS

At this stage, it would have been gratifying to weave the threads of this chapter into a coherent fabric: to summarize the conclusions drawn from the various considerations outlined in this chapter, to set forth the basic components needed for assessment of the hazard resulting from discharge of an organic compound to the aquatic environment, and to assemble the procedures whereby the resulting data may be synthesized into a decision. None of these aims has, in fact, been achieved; the simple fact is that not enough is known about the response of natural populations to alterations in their environment — whether brought about by natural causes or by the intrusion of man. Even for microorganisms, the mechanism and effect of mutation induced by exposure to xenobiotics are incompletely understood (Cairns et al. 1988; Hall 1990; Thomas et al. 1992).

This may seem a bleak conclusion after so many words: each situation will necessarily involve consideration of specific features viewed against a background of general principles. It can only be hoped that these have emerged through the murky details of the reality of the natural world. The presentation has attempted to provide an accurate and hopefully not too partisan account of the complexity of the problem and of the difficulty of attaining the objectives and the procedures involved; an impression of the magnitude of the many hitherto unresolved issues must surely have emerged. It seems unlikely indeed that the solution to a given problem will meet with approval from all who are concerned. At the very least, however, conscientious effort should be directed to obtaining a consensus which takes into account all the relevant issues and seeks a balanced judgment of those unresolved issues that possibly inevitably remain controversial.

It may, nonetheless, be valuable to summarize briefly the salient features that have emerged from this discussion on environmental hazard assessment.

1. A distinction should be made between procedures designed for predictive and for retroactive analysis.
2. For prediction of environmental effects in natural ecosystems, there are serious limitations in the information that can be provided by traditional procedures that rely exclusively on the results of tests for ready biodegradability, for acute toxicity, and surrogate methods for assessing bioconcentration potential. The developments of more sophisticated and more acceptable alternatives to all of these have been outlined in the appropriate chapters of this book.
3. Predictive hazard evaluations are provisional rather than absolute. As a safe-guard against errors in prediction, or the emergence of previously unknown factors that compromise the earlier conclusions, a monitoring program should be carried out. Its design should be carefully evaluated and analysis directed to specific compounds including possible transformation products from the compounds initially discharged.
4. Retroactive analysis — ecoepidemiology — is a valuable exercise since it has limited and defined objectives and accepts both the subjective nature of the evaluation and the possibility of many-to-one relationships between cause and effect. Such an open-ended approach enables an evaluation of all viable alternatives to be made and may, in addition, focus attention on hitherto unexplored facets of the problem.
5. Examination of different case studies reveals the limitations of many procedures that have been applied to assessing the environmental impact of organic compounds released into the aquatic environment; it seems clear that an integrated strategy should be developed and that this should include the sophisticated application of organic chemistry, microbiology, and toxicology. It would be

attractive to develop all of these procedures into a hierarchical system that could be applied to both initial and advanced hazard assessments.

6. In addition to these purely scientific aspects of hazard assessment, it should be appreciated that, in many cases, trans-scientific issues intrude and must be taken into account; among these, social and economic factors may, in some cases, be of determinative significance.

REFERENCES

Addison, R.F. 1989. Organochlorines and marine mammal reproduction. *Can. J. Fish. Aquat. Sci.* 46: 360-368.

Ames, B.N. 1991. Natural carcinogens and dioxin. *Sci. Total Environ.* 104: 159-166.

Ames, B.N., and L.S. Gold. 1991. Cancer prevention strategies greatly exaggerate risks. *Chem. Eng. News* 69(1): 28-32.

Ankley, G.T., K. Lodge, D.J. Call, M.D. Balcer, L.T. Brooke, P.M. Cook, R.G. Kreis, A.R. Carlson, R.D. Johnson, G.J. Niemi, R.A. Hoke, C.W. West, J.P. Giesy, P.D. Jones, and Z.C. Fuying. 1992. Integrated assessment of contaminated sediments in the lower Fox River and Green Bay, Wisconsin. *Ecotoxicol. Environ. Saf.* 23: 46-63.

Berry, R.M., C.E. Luthe, and R.H. Voss. 1993. Ubiquitous nature of dioxins: a comparison of the dioxins content of common everyday materials with that of pulps and papers. *Environ. Sci. Technol.* 27: 1164-1168.

Bringmann, G., D. Feineis, H. Friedrich, and A, Hille. 1991. Endogenous alkaloids in man — synthesis, analytics, and *in vivo* identification, and medicinal importance. *Planta Med.* 57: Supplement Issue 1: S 73-S 84.

Brouwer, A., P.J.H. Reijnders, and J.H. Koeman. 1989. Polychlorinated biphenyl (PCB)-contaminated fish induces vitamin A and thyroid hormone deficiency in the common seal (*Phoca vitulina*). *Aquat. Toxicol.* 15: 99-106.

Cairns, J., J. Overbaugh, and S. Miller. 1988. The origin of mutants. *Nature* 335: 142-145.

Crouch, D. 1986. Science and trans-science in radiation risk assessment: child cancer around the nuclear fuel reprocessing plant at Sellafield, U.K. *Sci. Total Environ.* 53: 201-216.

Dunster, H.J. 1984. Changes in public and worker attitudes. *J. Soc. Radiol. Prot.* 4: 118-121.

Fischhoff, B., S. Lichtenstein, P. Slovic, S.L. Derby, and R.L. Keeney. 1981. *Acceptable risk.* Cambridge University Press, Cambridge.

Fox, G.A. 1991. Practical causal inference for ecoepidemiologists. *J. Toxicol. Environ. Health* 33: 359-373.

Fox, G.A., S.W. Kennedy, RJ. Norstrom, and D.C. Wigfield. 1988. Porphyria in herring gulls: a biochemical response to chemical contamination of Great Lakes food chains. *Environ. Toxicol. Chem.* 7: 831-839.

Hall, B.G. 1990. Spontaneous point mutations that occur more often when advantageous than when neutral. *Genetics* 126: 5-16.

Harshbarger, J.C., and J.B. Clark. 1990. Epizootiology of neoplasms in bony fish of North America. *Sci. Total Environ.* 94: 1-32.

Hashimoto, S., T. Wakimoto, and R. Tatsukawa. 1990. PCDDs in the sediments accumulated about 8120 years ago from Japanese coastal areas. *Chemosphere* 21: 825-835.

Hawkins, W.E., W.W. Walker, R.M. Overstreet, J.S. Lytle, and T.F. Lytle. 1990. Carcinogenic effects of some polycyclic aromatic hydrocarbons on the Japanses medaka and guppy in waterborne exposures. *Sci. Total Environ.* 94: 155-167.

Hites, R.A. 1990. Environmental behavior of chlorinated dioxins and furans. *Acc. Chem. Res.* 23: 194-201.

Hovinga, M.E., M. Sowers, and H.E.B. Humphrey. 1993. Environmental exposure and lifestyle predictors of lead, cadmium, PCB, and DDT levels in Great Lakes fish eaters. *Arch. Environ. Health* 48: 98-104.

Iwata, H., S. Tanabe, N. Sakai, and R. Tatsukawa. 1993. Distibution of persistent organochlorines in the oceanic air and surface seawater and the role of ocean on their global transport and fate. *Environ. Sci. Technol.* 27: 1080-1098.

Jarman, W.M., M. Simon, R.J. Norstrom, S.A. Burns, C.A. Bacon, B.R.T. Simoneit, and R.W. Risebrough. 1992. Global distribution of tris(4-chlorophenyl)methanol in high trophic level birds and mammals. *Environ. Sci. Technol.* 26: 1770-1774.

Johnson, L.L., C.M. Stehr, O.P. Olson, M.S. Myers, S.M. Pierce, C.A. Wigren, B.B. McCain, and U. Varanasi. 1993. Chemical contaminants and hepatic lesions in winter flounder (*Pleuronectes americanus*) from the northeast coast of the United States. *Environ. Sci. Technol.* 27: 2759-2771.

Keilhau, B.M. 1831. Reise i Øst- og Vest-Finnmarken samt til Beeren-Eiland og Spitsbergen i Aarene 1827 og 1828. Christiania.

Kinloch, D., H. Kuhnlein, and D.C.G. Muir. 1992. Inuit foods and diet: a preliminary assessment of benefits and risks. *Sci. Total Environ.* 122: 247-278.

Kovacs, T.G., P.H. Martel, R.H. Voss, P.E. Wrist, and R.F. Willes. 1993. Aquatic toxicity equivalency factors for chlorinated phenolic compounds present in pulp mill effluents. *Environ. Toxicol. Chem.* 12: 281-289.

Kutz, F.W., D.G. Barnes, E.W. Bretthauer, D.P. Bottimore, and H. Greim. 1990. The international toxicity equivalency factor (I-TEF) method for estimating risks associated with exposures to complex mixtures of dioxins and related compounds. *Toxicol. Environ. Chem.* 26: 99-109.

Labare, M.P., and M. Alexander. 1993. Biodegradation of sucralose, a chlorinated carbohydrate in samples of natural environments. *Environ. Toxicol. Chem.* 12: 797-804.

Landner, L. (ed.) 1989. *Chemicals in the aquatic environment*. Springer-Verlag, Berlin.

Last, J.M. ed. 1983. *A dictionary of epidemiology*. Oxford University Press, New York.

Lewis, H.W. 1990. *Technological risk*. W.H. Norton and Company, Inc., New York.

Mac, M.J., and C.C. Edsall. 1991. Environmental contaminants and the reproductive success of lake trout in the Great Lakes: an epidemiological approach. *J. Toxicol. Environ. Health* 33: 375-394.

Malins, D.C, M.M. Krahn, M.S. Myers, L.D. Rhodes, D.W. Brown, C.A. Krone, B.B. McCain, and S.-L. Chan. 1985. Toxic chemicals in sediments and biota from a creosote-polluted harbor: relationships with hepatic neoplasms and other hepatic lesions in English sole (*Parophrys vetulus*). *Carcinogenesis* 6: 1463-1469.

Maltby, L., and P. Calow. 1989. The application of bioassays in the resolution of environmental problems; past, present and future. *Hydrobiologia* 188/189: 65-76.

McMaster, M.E., C.B. Portt, K.R. Munnkittrick, and D.G. Dixon. 1992. Milt characteristics, reproductive performance, and larval survival and development of while sucker exposed to bleached kraft mill effluent. *Ecotoxicol. Environ. Saf.* 23: 103-117.

Moors, P.J. 1986. Decline in numbers of rockhopper penguins at Campbell Island. *Polar Rec.* 23: 69-73.

Myers, M.S., J.T. Landahl, M.M. Krahn, and B.B. McCain. 1991. Relationships between hepatic neoplasms and related lesions and exposure to toxic chemicals in marine fish from the U.S. *West Coast. Environ. Health Perspect.* 90: 7-15.

Ottoboni, M.A. 1984. *The dose makes the poison*. Vincente Books, Berkeley, California.

Pearson, T.H. 1981. The Loch Eil Project: introduction and rationale. *J. Exp. Mar. Biol. Ecol.* 55: 93-102.

Pearson, T.H. 1982. The Loch Eil Project: assessment and synthesis with a discussion of certain biological questions arising from a study of the organic pollution of sediments. *J. Exp. Mar. Biol. Ecol.* 57: 93-124.

Peterman, P.H., and J.J. Delfino. 1990. Identification of isopropylbiphenyl, alkyl diphenylmethanes, diisopropylnaphthalene, linear alkyl benzenzenes and other polychlorinated biphenyl replacement compounds in effluents, sediments and fish in the Fox River system, Wisconsin. *Biomed. Environ. Mass Spectrom.* 19: 755-770.

Pochin, E.E. 1983. Sizewell Inquiry Transcripts, Day 151: 101 problems; past, present and future. *Hydrobiologia* 188/189: 65-76.

Remberger, M., P.-Å. Hynning, and A.H. Neilson. 1991. Chlorinated benzo-1,2-quinones: an example of chemical transformation of toxicants during tests with aquatic organisms. *Ecotoxicol. Environ. Saf.* 22: 320-336.

Rodricks, J.V. 1992. *Calculated risks.* Cambridge University Press, Cambridge.

Safe, S. 1990. Polychlorinated biphenyls (PCBs), dibenzo-*p*-dioxins (PCDDs), dibenzofurans(PCDFs), and related compounds: environmental and mechanistic considerations which support the development of toxicity equivalent factors (TEFs). *CRC Crit. Revs. Toxicol.* 21: 51-88.

Sarokin, D., and J. Schulkin. 1992. The role of pollution in large-scale population disturbances. I: aquatic populations. *Environ. Sci. Technol.* 26: 1476-1484.

Sergy, G.A., Ed. 1987. The Baffin Island Oil Spill (BIOS) Project. *Arctic* 40: Supplement 1: 1-279.

Smit, E., J.D. van Elsas, and J.A. van Veen. 1992. Risks associated with the application of genetically modified microorganisms in terrestrial ecosystems. *FEMS Microbiol. Revs.* 88: 263-278.

Smith-Somerville, H.E., V.B. Huryn, C. Walker, and A.L. Winters. 1991. Survival of *Legionella pneumophila* in the cold-water ciliate *Tetrahymena vorax*. *Appl. Environ. Microbiol.* 57: 2742-2749.

Smyth, H.F. 1967. Sufficient challenge. *Food Cosmet. Toxicol.* 5: 51-58.

Sobecky, P.A., M.A. Schell, M.A. Moran, and R.E. Hodson. 1992. Adaptation of model genetically engineered microorganisms to lake water: growth rate enhancements and plasmid loss. *Appl. Environ. Microbiol.* 58: 3630-3637.

Suflita, J.M., and M.R. Mormile. 1993. Anaerobic biodegradation of known and potential gasoline oxygenates in the terrestrial subsurface. *Environ. Sci. Technol.* 27: 976-978.

Susser, M. 1986. Rules of interference in epidemiology. *Reg. Toxicol. Pharmacol.* 6: 116-128.

Swain, W.R. 1991. Effects of organochlorine chemicals on the reproductive outcome of humans who consumed contaminated Great Lakes fish: an epidemiologic consideration. *J. Toxicol. Environ. Health* 33: 587-639.

Thomas, A.W., J. Lewington, S. Hope, A.W. Topping, A.J. Weightman, and J.H. Slater. 1992. Environmentally directed mutations in the dehalogenase system of *Pseudomonas putida* strain PP3. *Arch. Microbiol.* 158: 176-182.

Underdal, B., O.M. Skulberg, E. Dahl, and T. Aune. 1989. Disastrous bloom of *Chrysochromulina polylepis* (Prymnesiophyceae) in Norwegian coastal waters 1988 — mortality in marine biota. *Ambio* 18: 265-270.

Van Der Kraak, G.J., K.R. Munkittrick, M.E. McMaster, C.B. Portt, and J.P. Chang. 1992. Exposure to bleached kraft pulp mill effluent disrupts the pituitary-gonadal axis of white sucker at multiple sites. *Toxicol. Appl. Pharmacol.* 115: 224-233.

Visser, I.K.G., V.P. Kumarev, C. Örvell, P. de Vries, H.W.J. Broeders, M.W.G. van de Bildt, J. Groen, J.S. Teppema, M.C. Burger, F.G.C.M. UytdeHaag, and A.D.M.E. Osterhaus. 1990. Comparison of two morbilliviruses isolated from seals during outbreaks of distemper in North West Europe and Siberia. *Arch. Virol.* 111: 149-164.

Wadowsky, R.M., L.J. Butler, M.K. Cook, S.M. Verma, M.A. Paul, B.S. Fields, G. Keleti, J.L. Sykora, and R.B. Yee. 1988. Growth-supporting activity for *Legionella pneumophila* in tap water cultures and implication of hartmannellid amoebae as growth factors. *Appl. Environ. Microbiol.* 54: 2677-2682.

Weinberg, A.M. 1972. Science and trans-science. *Minerva* X: 209-222.

Welling, W., and J.W. de Vries. 1992. Bioconcentration kinetics of the organophosphate insecticide chlorpyrifos in guppies (*Poecilia reticulata*). *Ecotoxicol. Environ. Saf.* 23: 64-75.

Epilogue

Perhaps a few final comments may be allowed.

The problem of pollution is not new, and some conclusions from the historical past are as relevant today as they were more than a hundred years ago; indeed, the problem of pollution from by-products of the chemical industry is as old as the industry itself. Environmental concern should not, however, be directed only to the chemical industry, and it is well to remember in fairness the extent to which all of us are dependent on its products — agrochemicals, food preservatives, pharmaceutical products, plastics and adhesives, and dyestuffs.

Consider the following. In England, during the late eighteenth and a large part of the nineteenth centuries, the demand for alkali by manufacturers of glass and soap led to the gradual development of the heavy chemical industry. The oldest process for the production of sodium carbonate — the Leblanc process — used sodium chloride, sulfuric acid, and calcium carbonate as raw materials; the by-products were hydrogen chloride and calcium sulfide. Both presented serious problems, and the formation of the Alkali Inspectorate in England in 1863 was motivated by the disposal of hydrogen chloride; discharge of this into waterways had become so extensive and so serious that in some of the canals, iron barges could not be used because of corrosion to the hulls. That aspect was resolved eventually by the development of processes for the conversion of hydrochloric acid into chlorine which was used for the bleaching of cotton and which developed into an important industry in parts of Britain and elsewhere. The problem with calcium sulfide remained: the huge amounts accumulating as a by-product presented both a loss of the sulfuric acid used in the alkali production and an unsightly and malodorous waste. In this case, it was economic pressure to reduce the cost of alkali that led to the development of processes for recovery of the sulfur and its reconversion into sulfuric acid.

In the long run, the Leblanc process was superseded first by the more elegant and commercially attractive Solvay ammonia process and eventually by electrolytic production of alkali directly from sodium chloride. This, in turn, led to the recognition of two new environmental problems: the discharge of mercury into the environment, and when mercury electrodes were replaced by graphite electrodes for the production of chlorine, the discharge of highly chlorinated styrenes especially the octachloro congener. The toxicity of mercury compounds had, of course, been known for many years; it was the mechanism for its dissipation among environmental compartments in the form of the highly toxic methyl mercury that caused serious alarm. In addition, the ready availability of chlorine led to the industrial production of a wide range of organochlorine compounds. Some of them such as polyvinyl chloride are produced in vast quantities, and recognition of the toxicity of the monomeric vinyl chloride necessitated substantial effort to reduce worker exposure to acceptable levels. Many of the other compounds such as the PCBs and PCCs were in the long run seen as a major — and in many cases more subtle — environmental threats. Again, to keep the issue in perspective, it should be appreciated that it is not only organochlorine compounds that may cause environmental perturbations.

This little scenario reveals just how easy it is to solve one immediate problem and set in motion a whole train of circumstances with unforeseeable consequences; hindsight is, however, a valuable exercise if it succeeds in eliminating a repetition of past errors. Indeed, few apparent improvements have not brought in their wake serious problems that are as pressing today as they were more than a century ago. The complexities of evaluating and predicting environmental hazard will have become clear in Chapter 8, and they seem unlikely to diminish.

This is, in the final analysis, the justification for discussing at some length the procedures for assessing the distribution, the fate, and persistence, and the toxicity of compounds discharged into the aquatic environment that have been discussed in this book. There is no certainty that even the most extensive precautions can eliminate the possibility of undesirable perturbations; what is certain, however, is that if unconcern over the relevant issues prevails, or if a sense of false optimism and uncritical evaluation of their ultimate significance are accepted, disaster will surely follow sooner or later. Environmental hazard assessment requires a contribution from chemists, biologists, microbiologists, and mathematicians, but the final word must be left to those who draw together the often conflicting points of view, weigh them in a balance, and come to a decision on the available evidence.

INDEX